SURFACTANT BIODEGRADATION

SURFACTANT SCIENCE SERIES

CONSULTING EDITORS

MARTIN J. SCHICK

Diamond Shamrock Chemical
Company
Nopco Chemical Division
Newark, New Jersey

FREDERICK M. FOWKES

Chairman of the Department
of Chemistry
Lehigh University
Bethlehem, Pennsylvania

SURFACTANT BIODEGRADATION

By

R. D. SWISHER

MONSANTO COMPANY
INORGANIC CHEMICALS DIVISION
RESEARCH DEPARTMENT
ST. LOUIS, MISSOURI

1970

MARCEL DEKKER, INC., NEW YORK

MARCEL DEKKER, INC.
95 Madison Avenue, New York, New York 10016

LIBRARY OF CONGRESS CATALOG CARD NUMBER 79-107757

PRINTED IN THE UNITED STATES OF AMERICA

PREFACE

The waste surfactant problem, one small facet of the world's burgeoning pollution problem, developed in the 1950s and was corrected in the 1960s. Although the surfactants are only a small, specialized class among organic materials, the principles and techniques applied in that campaign are quite general and should be applicable likewise in attacking some of the larger pollution problems which we still face.

Accordingly, this book may serve two purposes in bringing together the information which has been developed by academic, governmental, and industrial scientists around the world during the past twenty years. First, it shows the current state of knowledge and understanding of surfactant biodegradation as of mid-1969; second, it provides a useful point of departure for future work on other pollution problems.

The area covered in this book lies at the intersection of many scientific and technological disciplines. I have tried to present enough background in the key ones to be minimally self-sufficient, so that the reader need not consult other sources. However, such consultation is recommended, and pertinent references are included for that purpose whenever possible. If it happens that my treatment of some particular point seems absurdly elementary to you, I beg your indulgence on behalf of those other readers (or at least of myself) who may find it less simple.

Although my discussion of the environmental-waste surfactant historical background is confined for the most part to the U.S. situation, there is no such limitation on the coverage of the pertinent scientific and technical literature. Geographically the coverage has been world-wide to the fullest extent possible, and chronologically to mid-1969 for the major journals. A few references which have subsequently come to hand have been added in press, in token form, as footnotes in the text. The most likely gap in coverage would be for the work in some countries of Eastern Europe. The recent comprehensive review by Pitter (1968e) should fill that gap, if there is one.

An important key to developing order out of the chaos of surfactant

v

biodegradation is the chemical structure of the surfactant. Many of the structural formulas used herein have been simplified by omission of hydrogen atoms, and sometimes by omission of bonds also, to bring out more clearly the molecular topology. For an appreciation of the actual topography, the best approach still seems to be the molecular model, properly scaled, studied during assembly as well as afterward.

A newly designed Bibliography/Author Index adds a new dimension to the indexing system by its two-directional nature. Each reference cited includes its title in addition to its author and journal location, and each is keyed back to the appropriate pages in the book. This amplifies the power of the author index considerably, since entry to the text via the author index can now be on a selective basis, with the title indicating the subject matter. Operations in the opposite direction are improved also, since the name and year citation in the text provides a much more informative referent than does a naked, anonymous reference number.

I wish to thank the individuals who have helped me, directly or indirectly, in this undertaking, but to avoid inadvertent omissions I refrain from listing them by name. However, there are two groups that I will mention specifically, because without them I could never have written this book. One is the international group of many hundreds of research workers who provided the basic data and the thought and the interpretation which make up the substance of the book; their names and works comprise the Bibliography/Author Index. The other is Monsanto's research management, who early recognized the importance of the waste surfactant problem and who provided the time, facilities, suggestions, and patience necessary for intelligent, constructive action upon it. And who provided the stimulation and encouragement for me to write about it.

R. D. Swisher

17 February 1970

CONTENTS

LIST OF ABBREVIATIONS

The following abbreviations have been used throughout the text, often without definition. See also special lists showing those used in the tables of Chapter 8 (Sections II–VI of Chapter 8) and in the Bibliography/Author Index (at its beginning).

ABS	Alkylbenzenesulfonate, usually with alkyl group in the detergent range, above C_6 or thereabouts
AMP	Adenosine monophosphate
AOS	α-Olefin sulfonate
APE	Alkylphenol ethoxylate; APE_n averages n mols of EO in the hydrophilic group
BOD	Biochemical oxygen demand
C	Celsius, centigrade; carbon
cm	Centimeter
cm^{-1}	Wave number, waves/cm
CMC	Critical micelle concentration
COD	Chemical oxygen demand
CPB	Cetylpyridinium bromide
CTAB	Cetyltrimethylammonium bromide
CTAS	Cobalt thiocyanate active substance; ethoxylate nonionics and related substances
d	Days
DNA	Desoxyribonucleic acid
DTAB	Dodecyltrimethylammonium bromide
E_n	Polyethylene glycol or ethoxylate nonionic averaging n mols of EO
EO	Ethylene oxide
F	Fahrenheit
GC	Gas chromatography
gm	Grams
h, hr	Hours

i	Iso
IR	Infrared
KBS	ABS with alkyl group derived from kerosene
LAS	Linear alkylbenzenesulfonate; ABS with linear secondary alkyl group
LPAE	Linear primary alcohol ethoxylate; $LPAE_n$ averages n mols of EO in the hydrophilic group
LPAS	Linear primary alkyl sulfate
LSAE	Linear secondary alcohol ethoxylate; $LSAE_n$ averages n mols of EO in the hydrophilic group
m^3	Cubic meter
MBAS	Methylene blue active substance; anionic surfactants, etc.
Me	Methyl
mg	Milligram
min	Minute
ml	Milliliter
MLSS	Mixed liquor suspended solids
mm	Millimeter
mmol	Millimol
mol. wt.	Molecular weight
n-	Normal
nm	Nanometer; millimicron
NMR	Nuclear magnetic resonance
NTA	Nitrilotriacetate
OPE	Octyl phenol ethoxylate; OPE_n averages n mols of EO in the hydrophilic group
PEG	Polyethylene glycol
ppm	Parts per million; used interchangeably with milligrams per liter
QBS	ABS with quaternary carbon atom in alkyl group
RNA	Ribonucleic acid
rpm	Revolutions per minute
s-, sec-	Secondary
SAS	Secondary alkane sulfonate
SDA	Soap & Detergent Association
SDS	Sodium dodecyl sulfate
SLS	Sodium lauryl sulfate
STCSD	Standing Technical Committee on Synthetic Detergents
SϕU	Sulfophenylundecanoic acid
t-, tert-	Tertiary

TBS	ABS derived from tetrapropylene
TLC	Thin layer chromatography
tp-	Tetrapropylene; polypropylene
U.K.	United Kingdom
U.S.	United States
UV	Ultraviolet
WPRL	Water Pollution Research Laboratory
λ	Microliter
μ	Micron
μg	Microgram
μmol	Micromol
Σ	Sulfonate Group; $O\Sigma$, sulfate ester
ϕ	Benzene ring with substituents as indicated

CHAPTER 1

BACKGROUND AND PERSPECTIVE

I. DETERGENT REVOLUTION—ENVIRONMENTAL REACTION

A. The Development of ABS

Study of surfactant biodegradation has assumed importance as a consequence of the chemical revolution which occurred in the detergent industry during the decade centering on 1950. At that time the replacement of soap in cleaning and laundry formulations by synthetic surfactants became technically and economically feasible as a result of the commercial development of alkylbenzenesulfonates (ABS). In a very few years the changeover was essentially complete in the United States, with the single exception of bar soaps. ABS became the surfactant in major use, first in the U.S.A. and soon afterward around the world wherever advanced chemical technology made it possible.

Soap, one of man's earliest chemical products, has been in commercial production for several thousand years. That it could be displaced from its dominant position in a single decade can be attributed to two factors. First, in hard water soap forms insoluble calcium and magnesium salts which have no detersive properties, and which furthermore deposit on the materials being cleaned. Second, the new products are not only far superior in this respect, but are also lower in price.

B. The Wastewater Problem and Surfactant Biodegradation

Despite the manifest advantages of the new products, they were viewed with certain misgivings by some operators of sewage treatment plants, who felt that their unique physical properties might upset their operation in any of several ways (Elton, 1949; Goldthorpe, 1949). These fears never materialized to any significant extent, thanks to the technical competence of the operators in developing improvements in their processes to meet the new conditions.

But a different problem gradually became evident. The synthetic surfactants began to be noticeable in wastewaters, treated sewage, and the receiving waters because of the same property which had led to their success—they retain their foaming properties in natural water at concentrations down to around one part per million, concentrations far below those at which other wastewater components, including soap, can be detected by casual observation. And the presence of ABS could be readily verified by simple and extremely sensitive analytical methods capable of detecting a few tenths of a part per million very easily.

Even though surfactants are essentially nontoxic to humans (Swisher, 1968b), and the concentrations normally met in public water supplies as a result of reuse of wastewater are very low, there is general agreement that their presence in drinking water is undesirable, if only from an aesthetic standpoint. It was for aesthetic reasons, for example, that the revised water standards of the U.S. Public Health Service (1962) set an upper limit of 0.5 ppm on ABS; foaming is not noticeable below that concentration.

Shortly after its recognition, the waste surfactant problem was found to derive primarily from the major surfactant in general use by that time, ABS derived from tetrapropylene, which we shall abbreviate as TBS. It contained some components which were quite resistant to microbial attack. One solution to the problem was accordingly the development of surfactants such as linear alkylbenzenesulfonate (LAS) which in addition to having all the superior functional and economic properties leading to the overwhelming acceptance of synthetic surfactants in the first place, could also be degraded by microorganisms. Study of the biological and biochemical aspects of the problem by investigators around the world has developed the knowledge and experience covered in this book.

C. The Wastewater Problem—Focal Points

The waste surfactant problem has primarily been felt in three areas—in sewage treatment plants, in rivers, and in groundwaters. It is in the context of conditions prevailing in those areas that most studies of surfactant biodegradation are made and interpreted. It will be helpful to review those conditions to provide frames of reference.

C.1. *Sewage Treatment Processes*

Sewage treatment plants provide *primary treatment*, which is simply removal of solid materials by mechanical means such as settling, skimming, or screening. In many installations this is followed by *secondary treatment*, which removes much of the dissolved organic materials by bacterial oxidation, after which the effluent water is usually discharged into a stream or river. The solids removed in the primary treatment and the excess bacterial floc which grows in the secondary treatment are usually further concentrated and decomposed by *anaerobic digestion*, and the remaining solids from this operation are disposed of in various ways, for example, as soil conditioning agents.

The secondary treatment is often accomplished by the *activated sludge* process. The sludge is actually a bacterial floc which develops naturally upon aeration of any nutrient solution under nonsterile conditions. It is mixed with the primary effluent under vigorous aeration, whereby the bacteria assimilate many of the organic components. This *mixed liquor* flows through the aeration tank with a retention time of several hours to several days, then to sedimentation vessels where the bacterial floc settles out and is returned to the aeration tank. Historic development of the activated sludge process has been discussed by Sawyer (1965), general principles by Hawkes (1963), and both by McKinney (1968).

Secondary treatment may also be carried out in a *trickling filter* plant. This is not a filter in the ordinary chemical sense, but is a bed of stones or prefabricated packing medium providing a large surface area over which the wastewater trickles, and providing voids through which air flows upward. The surfaces of the packing quickly become coated with a bacterial film or plaque, and the microorganisms residing therein oxidize the organic components of the sewage. The extent of oxidation is surprisingly great during the few minutes retention time available in passing through the filter bed, but several passages may be required to achieve the completeness characteristic of the activated sludge process. Hawkes (1963) has reviewed the basic principles and technology.

C.2. Surfactants in Sewage Treatment

The vigorous aeration used in the activated sludge process makes it a good foam generator if foaming components are present and if other conditions permit. Excessive foam is not welcome because it contributes to unsure footing and unsightly conditions in the plant, and when windblown can be a nuisance to the entire neighborhood.

Foaming problems were encountered before the days of surfactants, because proteinaceous material, always a component of sewage, is an excellent foaming agent under suitable conditions. After the introduction of synthetic detergents the new surfactants contributed further to the problem. By 1960 they were present in sewage in amounts ranging up to 10 to 20 ppm, and in sewage plant effluents to around 5 to 10 ppm. However, even though these concentrations were well into the foaming range, foaming problems in the plants were held to a minimum by control of operating variables. Maintaining relatively high concentrations of activated sludge (3000 to 4000 ppm) in the aeration tanks and avoidance of excessive aeration rates, or mechanical means such as spraying a fraction of the final effluent back onto the surface of the aeration tank, or sometimes, as a last resort, addition of chemical defoamants—all these measures helped to keep the foam under control.

The 5 to 10 ppm of ABS found in the final effluents from the sewage treatment plants in the early 1960s represented the more resistant of the molecular structures which made up the TBS being used at that time. In point of magnitude, however, this was only a minor component; the final effluents also contained (and still contain) up to 50 to 100 ppm of other organic materials resistant to biodegradation, compounds mostly of unknown structure and nature (Bunch, 1961; Murtagh, 1965) (Table 5.4). These compounds originate in the "natural" components of the sewage; they are either already present there or are formed from degradable components by bacterial action (Chapter 4, Section II). The composition of sewage itself is discussed in Chapter 5, Section III.A.

C.3. *Surfactants in Rivers*

The rivers of the world receive the sewage plant effluents and dilute them sufficiently so that ordinarily the major impurities in the effluent are not particularly evident. But the residual undegraded fraction of the TBS was much more visible, even though present in minute amounts. The situation as it existed in the U.S. up to the early 1960s is summarized in a report by Orsanco (1963).

The Ohio River, for example, averaged 0.16 ppm of ABS for the five years 1954–1959, and occasionally exceeded 0.3 ppm. This is somewhat below the foaming concentration, and one must look to other sources than domestic sewage for an explanation of the spectacular occasion in November 1963 when the river was covered with a 2-foot blanket of foam from shore to shore below a dam at Wheeling, West Virginia (Middleton, 1954; Todd, 1954). Perhaps it was the result of accidental or intentional dumping of large amounts of surfactant upstream. Whatever the reason, it was not to become a normal phenomenon on the Ohio River, and never recurred despite continued and greatly increased use of TBS.

C.4. *The Illinois River*

The Illinois has shown somewhat higher concentrations of residual ABS because it receives the effluent from a large Chicago sewage treatment plant. In 1959 the ABS content of the plant effluent ran about 1.6 ppm, and the ABS level was down to about 0.5 ppm by the time it reached the Pekin–Peoria region 160 miles downstream (Hurwitz, 1960; Vogel, 1962). Detailed study of concentrations and flow rates at intermediate points showed minor additions of ABS from smaller towns along the river, but an overall net decrease in concentration due in part to dilution by tributary streams and in part to removal of ABS by biodegradation or other natural means.

The Illinois represents an extreme case in that it receives sewage effluent from a very large city. In general the larger rivers of the country have shown only minor amounts of ABS, as have the smaller rivers and streams, unless they receive disproportionately large amounts of sewage along the way.

The success of the change from TBS to LAS is exemplified by the results from subsequent monitoring of the Illinois for methylene blue active substances (MBAS) (which include TBS and LAS, other anionic surfactants, and certain natural materials). In the Pekin–Peoria reach the MBAS content had averaged around 0.5 ppm all during the 1959–1965 period. After the change it dropped to around 0.2 ppm in 1965–1966 (Sullivan, 1968). By early 1968 it had dropped further to 0.05 ppm, barely detectable by the ordinary analytical procedure. The exact chemical nature of the MBAS at that time was uncertain, except that less than 20% of it was LAS. Thus the LAS level in the river was then below 0.01 ppm, compared to 0.5 ppm TBS in the earlier era (Sullivan, 1969).

The effects of changing from TBS to LAS are further covered in Chapter 5, Section X.E.

C.5. *The Neosho River*

An ultimate example is provided by the Neosho River at Chanute, Kansas, which in a sense was made up almost entirely of treated sewage for a period of about five months in 1956–1957. The river served as the town's water supply, but due to a prolonged drought it had dried up completely. As an emergency measure the effluent from their sewage treatment plant was returned to the river bed at the intake channel of the municipal water plant, was run through the usual steps of purification, and then back into the town's water mains. There was close attention to the health aspects and good collection of technical data, subsequently published (Metzler, 1958).

Before the river started flowing naturally again, Chanute's water had been recycled seven times. It was yellow in color and unpleasant in odor and taste, and very few people, if any, were using it for drinking. The total solids had risen to 1200 ppm and the content of resistant organic materials (measured as chemical oxygen demand, COD) to 50 ppm during the recycle operations. Included in the latter was 5 ppm of ABS, which undoubtedly contributed considerably to the pronounced foaming tendencies developed by the water as the recycling proceeded. Although the aesthetic quality of the water deteriorated markedly, no health problems became apparent as a result of the recycling.

The Chanute experiment sometimes is cited as a horrible example of the effects of detergent pollution, but it is nothing of the sort. If the townspeople had been using no synthetic detergents at all, the final water would still have been yellow in color, unpleasant in odor and taste, with total solids 1200 ppm and organic content in the neighborhood of 50 ppm. True, its ABS content would have been zero and its foaming properties undoubtedly less pronounced, but its health quality would have been the same and any superiority in aesthetic quality marginal at best.

C.6. *Groundwater*

The groundwater was the third general area where waste surfactants had made their presence known by the early 1960s, as a result of increasing encroachment by sewage. In some localities with high population densities and without community sewage treatment facilities the groundwaters by 1960 were receiving effluents from cesspools and septic tanks in quantities sufficient to build up ABS concentration above the foam threshold, sometimes as high as several parts per million.

Some of the people so afflicted had the feeling that if the surfactant were somehow removed, their sewage would assume pristine purity and become quite suitable for adding to their groundwater without deterioration of quality. This naive viewpoint was apparently shared for a time by some public officials, but it is by now generally recognized that such primitive treatment systems will degrade groundwater quality regardless of presence or absence of surfactants.

II. BIODEGRADATION FOR WASTE SURFACTANT CONTROL

The wastewater problem is becoming more pressing as population densities increase. Surfactants were a rather small factor in the total problem, but one potentially correctable at its source. As has been suggested above, the residual ABS which may appear as a component of water pollutants is resistant to bacterial attack. Studies of surfactant foods are concentrated in this manner—perhaps not the simpler, lower molecular weight ones, but colloidal ones like starch (Banerji, 1966). Surfactants are particularly susceptible to adsorption, and if the solids content (i.e., bacterial cells or floc) in such a system is high there may be a considerable removal of the surfactant upon its first addition (Chapter 4, Section V). Removal of further increments would not occur, however, once all the adsorption sites were occupied by surfactant molecules. With a biodegradable surfactant, as with the more usual foods, such blocking does not happen, because the adsorption sites are continually freed for further use as the adsorbed molecules are oxidized or transported into the cell (Chapter 5, Section VII.F.7).

Closely related to their adsorption on solid surfaces is the tendency of surfactant molecules to congregate at the surface of the solution, the liquid–air interface. Foam is made up almost entirely of such interfaces, and foam from a dilute surfactant solution contains a much higher surfactant concentration than the main body of the solution. Much of the surfactant content can be removed from the main body by generation and removal of foam.

G. Surfactant Removal by Chemical Means

Removal of surfactants from the usual biological systems through chemical reactions of nonbiological origin is theoretically possible, but is not an important factor with present-day surfactants. Three possible courses of such action are precipitation, hydrolysis, and oxidation. Precipitation is very important with soaps, since the calcium and magnesium ions normally present form very insoluble salts of the fatty acids. It was this property which made soaps vulnerable to replacement by the synthetic surfactants in the first place, so that synthetics having this property would probably not be able to gain any commercial importance.

Hydrolysis is an effective way to destroy alkyl sulfates, an important group of surfactants. But severe conditions are required for their chemical hydrolysis, and the hydrolysis of these materials which does often occur in biological systems is of biological rather than purely chemical origin. In other words, it is biodegradation. Similar considerations apply to removal by oxidation. All surfactants may be destroyed by chemical oxidation, but again the conditions (in terms of temperature or reagents, for instance) are much more drastic than those met in biological systems. Except for biooxidation, which is the principal route of biodegradation.

Removal of surfactants from wastewaters by nonbiological means has received considerable attention and many processes have been developed to accomplish it. These are of particular interest with nonbiodegradable surfactants in more concentrated systems such as laundromat wastes. They do not fit properly into the subject of surfactant biodegradation and will not be discussed further. Still, the possibilities of nonbiological removal must be kept in mind to the extent that they may be operative in the biological systems under examination, and if neglected these factors may lead to incorrect interpretation of the mechanisms involved in the surfactant removal.

III. BIODEGRADATION TEST METHODS AND THEIR LIMITATIONS

A. Typical Systems and Uncertainties

Study of surfactant biodegradation is exceedingly simple in principle. The surfactant is exposed to bacteria and its fate is observed. On the other hand, a simple quantitative expression of the results may be exceedingly difficult in view of the multiplicity of pertinent variables, of test methods, and of intermediate degradation products.

Since the day when Degens (1950) first put some Teepol into a complete aquarium and then analyzed the water at subsequent intervals to observe its biodegradation, test methods have been developed in limitless variety. Degens himself reports three in that one paper—the aquarium test, the canal water test (using brackish water from Amsterdam Harbor in 5 liter beakers), and the BOD test (wherein oxygen consumption is the indication that biodegradation has occurred).

An important reason for biodegradation study is to enable prediction of probable environmental consequences of widespread use of the surfactant in question. This suggests that the conditions of bacterial exposure should bear some convenient relationship to the conditions existing in nature or in sewage treatment processes, and further that the degradation products should meet criteria relating to their foaming properties, toxicity, or ecological effects in the receiving waters. A second major reason for the study is to gain insight into the biochemical reactions involved in the introduction of the surfactant into the bacterial metabolic reactions. This is less circumscribed with regard to conditions of exposure and range of degradation products to be studied, and may result in new molecular designs with improved biodegradability.

Some of the more important variables pertinent to biodegradation testing are discussed in Section II of this chapter. They include

Microorganisms—nature, acclimation, concentration

Food—nature, concentration

Toxic or bacteriostatic agents

Oxygen

Temperature

Surfactant concentration

Analytical method

Depending on these variables, and too often on others unknown or unrecognized, the biodegradation results may vary. This is a characteristic not only of surfactants, but of other organic compounds as well, including the more usual foods. Further, it is characteristic not only of laboratory tests, but also of the larger scale operations to which we are attempting to extrapolate. Because of such factors the efficiency of removal of organics in sewage treatment plants may vary from process to process, and for a given process from plant to plant, and for a given plant from day to day. In the wider and uncontrolled environment of nature the efficiency of removal of organics from surface or river waters can be expected to vary even more widely from place to place and from time to time.

Thus if biodegradation is to be the means for removing surfactants from our waste waters, it is important to study not only the removal in the various biological systems, but also the ease of removal relative to the other more natural components likely to be present. In other words, the biodegradability index (Section II.F, and Chapter 5, Section II.C).

The above circumstances make it obvious that the biodegradability of a surfactant (or of any other material) cannot be expressed in any scientific sense as a single number. In a legal sense it is done all the time. The German Government (1962) specifies an 80% minimum in their detergent law, and a 90% limit was agreed upon by U.S. manufacturers (SDA, 1965). Even though tied to specific test methods it is questionable whether these, or any, numbers could have been used successfully had not the manufacturers been able to develop surfactants with degradability far above the limits which were established.

B. Typical Results

A typical biodegradation experiment is shown in Fig. 1.1, using the river water technique of Hammerton (1955). For each of the three surfactants, a 7 mg sample was dissolved in 1 liter of river water and the solution was analyzed every few days by one of the methylene blue

methods. One of the surfactants was readily attacked by the microorganisms present in the river water, one was very resistant, while the
third fell in between.

These curves illustrate several points. First, they show an induction
period during which there is little change in surfactant concentration.

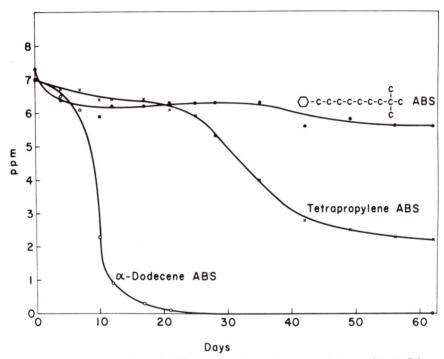

Fig. 1.1 Biodegradation of ABS varieties in a river water test, as detected by
methylene blue analysis [Swisher, 1963d].

This is the period of acclimation during which the appropriate enzymes
of the bacteria become adapted to oxidize the surfactant, a new and
unfamiliar food. A rapid disappearance of the α-dodecene LAS then
ensues, characteristic of a readily degradable surfactant. The quaternary
ABS at the upper right of Fig. 1.1 represents a different situation. Because
of its chemical structure it is much more resistant to biodegradation, and
the organisms failed to utilize it during the 2 months of exposure. The
TBS fell in between the two extremes. Compared to the C_{12}LAS it required
a considerably longer period of acclimation; its rate of disappearance,

once started, was slower, and its curve leveled off with about 30% of the original TBS still remaining undegraded.

C. Completeness of Degradation

Using a rather common terminology, the $C_{12}LAS$ may be said to have been completely degraded in the test in Fig. 1.1, the TBS only about 70% degraded. Note that this terminology is used with reference to primary biodegradatiôn. So far as can be determined from this particular experiment, we can say only that the $C_{12}LAS$ is completely degraded in its property of response to the methylene blue analytical method. Semantically, "complete" degradation could refer equally well to what we have defined as ultimate biodegradation to carbon dioxide and water, but we will not use it with such a meaning in this book. As shown in Chapter 7, Section III, $C_{12}LAS$ is indeed susceptible to ultimate biodegradation, but the proof involves techniques considerably beyond simple methylene blue analysis.

D. Multicomponent Surfactants

The behavior of TBS in Fig. 1.1 suggests strongly that it is not a pure compound (a fact well established by other means, as discussed in Chapter 2, Section II.D). If it were, all of its molecules would be identical and if 70% of them were degradable the other 30% should have been also. It might be argued that something had happened to inactivate the bacteria by the time the 70% had disappeared. Such is not the case, since addition of a second increment of TBS at that stage usually results in the disappearance of a like fraction of the new increment. Actually the TBS is a mixture of many different, closely related compounds, and the 30% undegraded is that fraction of the original material with molecular structures considerably more resistant to bacterial oxidation.

Note again that so far as this experiment is concerned, the fate of the 70% is indeterminate. It may have undergone ultimate biodegradation to carbon dioxide and water, or it may have been changed to some lesser extent, perhaps barely sufficient to make the molecules no longer responsive to methylene blue. As shown in Chapter 7, Section V, its actual fate may range from one of those extremes to the other, depending on the circumstances.

E. Variables

Operation of the several variables listed earlier might have altered the results of the river water experiments (Fig. 1.1) in several ways.

The bacteria responsible for the degradation were most likely those known as the common soil bacteria, species of which would be present in most of the world's rivers. Similar results would be expected with little regard to the geographical origin of the river water so long as unusual chemical or physical circumstances were not involved. River waters will ordinarily contain from a few hundred to a few hundred thousand bacteria per milliliter and within those limits the experimental results probably would not have differed greatly. At bacterial concentrations a million times higher, such as those met in sewage treatment processes, the degradation might have required only hours instead of days.

Effects of bacterial acclimation could have been demonstrated by adding a second increment of the $C_{12}LAS$ to the water just after degradation of the first was complete; it would have disappeared promptly, without the long induction period. If the river water had been obtained from a location where the bacteria might have been already exposed to LAS, they might have been already acclimated at that time.

The river water used in this experiment contained no food for the bacteria—its biochemical oxygen demand (BOD) was very low. If some food (such as carbohydrate, protein, or other oxidizable organic materials) had been added along with the surfactant at the beginning of the test, it might have been preferentially degraded by the bacteria, thereby delaying their attack on the surfactant. Similar effects might result if the original river water had itself contained foods, entering perhaps as sewage, incompletely utilized at the time the test was started. Many of the test methods used in surfactant biodegradability determinations provide for the addition of such food materials as a better simulation of natural conditions.

If the water had contained materials toxic to the bacteria, perhaps originating from natural sources, or from agricultural pesticide treatments, or from industrial wastes, the attack on the added surfactant might have been less vigorous. Presence of bacteriophages might likewise have interfered with the continued existence of the bacteria, and thus with their degradation of the surfactant. Absence of oxygen (i.e., anaerobic conditions) during the test, or prevalence of temperatures outside the preferred range of the bacteria in the sample would of course have affected their growth and propagation adversely, and their degradation of the surfactant as well.

In some cases the surfactant itself may be inhibitory or toxic to the bacteria; if 20 ppm of the $C_{12}LAS$ had been used at the start instead of 7 ppm, the induction period might have been several times as long and the subsequent degradation rate slower. If the longer chain C_{14}LAS had been used instead of the C_{12}, in the absence of acclimation the induction period would certainly have been longer. In highly concentrated bacterial systems such as activated sludge, much higher concentrations can be tolerated, up to at least several hundred parts per million in the feed, with no such difficulty, once acclimation has been achieved.

Finally, the shape and character of a biodegradation curve is dependent upon the analytical method chosen to follow the course of the degradation. The curves in Fig. 1.1 depict the methylene blue active substances (MBAS) present in the system as a function of time. If some other parameter had been chosen, say the total organic carbon content of the system, or its foaming capacity, or the amount of carbon dioxide formed, the resulting curves might have differed considerably.

CHAPTER 2

SURFACTANTS—THEIR
NATURE, BEHAVIOR, AND STRUCTURE

I. HYDROPHILIC HYDROPHOBES—THEIR PHYSICAL PROPERTIES

A. Surfactants and Detergents

Biodegradation is a chemical process carried out by biological agents, and a treatment of surfactant biodegradation may appropriately begin with a review of surfactant chemistry. In turn, surfactant chemistry derives from a characteristic and necessary feature of a surfactant—the presence of a strongly hydrophilic group and a strongly hydrophobic group linked together in the same molecule. Regardless of the exact chemical nature of these groups, surfactants have a certain set of physico-chemical properties in common, and we will briefly outline those properties before examining the range of chemical structures by which they can be achieved. Literature sources and further details are given in a recent review by Hutchinson (1967).

The word *surfactant* is a shortened form of the rather awkward term *surface active agent,* and denotes the outstanding property of these compounds: they tend to concentrate at the surface of an aqueous solution and to alter its surface properties. In distinction we apply the term *detergent* to a product or formulation designed for cleaning or laundering. Modern detergents usually contain about 10% to 30% of surfactant (often called the *active*), larger proportions of polyphosphate salts (the *builder*), and a number of other ingredients in small percentages. The cleaning efficacy of a properly formulated product is much greater than that of an equal weight of the pure surfactant.

B. Surface Phenomena

Because of the presence of the hydrophilic group a surfactant is more or less readily soluble in water. However, the hydrophobic group is repelled by water, so that there is a tendency for that portion of the molecule to leave the aqueous phase. This leads to a higher concentration at the surfaces or boundaries than in the main body of the solution. At the surface of the solution—the air–water interface—the surfactant molecules orient themselves with the hydrophilic groups in the water phase, the hydrophobic groups extending as far as possible in the other direction, still consistent with the molecular dimensions and geometry and with the intermolecular forces acting upon them. The result of this oriented surface film is the lowering of the surface tension of the water, and a greater tendency toward bubble and foam formation. In the presence of an immiscible liquid a similar layer tends to form at the liquid–liquid interface, hydrophilic groups oriented toward the water, hydrophobic toward the other liquid. This promotes dispersion and emulsification as droplets. At liquid–solid interfaces a similar phenomenon occurs.

In all of these cases an equilibrium exists between the surface molecules at the interface and the interior ones, with molecules constantly entering and leaving the two regions. In very dilute solutions, say at 1 ppm, the tendency toward the interface may lead to noticeable depletion of the solution if it is transferred to another vessel. Some of the surfactant is left behind on the walls of the old vessel, and some leaves the main body of solution to concentrate at the walls of the new one. Or if the area of the water–air interface is increased, as by blowing air through the solution and generating a foam, the concentration of surfactant in the main body can be markedly lowered.

On the other hand, if the surface area of the solution is held constant, the space available for surfactant molecules at the surface becomes a limiting factor. If the surfactant concentration is increased the surface approaches a state in which it is filled with a closely packed array of surfactant molecules. Beyond this state, as more surfactant is added it can only enter the main body, and the concentration there approaches that in the surface layer.

The simultaneous attraction and repulsion of the surfactant molecule by water may be pictured as the mechanism responsible for these phenomena. In many cases the resulting surface adsorption is a rather loose one. On the other hand, if the solid surface contains chemically active centers (as in activated carbon or ion exchange resins) they may exert a strong attraction for either the hydrophobic or the hydrophilic groups. In such a case the adsorption effects are very much greater than those exhibited by inert solids such as the walls of the container.

C. Micelles

Another phenomenon characteristic of surfactants in aqueous solution is the aggregation of their molecules into larger, oriented groups called *micelles*. In very dilute solutions, say 1 ppm, the individual single molecules are present, or their ions. Further increments of surfactant also dissolve to form separate molecules or ions up to a certain point, known as the *critical micelle concentration*, abbreviated CMC. Beyond this point the concentration of single molecules remains relatively constant. Much more surfactant may still be dissolved to give clear solutions, but the added increments form micelles in the solution instead of appearing as individual molecules. At some higher concentration, often much higher, the solubility limit for that particular surfactant may be reached. Further increments will no longer dissolve in the saturated solution, but instead form a new phase, solid or liquid, usually hydrated.

In the micelles of an aqueous solution the molecules are oriented with their hydrophobic portions clustered together, the hydrophilic ends extending outward. Such ordering results from repulsion of the hydrophobic groups by the water, further aided by attraction of the groups for each other. The micelles are in equilibrium with the solution, and one may picture relatively free passage of the surfactant molecules back and forth between the two.

Sizes and shapes of micelles can be determined by various physicochemical means. Depending on such factors as the chemical nature and architecture of the surfactant, the salt content of the solution, and the temperature, they may be spheres, ellipsoids, or cylinders, and may average tens or hundreds of molecules per micelle. Depending likewise on those factors the CMC may be around 1000 ppm or it may be an order of magnitude or so lower or higher.

Many organic materials which are insoluble in water but soluble in organic solvents may be solubilized in aqueous solution to a certain extent by the presence of surfactant micelles. The organic material is found to be molecularly dispersed in the internal region of the micelle as a quasi-solution in the clustered hydrophobes, one or more (or less) molecules per micelle. This solubilizing action is no longer exhibited below the CMC of the surfactant, and is to be distinguished from the emulsification or dispersion of insoluble materials in water by the aid of surfactants. In the latter case the insoluble material is present in the form of a second phase of macroscopic droplets or particles, restrained from coalescing or precipitating by the presence of the oriented surface layer of surfactant.

The main reason for bringing micelles into this discussion of surfactant biodegradation is to emphasize that micelle phenomema have little or nothing to do with the subject so far as is known at present. The surfactant concentrations of primary interest in biodegradation work are in the neighborhood of and below 10 ppm, while the critical micelle concentrations are usually ten to a hundred times higher.

D. Chemical Origins

The general properties of surfactants outlined above stem from their general structures—molecules in which a hydrophilic group is linked to a hydrophobic group. We shall now consider some of the specific chemical structures which embody that feature.

Hydrophobic groups, hydrophilic groups, and the means of joining them together are available to the synthetic organic chemist in almost limitless variety. The size of such compendia as Schwartz and Perry (Schwartz, 1949, 1958), Lindner (1964), and Schick (1967) attest the diligence with which the permutations and combinations of these have been explored since the 1920s. Such sources should be consulted for a detailed picture. Certain of these products have enjoyed widespread commercial use because of one advantage or another and those will receive our primary attention.

II. HYDROPHOBIC GROUPS AND THEIR SOURCES

By far the most prevalent hydrophobic group used in surfactants is the hydrocarbon radical having a total of from about 10 to 20 carbon atoms. Commercially there are two main sources of supply for such radicals in sufficient quantity and in a suitable price range: agriculture (and fishing) and the petroleum industry.

A. Fatty Acids

Agriculture contributes fats and oils, which are predominantly triglyceride esters of fatty acids and which are readily hydrolyzed to the fatty acids themselves [Eq. (2.1)]. A naturally occurring plant or animal

$$
\begin{array}{c}
\quad\quad\quad\;\; \overset{\displaystyle O}{\overset{\displaystyle \|}{R_a\!-\!C}}\!-\!O\!-\!CH_2 \\[4pt]
\overset{\displaystyle O}{\overset{\displaystyle \|}{R_b\!-\!C}}\!-\!O\!-\!CH \;\rightarrow\; 3\,\overset{\displaystyle O}{\overset{\displaystyle \|}{R\!-\!C}}\!-\!OH \\[4pt]
\overset{\displaystyle O}{\overset{\displaystyle \|}{R_c\!-\!C}}\!-\!O\!-\!CH_2
\end{array}
\qquad (2.1)
$$

fatty acid usually contains an even number of carbon atoms, so the group symbolized as R contains an odd number. The carbons are linked together in a straight chain with a wide range of chain lengths; those with 16 and 18 carbons are very common. They may be saturated, in which case the R group has the formula C_nH_{2n+1}, or there may be one or more double bonds along the chain. Hydroxyl groups along the chain are rather less common, and other substituents are rare.

Commercially the largest surfactant outlet for fatty acids is conversion to soap by neutralization with an alkali [Eq. (2.2)].

$$RCO_2H + NaOH \rightarrow RCO_2Na + H_2O \qquad (2.2)$$

In a strict sense this reaction is a synthesis, and soap might be considered a synthetic surfactant. Or a synthetic detergent, since it is often used as such without formulation. However, we will follow current usage and reserve the term "synthetic" for the more modern products which have been developed as improvements over soap and are obtained by rather more complex processes.

The carboxyl group of a fatty acid is chemically reactive and may be used to join its hydrocarbon chain to a desired hydrophilic group in many ways. Alternatively, the carboxyl group may first be reduced to an alcohol group [Eq. (2.3)] giving a fatty alcohol which can serve as an intermediate in the synthesis of still other types of surfactants.

$$RCO_2H + 2H_2 \rightarrow RCH_2OH + H_2O \qquad (2.3)$$

B. Paraffins

The hydrophobic groups contributed by the petroleum industry are principally hydrocarbons, deriving originally from the paraffins of crude oil. The chain lengths most suitable for detergent hydrophobes, C_{10} to C_{20}, occur in the crude oil cuts boiling somewhat higher than gasoline, namely kerosene and beyond. The main components of kerosene are saturates ranging from about $C_{10}H_{22}$ to $C_{15}H_{32}$, ordinarily containing 10% to 25% of straight chain, linear, homologs, **2.1**. There are larger amounts of branched chain isomers, of which **2.2** represents one of the hundreds of structures which may be present. In addition, quantities of saturated cyclic derivatives, naphthenes, may be present also, $C_{10}H_{20}$ to $C_{15}H_{30}$, exemplified by **2.3**, and minor amounts of polycyclic derivatives.

C—C—C—C—C—C—C—C—C—C—C—C

(2.1)

$$C-\underset{\underset{C}{|}}{C}-C-\underset{\underset{C}{|}}{C}-C-C-C-C-C$$

(2.2)

$$C-\underset{\underset{C}{|}}{C}-C\underset{C-C}{\overset{C-C}{<}}C-C-C$$

(2.3)

The paraffins have the disadvantage of being chemically unreactive so that direct conversion to surfactants is rather difficult. Instead, it is usually preferred to go by way of other intermediates, most commonly olefins **(2.4)**, alkylbenzenes **(2.5)**, or alcohols. These contain active

$$R-CH=CH_2 \qquad R_a-CH=CH-R_b$$

(2.4a)   **(2.4b)**

$$\underset{R_b}{\overset{R_a}{>}}C=CH_2 \qquad \underset{R_b}{\overset{R_a}{>}}C=CH-R_c \qquad \underset{R_b}{\overset{R_a}{>}}C=C\underset{R_d}{\overset{R_c}{<}}$$

(2.4c)   **(2.4d)**   **(2.4e)**

$$R-\langle \bigcirc \rangle$$

(2.5)

centers—the double bond, the benzene ring, or the OH group—which are more reactive than the paraffins themselves and are more easily linked to hydrophilic groups to make surfactants.

C. Olefins

Olefins with the desired C_{10} to C_{20} chain length are made by building up from smaller olefins (polymerization), or by breaking down larger molecules (cracking), or by production from paraffins of the same chain length.

A major example of olefin produced by the building process is tetra-propylene, the raw material for TBS. It is prepared by polymerization of propylene, a refinery by-product, under the influence of a phosphoric acid catalyst as indicated in Eq. (2.4). This is a very imperfect representation

$$C-C=C \rightarrow C_6H_{12} \rightarrow C_9H_{18} \rightarrow C_{12}H_{24} \rightarrow C_{15}H_{30} \rightarrow \text{etc.} \qquad (2.4)$$

of the actual reactions occurring. The conditions are drastic, so that

splitting and recombination of the polymer molecules occur, with formation of substantial amounts of intermediate molecules such as C_{10}, C_{11},
C_{13}, and C_{14} olefins. The final product comprises a great variety of highly
branched isomers and homologs with the double bond usually situated
internally as in **2.4d** and **2.4e**. The upper gas chromatogram in Fig. 2.1
indicates the complex makeup of commercial tetrapropylene. At least
50 components are evident as peaks and shoulders, and minor components
are undoubtedly present in larger numbers.

A second type of built-up olefin attained importance somewhat later—
the polymer obtained from ethylene in the Ziegler process using an aluminum trialkyl as catalyst [Eq. (2.5)]. These are predominantly linear

$$C{=}C \rightarrow C{-}C{-}C{=}C \rightarrow C{-}C{-}C{-}C{-}C{=}C \rightarrow etc. \qquad (2.5)$$

olefins with even carbon numbers; branched olefins of the type **2.6** are

$$
\begin{array}{c}
C{-}C \\
| \\
C{-}C{-}C{-}C{-}C{-}C{-}C{-}C{-}C{=}C \\
\textbf{(2.6)}
\end{array}
$$

formed also and are usually present in small amounts. The ethylene raw
material historically has been more expensive than propylene, the catalyst
is more expensive as well, and the reaction conditions are more sensitive
and critical, and give a very broad distribution of product chain lengths.
Thus this type of olefin is more expensive than the polypropylene discussed
above.

Production of detergent olefins from higher molecular weight precursors
is accomplished by the cracking process, which uses a drastic thermal
treatment, sometimes assisted by catalysts. Basically the reaction is the
splitting of a paraffin into two smaller molecules, a paraffin and an olefin,
as in Eq. (2.6). Actually, a wide range of products is obtained because the

$$C_{26}H_{54} \rightarrow C_{12}H_{26} + C_{14}H_{28} \qquad (2.6)$$

original molecule may be split at any spot along its chain, and because
the resulting products are themselves often split further. Each olefin
molecule which undergoes such further cracking gives two more olefin
molecules, so the paraffin ratio becomes progressively smaller. The olefins
produced are predominantly α-olefins, with the double bond at the end
of the chain (or more correctly at the beginning, the α-position). If the
original cracking stock is linear, the product olefins will be predominantly
linear; if branched or cyclic structures are originally present, such
structures will also appear in the product.

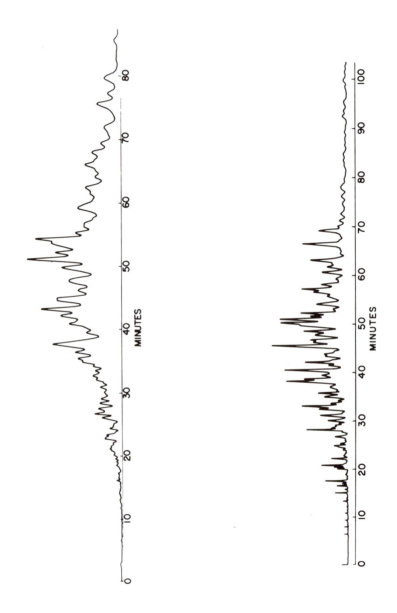

Fig. 2.1   Gas chromatograms of tetrapropylene (upper) and of tetrapropylene-derived alkylbenzene (lower) (Kaelble, 1963).

The third route to detergent range olefins is from paraffins of the same chain length. In principle one need only remove two hydrogen atoms from a pair of adjacent carbons along the chain [Eq. (2.7)], but the difficulties

$$C_{12}H_{26} \rightarrow C_{12}H_{24} + H_2 \tag{2.7}$$

of dehydrogenation are such that a two-step process of chlorination and dehydrochlorination [Eq. (2.8)] has been developed. In either process the reaction easily proceeds past the desired stage to give diolefin or dichloroparaffin, and beyond, as undesired by-products.

$$\begin{aligned} C_{12}H_{26} + Cl_2 &\rightarrow C_{12}H_{25}Cl + HCl \\ C_{12}H_{25}Cl &\rightarrow C_{12}H_{24} + HCl \end{aligned} \tag{2.8}$$

D. Alkylbenzenes

Alkylbenzenes made their appearance as surfactant intermediates in the late 1940s; they were prepared via chlorination of kerosene and Friedel-Crafts alkylation of benzene [Eq. (2.9)]. Their real dominance in

$$C_{12}H_{26} + Cl_2 \rightarrow C_{12}H_{25}Cl + HCl$$

$$C_{12}H_{25}Cl + \bigcirc \rightarrow C_{12}H_{25}\!-\!\bigcirc + HCl \tag{2.9}$$

the field began in the early 1950s with the appearance of tetrapropylene alkylbenzene, a better and cheaper product obtained in a one-step process involving addition of benzene at the double bond of the olefin [Eq. (2.10)]. A variety of Friedel-Crafts catalysts may be used, most commonly aluminum chloride or hydrogen fluoride, less often sulfuric acid.

$$C_{12}H_{24} + \bigcirc \rightarrow C_{12}H_{25}\!-\!\bigcirc \tag{2.10}$$

The alkyl group of the alkylbenzene might be expected to have a carbon skeleton identical with that of the olefin from which it was derived. This is probably true in most cases, but not necessarily so because the alkylation catalyst may cause rearrangement of the carbons in the chain. Furthermore, the reaction of the benzene ring with the double bond of the olefin involves a number of intermediate steps during which rearrangements may occur, so that the benzene may finally link on elsewhere than at the original position of the double bond. Thus each of the many species which make up the olefin tetrapropylene may give rise to several isomeric alkylbenzenes, and the resulting tetrapropylene alkylbenzene is an even more complex mixture than the original olefin. The lower gas chromatogram in Fig. 2.1 shows about a hundred components at least partially resolved, and undoubtedly many more are present in smaller amounts. Generically the commercial products would be better called polypropylene than tetrapropylene alkylbenzenes since alkyl groups other than C_{12} are present. Some specialty products designed to meet particular requirements have averaged as high as C_{13} or C_{14}.

Polypropylene derivatives such as TBS are obsolescent because of the exceptional biological stability of some of the components. During 1965, for example, the U.S. detergent industry abandoned TBS in favor of the corresponding straight chain product LAS. Similar changes were made in Germany and the United Kingdom a little earlier and a little later. The linear alkylbenzenes may be prepared either from linear paraffins via the intermediate chloroparaffins, or from linear olefins. Here, too, the chain length may range from C_{10} through C_{15}, variously distributed according to the properties desired in the final surfactant.

Alkylation of benzene with a linear olefin ordinarily results in a mixture of all possible linear secondary alkylbenzenes of that chain length, even though the original olefin may have been a single pure compound. (A *secondary* alkylbenzene is one in which the ring is linked to a carbon atom which in turn is linked to two other carbons.) During the alkylation reaction a rearrangement occurs so that the ring may join any of the carbons along the chain except the end ones [Eq. (2.11)]. *Primary*

$$
\begin{array}{l}
\text{C=C—C—C—C—C—C—C—C—C—C—C} \\[2pt]
\quad\quad\quad\quad\quad\quad\downarrow \\[2pt]
\text{C—C—C—C—C—C}\!\!-\!\!\text{C—C—C—C—C—C} \\
\quad\quad\phi \\[2pt]
\quad\quad\quad\quad\quad\quad + \\
\text{C—C—C—C—C—C}\!\!-\!\!\text{C—C—C—C—C—C} \\
\quad\quad\quad\phi \\[2pt]
\quad\quad\quad\quad\quad\quad + \\
\text{C—C—C—C—C—C}\!\!-\!\!\text{C—C—C—C—C—C} \\
\quad\quad\quad\quad\phi \\[2pt]
\quad\quad\quad\quad\quad\quad + \\
\text{C—C—C—C—C—C}\!\!-\!\!\text{C—C—C—C—C—C} \\
\quad\quad\quad\quad\quad\phi \\[2pt]
\quad\quad\quad\quad\quad\quad + \\
\text{C—C—C—C—C—C}\!\!-\!\!\text{C—C—C—C—C—C} \\
\quad\quad\quad\quad\quad\quad\phi
\end{array}
\tag{2.11}
$$

alkylbenzenes (linkage to a carbon attached to only one other carbon) are not formed. These same five isomers would have been obtained [Eq. (2.11)] regardless of the original location of the double bond. Indeed, any of the five phenyldodecane isomers may be converted into a mixture of all five by treatment with $AlCl_3$, the catalyst often used in the alkylation

(Swisher, 1961). Figure 2.2 shows gas chromatograms of such mixtures from alkylation and from isomerization.

A similar mixture is obtained using the chloroparaffin process. Since there is not much difference in the reactivity of the 26 hydrogen atoms in

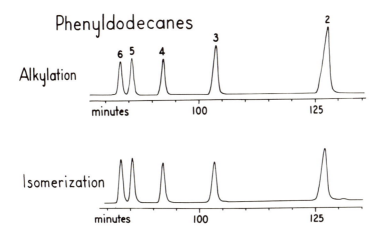

Fig. 2.2   Gas chromatograms of linear C_{12} alkylbenzene (Swisher, 1961). Upper: From alkylation of benzene with α-dodecene. Lower: From isomerization of 3-phenyldodecane.

dodecane, all possible monochloro isomers are present in the intermediate chlorododecane and all possible phenyl isomers are present in the resulting phenyldodecane mixture, including the 1-phenyl isomer.

A gas chromatogram of a typical linear alkylbenzene mixture is shown in the upper part of Fig. 7.9.

E. Alcohols

Long chain alcohols have been used as a source of hydrophobes since the earliest days of synthetic detergents. Linear alcohols have been used since the beginning, branched alcohols more recently.

The classical route to a linear alcohol is by reduction of the carboxyl group of a natural fatty acid [Eq. (2.12)]. Actually an ester of the carboxylic

$$RCO_2H + 2 H_2 \rightarrow RCH_2OH + H_2O \qquad (2.12)$$

acid is usually used since the carboxyl group itself is rather sluggish. Lauryl and tallow alcohols are two that are commonly used for surfactants. The first is derived from lauric acid, predominantly C_{12}, but usually containing some amounts of lower and higher homologs. The tallow alcohol averages around C_{18}; depending upon the reduction process used it may contain some unsaturated alcohols derived from the unsaturates in the original tallow fatty acids.

These are linear *primary* alcohols, in which the OH group is joined to a carbon atom which has two hydrogen atoms attached (RCH_2OH); a *secondary* alcohol has only one hydrogen on the alcoholic carbon (R_2CHOH), and a *tertiary* alcohol has none (R_3COH).

Since the early 1960s linear primary alcohols have also been available from petroleum sources, namely ethylene. The process is similar to the Ziegler process for linear olefins, except that the last step is an oxidative one yielding the alcohol instead of the olefin. Being polymers of ethylene, the Ziegler alcohols come in even-numbered chain lengths. The average chain length and distribution of homologs can be controlled somewhat by the reaction conditions, and completely by subsequent distillation. The OH group is at the end of the chain, so they are identical with the alcohols derived from the saturated fatty acids, for example, **2.7**. However the two

$$C—C—C—C—C—C—C—C—C—C—C—C—OH$$

$$(2.7)$$

products may differ in minor ways because the minor components or impurities may differ.

Branched chain alcohols were also used extensively for surfactant manufacture prior to the changeover to more readily degradable products. They were usually derived from polypropylenes by the oxo process, which involves catalytic addition of carbon monoxide and hydrogen to the double bond in a sequence of reactions which add up to Eq. (2.13). Thus tetrapropylene gives a C_{13} alcohol, as highly branched as the original raw material, and primary.

$$C_{12}H_{24} + CO + 2 H_2 \rightarrow C_{12}H_{25}CH_2OH \qquad (2.13)$$

If a linear α-olefin is used in the oxo process, addition may occur at either end of the double bond to give a mixture of linear and singly branched primary alcohols as in Eq. (2.14). Substitution further down

$$C—C—C—C—C—C—C—C—C—C—C{=}C$$
$$\downarrow$$
$$C—C—C—C—C—C—C{-}C—C—C—C—C—OH$$
$$+$$
$$C—C—C—C—C—C—C{-}C—C—C—C—C \qquad (2.14)$$
$$\vert$$
$$C—OH$$

the chain occurs in only minor amounts, and with proper choice of reaction conditions the proportion of linear primary alcohol may reach 80% or more.

The development of linear paraffin supplies for production of LAS also made linear secondary alcohols feasible as surfactant hydrophobes. Here the OH group may be introduced by reaction of the paraffin with oxygen, or by chlorination and subsequent hydrolysis. In either case all possible isomers are formed and the OH group is found on any of the carbons along the chain.

F. Alkylphenols

The alkylphenol hydrophobes are produced by addition of phenol to the double bond of an olefin as in Eq. (2.15). The alkyl group may be

$$C_{12}H_{24} + \langle \rangle{-}OH \longrightarrow C_{12}H_{25}{-}\langle \rangle{-}\text{-OH} \qquad (2.15)$$

linked to the ring either ortho, meta, or para to the OH group. The earlier commercial products were derived from branched olefins, for instance, octylphenol from diisobutylene and nonylphenol from tripropylene. More recently linear alkylphenols became available with the development of linear olefins for LAS in 1965.

G. Polyoxypropylenes

As an example of hydrophobes which are not derived from hydrocarbons we can cite the polyoxypropylenes, polymers of propylene oxide [Eq. (2.16)]. The importance of such products stems from the possibility

$$H_2O + nC{-}C\overset{O}{\triangle}C \rightarrow HO{-}C_3H_6O{-}C_3H_6O{-}\cdots C_3H_6OH \qquad (2.16)$$

of exercising control over the hydrophobic character, which is intensified as the degree of polymerization, n, is increased. The character of other hydrophobes such as alcohols and alkylphenols may also be modified by addition of propylene oxide as in Eq. (2.17).

$$ROH + nC{-}C\overset{O}{\triangle}C \rightarrow R{-}(O{-}C_3H_6)_n{-}OH \qquad (2.17)$$

III. HYDROPHILIC GROUPS

The hydrophilic groups of today's surfactants are of two classes—those which ionize in aqueous solution and those which do not. The hydrophilic property of the former derives from a strongly acidic or basic character which permits the formation of true, highly ionizing salts upon neutralization. On the other hand the nonionic hydrophilic groups have a multiplicity of elements which are individually rather weak hydrophiles but which have a cumulative effect; increasing their numbers in the group increases the hydrophilicity of the aggregate to the degree desired.

The more common hydrophilic groups include:

Sulfonate ($-\Sigma$) $-SO_3^-$
Sulfate ($-O\Sigma$) $-OSO_3^-$
Carboxylate $-CO_2^-$
Quaternary Ammonium
      $-R_3N^+$
Polyoxyethylene ($-E_n$)
      $-O-CH_2-CH_2-O-CH_2-CH_2-\cdots-O-CH_2-CH-OH$
Sucrose
      $-O-C_6H_7O(OH)_3-O-C_6H_7O(OH)_4$
Polypeptide
      $-NH-CHR-CO-NH-CHR'-CO-\cdots-NH-CHR''-CO_2H$

The first four are ionic and the last three nonionic, in the terminology of surfactant chemistry.

The above list introduces two abbreviations which will be used throughout this book. The symbol Σ represents the sulfonate group (SO_3^-, SO_3H, SO_3Na, etc.) ($O\Sigma$ is the sulfate ester group), while E_n represents the polyoxyethylene group derived from n mols of ethylene oxide. (For other abbreviations see list of abbreviations.)

IV. SURFACTANTS

A. The Three Types

The hydrophobic groups, the hydrophilic groups, and the modes of linking them together may be permuted and combined to give surfactants in unlimited variety. Most of these are mere chemical curiosities. However, certain ones have been extremely important for scientific or commercial reasons. The following discussion gives some indication of their scope, but is not intended to be exhaustive.

Anionic surfactants are those which give negatively charged surfactant ions in aqueous solution, usually originating in sulfonate, sulfate, or carboxylate groups [Eq. (2.18)]. Commercially these are very important

$$RSO_3Na \rightarrow RSO_3^- + Na^+ \qquad (2.18)$$

and represent the major fraction of surfactants in use today. For the most part they are produced by sulfonation or sulfation of the desired hydrophobe and various representatives are described in Section IV.B.

Cationic surfactants are those which give a positively charged surfactant ion in aqueous solution, for example, quaternary ammonium derivatives [Eq. (2.19)]. These are of interest in the detergent industry principally

$$RN(CH_3)_3Cl \rightarrow RN(CH_3)_3^+ + Cl^- \qquad (2.19)$$

because of their bacteriostatic or germicidal properties. Their performance as detergents is rather poor, especially so considering their price, and they represent only a very minor fraction of the surfactants being used.

Cationic and anionic surfactants neutralize each other when present in the same solution together. The oppositely charged surfactant ions join each other to form a water-insoluble salt, and if stoichiometrically equivalent quantities are present, no surfactancy properties remain evident. If an excess of one type is present, the excess continues to exhibit its usual properties.*

Nonionic surfactants contain hydrophilic groups which do not ionize appreciably in aqueous solution. The ones of greatest commercial importance contain a polyether hydrophobe group derived from ethylene oxide (EO). They comprise perhaps 20% to 25% of the total volume of surfactants produced.

The polyoxyethylene group can be introduced into any hydrophobic group which contains an active hydrogen atom, for example a hydrogen atom joined to an oxygen atom as in an alcohol. EO reacts with the active hydrogen to form a hydroxy ether [Eq. (2.20)]. This product contains its

$$ROH + \overset{O}{\overset{/\,\backslash}{CH_2-CH_2}} \rightarrow ROCH_2CH_2OH \qquad (2.20)$$

own active hydrogen atom, so that it too reacts with the EO [Eq. (2.21)], giving another product with an active hydrogen, and so on. A polyglycol

$$ROCH_2CH_2OH + \overset{O}{\overset{/\,\backslash}{CH_2-CH_2}} \rightarrow ROCH_2CH_2OCH_2CH_2OH \qquad (2.21)$$

* The interaction is not quite so simple, according to Papenmeier (1969).

chain of any desired length can be built up simply by continued introduction of EO into the reaction mixture. The products, or the hydrophilic
groups thereof, are variously termed polyglycol derivatives, polyoxyethylene derivatives, ethoxylates, ethoxamers, or the like.

The product of such a reaction is a mixture. If 10 mols of EO are added
per mol of hydrophobe, molecular species will be present ranging from
perhaps 5 to 15 polyether units or broader, in something like a Poisson
distribution averaging 10 units per molecule and with the population
peaking around 10 units per molecule. Depending upon the exact properties desired and the nature of the hydrophobic group, the average number
of EO units from product to product may range from 8 or 9 to 20 or more.
Typical examples are shown in Section IV.C.

B. Anionic Surfactants

Some of the more common representatives of this class are cited in the
following paragraphs.

Alkylbenzenesulfonates are obtained by reaction of the parent alkylbenzene with sulfuric acid or sulfur trioxide to give the sulfonic acid,
which is then neutralized to give the desired salt, often the sodium salt
[Eqs. (2.22)]. The reactions are smooth and the yields quantitative. The

$$R\phi + H_2SO_4 \rightarrow R\phi SO_3H + H_2O$$
$$R\phi SO_3H + NaOH \rightarrow R\phi SO_3Na + H_2O \tag{2.22}$$

sulfonate group enters the ring predominantly at the para position. Minor
amounts of the ortho sulfonate may be formed also, particularly if the
geometrical structure of the R group is such that steric hindrance at
the ortho positions is minimized (Gray, 1955). Alkylbenzenesulfonates are
the most widely used surfactants because of their excellent detersive
properties, their low cost, and the attractive physical properties of their
formulations.

Alkane sulfonates (aliphatic sulfonates or paraffin sulfonates) cannot be
formed by direct sulfonation since paraffins are relatively inert toward
sulfuric acid. Commercial production is feasible via the sulfoxidation
reaction; the paraffin is reacted with sulfur dioxide and oxygen to give
the sulfonic acid which is then neutralized with the desired base [Eq.
(2.23)]. The reaction can occur at any of the hydrogens along the paraffin

$$C_{16}H_{34} + SO_2 + \tfrac{1}{2}O_2 \rightarrow C_{16}H_{33}SO_3H \tag{2.23}$$

chain, giving a mixture of isomeric sulfonic acids. By-products include
di- and polysulfonic acids, which are formed in increasing amounts as the

reaction is driven toward completion. To date the alkane sulfonates have not had much success in competition with alkylbenzenesulfonates, presumably because of a rather unfavorable price/performance relationship.

Primary alkane sulfonates have been of considerable importance in scientific studies of surfactancy phenomena since they are readily synthesized [Eq. (2.24), for example], easily purified, and can be designed with varied molecular architecture. Reaction (2.24) has no present commercial significance, but in suitable circumstances these products might become commercially available through addition of a bisulfite to linear α-olefins [Eq. (2.25)].

$$C_{16}H_{33}Br + Na_2SO_3 \rightarrow C_{16}H_{33}SO_3Na + NaBr \qquad (2.24)$$

$$C_{14}H_{29}CH{=}CH_2 + NaHSO_3 \rightarrow C_{14}H_{29}CH_2CH_2SO_3Na \qquad (2.25)$$

Olefin sulfonates now coming into commercial use result from the action of sulfur trioxide on linear α-olefins, with subsequent neutralization. The reaction is complex and follows several paths, giving two major products: alkenesulfonate **(2.8)** and hydroxyalkane sulfonate **(2.9)**. Other isomers

CCCCCCCCCCCC=CCΣ   CCCCCCCCCCCCCCCΣ
                                                          |
                                                          OH

(2.8)   **(2.9)**

of each are also formed, and disulfonates as well.

Ester and *amide sulfonates* may be produced by reacting a fatty acid chloride with short chain hydroxy sulfonic acids or amino sulfonic acids [Eqs. (2.26) and (2.27)]. These products, originally known as Igepon A and Igepon T, respectively, have been in commercial production for many years for relatively small specialty uses where certain desired properties offset their higher cost.

$$RCOCl + HOCH_2CH_2SO_3Na \rightarrow RCO_2CH_2CH_2SO_3Na \qquad (2.26)$$

$$RCOCl + \underset{\underset{CH_3}{|}}{HNCH_2CH_2SO_3Na} \rightarrow \underset{\underset{CH_3}{|}}{RCONCH_2CH_2SO_3Na} \qquad (2.27)$$

Sulfo fatty acids can be obtained by direct sulfonation with sulfur trioxide. The sulfonic group enters at the α-position [Eq. (2.28)]. The sulfonate and carboxyl groups may both be neutralized, or the carboxyl group may be reacted with an alcohol to give a sulfo fatty ester. Commercially, for the near future these products will probably be limited to small volume specialty uses because of price.

$$C_{16}H_{33}CH_2CO_2H + SO_3 \rightarrow \underset{\underset{SO_3H}{|}}{C_{16}H_{33}CHCO_2H} \qquad (2.28)$$

Primary alkyl sulfates are formed by reaction of primary alcohols with sulfuric acid and neutralization of the acid ester with the desired base [Eqs. (2.29)]. These products have been made in rather large volume since

$$RCH_2OH + H_2SO_4 \rightarrow RCH_2OSO_3H + H_2O$$
$$RCH_2OSO_3H + NaOH \rightarrow RCH_2OSO_3Na + H_2O \qquad (2.29)$$

the earliest days of synthetic detergents. Hydrophobes are usually linear, such as lauryl alcohol, tallow alcohol, Ziegler alcohols, and linear oxo alcohols. The primary alkyl sulfates are stable in neutral or alkaline solution, but in the presence of acids are readily hydrolyzed [Eq. (2.30)]. Since the sodium bisulfate produced is strongly acidic, once hydrolysis starts it is self-acclerating.

$$RCH_2OSO_3Na + H_2O \rightarrow RCH_2OH + NaHSO_4 \qquad (2.30)$$

Secondary alkyl sulfates are not usually produced from the corresponding alcohols since they are more readily formed by direct reaction of the corresponding olefin with sulfuric acid. The sulfate ester group does not necessarily enter only at the position of the double bond; for example, with a linear olefin it is found at all positions except the primary ones at the end of the chain [Eq. (2.31)]. (This indicates the formation of a carbonium ion as an intermediate in the reaction.)

$$C-C-C-C-C-C-C-C-C-C-C-C-C=C$$
$$\downarrow H_2SO_4$$
$$C-C-C-C-C-C-C\underbrace{-C-C-C-C-C-C}_{O\Sigma}-C \qquad (2.31)$$

Secondary alkyl sulfates derived from cracked wax olefins have been marketed in Europe for many years under the name Teepol, mostly as an unformulated aqueous solution. Like the primary alkyl sulfates, they are prone to hydrolysis by acids, and in addition are sensitive to heat, reverting to olefin and sodium bisulfate.

Sulfated nonionics are primary sulfate esters derived from polyoxyethylene ethers **(2.10)**. Ordinarily the number, n, ranges below 6 or 7, rather fewer than the number required in a satisfactory hydrophilic group in a nonionic surfactant.

$$R(OCH_2CH_2)_nOSO_3Na$$
(2.10)

C. Nonionic Surfactants

The nonionic products which have seen large scale commercial use fall into four general types.

Alcohol ethoxylates are derived from aliphatic alcohols by reaction with EO [Eq. (2.32)]. The branched chain product prepared from tetrapropylene C_{13} alcohol has declined in importance because of its greater resistance to biodegradation. It is being supplanted by the linear alcohols derived by Ziegler polymerization of ethylene and by oxonation of linear olefins.

$$ROH + nC_2H_4O \rightarrow R(OC_2H_4)_nOH \qquad (\text{or } ROE_n) \qquad (2.32)$$

Alkylphenol ethoxylates are made according to Eq. (2.33). Those derived from polypropylene alkylphenols are obsolescent in the detergent industry for reasons of biodegradability.

$$R\phi OH + nC_2H_4O \rightarrow R\phi OE_n \qquad (2.33)$$

Polyoxyethylene esters are formed by ethoxylation of carboxylic acids as in Eq. (2.34). The acid is usually a natural fatty acid or the mixture of fatty and rosin acids occurring in tall oil.

$$RCO_2H + nC_2H_4O \rightarrow RCO_2E_n \qquad (2.34)$$

Polyoxyethylene-polyoxypropylene derivatives are mixed polymers with hydrophobic groups derived from propylene oxide, further reacted with EO until the desired properties are attained. Ordinarily they have rather high molecular weight, often with much more than the usual 8 to 15 mols of EO characteristic of the other nonionics.

D. Mixed Surfactant Types

Products are also in use which do not fit precisely into one of the three main categories, anionic, cationic, or nonionic, or which belong to more than one. These have been designed with a variety of special properties for specialty uses, and are produced in relatively small quantities. They are thus of quite minor significance in wastewater problems.

CHAPTER 3

THE ANALYTICAL METHODS

I. LIMITATIONS AND PRECAUTIONS

Methods in limitless variety can be used for estimating the progress and extent of biodegradation of a surfactant in the system under observation. We may make use of physical or chemical properties imparted by the surfactant itself or by its intermediate degradation products. Or, we may measure related functions such as the amount of oxygen absorbed or of carbon dioxide evolved by the organisms in the system. Each method has advantages and none is free of limitations. Fischer (1962) has reviewed methods for determination of the surfactant, directly or indirectly, while Allred (1964b) compares several analytical techniques, illustrating how they supplement each other in measuring surfactant biodegradation.

A. Sample Preservation

In many cases analysis of samples from biodegradation experiments cannot be carried out immediately upon collection, and steps must be taken to minimize change during storage prior to analysis. Generally, chemical preservations are used, selected with consideration of possible chemical changes which they may cause in the sample or of possible interference in the subsequent analysis. Loehr (1967) has reviewed the usual chemical methods.

Acidulation is perhaps the simplest. For instance, Renn (1964a) added enough sulfuric acid to bring the sample to pH 2, and subsequently neutralized for making BOD determinations. Allred (1964b) added 2 ml of 12 N HCl to a 12 ml sample.

Based on a comprehensive interlaboratory study with 16 participants, the Institute of Sewage Purification recommended that if not analyzed immediately after collection, samples should be preserved by addition of 50 ppm of mercuric chloride. Without preservation, losses of MBAS content averaged some 25% for TBS and 30% for LAS (Gameson, 1962).

Urban (1965) noted persistence of some viable bacteria in sewage and effluents at 10 ppm of mercuric chloride, but complete sterility at 50 ppm. Mercuric chloride has also been recommended by Eden (1962), Fischer (1965), and Klein (1965a, b), and is probably the most generally used method of preservation. Howe (1969) reports detailed comparisons with sulfuric acid.

Hellwig (1964) concluded that 60 to 80 ppm of mercuric chloride was the best of several methods for 2 week preservation of water samples in pollution studies, so far as 19 quality parameters were concerned. Chloroform, thymol, formalin, potassium cyanide, and sulfuric acid were rejected either because of poor preservation or because of interference with analytical methods. Subsequently Hellwig (1967) described the use of mercuric chloride at much higher levels, 900 ppm, to preserve sewage samples. Precipitation of the mercury as sulfide before the eventual analysis avoided interference.

Since chloroform is used in the MBAS analytical procedure, it is convenient to use also as preservative. Fairing (1956) states that addition of 2000 to 4000 ppm and refrigeration at 4° to 10°C permits storage for at least 2 weeks without loss of TBS content. Walker (1962) says that 2000 ppm of chloroform is better than 4000 ppm of hydrochloric acid or 200 ppm of mercuric chloride.

Formaldehyde has also been popular as a preservative. For example, House (1956) used $\frac{1}{2}\%$ of a 40% solution, while Sweeney (1964a), Knapp (1965), and Bunch (1967a) used 1%.

Low temperature is an alternate to chemical preservation. Bacterial action may still occur slowly even near 0°C if the sample is still liquid, and complete freezing is somewhat surer (Allred, 1964b). It is generally accepted that samples can be stored indefinitely if frozen, although comprehensive experimental investigations in support of this opinion seem to be scarce. Storage of raw sewage in the liquid state has been investigated by Loehr (1967) with regard to stability of COD, BOD, dissolved oxygen, pH, and suspended solids, with the conclusion that those properties remain unchanged for at least 6 days at 1°C. At 10° to 20°, storage for 12 to 24 hr produced no important variations in BOD, COD, or suspended solids, and no major changes in these occurred during 6 to 12 hr at 37°C.

Heat sterilization of raw sewage has been found to increase the dissolved solids, volatile dissolved solids, and reducing sugars (Jones, 1968c).

Berg (1966) describes a variant of heat and chemical sterilization which involves heating at 80°C for 1 hr, followed by addition of the antibiotic polymyxin B sulfate (1.5% to 3% of the COD) to take care of spore-forming organisms. The COD of the sewage was stabilized for about 40 days or longer.

B. Problems of Representative Sampling

When working with surfactants at the low concentrations usually met in biodegradation research, one universal limitation must always be kept in mind. Surfactants tend to concentrate or adsorb at interfaces (Chapter 2, Section I.B), and this calls for appropriate precautions. Foam, if present, will contain a much higher concentration of surfactant than the main bulk of solution, which may be significantly depleted thereby. The difficulty of obtaining a representative sample for analysis under such conditions is manifest; efforts should be made to avoid foaming conditions, or at least the foam should be allowed to subside before sampling.

Adsorption onto the walls of pipets, graduates, optical cells, and other equipment may also occur, causing depletion and cross-contamination of dilute solutions. Weber (1962) has reported that 0.01 M KH_2PO_4 solution will desorb the surfactant, so that its use as a medium and rinse in optical measurements will minimize errors from that source. Fairing (1956) took a different approach, recommending that all glassware be coated with a silicone water repellent; additionally the glassware was rinsed at certain stages with ammoniacal methanol to recover adsorbed surfactant.

C. Desorption Procedures

Biological growth, for example pure bacteria or activated sludge, may also adsorb significant quantities of surfactant (Chapter 4, Section V). Such adsorption may be the first step in biodegradation, but if the surfactant is not degraded, significant amounts may accumulate on the sludge—up to several percent of its dry weight in some conditions. If this factor is neglected and only the liquid phase analyzed, erroneous interpretations may ensue.

Desorption of ABS, and presumably of other surfactants as well, from sewage sludge, activated sludge, and the like, can be accomplished by extraction. Hot water (Hartmann, 1963a; Kelly, 1965; Tomiyama, 1968), ethanol (AASGP, 1961; Maurer, 1965), and methanol (Roberts, 1958; Fischer, 1962; Phillips, 1963, pp. 33, 107–108; Bruce, 1966) have all been used. After a comparative study Pitter (1967b) concluded that two methods were most useful and developed improved versions:

(i) For accurate, precise results (over 98% recovery) the sludge is dried, powdered, extracted 20 hr with methanol or ethanol, and an aliquot of the ethanolic extract is analyzed by the methylene blue procedure after dilution with water (adapted from Roberts, 1958).

(ii) For rapid but less accurate results (90% to 95% recovery) the wet sludge is mixed with methanol or ethanol, agitated 30 min and an aliquot of the supernatant is analyzed as above (adapted from Maurer, 1965).

Fairing (1956) points out that acid conditions increase the anionic surfactant binding power of proteins and hence are to be avoided. Meinck (1961) used a preliminary acid treatment, acidifying the sludge with HCl and boiling for 15 min under reflux, after which the excess acid was removed by evaporating almost to dryness and air-blowing. Excess alkali was then added and the mixture refluxed an hour to eliminate protein interferences before methylene blue analysis. Allred (1964b) also recommends preliminary hydrolysis of biodegradation samples by boiling with HCl in order to liberate surfactant from within the bacterial cells and from the cell walls. Using LAS randomly labelled with ^{14}C he showed that without hydrolysis a sizable fraction resisted adsorption by charcoal, presumably because of retention by the cells. After hydrolysis the charcoal captured substantially all of the ^{14}C activity.

Gould (1962) developed an alkaline aqueous acetone solvent highly active in desorbing anionics, while Huber (1962, 1968) used methanolic sodium hydroxide for activated sludges. Use of the organic solvent is probably more efficient than the hot aqueous NaOH procedure of Degens (1955).

For anionics Schönborn (1966, p. 93) preferred Soxhlet extraction with methanol instead of the methanolic sodium hydroxide reflux method, because it gave a cleaner extract. However, he did use the latter for desorbing nonionics from activated sludge prior to his cation exchange resin–mercuric iodide analysis.

II. SURFACTANCY METHODS

A. Foam

A.1. *Applicability and Restrictions*

The major impetus for study of surfactant biodegradation came from the residual foaming properties of incompletely degraded surfactants. Accordingly it is both fitting and pertinent that foam be used as a measure of the progress of biodegradation in a test system. The early work of Bogan (1956) demonstrated the validity of this approach.

However, the foaming phenomenon is a complex and transient one, deriving from the interaction of several physical properties and describable only in terms of multiple parameters—even then only incompletely and with a sorry lack of precision. Thus foam may be characterized by its volume, its texture, its persistence. For a given surfactant these will be dependent upon the concentration of the surfactant and of other materials present (which may either enhance or inhibit), the means used for generating the foam, the size and shape of the container, the previous history of the solution, temperature, humidity, and undoubtedly other factors as well.

If a system which initially gives a copious foam is subsequently found to be incapable of generating any significant amount after exposure to the chosen biodegradation conditions, a tentative conclusion would be that the surfactant has been degraded to some nonfoaming product. Often a system will show foaming properties without added surfactant, because many "natural" products have their surfactant properties also, particularly the proteinaceous products of biological origin—consider beer for example. In any event, there is a need for defining what amount of foam is a "significant" amount, and this implies a quantitative scale for describing foaming phenomena. Since a multidimensional system would be required, a one-dimensional ranking of test solutions or surfactants in an order of merit would be impossible.

Foaming is not a linear function of surfactant concentration. At a few tenths of a percent as used in laundering, the nonionics are often quite low in foaming tendency compared to the anionics. At a few parts per million the foaminess is not as pronounced, but the nonionics may equal or exceed the anionics in what foam there is. Such discrepancies are also possible among the partial degradation products of the surfactants as well. It is conceivable that one surfactant might be much more extensively degraded than another on a weight basis, yet might then show higher foaming properties if its degradation intermediates were better foamers.

Determination of surfactant concentration by foam measurements in waters of unknown composition is, of course, impossible. Foamability of a sample of sewage, treated effluent, river water, or the like certainly implies the presence of surfactant, by definition. But the surfactant may be of natural origin, such as protein, or of nondetergent origin, and useful conclusions cannot be drawn without other sorts of information.

A.2. *Determination of Foaming Potential*

Despite the problems of quantitative estimation, foaming properties have been used by numerous workers for monitoring the biodegradation process, particularly in the field of nonionic surfactants where chemical analysis is difficult. Thus in the SDA's Biodegradation Subcommittee procedure a 50 ml sample is shaken in a 100 ml glass-stoppered graduate and the volume of the resultant foam measured 15 and 60 sec later (SDA, 1969b). To increase reproducibility and sensitivity Bacon (1966) suggests rotating mechanically a 600 ml sample in a 1 liter graduate; the graduate is used as the container for carrying out the biodegradation test (river water), so that transfers need not be made.

Taylor (1965) recommends the "quart bottle foam test" which also is done in the original container without transfer. The 800 ml sample in a quart bottle is shaken vigorously ten times through an arc of about a foot in about 4 to 5 sec. The percent area of the liquid surface covered with foam is estimated after 1, 10, 60, and 120 min. Foam height is not measured, an amount of foam that covers the whole surface being considered environmentally unsatisfactory anyway.

Nelson (1961) describes a test wherein 150 ml of the liquid is placed in a 7 cm diameter tube 50 cm high and aerated by 100 ml/min of air introduced at the bottom through a sintered glass diffuser. The foam height is measured after 5 min. The South African Institute for Water Research (Urban, 1965, p. 32) uses a 2 liter sample, a 4 inch tube, 80 ml/min of air, and measures the maximum foam height before its first collapse. Studying variables involved in the foaming of activated sludge systems, Polkowski (1959) also used an air flow device for dynamic foam measurement as well as several simpler foam-generating procedures for static measurements.

Feng (1962) selected an arbitrary endpoint and determined what dilution of his sample was necessary to reach it. For instance a 4 ppm solution of TBS in distilled water would just start to form a clear spot at the center of the surface 15 sec after shaking. Unknown concentrations of TBS could be estimated by serial dilutions. Although inapplicable to determination of the concentration of unknown surfactants in unknown systems, their relative foaming tendencies may be compared by this method.

A.3. *Preconcentration by Foam Stripping*

Foaming is useful in preconcentration prior to analysis as well as in the analytical estimation of surfactants. Since the surfactant content of biodegradation systems and wastewaters is usually low and the content of other and potentially interfering materials high, a preconcentration or prepurification step is often applied to improve the response in the final analysis. This may be done by extraction or by adsorption and desorption under appropriate conditions, chosen to accept the surfactant and reject the other solutes. The foam-stripping procedure accomplishes these objectives by using some of the characteristic and unique surfactant properties described in Chapter 2, Section I.B. Thus it can in principle be used to concentrate the surfactant molecules from the system regardless of their origin or chemical nature, and without the prior knowledge of their other properties so necessary in designing other separation procedures.

If air is bubbled through a dilute surfactant solution, the concentration of surfactant in the resulting foam is much higher than in the main body of the liquid because of surface adsorption at the air–liquid interfaces which make up the foam. If the foam is continually removed from the system as it is formed, the solution is progressively depleted in surfactant content to a point, often below 1 ppm, at which it will no longer maintain a stable foam. The removed foam will contain most of the surfactants of the system regardless of their exact chemical nature, and only minor amounts of the other solutes. When it collapses (very quickly in the presence of an organic liquid such as ether or chloroform), the volume may be as little as one-hundredth of the original, and the subsequent analysis or examination is often greatly facilitated.

The operation is usually carried out in a vertical column, tall enough so that the foam can drain free of a large proportion of the solution phase and consist as much as possible of surface film. Bolton (1961) describes such equipment used to isolate TBS and LAS for IR determination of their relative proportions in sewage samples during the Luton test. Osburn (1966) used a somewhat different design in isolating nonionic surfactants, either intact starting material or intermediate degradation products, in biodegradation studies. Those intermediates which had lost surfactancy were largely left behind in the main body, from which they could be isolated later by extraction.

This same principle was applied by Sharman (1964b) and House (1965b) in their process for biodegradation of TBS in an activated sludge system. The TBS was continuously removed from the effluent by foam-stripping and recycled back to the aerator, where it built up to a steady state concentration high enough so that it could degrade at the same rate as introduced via the incoming sewage.

B. Surface Tension

B.1. *Origin and Limitations*

In contrast to the complexities of describing foam, surface tension is a specific, readily measured physical property describing the strength of the surface layer of a body of liquid. The surface tension of water at room temperature is around 72 dynes/cm; i.e., a force of 72 dynes is required to rupture a 1 cm length of the water surface. This cohesiveness of the surface film results from the nonisotropic distribution of forces between the molecules at the surface. In the main body of the liquid the molecules are surrounded by others on all sides and their forces of attraction are evenly balanced. In contrast, at the surface the molecules are attracted by adjacent ones in the surface and in the body but not nearly as much by the fewer molecules in the gas phase; this results in surface tension.

In the presence of a surfactant at concentrations of a few parts per million the surface tension of water is lowered significantly. It decreases further, although not linearly, with increasing concentration up to the critical micelle concentration, or CMC. From that point on there is little further change with further increase in concentration. With ordinary detergent-type surfactants the surface tension is in the range of 25 to 35 dynes/cm and the surfactant concentration in the range of 50 to 200 ppm when the CMC is reached. High concentrations of salts may lower the CMC, perhaps to 5 to 50 ppm.

The pronounced effect of surfactant upon surface tension at such low concentrations comes from the tendency of surfactant molecules to congregate at the surface (Chapter 2, Section I.B). At a specific concentration in the body of the solution, the surfactant will establish a certain specific, rather higher, concentration at the surface, giving rise to a characteristic value for the surface tension. Unfortunately, the equilibrium between body and surface may require a significant amount of time to become established, and therein lies a difficulty. Consider a surfactant solution having an equilibrium surface tension of say 55 dynes/cm. If the surface layer is decanted, simply by pouring off some of the solution, the remaining liquid may show around 68 to 70 dynes/cm, and it may take an hour or more before the equilibrium value of 55 is reestablished. A similar delay may occur in a sample transferred by pipet. Kloubek (1969) shows that the equilibration time increases with decreasing surfactant concentration and that 5 hr or more may be required.

Thus considerable care must be exercised if surface tension is expected to be a valid measure of surfactant concentration. Even then, as pointed out by Blankenship (1963), it can be only a qualitative measure in biodegrading systems because different surfactants differ in their degree of lowering surface tension (and likewise their various intermediate biodegradation products). The SDA (1969a, b) Biodegradation Subcommittee found that interferences caused poor reproducibility in measuring surface tension in biodegradation systems, and preferred determination of foaming potential.

B.2. *Procedures*

Of the many methods available for measuring surface tension, the duNouy tensiometer has been used most frequently, for example, by von Riesen (1955), Huddleston (1963), and Vath (1964). A circular wire ring, about 2 cm in diameter, is submerged horizontally below the surface, and is then lifted through the surface, with simultaneous measurement of the force applied. The maximum force comes at the instant the surface film is broken, and from this and the ring dimensions the surface tension is calculated. In addition to the errors of nonequilibrium discussed above, errors may also be introduced by the presence of particulate matter if it causes premature rupture of the surface film.

The hanging drop method should be less sensitive to particulate matter and probably would require less time for surface equilibration, but requires more time on the part of the operator. The surface tension is calculated from the shape of the hanging drop—the lower the surface tension the greater is the distortion from spherical by the force of gravity.

The capillary rise method for measurement of surface tension appears to be inapplicable in this field according to House (1965a); adsorption of the surfactant onto the walls of the capillary leaves the rising solution depleted, resulting in values much too high. Schwen (1966) found that such systems did eventually equilibrate, but that several days was required.

III. SPECIFIC CHEMICAL METHODS

A. Methylene Blue

A.1. *Chemical Principle*

For following the progress of biodegradation of anionic surfactants the methylene blue method is by far the most widely used. It is extremely sensitive, and in some of its modifications extremely simple to use. Methylene blue (**3.1**) is a cationic dye which, in the form of an inorganic

$$(3.1)$$

salt such as the chloride or sulfate, does not extract from water into an organic liquid such as chloroform. But if anionic surfactant is present, a salt of much lower water solubility is formed with the surfactant anion, a salt which is readily extractable into organic solvents, in much the same manner as salts between cationic surfactant and anionic surfactant.

Measurement of the intensity of blue color in the solvent gives a measure of the amount of anionic surfactant present, one molecule for each molecule of methylene blue. Its intense color makes for adequate sensitivity: 10 μg of ABS is readily detected, corresponding to a 100 ml sample of 0.1 ppm concentration.

A.2. *Interferences*

The reaction is responsive not only to anionic surfactants, but generally to any materials containing a single strong anionic center, strong enough to form a stable salt with the methylene blue cation, and at the same time containing a hydrophobic group sufficiently lipophilic to bring the salt preferentially into the organic layer, from the aqueous phase.

Although soaps are sufficiently hydrophobic, the carboxylate anion is not strong enough for salt formation under the acidic conditions usually used, and soaps are unresponsive in the method.

On the other hand, disulfonates, which may be present as impurities in some anionic surfactants, are generally overly hydrophilic and tend to remain in the aqueous phase, giving low results. Such a trend is quite evident in the analyses reported by Ōba (1968b) (Table 3.1). In a series of

TABLE 3.I

Methylene Blue Response of Disulfonates in α-Olefin Sulfonates

| | | MBAS, ppm[a] | |
| | Disulfonate | --- | --- |
Sample	content, %	Found	Calcd. for mono content
AOS(C)	<4	36	36.5
AOS(D)	15	31	32.4
AOS(E)	50	23	19.0

[a] Calculated from data of Ōba (1968b) assuming concentration 30 ppm in all cases; molecular weight of AOS(C) stated to be 353. MBAS values in terms of Manoxol OT, mol. wt. 444.

C_{15-18} α-olefin sulfonates the methylene blue active substance (MBAS) found was not far different from that calculated for the monosulfonate content, regardless of the disulfonate content. In other words, the response of the disulfonate was very much less than stoichiometric.

High concentrations of chlorides force some amounts of methylene blue into the chloroform layer, and hence cause positive interference; thus the method is inapplicable in seawater without precautionary measures. For instance, Evans (1950) reported that 1.79% NaCl solution at pH 1.8 gave the same response as 10 ppm of alkyl sulfate surfactant, as also did 1040 ppm of nitrate or 40 ppm of thiocyanate. He found that the methylene blue response of these ions decreased with increasing acidity, while that of the surfactant remained constant. He was able to compensate for the interference by making determinations at pH 3.25 and 0.7; linear extrapolation to pH −2 gave the amount of surfactant alone. Degens (1953) was able to minimize such interferences by special adjustment of dilution and vigor of extraction.

Evans was also able to use his extrapolation technique to eliminate the interference of the organic substances, possibly sulfate esters, which cause urine to give responses corresponding to 100 ppm or more of MBAS.

Amines and other cationic substances such as proteins may give rise to negative interference by competing with the methylene blue for salt formation by the surfactant anion. Fairing (1956) has examined this problem in considerable detail and developed means of avoiding it by a prepurification procedure involving extractions in the presence of 1-methylheptylamine. Bolton (1961) reports both negative and positive interferences by materials present in sewage, and that ABS extraction can be made quantitative by operation at pH 10.

A.3. *Procedures*

The fundamental methylene blue procedure is detailed in *Standard Methods* (APHA, 1965, p. 297). Considerable effort has gone into the investigation of various other combinations of cationic dye and immiscible solvent to minimize interferences (for example Edwards, 1954; LGC, 1962, p. 37) but none appear to offer enough advantages to gain much popularity. After methylene blue–chloroform, perhaps the most widely used is the Moore-Kolbeson methyl green–benzene method (Moore, 1956).

Of the methylene blue modified methods, the most successful is that of Longwell and Maniece (Longwell, 1955; ABCM-SAC, 1957), in which the methylene blue extraction is carried out twice, first at pH 10, with subsequent treatment of the chloroform layer at the Standard Method pH of 2. Among the potentially interfering substances which this variant removes are carboxylated sulfonates. These are formed as intermediates during the biodegradation of sulfonate surfactants. The carboxyl group is not ionized at the acidic pH of the Standard Method, so it behaves in a hydrophobic manner and the molecule responds to the test under those conditions. In the Longwell-Maniece method the carboxylate is ionized in the alkaline extraction step, giving a molecule which resembles the disulfonates in resisting extraction, and hence its interference is avoided (Allred, 1964b; Swisher, 1964a). The methyl green method rejects these materials partially, according to Allred (1964b), who found that sulfophenylundecanoic acid gave only about one-third of the stoichiometric response.

Automation of the methylene blue method has been accomplished using the Technicon Autoanalyzer. A series of samples is introduced into a continuous-flow system providing for addition of methylene blue reagent, mixing, extraction with chloroform, measuring, and recording the color intensity of the chloroform stream. Application to biodegradation analyses has been discussed by Barnhart (1963a), Huddleston (1964a), Renn (1964a), Setzkorn (1964), Testa (1964), and Kelly (1965), while de Bolt (1965), Södergren (1966), and de Jong (1969) have described it in detail.

The Hellige method is a much simplified variant of the Standard Method; the color is measured by visual comparison with a set of glass standards instead of with a spectrophotometer. A single chloroform extraction is used, and is sufficient to recover surfactant quantitatively in most cases. Thus only a portion of the organic layer need be taken for color measurement, eliminating the trouble of breaking any emulsion for complete recovery, as is necessary in the Standard Method (Swisher, 1964a). Michelsen (1961) developed a similar single-extraction procedure using copper–cobalt solutions instead of glass standards for color comparison.

A.4. *Molecular Weight, Stoichiometry, and Calibration*

Complete removal of the surfactant–methylene blue salt from 65 ml of aqueous phase into 10 ml of chloroform phase in a single extraction, as in the Hellige procedure (above), indicates an extremely favorable distribution coefficient for the methylene blue salt. As the size of the hydrophobe group is diminished, however, the salt becomes more hydrophilic and the extraction less complete. This effect is first noticeable in the octylbenzenesulfonates; it is quite pronounced in the hexylbenzenesulfonates and their removal is incomplete even with three extractions. The decrease in methylene blue response parallels the decrease in surfactancy as the hydrophobe group is shortened, and methylene blue thus gives a good measure of any sulfonates with surfactancy properties present in an unknown system.

Above octylbenzene the response of the sulfonates is close to stoichiometric, each molecule of surfactant carrying with itself one molecule of methylene blue into the chloroform (Laws, 1959; Hammerton, 1962; Swisher, 1964a). Therefore a single calibration curve is ordinarily sufficient to give results on a molar basis for all detergent range surfactants. This can be converted to weight basis by applying a molecular weight factor, if the molecular weight is known.

Despite the extremely favorable distribution coefficient mentioned above, there are indications that the rate of extraction into the chloroform phase may vary with the type of sulfonate. Specifically, LAS types may be slower to extract than TBS, even though they reach the same equilibrium value. This factor is important in an automatic analysis system which may derive its results from nonequilibrium values of the color intensity in the chloroform solution. In such a case, different calibration curves may be required for TBS and for LAS even with identical molecular weights.

The relationship between methylene blue response and chain length discussed above relates to the intact surfactants and is not particularly pertinent in the detection of intermediate biodegradation products. The intermediates are predominantly carboxylated derivatives, and their response versus chain length has not been well explored. In the Standard Method sulfophenylundecanoic acid, the 11-carbon homolog, is somewhat less responsive than C_{11}LAS (Swisher, 1964a), so it may be surmised that the cutoff point for the shorter intermediates might be somewhat above the 5- or 6-carbon chains noted for the unsubstituted ABS.

A.5. MBAS—Methylene Blue Active Substances

Since the methylene blue reaction is so lacking in specificity, the entities detected by it in unknown waters are properly referred to simply as MBAS in the absence of further identification.

Ōba (1965c, d) approached this problem using thin layer chromatography. Using a 5:1 chloroform:ethanol developing solvent the ABS–methylene blue salt remained at the starting line while certain interfering MBAS migrated. Development with 99% ethanol then dissociated the ABS salt and the free ABS migrated to its characteristic location.

A.6. Micromols, Milligrams, and Standards

The methylene blue response of detergent-type surfactants is fundamentally an indication of the number of mols present, and information as to its molecular weight is necessary before the results can be expressed in terms of weight, such as parts per million. In Britain the usual practice is to express the result as parts per million of a standard surfactant, Manoxol OT, giving an equivalent color (Bolton, 1961). Manoxol OT, also known as Aerosol OT, is the sodium salt of bis(2-ethylhexyl) sulfo-succinate (3.2), mol. wt. 444. It would seem more appropriate in the case of wastewaters to use a $C_{12}TBS$, mol. wt. 348, as a standard; 3.48 ppm of that would correspond to 4.44 ppm of Manoxol OT.

$$NaO_3S-C-CO_2C-\overset{\overset{\textstyle C-C}{|}}{C}-C-C-C-C-C$$
$$C-CO_2C-C-C-C-C-C$$
$$\underset{\underset{\textstyle C-C}{|}}{}$$

(3.2)

In France sodium myristyl sulfate has been used as a standard, while the Comité Internationale des Dérivés Tensio-Actifs used purified sodium lauryl sulfate (Reid, 1967, 1968). Another compound suitable as a primary standard is 1-phenyldodecane-p-sulfonate sodium salt (Gray, 1955). It is readily recrystallized to a high degree of purity by virtue of its very low solubility (200 to 300 ppm) in cold water compared to hot, and it is chemically stable and nonhygroscopic.

House (1954), Gardner (1967), and Reid (1968) all recommend use of TBS or LAS sulfonic acid as standard. It is prepared by ether extraction of an aqueous mixture of the sodium salt and hydrochloric acid, recovered from the ether, and standardized by titration with sodium hydroxide. Errors would, of course, be introduced if the extract contained other acids than the ABS acid, or if it contained salts of the ABS.

Methylene blue itself might be a desirable primary standard except for difficulties of purification, and also the instability toward oxidation by atmospheric oxygen detailed by Abbott (1962).

A.7. *Precision and Accuracy*

Pitter (1960) concluded that the methylene blue method was suitable for analysis of surface waters but that errors might be expected in sewage analysis. Gameson (1962) details a study by the Institute of Sewage Purification covering Manoxol OT and Dobane PT and JN sulfonates at several levels in several media, including settled sewage and river water. Results from the 16 laboratories involved ranged from about 7 to 13 ppm on a 10 ppm sample and from 1 to 3 on a 2 ppm sample. Standard deviations were about 10% to 20% of the mean value.

Standard Methods (APHA, 1965, pp. 298–299) reports a standard deviation of ± 10–15% in an interlaboratory study involving 57 analysts.

A later study with 111 participants using the SDA standard LAS gave the same result (Lishka, 1968). The Fachkommission "Abwasser" of the German Surfactant Committee (DAGS, 1961) has reported similar studies, as has Bolton (1962), while Wayman (1965) found adequate agreement between the Hellige and the Standard Method.

It is safe to say that whatever the uncertainties introduced into biodegradation results by any of these methylene blue analysis methods, they are on the average considerably smaller than those of biological origin. Setzkorn (1964) agrees.

A.8. *Two-Phase Titration*

Although usually used at higher levels of surfactant concentration, the two-phase titration method may have some potential use at the lower levels found in biodegradation work. There are several modifications, usually using methylene blue (or other cationic dye) as the indicator, because of the same properties discussed above. These methods, variants and improvements of one developed by Epton (1948)—for examples see Weatherburn (1951), Reid (1967)—are capable of high precision. The anionic surfactant present is titrated with a standard solution of cationic surfactant such as Hyamine 1622. This binds the anionic more strongly than does the methylene blue, which is added in small amount.

The titration is carried out in mixed aqueous chloroform medium, and the presence of any unneutralized anionic is signalled by the blue color of its methylene blue salt in the chloroform phase. At the beginning all of the methylene blue is in the chloroform in that form; near the endpoint the cationic titrant, having neutralized all the free anionic, begins to abstract it from its methylene blue salt and the blue color moves into the aqueous phase. The theoretical endpoint (colorless chloroform) at the completion of this process is difficult to detect precisely, so a slightly earlier, arbitrary but accurately reproducible, endpoint is usually taken when the colors of the two phases are equally intense.

The Analysis Commission of the Comité Internationale des Dérivés Tensio-Actifs has developed a standardized titration method and has reported on the use of various primary standards and on the reproducibility (Reid, 1967, 1968).

The STCSD (1966, Appendix IIIB) describes a method said to be suitable for relatively unpolluted samples, in which 1 ml of titrant corresponds to 1 ppm of anionic in a 100 ml sample. Edwards (1954) had earlier developed a procedure for use with sewage wherein the interfering emulsions were broken by centrifugation. Both of these methods use the anionic bromphenol blue, carried into the organic layer by the excess cationic titrant which begins to appear beyond the endpoint.

B. Sulfate

Formation of inorganic sulfate from sulfonate or sulfate type surfactant means that a corresponding amount of the surfactant has been destroyed, at least so far as its surfactancy properties are concerned. Two methods have been used frequently: (i) turbidimetric determination as $BaSO_4$ and (ii) radiotracer techniques involving ^{35}S-labeled surfactant.

A turbidimetric procedure is recommended by *Standard Methods* (APHA, 1965) for sulfate concentrations from 10 down to 1 ppm, which is stated to be the minimum detectable concentration. The $BaSO_4$ is precipitated by adding $BaCl_2$ under controlled conditions in the presence of $\frac{1}{2}\%$ to 1% of glycerol, ethanol, HCl, and NaCl, and the resulting turbidity is measured with a photometer. Ryckman (1956) checked the method on knowns containing (after dilution) 20 ppm of sulfate and 30 ppm of various alkylbenzenesulfonates. Short chain alkyl groups had no effect, but in the range of C_{10} to C_{14} alkyl the turbidity was decreased and the apparent sulfate was only 7 to 15 ppm. He minimized this interference by using smaller samples and thus precipitating the $BaSO_4$ at lower concentration.

Cordon (1968b) considered this method unsuitable in the presence of surfactants and developed one in which the precipitation was done in 50% aqueous ethanol containing gelatin. The ethanol promoted complete precipitation of the $BaSO_4$, which otherwise was solubilized to some extent by the surfactant. Some sulfonate surfactants themselves gave turbidity in the absence of sulfate ion, until primary biodegradation had occurred, necessitating a correction factor calculated from a parallel MBAS analysis.

Radiotracer techniques can detect even minute amounts of sulfate formed from ^{35}S-labeled sulfonates. House (1956) gives a detailed procedure in which ordinary Na_2SO_4 is added to the sample as carrier, precipitated as $BaSO_4$, and any ^{35}S therein detected by counting. Michael (1968) isolated the sulfate as Na_2SO_4, precipitating it from aqueous solution by addition of ethanol.

C. Polyethoxylate Nonionic Problems

The determination of the polyethoxylate surfactants at low levels in biodegradation media or in sewage or natural waters is considerably more complex than for the anionics, where the methylene blue method has found almost universal acceptance. The polyethoxylates have a wider variety of chemical structures, their biodegradation intermediates are not as sharply delimited from the intact surfactant in analytical response, and interference by naturally occurring materials is more troublesome.

These factors make some sort of prepurification almost mandatory, and the conditions used for it can profoundly affect the analytical results. The prepurification may reject biodegradation intermediates, for example carboxylated ethoxylates, which might otherwise respond to the analytical reactions and which might still be environmentally objectionable because of foaming or other properties. If the analytical reaction itself is not sensitive to surface active intermediates, this same difficulty is encountered, as pointed out by Huddleston (1965b). For these reasons the SDA Biodegradation Subcommittee has tended toward the use of foam detection in measuring nonionic biodegradability, rather than employing specific chemical procedures.

Heinerth (1966), Schönborn (1966), and SDA (1969b) have reviewed in detail the chemical analytical procedures applicable for nonionics in biodegradation and environmental studies. They usually depend on some reaction characteristic of the polyethoxylate chain—formation of complexes with compounds of cobalt, bismuth, tungsten, molybdenum, mercury, and the like, which are discussed in the following sections. Wickbold (1966) has also reviewed these methods, more briefly, in conjunction with development of one of his own.

Han (1967) devised an entirely different method wherein the polyethoxylate, after isolation by chloroform extraction, is converted to the EO sulfate by addition of chlorosulfonic acid. It is then determined by the methylene blue method.

D. Cobalt Thiocyanate

D.1. *Principles and Procedures*

Like the methylene blue method for anionics, the cobalt thiocyanate method has been the subject of considerable study and several variants have been developed. It is based on the formation of a complex with the polyethoxylate hydrophilic group, presumably held together by secondary valence forces bonding the ether oxygens to the central cobalt atom. The hydrophobic group confers lipophilic properties upon the complex, and it can be extracted from the aqueous phase into an organic solvent. There its amount can be estimated from the intensity of its absorption bands in the UV around 320 nm, or, with less sensitivity, at 620 nm in the visible. Quantities down to and below 100 μg can be determined, corresponding to a 100 ml sample of 1 ppm.

This method seems to be responsive for hydrophilic groups as short as E_5 or E_6. Below this limit complex formation is incomplete, presumably because of an insufficiency of ether oxygens in the molecule and resultant weakening of bonding in the complex. The method has been applied to the analysis of nonionic biodegradation systems by many, including Crabb (1964, 1968), who preextracted the nonionic with ether, and by Greff (1965), who formed the complex directly in the biodegradation medium. Huddleston (1965b) recommends the preextraction with ether. Sebban (1968) increased the sensitivity of the Greff method for ethoxylates above E_{12} by extracting with chloroform instead of benzene.

The preextraction step which is usually used serves two purposes: (i) it may reject interfering substances, and (ii) it preconcentrates the surfactant and thereby increases the sensitivity.

Lashen (1966) reports an increase in sensitivity achieved by a different means. The amount of complex formed was measured not by its UV absorption but by its thiocyanate content, determined by a very sensitive color reaction. Such a variant was also recommended by Watts (1968). Alternatively, Morgan (1962) analyzed for the cobalt content by its color reaction with nitroso-R salt.

D.2. *Intermediate Biodegradation Products*

The possible presence of biodegradation intermediates must be taken into consideration in the interpretation of results. Frazee (1964b) and Osburn (1966) have advanced considerable evidence that nonionic biodegradation may proceed through either end of the molecule, (i) by carboxylation and subsequent shortening of the hydrophobic group, or (ii) by hydrolysis and shortening of the polyethoxylate hydrophilic group. Although direct experimental evidence seems to be lacking, we may presume that the introduction of a carboxyl group would per se have little influence on formation of the complex and its subsequent extractability. But shortening of the hydrophobic chain by continued biodegradation would at some point so reduce its lipophilic properties as to render the complex unextractable. Before that limiting point the carboxylated intermediates would probably be measured along with any of the original surfactant still remaining—unless the carboxylates have already been separated by some preliminary purification step such as extraction under alkaline conditions.

Decline in CTAS (cobalt thiocyanate active substances) response with shortening of the EO chain may be a factor leading in some cases to large discrepancies in the apparent extent of biodegradation. For example, the CTAS may drop to zero while the foaming properties are still high (Huddleston, 1965b; SDA, 1969a, b), or while IR analysis shows significant amounts of nonionic remaining (Osburn, 1966).

E. Bismuth Iodide–Barium Complexes; Mercuric Iodide

Among the many substances which form precipitates with Dragendorff reagent (a preparation containing bismuth iodide) are the nonionic surfactants, presumably through interaction with the ethoxylate oxygen atoms as in the cobalt thiocyanate complexes. Bürger (1963a) has developed this into a useful method for biodegradation analysis in which the nonionics are preconcentrated by butanone extraction and precipitated with a barium-modified Dragendorff reagent. The amount of precipitate is measured by centrifuging into a capillary tube under controlled conditions and measuring the length of the column. The method is said to respond to ethoxylate nonionics from E_2 to E_{20}, certain modifications being required for the E_2 to E_4 derivatives because of their limited water solubility (Bürger, 1964).

Presence of the hydrophobe group is not necessary for formation of the precipitate; the polyethylene glycols also give a positive reaction. Since such products may be formed by biological removal of the hydrophobe group from the nonionic surfactant, and since the biological systems may develop other precipitatable components even in the absence of surfactants, Bürger's method includes a preliminary extraction of the nonionic into butanone, leaving any polyglycols behind in the aqueous phase. Sensitivity is comparable to that of the cobalt thiocyanate method, and 100 ml of 1 ppm solution is a suitable sample size.

Wickbold (1966) further modified Bürger's method. He preextracts from aqueous HCl solution with n-butanol to separate the polyethoxylate nonionics from the polyglycols, and measures the amount of Bürger reagent precipitate by colorimetric reaction of the bismuth with pyrrolidine dithiocarbamate. He found that carboxylated intermediate biodegradation products go along with the intact nonionics in the acid extraction and subsequently respond to the bismuth iodide if they contain long enough polyglycol groups. They can be rejected by extraction from bicarbonate solution, where they remain in the aqueous phase as the sodium salts.

In the course of this work Wickbold observed that many polyethoxylate nonionics ranging from E_7 to E_{20} from precipitates with complex barium salts, averaging about 10 ethoxylate oxygens per atom of barium. This relation held for barium silicotungstate, phosphotungstate, tetraphenylboronate, tetraiodobismuthate, and cobaltothiocyanate.

It is not evident whether the potassium mercuric iodide–barium chloride method used by Schönborn (1966, p. 42, 62) falls in this category or not. He found it advantageous to use a prepurification step, either (i) adsorption onto a cation exchange resin followed by elution with methanol or (ii) extraction with butanone.

F. Phosphotungstate, Phosphomolybdate

Pitter (1962a, b, 1966a, 1967a) complexed polyethoxylates with phosphotungstic acid for analysis, recovering the complex as a precipitate by centrifuging and then measuring the color developed by treatment with hydroquinone in concentrated sulfuric acid. He reports a reproducibility of ± 1 ppm in analyzing a 3 ppm solution. The response falls off with lower ethoxylates such as lauryl alcohol E_4. Prior deproteination is required if activated sludge effluents are to be analyzed, since proteins also precipitate with phosphotungstic acid (Pitter, 1963a).

Another phosphotungstate procedure has been developed by Burttschell (1966). Interfering materials are removed by centrifugation followed by mixed bed ion exchange. The phosphotungstate complex is then formed and extracted into butanone at pH 5, under which conditions the excess uncomplexed phosphotungstic acid decomposes and is not extracted. The tungsten in the complex is determined by $TiCl_3$ reduction and color formation with toluene-3,4-dithiol. A lower limit of about 10 μg of nonionic (corresponding to 100 ml of 0.1 ppm solution) is imposed by the fairly large blank when the ion exchange resin is used. IR spectroscopy was used to prove that the material complexed from sewage in this procedure was indeed nonionic surfactant.

Polyethoxylate nonionics also form complexes with phosphomolybdic acid. Stevenson (1954) dissolved the resulting precipitate in concentrated sulfuric acid, developing a violet to pink color which was measured at 520 nm and compared to known standards. Stevenson developed this method for Lissapol N, an alkylphenol ethoxylate, and it is not clear whether the color is formed with aliphatic derivatives as well. However, an alternate procedure is also described, in which the molybdenum content of the precipitate is determined by measurement of the yellow color formed with thiocyanate (470 nm); this should be applicable to either aliphatic or aromatic derivatives. Huyser (1960) used one or the other of these methods in his early work on polyethoxylate biodegradation. Sheridan (1969) estimates the molybdenum by atomic absorption spectroscopy. The amount of nonionic is calculated using an empirical factor which varies with the nature of the nonionic and the system being analyzed.

IV. PHYSICOCHEMICAL METHODS

A. Thin Layer Chromatography (TLC)

A.1. *General Principles*

TLC and its companion, paper chromatography, have been useful particularly in the study of nonionic biodegradation and the intermediate products thereof. The analytical sample is usually preconcentrated by extraction into an appropriate solvent for deposition on the chromatographic medium, where it is developed with appropriate solvent combinations and the resulting spots are then made visible by spraying with a suitable reagent.

The nonionic surfactants may be separated from the corresponding polyglycols (either present as impurity in the initial surfactant or formed by biodegradation) by proper choice of extraction conditions or of the developing solvents. Further, their spots may be distinguished from each other by differences in their color reactions with certain spray reagents.

Both nonionics and polyglycols are usually mixed products comprising a broad range of EO mole ratios. These may be separated into their individual components by use of certain developing solvents, giving a series of spots in the chromatogram. Two-dimensional techniques are often useful, developing first in such a manner as to separate the general classes and then at right angles with a different eluent to resolve the individual components.

The lower the EO content of the nonionic, the more readily is it carried along by the organic solvent during development. A solvent of low polarity such as benzene:chloroform will develop only the lower ethoxylates, leaving those higher than E_3 or E_4 back at the starting line. The higher components can be carried forward to the desired degree by use of more polar developing solvents containing suitable amounts of alcohols, ketones, or water. The free polyglycols are less mobile than the nonionics, and still higher water content is required in the developer if they are to be resolved.

A.2. *TLC Procedures*

Bürger (1963b) used butanone as a preliminary extractant for the nonionics, leaving free polyglycols behind in the aqueous phase. He used a butanone:water developer with his silica substrate, and a modified Dragendorff bismuth iodide solution as a spray reagent for coloring the spots. He found butanol:acetic acid:water to be more suitable for developing the free polyglycols.* Skelly (1966) used alumina medium instead of silica and was able to resolve nonionics as high as nonylphenol E_{18-22}, while Hayano (1968) with silica and butanone:water resolved an average alkylphenol E_9 mixture into each individual component from E_1 to E_{15}.

Mansfield (1964) developed with benzene:acetone, using higher acetone content for products with higher E_n, and used permanganate as the spray reagent. Borecký (1965) devised a whole series of developing solvents for use with nonionics ranging from low to high E_n, using either Bürger's Dragendorff reagent or formaldehyde:sulfuric acid for spraying. Thoma (1965) made a gross separation of the nonionics by butanol:ethanol: ammonia, and then developed at right angles with butanol:water for resolving the individual ethoxylates. Alternatively he used chloroform: methanol:water in the second dimension to resolve the polyglycols. Nonionics were distinguished from polyglycols by iodine and starch spray reagents.

The foregoing workers deposited samples of 20 to 200 μg for development of the chromatograms. Patterson was able to apply these techniques to the detection and identification of 1 to 5 μg quantities of nonionics in sewage (1966a) and river waters (1966b), and to the study of nonionic biodegradation. By using ethanol:acetic acid:water 40:30:30 the nonionic could be separated as a single spot suitable for quantitative estimation by comparison with adjacent knowns. Separation of the individual components for qualitative examination was accomplished by a 70:16:15 solvent ratio. Further information regarding the development of the method is reported in LGC (1965, pp. 105–108), while Ellerker (1968, p. 550) gives further operating suggestions. WPRL (1965, p. 192) reports that the color intensity of the nonionic single spot varies from one product to another; in six compounds tested it ranged from 0.35 to 1.4, with some indication that the color intensity increased with increasing E_n number.

Thin layer and paper chromatography are also adaptable to anionic surfactants, but little application has been made in biodegradation studies, presumably because of the adequacy of simpler methods. Franks (1955, 1956) was able to resolve the individual chain lengths in alkyl sulfate, alkyl aryl sulfonate, and cationic surfactants, while Püschel (1968) studied alkyl sulfates and alkane, alkene, hydroxyalkane, and ketoalkane sulfonates. Procedures have been developed by Drewry (1963, 1964), Takagi (1964), Bey (1965), Loeser (1965), and Borecký (1966) for separation and identification of all three classes—anionic, cationic, and

---

* Bürger (1967) later showed the utility of his TLC method in studying biodegradation of APEs.

nonionic—using TLC and paper techniques. Ōba (1965c, d) has applied TLC to the examination of MBAS as described in Section III.A.5.

B. Gas Chromatography (GC)

B.1. *ABS—Desulfonation–Gas Chromatography*

Anionic surfactants, being salts, are not volatile enough for useful results with gas chromatography. Pyrolysis GC (Liddicoet, 1965) might be applicable to the problems of biodegradation, but such use does not seem to have been reported in the literature.* However, the ABS surfactants are susceptible to desulfonation by boiling with concentrated phosphoric acid at 200°C (Knight, 1959). The unsulfonated alkylbenzene is liberated and can be isolated from the distillate. Microdesulfonation procedures have been developed by Setzkorn (1963) and Swisher (1963b) suitable for use with biodegradation samples, giving alkylbenzenes for identification and measurement by GC. Capillary columns are particularly useful because of their high resolving power and the minute amount of sample required, if used in conjunction with sensitive detectors.

GC is especially suitable for LAS because of its very characteristic and regular pattern of peaks of the individual isomers and homologs, not easily obscured by other waste components or bacterial by-products which may be present (Fig. 7.9, upper). A satisfactory chromatogram can be obtained from 1 μg or less of LAS, but problems of handling are minimized if samples on the order of 0.1 to 1 mg are available, such as might be obtained from a few hundred milliliters of 1 ppm solution. Since ABS is not volatile and since interference is not often a problem, it is usually sufficient to simply evaporate the sample to dryness for the desulfonation, thus eliminating any possibility of loss in prepurification. If some sort of separation of the ABS does prove to be necessary, this is readily accomplished by running a large scale methylene blue type separation (Swisher, 1966b), or somewhat less easily by the prepurification procedure developed for the IR method (Section IV.C.1).

The desulfonation–GC technique has been vital to the elucidation of the biodegradation process in LAS and other ABS compounds, by measuring their relative rates of disappearance in mixtures (Huddleston, 1963; Swisher, 1963a, b; Allred, 1964b; Hughes, 1968b), and by detecting the intermediate biodegradation products of LAS (Swisher, 1963c, 1964b). It is also indispensible for unequivocal proof of the presence or absence of LAS in wastewaters and effluents (Swisher, 1966b) and in natural waters (Sullivan, 1969).

* Van Cauwenberghe (1969) finds that PGC is unsatisfactory for ABS but useful for examining alkylbenzenes.

B.2. *Aliphatic Sulfates and Sulfonates*

Payne (1965) used GC in studies of the fatty alcohols liberated in the biodegradation of alkyl sulfates. Application to aliphatic sulfonates is more difficult. In contrast to the smooth desulfonation of the aromatic sulfonates, aliphatic sulfonates undergo extensive decomposition when desulfonation is attempted. Tomiyama (1968) was able to prepare derivatives sufficiently volatile for GC by converting the sulfonate to the sulfonyl chloride. This was achieved with both the alkene sulfonate and the hydroxyalkane sulfonate fractions of α-olefin sulfonates, making it possible to study their relative biodegradation rates.

B.3. *Nonionics*

Nonionic surfactants with low EO content can be separated by GC, as shown by Nadeau (1964), but products over E_8, the range of greatest interest for surfactants, were not volatile enough to show up using his method. Volatility has been improved by conversion to the trimethylsilyl ethers (Suffis, 1965) or acetate esters (Gildenberg, 1965), but applicability in biodegradation research has not yet been reported.

C. Infrared Spectroscopy

C.1. *General Applications*

Most organic compounds show characteristic and complex IR absorption spectra due to vibrations of their component atoms. The specific frequencies are dependent on the chemical bonds linking the atoms together in the molecule and upon other molecular features. Surfactants are no exception; they exhibit a wealth of detail in their IR spectra. This can be applied to biodegradation studies, with proper attention to the fact that just about every other molecular type in the vicinity will contribute an interfering spectrum of its own.

Thus Ryckman (1956, 1957) used the IR spectra of LAS and of linear primary ABS to demonstrate their degradation in activated sludge systems; persistence of the TBS spectrum in parallel experiments showed that it was not degraded, and validated the applicability of the IR technique to the problem as well. IR spectroscopy was used by Frazee (1964b) and Osburn (1966) in their studies of nonionic biodegradation. Samples were concentrated and purified by charcoal adsorption, chloroform extraction, or foam stripping. From the spectra were estimated the amounts of aliphatic ether oxygen ($1120\ cm^{-1}$), aromatic ether oxygen ($1250\ cm^{-1}$), hydroxyl ($1040\ cm^{-1}$), and carboxyl ($1700\ cm^{-1}$) groups present.

C.2. *ABS in Waters and Wastewaters*

The IR technique has found its main utility as a referee method for confirming the presence of ABS in waters where it has already been presumptively indicated by the less specific methylene blue method. The procedure was developed by AASGP (1956) and adapted as a Standard Method by the APHA (1965, p. 299). The key operations are a lengthy series of prepurifications designed to concentrate the ABS and eliminate all other substances. Briefly, these involve charcoal adsorption, desorption with methanol:chloroform:ammonia, hydrolysis, solvent treatment, conversion to an amine salt, and extraction, all in preparation for determination of the spectrum. Comparison with the spectrum of a known ABS constitutes the identification.

Simplifications in the charcoal prepurification steps were introduced by Ogden (1961) and AASGP (1961). Hughes (1968b) used in anion exchange resin instead of charcoal, desorbing with a small volume of methanol:HCl.

C.3. *Differentiation of TBS and LAS*

Identification was once a simple matter in the days when TBS was the only sort of ABS likely to be present. But after the introduction of LAS there arose the problem of distinguishing the two, since the prepurification procedure admits all types of ABS indiscrimately. Although spectra of all varieties of ABS resemble each other closely, there are minor differences depending on the exact chemical structure. TBS and LAS exhibit such differences, for example, in the region of 7.1 to 7.3 μ (1365–1410 cm^{-1}), and their relative amounts can be calculated from the intensities of particular bands. Known synthetic mixtures of the two in pure form are used for comparison, assuming that these are the only components present in the unknown.

Contributions in this area have been made by Bolton (1961), Ogden (1961), Frazee (1964a), Urban (1965, p. 36), Maehler (1967), Ōba (1968a), and Hughes (1968b). The spectral differences are slight and estimation becomes uncertain when one or the other component is below the range of 10% to 20%. In such circumstances the desulfonation–GC technique (Section IV.B.1) is particularly useful for unequivocal identification of LAS and estimation of its amount.

D. Ultraviolet Spectroscopy—Benzene Rings

The benzene ring exhibits three characteristic and intense absorption bands in the UV at wavelengths which depend somewhat upon the nature of the substituent groups (Jaffe, 1962). Weber (1962) used this to determine the ABS content of pure aqueous solutions at low levels. In LAS the weakest of the three bands is in the neighborhood of 260 nm; the second, about 20 times as intense, is at 223, and the third, four times stronger yet, is at 193 nm (Swisher, 1967b). These bands are easily detectable in aqueous solutions at concentrations of 1 ppm or lower, and provide an excellent means for studying biodegradation of the rings in benzenoid surfactants such as ABS and the alkylphenol ethoxylates.

Since the molar absorbance of these bands appears to be fairly constant from one benzene compound to another, UV spectroscopy is not applicable to the study of primary biodegradation of these compounds. The bands persist substantially unchanged as long as the rings are present in the system, regardless of whether in the intact original molecules or in extensively degraded intermediates. By the same token the method is particularly suited to study of the later stages of biodegradation of benzenoid surfactants, since weakening or disappearance of the bands implies destruction or removal of the rings. Such application has been made by Setzkorn (1965) for aromatic sulfonate hydrotropes with short alkyl groups, and by Hammerton (1962), Kölbel (1967), and Swisher (1967b, c, 1968a, 1969) for LAS and other ABS.

The alkylphenol nonionics are also open to UV spectroscopy by virtue of their benzene ring. Griffith (1957) used the 278 nm band in his adsorption studies, reporting that concentrations as low as 50 ppm could be determined. Sensitivity could probably be increased manyfold by using the more intense bands at shorter wavelengths. Osburn (1966) supplemented UV with IR in his work on the biodegradation of alkylphenol ethoxylates. UV has also been applied to the detection of these surfactants in water and sewage, after extensive prepurification (SDA, 1964).

E. Tracer Techniques

Compounds tagged with stable or radioactive isotopes can provide great insight into the course of a complex reaction and assist in analysis of complex systems. Some use has been made of the technique in surfactant biodegradation work, and there should be considerable potential for further application.

It is a relatively simple matter to prepare an ABS tagged with ^{35}S in the sulfonate group by sulfonating the precursor alkylbenzene with $H_2^{35}SO_4$ (House, 1956). Such material can subsequently be detected in a system to which it has been added for biodegradation study, by isolation in an appropriate manner and counting the radioactivity present. The method is capable of extreme sensitivity and freedom from interference. Biodegradation of any specific variety of ABS can be investigated in sewage and field systems by this means without regard for the "natural" ABS which may be present.

Not only can the amount of intact original material remaining at any given time be found, but also the amount of inorganic sulfate formed in the degradation can be determined by isolating all the sulfate present in the system and counting its radioactivity (Section III.B). If intact surfactant and the derived sulfate do not add up to 100% of the amount originally fed, the difference represents the sulfur present in the form of intermediate degradation products. These may be studied more closely through separation into fractions by chemical or physical means and further counting. These radiosulfur techniques have been used very effectively by House (1956), McGauhey (1959a, b), Robeck (1963), Straus (1963), Klein (1964a, p. 75), Sharman (1964a), Sweeney (1964a), Williams (1964), Payne (1965), and undoubtedly others.

Alternatively the surfactant may be tagged by ^{13}C or radiocarbon. Introduction of the tracer is considerably more difficult in this case, particularly if it is to be in a less accessible location in the molecule. Carbon tracing is very useful in studying the chemical mechanisms of the biodegradation process, and has been applied to that end by Huddleston (1963), Vath (1964), and Lashen (1966). Inversely, Ludzack (1964) tagged the bacteria themselves to gain better insight into the biodegradation process, a technique applied to surfactant studies by Brink (1966).

Hydrogen tracing by means of deuterium or tritium in the surfactant molecule is also possible. For instance, Lashen (1966) used a doubly tagged OPE_{10} nonionic to demonstrate that losses of ^{14}C from the test system were due to biodegradation rather than adsorption; the tritium was completely recoverable in the effluent.

V. METABOLIC AND OTHER NONSPECIFIC METHODS

In the absence of specific analytical methods for the surfactant under test, or for its intermediate degradation products, nonspecific methods have been called upon to monitor the course of biodegradation. Thus the oxygen uptake or the carbon dioxide production by the organisms of the test system may be used as a measure of the assimilation of the test compound. Since the bacteria maintain a fairly high endogeneous metabolic activity in the absence of added foods, this must be corrected for. It is usually accomplished by setting up a control system, identical to the test system in all respects except that the test surfactant is not fed to it, and subtracting the control results from the test results. Generally such measures must be taken in using any of the nonspecific methods.

Two major disadvantages weaken such a procedure. First, the net number is often the difference between two considerably larger numbers, with resulting magnification of relative error. Second, it is often very difficult to prove that the reactions occurring in the control system occur to the same extent in the test system, and thus the validity of subtracting the control values may be open to question. These disadvantages are inherent, to a greater or lesser extent, in all of the methods discussed in this section.

A. Organic Content

A.1. *Interferences and Uncertainties*

Several useful analytical methods have been developed to measure (or to estimate) the total organic carbon content of the medium, ignoring its exact chemical nature. Bacterial foods, metabolic products, and the bacteria themselves are, like the surfactant, all made up of organic carbon compounds. This gives a background against which the change due to degradation of the surfactant is often difficult to distinguish.

Interference by the organisms themselves and by cellular debris may be eliminated by centrifugation or by filtration through a membrane capable of retaining particles of such size. Adsorption of solutes or contribution of solubles by the membrane must be guarded against.

But soluble organic materials are also formed by the bacteria—natural metabolic products resistant to further degradation. Thus an exhaustively degraded activated sludge effluent may contain as much as 50 ppm of such organics arising from natural foods in the absence of surfactants. This is discussed further in Chapter 4, Section II. It is unfortunate that these factors interfere in applying the carbon methods in their potentially most useful area, namely in the study of ultimate biodegradation. In spite of this they have been used with considerable effectiveness since the time Nelson (1961) employed COD analysis for that purpose.

Most workers have preferred to use the COD analysis (material oxidizable by dichromate) or the combustion method (amount of CO_2 formed). Janicke (1968a, c) compared biodegradation results on an alkylphenol nonionic by these methods as well as by two methods specific for the nonionic, with the rather discordant results shown in Table 3.2. Poor agreement is often encountered in biodegradation tests on resistant surfactants, and that is exemplified by the 14% and 34% results for the German dichromate method. But the results in each vertical column presumably all come from the same effluent, and discrepancies there cannot be attributed to bacterial vagaries. Rather, they must reflect

TABLE 3.2

Comparison of General and Specific Analytical Methods in
Biodegradation of an Alkylphenol Nonionic[a]

Analytical method	Percent removal	
	1968a	1968c
COD (German dichromate method)	14	34
COD (U.S. dichromate method)	—	50
Combustion (total organic carbon)	24	—
Permanganate oxidation	—	29
Bismuth iodide (specific) (Wickbold, 1966)	—	13
Mercuric iodide (specific) (Schönborn, 1966)	—	28

[a] tp-C_9APE_7 in official German continuous activated sludge test,
Janicke (1968a, c).

varying response of the initial surfactant and its intermediate degradation
products to the different oxidation conditions in the several analytical
methods.

A.2. Chemical Oxygen Demand (COD)

This widely used procedure is covered in *Standard Methods* (APHA
1965, p. 510). It comprises a wet oxidation by boiling with excess standard
dichromate solution and back-titration with ferrous ammonium sulfate. It
responds to most organic matter, which generally is oxidizable by
dichromate with consumption dependent upon the initial oxidation state
of the organic. Exceptional structures may be unreactive. Interference by
noncarbonaceous reducing substances is not a problem, since such
materials are not ordinarily present in biological systems to any great
extent.

The analytical result is expressed as the amount of oxygen equivalent
to the dichromate consumed. For an example, consider the theoretical
oxidation equations for glucose (3.1) and a $C_{12}ABS$ (3.2). A solution

$$C_6H_{12}O_6 + 6\ O_2 \rightarrow 6\ CO_2 + 6\ H_2O \tag{3.1}$$
$$\phantom{C_6H_{12}O_6}\ 180 \qquad 192 \qquad 264 \qquad 108$$

$$C_{18}H_{29}SO_3Na + 25\tfrac{1}{2}\ O_2 \rightarrow 18\ CO_2 + 14\ H_2O + NaHSO_4 \tag{3.2}$$
$$\phantom{C_{18}}348 \qquad\quad 816 \qquad\quad 792 \qquad 252 \qquad 120$$

containing 180 ppm glucose and 3.48 ppm of the ABS should theoretically
show a COD of 192 + 8.16, or about 200 ppm. In actual practice glucose

and ABS are found to exhibit CODs in reasonable agreement with those calculated, as do most other organic materials.

The chemical operations of the test are subject to the usual limitations on precision and accuracy, which may introduce uncertainties on the order of several parts per million. Thus the method is of questionable applicability for systems containing below about 5 to 10 ppm of surfactant, the range particularly pertinent to biodegradation. Ballinger (1962) reports a standard deviation of about $\pm 8\%$ in an interlaboratory comparison on a standard sample of glucose–glutamic acid mixture. The mean value found for the COD was 92% of the theoretical.

Oxidation with permanganate instead of dichromate has been practiced quite extensively in Europe for estimating the organic content of a system. It is subject to similar difficulties and appears to offer no particular advantages. Janicke (1968c) compares biodegradation results using the permanganate method and two dichromate methods as shown in Table 3.2.

A rapid "combustion" method has been developed by Stenger (1967), wherein a tiny aqueous sample is injected into a stream of carbon dioxide and pyrolyzed. Carbon monoxide is formed in proportion to the COD of the sample and is measured by its IR absorption as the gas stream passes through a spectrometer. This method has not yet been used widely enough to allow full assessment of its capabilities.

A.3. *Combustion*

An alternate approach different from COD is increasingly prevalent in the estimation of organic content. It involves complete oxidation of the sample by injection into a stream of air or oxygen which then passes over a catalyst. The organics yield carbon dioxide which is measured by its IR absorption (Menzel, 1963; Van Hall, 1964, 1967; Schaffer, 1965; ASTM, 1968). The system is available commercially; for aqueous solutions in the range of 1 to 100 ppm a $20\ \lambda$ sample is sufficient, and results are obtained less than a minute after injection.

Inorganic carbon (carbon dioxide and carbonates) present in the sample will also contribute to the CO_2 measured. Since organic carbon is in most cases the item of interest, the inorganics are usually removed (by acidulation and expelling CO_2) before the analysis, or are determined separately on another sample (evolution of CO_2 using an acid catalyst and lower temperature) and subtracting from the total (van Hall, 1967).

It must be noted that the carbon combustion method, corrected for inorganic carbon content if necessary, tells the total organic carbon content, regardless of its oxidation state. This is not quite the same information as is given by the COD method, where the COD value is higher for a lower oxidation state, even though the total carbon content may be the same. True COD values are given by the pyrolysis method of Stenger (1967) (Section V.A.2).

Busch (1966) has discussed the general principles, applicability and interpretation of the combustion method in comparison with COD. In another comparison of the two, Janicke (1968a) found a precision corresponding to around ± 1 ppm for the carbon and ± 10 ppm for the dichromate (equivalent to ± 1.6 ppm of COD). However, that is analytical precision; precision of the biodegradation results may be very much poorer, as suggested by Table 3.2.

Rickard (1965) has demonstrated the suitability of the combustion method not only for analyzing the organic content of the liquid phase during biodegradation, but also for measuring the growth of the bacterial cells themselves, isolated by centrifuging and washing. Young (1968) included such determinations in a proposed standard biodegradation test procedure. The combustion method has lately become preferred for continuous automatic monitoring of effluents and receiving waters.

A.4. Gravimetric Determinations

Payne (1963a, b) took a direct approach to measurement of the organic content by simply evaporating the cell-free filtrate to dryness; the loss of weight upon subsequent ashing was a measure of the organics. Instead of ashing the residue, Borstlap (1967a) determined its content of n-butanol-soluble materials by extraction and subsequent evaporation of the extract. In either case corrections must be applied by subtraction of corresponding results from control cultures which have not been fed the test compound. Presumably these corrections would be much larger in Payne's method than in Borstlap's. But the latter method would miss any butanol-insoluble biodegradation products, if present.

B. Oxygen—Biochemical Oxygen Demand (BOD)

B.1. Difficulties of Interpretation

Biodegradation is principally an oxidative process. The theoretical amount of oxygen required for complete oxidation of the compound is of course the same regardless of the mechanism—chemical or biochemical. It is calculated as described for COD (Section V.A.2). In principle, then, the biodegradability of a surfactant (or of any other compound) should be measurable as its BOD, the weight of oxygen taken up per unit weight of the compound in an appropriate biological system. Lamb (1952), Heukelekian (1955), and Ludzack (1960b) have tabulated BOD data for many organic compounds.

In practice the interpretation of such results is far from straightforward because of the extreme complexity of the metabolic processes involved. Oxygen is utilized by the microorganisms for a multitude of reactions, not simply for oxidation of the added food or test compound. Correction for this by subtracting a blank value determined from a control culture is often highly uncertain.

Even more troublesome is the fact that the organisms do not oxidize any food completely to carbon dioxide and water; living protoplasm is synthesized, and inert metabolic by-products are also formed, in amounts which may vary widely from one circumstance to another. Thus the proper interpretation of oxygen uptake measurements in terms of percent biodegradation is exceedingly difficult if not impossible, and only qualitative at best.

These problems are discussed in detail in Chapter 4, Section II and Chapter 5, Section IV, and we will direct our attention at present primarily to the oxygen determination itself.

B.2. *Closed Bottle Techniques*

Measurement of oxygen uptake by the classical BOD method is detailed in *Standard Methods* (APHA, 1965, p. 415). A bottle is completely filled with an aqueous solution containing known amounts of dissolved oxygen and the test compound, along with a bacterial inoculum. After standing at the desired temperature for the desired time, the amount of oxygen remaining is determined by any of several chemical methods (APHA, 1965, p. 405), and subtracted from the initial value to find the uptake by the organisms. A control run is also made, identical in all respects except that the test compound is omitted, and its uptake is subtracted to give a net value for the compound.

The initial oxygen content is limited by its solubility to around 8 to 10 ppm. The amount of test compound must be chosen so as to react with a substantial amount of this oxygen and yet to leave a significant amount, say 1 or 2 ppm, unreacted at the end. This is ordinarily accomplished by running several bottles with various amounts of the compound, hoping to bracket the desired range. Alternatively, large volume techniques have been developed whereby many samples can be withdrawn for analysis, and with provision for replenishment of oxygen as it is depleted. Such a technique, developed by Orford (1953) and later used by Ryckman (1956, 1957), has been further recommended by Young (1968).

A standard time of 5 days is often used in wastewater chemistry, but for research purposes the determinations may extend to 20 days or more. Ballinger (1962) reports a comparison of results from 34 laboratories on a standard sample of glucose–glutamic acid. For the 5 day BOD the oxygen uptake was 62% of theoretical and extrapolation indicated that the ultimate BOD would average around 70%. The standard deviation was around $\pm 20\%$ of the value determined.

A more comprehensive picture of the biodegradation may be obtained if, instead of a single determination at 5 days, a curve showing oxygen uptake as a function of time is developed. A separate pair of bottles, compound and blank, is required for each point on such a curve, unless a large volume method is used.

With the development of oxygen-sensing electrodes it has become possible to develop BOD curves more simply. The oxygen content can be determined at intervals—or continuously—over the desired period with little if any disturbance of the system. Current state of the art is indicated in reviews by the Water Pollution Research Laboratory (WPRL, 1964; 1967, pp. 191–192) and Montgomery (1967), and in papers by Gannon (1965), Downing (1965), Eye (1966), and Eden (1967).

B.3. *Respirometry*

In the closed bottle BOD measurement the entire amount of oxygen required for the degradation must be dissolved in the water at the start, thus limiting the initial concentration of the test compound to a few parts per million and limiting as well the number of organisms which can be allowed in the system. Such limitations are avoided by use of respirometric techniques, of which the most common is the Warburg, described in detail in an earlier edition of *Standard Methods* (APHA, 1960, p. 396).

Here the oxygen is supplied to the system from the gas phase, where it can be made available to the full extent necessary. Transport into the liquid phase containing the test compound and the inoculum is facilitated by agitation. Oxygen uptake may be measured at any time by the decrease in volume or pressure of the gas phase. The carbon dioxide produced by the bacteria would interfere with this means of oxygen measurement, so it must be continuously removed, for example, by absorption into a small reservoir of potassium hydroxide. (But care must be taken to avoid complete absence of CO_2 lest bacterial metabolism be inhibited thereby.)

With the Warburg technique the precision of the oxygen demand determination may be considerably improved over that of the BOD bottle, but the difficulties of interpretation remain. The metabolic reactions remain complex, and the suitability of the control run remains uncertain.

Many other apparatus systems have been devised for respirometric studies. These have been discussed in reviews by Ludzack (1963a) and Montgomery (1967). Tool (1967) reviews the shortcomings of both BOD and Warburg methods and describes the Hach modification, designed to simplify the procedures and avoid the shortcomings. Discussing the Hach (manometric) as well as manostat-type apparatus, Fuhs (1968) points out that kinetic data (i.e., the shape of the oxygen uptake curve) will be influenced not only by the characteristics of substrate and microorganisms, but also by extraneous factors such as carbon dioxide deficiency [which inhibits the biochemical processes (Gaffney, 1965)], inadequate oxygen transfer, and light (whichmay cause photosynthetic formation of oxygen).

For the most part these newer variants have not as yet been applied to study of surfactant biodegradation, except by Pitter (1963b), who developed a large volume respirometer which he has used extensively in his surfactant work.

C. Carbon Dioxide

Measurement of the carbon dioxide formed during biological oxidation of the test compound should be about as informative as measurement of the oxygen uptake. Porges (1952) devised a system for so doing, after having observed in Warburg runs on skim milk, lactose, and casein that about 1 mol of CO_2 was produced per mol of oxygen absorbed. He swept CO_2-free air through activated sludge and then through scrubbers charged with barium hydroxide to pick up the metabolic CO_2 for determination.

He showed that some of the CO_2 might remain in the mixed liquor, to the extent that alkali was generated therein (for example, by oxidation of sodium salts of organic acids). This carbonate CO_2 could be liberated at the end of the run by acidification; correction should be made for any carbonates present initially. Porges further pointed out that endogenous respiration in the absence of added food must be taken into consideration, and that CO_2 formation would be diminished by synthesis of cells. Sludge plus 200 ppm of glucose gave 190 ppm of CO_2 in 5 hr, but it gave 130 ppm in the absence of glucose. The theoretical amount from 200 ppm glucose is 284 ppm, so the net CO_2 production in this experiment was only about 20% of theoretical. Porges did not attempt to push degradation to completion in this work, but proposed the method as a means of determining the oxidative activity of a sludge.

Ludzack and his associates have adapted the CO_2 measurement technique to biodegradation studies (for example, Ludzack, 1957, 1959, 1963b; see also Pahren, 1961), but it has been used only rarely in surfactant biodegradation work. Conway (1966) did use such a procedure in his studies on nonionic surfactants, working at 15 ppm in river water. Interpretation and evaluation of the results were facilitated by comparison with a control system containing diethylene glycol, chemically related to the nonionics and having known biodegradation properties.

Ludzack (1964) studied the variations in the endogenous respiration of ^{14}C-labeled activated sludge, using a closed system. The aeration gas was recycled through a barium hydroxide absorber and back through the culture. Total CO_2 evolution was determined by decrease in barium hydroxide alkalinity, sludge CO_2 by radioactivity of the precipitated $BaCO_3$, and oxygen utilization by volume change in the constant pressure system.

D. Bacterial Growth

If bacteria multiply significantly in a system, it can be concluded that some components of the system are being used as food, i.e., are being biodegraded. In the presence of sufficient food in a liquid medium, multiplication of the bacterial cells can be easily observed qualitatively as an increasing opalescence or turbidity. Measurement of optical absorbance provides a basis for quantitative estimation of numbers or weight of bacteria present, provided that they are evenly distributed in the medium instead of clumping or settling.

If a surfactant is used as the sole food in such a system, bacterial growth indicates that some degradation of the surfactant must have occurred, but not necessarily ultimate biodegradation to carbon dioxide and water. For instance, Anderson (1964) observed growth by one of his bacterial strains on TBS, and at a rate comparable with that on nutrient broth, even though TBS is quite poorly degradable. On the other hand, lack of growth does not necessarily prove that the surfactant is not biodegradable; the conditions may be unsuitable for growth for some entirely different reason.

Relatively high concentrations of food are necessary if visible growth is to be achieved, concentrations at which some surfactants may be bacteriostatic. This factor doubtless influenced the results of von Riesen (1955), who worked at 1000 ppm. Nevertheless Payne and co-workers have made good use of the technique, although for quantitative estimations they had to use cell count instead of turbidity under some circumstances (Prochazka, 1965, 1967).

Pipes (1963a) measured the cell count by direct microscopic examination, while Vath (1964) measured the bacterial growth gravimetrically and Rickard (1965) estimated it from the organic carbon content of the centrifuged, washed cells. Pipes established that maximum cell count attained with readily assimilable foods was proportional to the total COD or theoretical oxygen demand of the food.

The bacterial cell count in a growing system is not a linear function of the optical density, cell mass, or cell carbon content. Rickard (1965) and Riley (1965) present data showing that when resting cells were introduced into a nutrient medium they underwent considerable growth without dividing, increasing perhaps eight- to tenfold in mass during the first 1 or 2 hr before division began. The cell count increased rapidly during the next several hours, much more quickly than further increase in total cell mass. Thus the average mass of a single cell went through a maximum and then slowly dropped back toward the original value when no more food was available.

Gaudy (1964a) found a relatively constant ratio of cell COD to cell mass, but with a considerable random variation which ranged from 1.2 to 1.7 mg COD/mg dry solids for young cells and 1.1 to 1.5 for old ones.

Data on bacterial cell counts during biodegradation are presented by Husmann (1963b, pp. 44–46), Brebion (1966), and Young (1968). With good growth conditions the populations usually increased by a factor of 5 or 10 to a peak of perhaps 1 to 10 million/ml at the 3rd to 5th day. Subsequent decline was at a slower rate. Wayman (1963a) gives examples of more rapid growth, up to 100 million/ml or more within a day or two.

E. Toxicity

Inasmuch as toxicity is a major point in assessing the environmental acceptability of a material, toxicity should be, in principle, a good property for use in monitoring biodegradability. It would have the added advantage of safeguarding against the improbable situation wherein an innocuous material might give toxic intermediate biodegradation products.

In actual practice, monitoring in this manner is rare indeed, because of the effort required in making the toxicity determinations and because the information so acquired is mainly qualitative and not very informative about any other aspects of the biodegradation process. Usually the biodegradation is monitored and studied by chemical or physical means to the full extent possible, with the toxicity questions being checked in associated studies under the most pertinent conditions. For example Swisher (1964a) confirmed that biodegradation of LAS did in fact destroy its toxicity to fish along with its surfactancy properties. The effluent from a laboratory continuous-flow activated sludge unit was directed through a small aquarium, where it served as the sole medium for the test fishes. Even when influent LAS concentrations were brought up to and beyond a hundred times the nominal toxic level, there was little evidence of any toxicity in the effluent. Bringmann (1963) had earlier reached a similar conclusion on exposing *Daphnia magna* to the effluent from a trickling filter which was degrading Marlon BW1043, a commercial-type LAS.

Aubert (1969) reports biodegradability studies monitored by toxicity measurements, here again with the major aim of demonstrating the alteration of toxicity by biodegradation rather than using the toxicity values to demonstrate biodegradation. Five test organisms were used, *Diogenes* sp. (alga), *Asterionella japonica* (diatom), *Artemia salina* (microcrustacean), *Carassius auratus* (fish), and *Mytilus edulis* (mussel). The test compounds were held in seawater for 0, 3, 6, and 9 days for biodegradation before the test organisms were introduced; their subsequent fate was observed over the next 9 days in comparison with controls. The initial surfactant concentration was twice the toxic threshold (defined as the level above which there was noticeable effect on growth or development), ranging from 15 to 600 ppm. These levels were necessary so that any effect on toxicity could be detected, but were higher than desirable for biodegradation studies because of possible inhibitory action on the bacteria. Of course they were much higher than would be expected to occur in cases of actual pollution by sewage.

Aubert's results indicated loss of toxicity in some instances and not in others, but it is difficult to judge how much of a part biodegradation played in this in the absence of any supplementary analytical determinations. Among the five materials examined were three surfactants; their chemical composition was not revealed for proprietary reasons, because they were commercial products.

CHAPTER 4

THE BIOLOGICAL BACKGROUND

We turn now to the biological agents which may bring about surfactant biodegradation. Their normal life processes, capabilities, and limitations will be considered, providing the background necessary for proper interpretation of results obtained when they degrade surfactants. Further, their interactions with surfactants other than by biodegradation will be reviewed, since those interactions may alter or obscure the biodegradation reactions if not properly controlled or taken into account.

It will become evident that simply mixing some bacteria with a compound—be it surfactant, natural food, or whatever—may not per se give useful information regarding its biodegradability. Choice of reaction conditions in laboratory systems may have a profound effect on the results; however certain they may be, they may completely contradict equally clear-cut results from another set of conditions. Out in the field, in treatment plants and in natural waters, the conditions are not rigidly fixed from one moment to the next, but may be continually changing. This has an averaging effect, so that the overall result there is more definitive and dependable.

I. THE MICROORGANISMS

A. Nonbacteria

In the systems pertinent to our subject, most of the biodegradation is accomplished by bacteria, although Wurtz-Arlet (1964) and Klein (1964a) have reported that ABS is degraded by algae. Davis (1967) found that TBS, LAS, alcohol E_7, and alkylphenol E_9 were all degraded by algae in pure culture, with the extent varying widely from one species to another. And Klein (1962, 1963a) found that sunflowers can degrade TBS quite extensively. In the usual systems also, nonbacterial forms of life, monocellular and multicellular, are commonly present, feeding on the bacteria and on each other if not on the chemical foods. Their presence is not absolutely necessary for the biodegradation to proceed, even though they may be useful indicators of a healthy, balanced ecology, particularly in activated sludge and trickling filter systems.

McKinney (1956c) and Brown (1965) have studied the protozoa of activated sludge and Hawkes (1963) has reviewed and discussed these questions in detail. Curds (1968) has demonstrated quite convincingly a marked improvement in clarity of activated sludge effluents following introduction of ciliated protozoa into the system, attributable either directly or indirectly to their presence, while Calaway (1968) has emphasized the importance of the rotifers.

A major factor affecting the distribution of both bacterial and protozoan species in a system is the suitability of the bacteria as food for the protozoa. Coler (1969) showed that a *Colpoda* species, a soil ciliate, could thrive on certain bacterial species, particularly *Escherichia coli* and *Aerobacter aerogenes*. In contrast, growth was very limited on a number of *Arthrobacter* species, apparently because of a toxic cytoplasmic component. Coler points out the probable relation of these observations to the ubiquity of *Arthrobacter* and to the rapid disappearance of coliforms from soil and surface waters.*

B. Bacterial Habitats and Species Distribution

Few problems exist in obtaining bacterial species suitable for surfactant biodegradation; they are present everywhere in our environment—in the soil, in natural waters, and floating in the air. And in sewage.

Harkness (1966) has reviewed the literature on the bacterial species found at all stages of sewage treatment, and Baars (1965) has discussed their ever-changing population distribution, their structure, and their biochemistry. Brebion (1966), working with samples of domestic sewage, identified nine predominant bacterial genera and investigated the effects of several nutrient media upon the population distribution during their subsequent growth.

In his examination of activated sludge from varied sources McKinney (1953) isolated 72 bacteria representing 14 genera. Van Gils (1964) investigated the bacteriology of activated sludge in detail, and Dias (1964, 1965) isolated over 300 bacterial strains from it; he noted certain differences in the distribution of the strains in activated sludge compared to those found in sewage. Shaposhnikov (1968) reports isolation of 132 pure strains of bacteria and fungi from activated sludge derived from petroleum-containing sewage.

The genus *Pseudomonas* occupies a prominent place among these environmental bacteria. Its members have been involved in surfactant

---

* The WPRL (1968, p. 151–7) gives a census of ciliated protozoa species found in British sewage treatment plants, and confirms their improvement of effluent quality and their extensive predation on *E. coli* and other bacteria.

biodegradation since the earliest days, as when Williams (1949) observed accidental bacterial growth in a proprietary shampoo and tentatively identified the organism as a *Pseudomonas* species. He then tested 12 different pure bacterial cultures and found that 7 would grow on the shampoo while 5 did not. Ten years later McKinney (1959a) reported the isolation of 34 pure cultures able to live on TBS. Fifteen of these belonged to the genus *Pseudomonas*, 10 to *Alcaligenes*, and the rest were distributed among 5 other genera. Like McKinney, Schönborn (1962b) isolated 34 strains by similar enrichment techniques on TBS and LAS. One of these was a fungus type, one a Gram-positive bacterium, and all the rest Gram-negative bacilli. Anderson (1964) studied the growth rates of 10 cultures isolated via TBS enrichment; one of these (an *Alcaligenes* species) grew as well on 5000 ppm TBS medium as on nutrient broth, the others less readily. None would utilize unsubstituted sodium benzene-sulfonate without an alkyl group.

Ilişescu (1966) isolated 10 *Pseudomonas* species and one each of *Aerobacter*, *Alcaligenes*, *Achromobacter*, and *Escherichia* from culture media containing alkyl sulfate or ABS as sole carbon source, after inoculation with acclimated activated sludge. Skinner (1959) isolated a soil bacterium capable of living on the secondary alkyl sulfate Teepol, primary alkyl sulfate, and fatty alkanolamides. It was a *Pseudomonas*, species unidentified. Likewise was Payne's (1963a) isolate "C12B," which lived on ABS (see Feisal, 1966, pp. 25–29). Hsu (1965) obtained six *Pseudomonas* strains from sewage, each capable of degrading alkyl sulfates. In contrast to all these, Minami (1958, 1959) identified his Tween 80(oleoyl sorbitan E_{20})-degrading soil organism as a *Micrococcus*, christened *M. tweenis*.

Robeck (1963) examined the microorganism populations at several locations in a 2 year old soil lysimeter which had been used in TBS degradation experiments. *Pseudomonas* species accounted for about 5% of the organisms found near the top, 0.5% of those near the bottom, and almost 50% of those in the effluent. A dozen species of fungi were present also. Jones (1968b) enumerates many bacterial and fungal species found in moorland soil at various depths, and reports that 15% to 20% of them were capable of growing on hydrocarbons.

The bacteria involved in anaerobic systems such as sewage sludge digesters have been discussed by Burbank (1964), Cookson (1965), Harkness (1966), Hattingh (1967), Toerien (1967a, b; 1969) and Siebert (1969). Some of the common aerobic bacteria already mentioned (including *Pseudomonas*) are found here also, either because the systems may not have been completely devoid of oxygen or because some of those aerobic

TABLE 4.I

Some Microorgansim Genera Commonly
Encountered in the Environment

Achromobacter	*Lophomonas*
Aerobacter	*Micrococcus*
Alcaligenes	*Mycobacterium*
Arthrobacter	*Neisseria*
Bacillus	*Paracolobactrum*
Citrobacter	*Proteus*
Corynebacterium	*Pseudomonas*
Empedobacter	*Serratia*
Escherichia	*Sphaerotilus*
Flavobacterium	*Streptococcus*
Klebsiella	*Zoogloea*

species are also facultative anaerobes. Others are obligate anaerobes and are not found in oxygenated systems at all. Ōba (1965f) reports isolation of *Citrobacter, Cloacae, Hafnia,* and *Klebsiella* species from a cesspool in which LAS degradation studies were being made.

Hoadley (1965) isolated 11 types of lake and stream bacteria and reports that the predominant ones were *Pseudomonas* species. Stumm-Zollinger (1968) accomplished the separation of river water isolates into numerous classes by their morphology and physiology although they were not identified taxonomically.

Taxonomy of 267 *Pseudomonas* strains has been comprehensively studied by Stanier (1966), while de Ley (1964a) has reviewed *Pseudomonas* and many related genera. Along with *Pseudomonas* the names of certain other genera have recurred again and again throughout these studies and elsewhere in surfactant biodegradation work. Some are listed in Table 4.1. They are ubiquitous in our environment and are there, readily available, whenever needed.

II. BACTERIAL DEVELOPMENT, DECLINE, AND END PRODUCTS; OR, HOW ULTIMATE IS ULTIMATE BIODEGRADATION?

A. An Idealized Picture

A.1. *Starvation to Plenty to Starvation*

Consider a bacterial system which is without food but meets all other requirements for growth. Oxygen absorption proceeds continually at a slow rate, the rate of *endogenous respiration*. The cells are in a resting state, generating energy for their life processes at this minimal rate by oxidation of certain of their own cell components or by oxidation of organic materials liberated in the breakup and dissolution of dying companions. In his review of endogenous metabolism Dawes (1964) enumerates many cell components which can undergo oxidation in such a case—amino acids, proteins, glycogen, polyhydroxybutyric acid, RNA, and possibly lipids and DNA—and also discusses the biochemical reactions involved.

Now let us add an increment of food, say glucose, to our resting system. Under certain circumstances such an addition (or indeed, addition of any other nutrient or growth factor withheld during the starvation period) may, strangely, result in death (Postgate, 1963; Dawes, 1964). Ordinarily, however, when the food is added it will be utilized rather quickly. In our example, analytical methods specific for the intact glucose will show that it rapidly disappears. Simultaneously with the feeding, the oxygen absorption rate of the cells increases abruptly as glucose oxidation begins, but the immediate amount of oxygen absorbed is far less than stoichiometric. By the time the glucose is all gone perhaps only 20% or 30% of the theoretical oxygen uptake has occurred, of the amount necessary for complete oxidation of the glucose to carbon dioxide and water.

The 70% to 80% discrepancy arises from conversion of the glucose to incompletely oxidized products (i) present either in the medium or in the cells (or associated with them), and (ii) present either in dissolved or undissolved state (perhaps highly polymeric). Such products eventually might range from stored foods, perhaps changed hardly at all from the original glucose, to small molecular fragments actively involved in the metabolic processes, to cell protoplasm, to new cells, to dead cell walls, to inert metabolic products incapable of being degraded further.

Metabolic oxidation of many of these intermediates continues, so the oxygen uptake continues also, but at a decreasing rate which eventually levels off at a new endogenous rate when all of the added food values have been utilized. The overall time required for these processes may range from a few hours in a highly concentrated bacterial culture, say 1000 or 10,000 ppm activated sludge, to many days in a dilute system, as upon inoculation with a few thousand or a few million cells per milliliter. Dean (1964) describes in detail the changes undergone by the cells during such a cycle.

A.2. *And Beyond Starvation*

The endogenous respiration rate of the foodless system is approximately proportional to the number of living cells present. When the increment of glucose is added the cells grow and multiply. By the time the increment is completely metabolized and the endogenous state is again entered, the number of cells present is thus higher than originally, and the endogenous respiration rate of the system is correspondingly higher. At this point the degradation of the glucose increment has still not proceeded to the ultimate extent. Those extra cells are made in part of carbon from the glucose, and that carbon is susceptible to still further oxidation.

In principle, the oxidation of the glucose increment should reach its ultimate at some later time, $t = U$, when those new cells (or an equivalent number of companions) have all died and their substance has been metabolized to the fullest possible extent by the survivors. Now the endogenous rate has dropped back to its initial value at time $t = 0$, when the increment was originally added.

The amount of oxygen absorbed from time zero to time U may be viewed as that required to oxidize the glucose increment. However, there is some disagreement as to whether a correction should be made for endogenous respiration during the entire period. In other words, does endogenous respiration proceed at its usual pace during and in addition to the oxidation of the added food? Usually the assumption is made that it does, and the observed oxygen uptake is corrected by subtracting the endogenous amount, either calculated by extrapolation of the original rate or more often determined directly by observation of a parallel control culture which received no food. (The risks of such an assumption are discussed in Chapter 5, Section IV.A.2.)

When this endogenous correction is made it is found that the total amount of oxygen absorbed is significantly less than the amount calculated for complete oxidation of the food to carbon dioxide and water, perhaps only 70% to 90% of it. The difference from 100% represents those resistant metabolic products and cellular debris which cannot be degraded further biologically. Thus we see that by this criterion glucose, one of the most readily assimilable of natural foods, is not completely biodegradable to carbon dioxide and water, nor is there indication that any other food should exist which would be superior in this respect.

B. The Experimental Observations

The above conclusion that ultimate biodegradation must fall somewhat short of complete oxidation may be open to some question because of the uncertain validity of the endogenous correction. However, essentially the same conclusion is reached by direct measurement.

Thus Porges (1958) and McWhorter (1962) have observed that, after feeding glucose, a considerable amount of organic matter still remains dissolved in the medium after the glucose is gone. This subsequently disappears at a slower rate, some of it being converted to insoluble cellular or metabolic products. In some cases the conversion takes place more rapidly with only minor formation of the dissolved organic intermediates (Painter, 1968). Van Gils (1964) presents extensive data on the oxidation, assimilation, and storage of glucose by various activated sludges and pure bacterial cultures. Wide variations are shown, perhaps reflecting differences in acclimation and hence in reaction rates.

There is general agreement that rather more than half, perhaps as much as 60% to 65%, of a readily utilizable food can be converted to cell substance at the peak of growth. Since destruction of some of the old cells can be occurring before completion of synthesis of the new, the maximum reached may depend to some extent upon the experimental conditions; the more the two processes overlap, the lower the maximum. Upon subsequent starvation the amount of cell substance remaining will diminish, going down as low as 10% to 20% of the food increment but not much lower (Kountz, 1959; Busch, 1961; Washington, 1962; McKinney, 1963; Hetling, 1964; Simpson, 1964).

This remaining material includes rather insoluble, particulate cellular debris which cannot be further metabolized by the surviving organisms. In an extensive study using both radiotracer and conventional techniques Washington (1962) found that the inert solids were formed not only from glucose but also from acetate (representing fatty foods) and glycine (representing proteins) in similar amounts of 5% to 15%. Rough estimates suggested that the inert material was somewhat over half polysaccharide, somewhat less than half proteinaceous, with small amounts of fatty materials, and that it may have originated in cell capsular material and extracellular slime. Although Washington (1964) did obtain some evidence that further degradation of the resistant material, perhaps half of it, could be accomplished by a specially adapted organism resembling *Pseudomonas fluorescens*, the effects were not reproducible and must remain questionable until the controlling variables are clarified.

Formation of nonbiodegradable products may be much greater than 10% to 20% in nitrogen-deficient systems, according to Symons (1958). In the presence of excess nitrogen only a minor fraction went to biologically inert polysaccharide. In nitrogen-deficient medium, protoplasm synthesis was limited to an amount corresponding to the nitrogen available and a much larger fraction of the food was converted to the inert polysaccharide.

In addition to the insoluble cellular products, formation of soluble organic material resistant to further biodegradation can also occur.

McWhorter found that material corresponding to some 5% to 15% of the food increment persisted up to 25 days in his experiments, detected by COD analysis of filtrates. Similarly Bhatla (1966) found 10% to 20% of the initial COD remaining after 5 days in glucose-fed systems. This was an increase from a minimum of 2–10% at around 1 to 2 days. A decrease in cellular mass occurred along with the COD increase, suggesting release of soluble cell components upon disintegration. Glucose content of the solution was estimated by the anthrone method. The anthrone curve exactly paralleled the COD curve through the minimum and the subsequent rise, indicating that the COD was probably carbohydrate in nature.

McKinney (1963) in his discussion of Busch (1963) lists a set of empirical but useful relationships between food uptake, synthesis of cells, and endogenous metabolism in these systems:

(i) One-third of the organic matter is oxidized to convert the other two-thirds into cellular protoplasm.

(ii) The oxygen equivalence [theoretical oxygen demand] of the volatile solids [cells and their associated solids] is 1.42 times their dry weight.

(iii) The ultimate oxygen equivalence of a mixed waste is 1.5 times the 5-day BOD determined with an acclimated bacterial seed.

(iv) The rate of endogenous metabolism is estimated at 38% per day based on active mass of activated sludge; the corresponding oxygen uptake rate is 24 mg/hr/gm of active mass.

(v) About 20% of the active mass of microorganisms is undegradable and remains after the endogenous reaction is complete.

Detailed investigation, discussion, and review of these aspects of bacterial growth and decay have been provided by McWhorter (1962), Washington (1962), Hetling (1964), Simpson (1964, 1965), Gaudy (1965, 1966), Bhatla (1966), and Krishnan (1966). Beyond question ultimate biodegradation, the fullest extent attainable in a biological system, falls significantly short of complete oxidation to carbon dioxide, water, and inorganic products even when the substrate is a natural food. This is taken into account in the definition of ultimate biodegradation presented in Chapter 1, Section II.A.

III. ACCLIMATION AND DEACCLIMATION

Metabolic versatility is a major bacterial characteristic, but even so there are limitations. Acclimation is an extremely important factor in determining those limitations, and is equally important in any biodegradation research. From the standpoint of suitability as food (i.e., susceptibility to biodegradation), organic compounds, including surfactants, may be divided into three categories: (i) readily utilizable, (ii) utilizable after acclimation, and (iii) not utilizable. There is no sharp dividing line from one category to the next, and furthermore the classification of a given organic compound may vary from one situation to another.

Many of the readily utilizable compounds are those directly involved in the fundamental metabolic cycles of the bacterial life processes, or closely related thereto. These would include the smallest building blocks such as acetate ion, larger molecules such as certain simple sugars, amino acids and fatty acids, and much larger ones such as certain polysaccharides, proteins, and lipids. These molecules are continually being synthesized and broken down in the chemical reactions which in the aggregate make up the normal life process of the cell. The enzymes which catalyze these reactions, the *constitutive enzymes*, are usually present at significant concentrations at all times. Any organic molecules structurally matched to those enzymes can thus enter the metabolic cycle without delay, with resultant assimilation.

A. Induction of Enzymes

Many organic compounds not capable of immediate assimilation can nevertheless serve as food equally well if acclimation occurs. This involves the development (or induction) of *adaptive* (or *induced*) *enzymes* by the bacterial cells, enzymes matched to the chemical structure of the new compound and capable of modifying it or breaking it down to products suitable for entering the cell's fundamental metabolic reactions.

Acclimation is a rather unpredictable process. A certain amount of time is needed for the development of the adaptive enzymes. Also other environmental factors may enter into the process; for example, the absence of natural foods may provide a driving force for acclimation to an exotic compound. Or a mixed bacterial culture may show a greater facility in developing acclimation, since bacterial species differ in their chemical capabilities and a given species may be quite unsuited for initiating the attack on a given compound.

Such mixed systems form the basis for the *enrichment culture* technique, a very powerful means of isolating or developing bacterial strains capable of assimilating or degrading any particular compound of interest. Mixtures of organisms from the natural environment, from soil for example, are simply cultured in the presence of the compound. After repeated transfer and growth under such conditions, organisms with the desired properties often show up and predominate.

These factors make it difficult to place a compound in the third category, "not utilizable," except for specifically designated conditions. There is always the possibility that under other conditions or with other bacterial species acclimation will occur. This may eventually happen even under the specifically designated conditions if unrecognized or uncontrollable variables are involved.

Delay in onset of degradation may be due to other factors besides necessity for acclimation. For example, Gaffney (1965) points out a requirement for carbon dioxide, particularly in systems with low concentration of organisms and high concentration of food. Uptake may be quite slow until sufficient metabolic carbon dioxide is formed. He reports a 2 day induction period for utilization of glucose in a CO_2-free system while it began immediately in a CO_2-enriched system.

Acclimation is often a key factor in surfactant biodegradation. Thus Huyser (1960) found that 5 to 6 days were needed for disappearance of 8-phenylpentadecane LAS in unacclimated river water, but only 1 to 2 days upon addition of acclimated river water in which the compound had already been degraded. Similar patterns have recurred again and again in subsequent studies on LAS and other surfactants. Even more striking are the results on the nonionic OPE_{10}, reviewed by Lashen (1967b) and summarized in Chapter 6, Section IX.C.2. Often quite resistant in routine laboratory tests, this material is nevertheless readily degraded in the field at locations where long-continued industrial use has provided adequate opportunity for acclimation.

But even after bacteria have become acclimated to a surfactant under one set of conditions there is no guarantee of immediate action under other conditions. Thus an inoculum from an acclimated semicontinuous activated sludge culture into a shake flask test may sometimes be no more effective than an unacclimated inoculum. It may need just as much acclimation to the new conditions; several transfers may be required, and sometimes the degradation may become even poorer before improvement sets in. Likewise, an acclimated semicontinuous sludge may do poorly for a week or so upon transfer to a continuous-flow system.

The experience of Cook (1968) with the slope culture technique is a further example. The aim of that procedure was to develop an acclimated inoculum for the British STCSD aeration test by prior growth on an agar medium containing the surfactant under test. She found that the control cultures, grown without surfactant in the agar, were just as effective as the "acclimated" ones.

Even so simple a change as an increase in influent surfactant concentration in a continuous-flow activated sludge test may require noticeable reacclimation (Huddleston, 1964a). Indeed, this may occur with any substrate, not only surfactant, as demonstrated by Storer (1969) for glucose (Chapter 5, Section VII.F.5).

The acclimation process is ordinarily a reversible one. If the exotic food is no longer supplied, the adaptive enzymes eventually disappear. Gaudy (1962) showed that this loss of acclimation was not necessarily due to a destruction of the adaptive enzyme. He worked with an activated sludge acclimated to lactose and found that after removal of lactose from the feed, the induced enzyme, β-galactosidase, disappeared on a curve which matched that calculated from simple dilution by the incoming feed.

Upon renewed exposure to the exotic food at a later time, reacclimation may be necessary before utilization again occurs. Since this happens in pure cultures as well as in mixed ones, it indicates that acclimation does not necessarily involve mutation or genetic alteration of the bacteria; i.e., development of a new strain or species is not required.

But acclimation by mutation can and does occur also. In his studies on the enzyme systems involved in the degradation of fatty acids by *Escherichia coli*, Overath (1969) produced numerous mutants with differing capabilities and deduced the locations of the responsible genes on the chromosome. The wild-type organism was unable to grow on C_4 and C_6 fatty acids but grew well on C_{12} and higher, while the mutants exhibited markedly different patterns. For example, some grew well on butyrate and others were unable to grow on oleate.

Acclimated bacterial cultures have been used extensively in exploring the metabolic pathways involved in the biodegradation of exotic compounds, making use of the phenomenon of simultaneous adaptation (*sequential induction*) described by Stanier (1947). Suppose a culture has been acclimated to compound A and now degrades it readily. For some (or no) reason, compound B is suspected to be an intermediate in the degradation. If B is fed to the culture and found to be utilized immediately at a rate comparable to that of A, this is good evidence that B is indeed an intermediate and lies on the pathway (unless B is also readily degraded by the organisms before acclimation to A). Valuable information on the biodegradation may also be obtained by observing the speed of response upon feeding other related compounds such as isomers or structural analogs of A.

B. Repression of Enzymes

Acclimation may be counteracted by *repression*: addition of a readily utilized substrate such as glucose may result in cessation of attack on adapted foods in the system until the glucose has been metabolized. Evidently the repressant (or a metabolic product thereof) interferes with the synthesis of the adaptive enzyme, or with its subsequent action, or with the transport of the adapted food across the cell membrane.

In the specific case of a tyrosine-adapted *Pseudomonas fluorescens* Jacoby (1964) determined that glucose repressed the formation of hydroxylase and oxygenase enzymes necessary for the early stages of tyrosine oxidation, but that interference with tyrosine transport was apparently not involved. He found that in this particular system several other compounds (e.g., citrate, fumarate, acetate) were even more effective repressants than glucose.

Hsie (1967) implicated glucose-6-phosphate, an early metabolite, rather than glucose itself as a probable primary factor in repression of several enzymes by a strain of *Escherichia coli*.* Stokes (1967) found that the apparent inability of a *Sphaerotilus* species to oxidize organic compounds was actually a case of enzyme repression due to glucose in the medium used for growing the cells prior to testing (5000 ppm peptone plus 2000 ppm glucose). Omitting the glucose from the propagation medium gave cells with markedly better capability for oxidizing amino acids, acetate, benzoate, and even glucose.

Gaudy (1963) showed that such phenomena can also be observed in mixed bacterial systems such as activated sludge. When a mixture of glucose and sorbitol was fed to a sorbitol-acclimated sludge, the sorbitol remained unchanged until the glucose had disappeared, provided that the sludge consisted of physiologically young cells. With old cells acclimated to sorbitol, both glucose and sorbitol disappeared together. With glucose-acclimated sludge, glucose always disappeared before sorbitol removal began, whether the cells were young or old. Later Gaudy (1964b) demonstrated similar phenomena with glucose and mannitol. Further he showed that in the absence of a nitrogen source the attack on glucose by sorbitol- or mannitol-acclimated sludges (old cells) was hindered greatly; without the nitrogen the glucose enzymes could not be synthesized in adequate quantity. Similar results were obtained with a wider range of sugars and sugar alcohols (Komolrit, 1966b) and also in continuous-flow systems. There the effects were less noticeable, but could be magnified by higher concentrations of glucose and/or by nitrogen deficiency (Komolrit, 1966a).

Gaudy (1966) has reviewed all this work and emphasizes the importance of these phenomena in waste treatment, where the influent composition may change rapidly. Likewise they may have a direct bearing on our present subject and must be kept in mind in designing test procedures for

---

* De Crombrugghe (1969) reports that enzyme synthesis appears to be regulated by the intracellular level of cyclic AMP, which is lowered by glucose.

measuring surfactant biodegradability and in the interpretation of the biodegradation data.*

Stumm-Zollinger (1968) reported somewhat different results with mixed organisms from river water and from activated sludge. In four out of five experiments, glucose had little or no effect on the assimilation of galactose and both substrates disappeared concurrently, even though the cells used were young and well dispersed. She suggested that glucose might have a repressing action only on certain bacterial species, for example the coliform group, and not on others, for example the slow-growing species encountered in natural water habitats.

Sequential uptake of mixed substrates can also occur in continuous-flow systems with either pure or mixed cultures. Mateles (1969) describes such phenomena with glucose–fructose, glucose–lactose, and glucose–butyrate mixtures, with evidence that the glucose, in addition to being degraded more rapidly, may retard the degradation of the other component.

C. Degree of Acclimation—Biodegradation Rates

In view of the complex interrelations and interactions outlined in the two preceding sections, the data and observations to be cited in this section are necessarily largely empirical and descriptive in nature. Yet there is little question but that a broad fundamental quantitative picture can eventually be developed, into which the fragmentary observations now available will fit.

Details of the biochemical mechanisms involved in acclimation and in enzyme induction and repression are not completely known, and in any case are rather beyond our scope. Dean (1964) has reviewed these phenomena in the general context of cell growth and regulation. He concluded that they are explainable in terms of the fundamental laws of physicochemical kinetics and equilibria as applied to the reactions involved, namely, synthesis of and by enzymes. All the same, Hughes (1968a) has shown that cells derived from a single parent cell may differ greatly in their degree of acclimation, which implies that the laws may have to be applied at the individual rather than at the statistical level.

Basically the rate at which any particular biodegradation reaction progresses depends on the effectiveness of the enzyme molecule as a catalyst for the reaction in question, on the number of enzyme molecules in the system, on the accessibility of the enzyme molecules to the substrate

* Contrariwise, Prakasam (1967b) found that sequential uptake of substrate could be a mere artifact of acclimating the sludge to a constant synthetic medium, not occurring with "natural" sludge.

molecules, and on the removal of the product molecules from the vicinity. As a first approximation, the highest rate should be attained when all the cells in the system contain a maximum amount of the enzyme. This maximum rate may not be attainable in a mixed bacterial system if some of the species present are not able to synthesize that particular enzyme. And the rate will be still lower if the species that are capable do not have each cell loaded with the enzyme to its full capacity. Such states represent lower degrees of acclimation.

Painter (1968) found that activated sludge from different sewage treatment plants exhibited different degrees of acclimation to glucose and other sugars, as judged by the speed of removing them when fed under standard conditions. For example, sludge from plant A removed up to several hundred milligrams per gram of sludge (dry weight) per hour; other sludges ranged from 30 to 100 mg/gm per hour, and the normal rate seemed to be around 50. This normal glucose removal rate could be increased severalfold by feeding the sludge glucose over prolonged periods, the rate rising slowly during several weeks and slowly declining after glucose feeding was stopped.

If the sludge was overfed by adding more glucose immediately after each preceding increment was gone, without allowing time for complete oxidation of the intermediate degradation products, the superacclimation disappeared and the glucose removal rate dropped back to the normal 50 mg/gm per hour. No interpretation of this unusual deacclimation could be advanced. In contrast, pure cultures of *Escherichia coli* and *Streptococcus faecalis* showed much higher rates, 950 and 550, respectively, which did not decline on overfeeding.

Stumm-Zollinger (1968) observed glucose removal rates in the range of 200 to 300 mg/gm per hour for activated sludge and for mixtures of river water bacteria, without necessity for specific acclimation.

Painter mentions in passing that the removal rate of an anionic surfactant (of unspecified nature) was around 1 mg/gm per hour, much lower than the normal rate for sugars. Removal rates of LAS can be much higher under suitable circumstances. For example, the miniature activated sludge units described by Swisher (1964a) have been operated over periods of many months with substantially complete degradation of 100 ppm of LAS in the feed by 3500 ppm of mixed liquor suspended solids during a 3 hr retention time; this corresponds to LAS removal of 10 mg/gm per hour.

In static tests on sludge grown in large semicontinuous units fed ordinary domestic sewage, WPRL (1967, pp. 175–176) measured LAS degradations of 1 to 2.5 mg/gm per hour. Existence of unknown factors affecting acclimation was strikingly evident in this work, since removal rates were only 0.54 and 0.35 using sludges grown in parallel experiments on a detergent-free domestic sewage spiked with 12 ppm of LAS or no LAS, respectively. No difference in LAS adsorption was evident between the acclimated and unacclimated sludges. Although the primary biodegradation rate of the LAS (MBAS analysis) was not affected by the presence of glucose or several other organic materials, biodegradation of the LAS benzene ring was more rapid with the acclimated sludge.

On the basis of degradation rate measurements of this sort, Painter (1967) has estimated that activated sludge or domestic sewage contains the proportions of various bacteria enumerated in Table 4.2.

TABLE 4.2

Proportions of Certain Acclimated Bacterial
Species Estimated to be Present in Activated
Sludge Developed on Domestic Sewage[a]

	Percent
Nitrifying (*Nitrosomonas*)	<1
Glucose-oxidizing	5
Acetate-oxidizing	10
ABS-degrading	3–7

[a] Painter (1967).

Hartmann (1968) gives quantitative data on development and loss of acclimation to phenol which may exemplify what might also happen in a surfactant biodegradation system. He first included phenol in and then later withdrew it from the feed to a (presumably) continuous-flow activated sludge unit. The state of acclimation was determined from Warburg respirometric curves showing oxidation of phenol by samples of the sludge, by comparison with a theoretical curve representing maximum acclimation. This theoretical curve was calculated from the Michaelis-Minton enzymatic reaction rate theory, using values for the Michaelis constant and saturated velocity determined from the respirometric runs themselves.

Some of Hartmann's results are summarized in Table 4.3. The acclimating feed to the sludge unit included 200 ppm of phenol and 500 ppm (as BOD) of Liebig's meat extract. After 70 hr the feed was changed to meat extract alone and continued for another 70 hr to study deacclimation. A sample taken at time zero, representing the original unacclimated sludge, required about 12 hr for 95% oxidation of 50 ppm of phenol in the

TABLE 4.3

Acclimation and Deacclimation of Activated Sludge to Phenol[a]

Sludge exposure time, hr			95% oxidation time in Warburg, hr		
Accl.	Deaccl.	Total	20 ppm[b]	50 ppm[b]	200 ppm[b]
0	—	0	—	12	35
15	—	15	—	1.0	6
70	—	70	0.7	1.1	3.7
—	22	92	—	—	4.9
—	47	117	1.0	1.1	[c]
—	70	140	1.3	3.6	[d]
Full acclimation[e]			0.6	0.8	0.9

[a] Continuous-flow, fed 200 ppm phenol first 70 hr, 500 ppm meat extract entire 140 hr (Hartmann, 1968).
[b] Phenol concentration used for acclimation test in Warburg.
[c] 80% oxidized at 5 hr.
[d] 10% oxidized at 7 hr.
[e] Calculated theoretical values.

Warburg; this included a lag period of about 10 hr for development of acclimation in the Warburg flask. A sample taken after 15 hr on the phenol feed accomplished 95% oxidation in 1 hr in the Warburg, and no further improvement was shown in a sludge sample taken at 70 hr; the theoretical time was calculated as 0.8 hr for a completely acclimated sludge.

After 22 hr without phenol in the feed, the sludge could still accomplish the phenol oxidation in about an hour in the Warburg, but took 3.6 hr to do it after 70 hr deacclimation. Using only 20 ppm of phenol in the Warburg the sludge approached the theoretical more closely both at 70 hr acclimation and 70 hr deacclimation. On the other hand, the sludge was much less effective, compared to theoretical, in handling 200 ppm of phenol in the Warburg, even though that same concentration was used in the acclimation feed. This latter is not surprising, since the phenol concentration in the continuous-flow mixed liquor (i.e., in the environment to which the sludge was acclimating) would be much lower than in the feed itself, if phenol degradation is occurring.

These results show clearly that acclimation is not an all-or-nothing phenomenon—there can be various degrees of acclimation. And further, that the degree of acclimation of a given sample of sludge or culture may not be a single characteristic value, but will depend on the conditions in the acclimation measurement itself. Furthermore, Kurz (1969) has shown that although the degree of acclimation to a substrate (e.g., to phenylacetic acid) may remain constant during the life cycle of the cell, acclimation to intermediate degradation products of the substrate (e.g., hydroxyphenylacetic acid) may vary markedly from one stage of the cycle to another.

Earlier, Hartmann (1967) had described some acclimation experiments involving anionic and nonionic surfactants. The Warburg oxygen uptake was markedly poorer after acclimation than before. For explanation he advanced the tentative hypothesis that two classes of bacteria were present in the seed. Before acclimation, type B are presumed to be killed by the surfactant upon its addition, liberating their cell components which provide food for type A with accompanying oxygen uptake. After acclimation type B are less sensitive and are not killed, or perhaps are no longer present, and hence there is less oxygen uptake. Whatever the explanation, these results are not typical of those often found in surfactant acclimation studies.

The work of van Eyk (1968) provides further examples of enzyme induction and repression, in paraffin oxidation systems (Chapter 7, Section I.A).

D. Thermal Acclimation and Temperature Effects

As pointed out in Chapter 1, Section II.D, at different temperatures the species distribution may be quite different in bacterial populations open to the general environment. Thus acclimation of the usual biodegradation test system following a temperature change (which we may term *thermal acclimation*) will involve two sets of changes: (i) redistribution of species or strains toward a preponderance of those whose growth is favored at the new temperature, and (ii) acclimation of those species to the test compound.

The temperature preference of a given bacterial strain is determined by genetic factors, at least in some instances, and may be altered by appropriate means. Thus Olsen (1968) was able to bring about such changes in certain mesophilic strains of *Pseudomonas aeruginosa* in two ways— ultraviolet irradiation and nucleic acid transfer. Upon ultraviolet treatment of intensity such that less than 1% of the organisms survived, about 1 in a million of the survivors had mutated to psychrophilic types; their limiting temperatures for growth became 0° and 32°C compared to 11° and 44° before mutation. Similar changes were accomplished by nucleic acid transfer via a *Pseudomonas* bacteriophage grown on a psychrophilic strain of *P. fluorescens* and subsequently acting on a mesophilic *P. aeruginosa*.

Palumbo (1969a) reported that a psychrophilic strain of *P. fluorescens* used the same metabolic pathways for glucose at 8°C as at 30°C, when the glucose concentration in the medium was limited to an amount which was completely utilized in a continuous-flow culture. At 8° this minimal concentration was 100 ppm compared to 330 ppm at 30°, and the growth rate was much slower, generation times being 13 and 2 hr respectively. Any excess glucose over the minimal was metabolized via a different pathway; thus the overall metabolic picture was temperature-related indirectly, in that the minimal concentration was governed by temperature. Material balance experiments (Palumbo, 1969b) determined that the cell yield was 0.39–0.46 mg dry weight per mg glucose at 8° compared to 0.65–0.67 at 30°.

Ludzack (1961) gives detailed data on performance of laboratory activated sludge units at 5° and 30°C, while Hunter (1966) reports little difference in the range 20° to 45°C but poorer efficiency at 4° and 55°. Hawkes (1963) has reviewed effects of temperature upon waste treatment systems in general, and Montgomery (1967) in respirometric tests.

Even with complete thermal acclimation, some reduction in specific biodegradation velocity is to be expected with reduction in temperature; after all, the metabolic reactions are chemical reactions. Lower temperature, other things being equal, will result in slower biodegradation of the more natural foods as well as of surfactants. Accordingly, comparison of the two under corresponding test conditions is necessary if a true judgment of a particular surfactant's low temperature biodegradability potential is to be reached. Few efforts have been made toward reaching such judgments, and those which have been reported do not seem to have given much attention to the above considerations. Fortunately the specific heat of water is large, so that the range of operating temperatures actually met in sewage treatment is much narrower than the extremes of atmospheric temperature which may be experienced.

Roberts (1960) observed that LAS biodegradation was faster at 37°C than at room temperature; Wayman (1963a) found it slower at 10°C. It is difficult to assess whether this reflected only the kinetic effect of temperature, or whether thermal acclimation was also involved.

In the course of studies on a sewage lagoon, Halvorson (1969b) collected organisms from under the ice at 2°C and found them unable to degrade C_{14}LAS at 2° in the laboratory for at least 12 days. The bacteria were undoubtedly thermally acclimated, but were obviously not acclimated to the LAS; it is an open question whether LAS acclimation could have been achieved if special efforts had been made. LAS degradation did occur at 10°, at about half the 25° rate, using summer organisms. Similar experiments with more natural substrates—palmitate, acetate, casamino acids, glucose—also showed very nearly zero utilization by winter organisms at 2°C in 2 hr Warburg runs (Halvorson, 1968). These substrates and others were eventually removed upon prolonged exposure, except for the casamino acids, still essentially unattacked in 6 days (Halvorson, 1969a). Thus even the natural foods may resist bacterial attack at low temperatures under laboratory conditions.

Krone (1968) found that biodegradation of sodium alkane sulfonate (SAS) and LAS was markedly poorer in the laboratory at 6° than at 25°C, and that LAS was influenced at 15° as well (Table 4.4). However, the

TABLE 4.4

Biodegradation at Low Temperatures[a,b]

	Temperature, °C		
	6°	15°	20°
SAS (secondary alkane sulfonate)	25	99	99
LAS	25	40–95	96

[a] Krone (1968).
[b] MBAS removal, %, in official German activated sludge test.

effects were less marked in the field, where SAS and LAS removals across a trickling filter during winter operation were 85% (8°C) and 76% (10°C), respectively. Perhaps thermal acclimation, more likely to have been achieved in the field, accounts for the difference from the laboratory results.

In a fresh soil lysimeter Robeck (1964) was unable to initiate TBS biodegradation in 60 days at 40°F, but acclimation occurred rapidly upon raising the temperature to 70°. A parallel unit started at 70° was removing about 60% within a few days. Lowering the temperature of this unit to 40° at day 65 resulted in a drop in biodegradation, which reached zero removal at day 110 and then climbed back to 40% by day 180. It was then brought to 70° again and attained 80% removal by day 235. Cooling back to 40° then caused a rapid drop to 45%, followed by a rise to 60% by day 250, when the run was stopped. The relatively high removals of the TBS at 40°F, 40% to 60%, would undoubtedly have been exceeded significantly if LAS had been used.

IV. SURFACTANT–BACTERIA INTERACTIONS

Two interactions with bacteria are particularly important in the study of surfactant biodegradation. First, high concentrations of the surfactant may inhibit or damage the bacteria so that they are unable to accomplish a degradation which may occur readily at lower concentrations. Second, the bacteria may remove surfactant from solution by adsorption; to mistake this for biodegradation leads to confusion and possibly to later disappointment.

Information on bacteriostatic and toxic effects, outlined in the following section, is mostly descriptive. But beginnings have been made toward an understanding of some of the mechanisms involved, exemplified in the studies of interactions with proteins and with enzymes, summarized in the succeeding sections. These in turn lead to the phenomena of adsorption, covered in Section V.

A. Surfactants as Bacteriostats and Germicides

Not only do bacteria degrade surfactants; surfactants have been known to retaliate on occasion—to degrade bacteria, in a manner of speaking. Much of the early work on surfactant–bacteria interactions was directed toward development of antiseptic and disinfectant systems; some of the surfactants are quite powerful in that respect, under proper conditions. Even so, under other conditions those same surfactants may be readily attacked by bacteria and used by them as a source of food. Such apparent paradoxes are well known. A classical example is phenol, used as a germicide ever since the relationship of bacteria to infection and disease was discovered. Yet phenol is readily removed from industrial wastewaters by bacterial action, the preferred method of treatment being the activated sludge process. As a further example, soaps and fatty acids show antibacterial activity just as do the synthetic surfactants, and they too are readily degraded by bacteria under other conditions.

A.1. *Conditions and Threshold Concentrations*

Surfactant effects on bacteria have been reviewed by Glassman (1948), Putnam (1948), Fischer (1958), James (1965), and Dychdala (1968), among others. At neutral pH cationic surfactants are generally the most toxic; their effectiveness is increased at higher pH and decreased at lower. In contrast, the anionics are less effective than the cationics in neutral solution but their activity is enhanced at lower pH. Nonionics are in general inactive against bacteria.

The antibacterial action of surfactants may be considerably diminished by the presence of proteins or other organic matter. Bacteria of different species may show widely differing sensitivity to a given surfactant. In general Gram-positive species are more susceptible to anionics than are Gram-negatives, while cationics act against both types. Increased chain length of an anionic appears to make it more effective. Voss (1963) found that Ca^{2+} and other divalent ions enhanced the antibacterial action of TBS and other anionics, presumably by promoting adsorption by and penetration of the cell wall.

With increasing use of surfactants by the detergent industry and their resultant appearance in sewage, a somewhat different consideration emerged. Rather than a maximum of antibacterial action, here the minimum of such effects was to be desired, to minimize possibility of interference with the bacteria involved in the sewage treatment processes. Thus Waddams (1950) investigated Teepol (a secondary alkyl sulfate) and reported that several bacterial species of interest were unaffected at concentrations up to 500 ppm, although it was somewhat bactericidal toward *Staphylococcus aureus* at concentrations as low as 250 ppm. The nonionic Lissapol N was found to have no effect on bacteria.

In contrast, Leclerc (1952) found that the contemporary commercial ABS (C_{12}) inhibited oxygen uptake in the BOD test at 60 ppm, although an alkylphenol nonionic showed no such effect even at 300 ppm. Further, Manganelli (1953, 1956) reported that Nacconol NR (a KBS) and Ceepryn (a cationic) markedly reduced the oxygen uptake of activated sludge at concentrations as low as 10 ppm. There were indications that they might reduce the efficiency of sewage treatment processes in several other respects as well. Incomplete acclimation was presumably a major factor in Manganelli's results, since these undesirable effects were not particularly evident in actual practice later, when ABS levels in incoming sewage approached 10 to 20 ppm (TBS or LAS) all around the world. Bringmann (1963) found that LAS (Marlon BW1043) did not inhibit the bacterial action in a trickling filter at feed levels below 300 ppm, although protozoa were disturbed above 160 ppm, causing cloudy effluents.

Generally the inhibitory effects of anionic surfactants become evident in the range from 100 to 1000 ppm. The threshold varies widely depending on the physiological criteria used and on the specific conditions—the type of surfactant, the presence of foods or other materials, the particular bacterial species, the bacterial concentration, and their recent history, i.e., their degree of acclimation. In some circumstances the upper limit can be extremely high. Payne (1963a) reports optimal growth of *Pseudomonas*

C12B in 0.1 M KBS (kerylbenzenesulfonate) and 0.015 M SLS (sodium lauryl sulfate), over 30,000 and 4000 ppm, respectively. Earlier, McKinney (1959b) had used 1000 ppm TBS medium in isolating TBS-degrading species, and later Anderson (1964) reported good growth in 5000 ppm TBS.

Inhibitory effects of surfactants on bacteria have usually been detected by observation of changes in their respirometric oxygen uptake. In an early study Baker (1941) found that at 300 ppm some anionics (Igepon AP, for example) inhibited the respiration of some bacteria almost completely, but actually stimulated respiration of others to the extent of 150% or more. Some species (*Sarcina lutea*, for example) ranged from 95% inhibition to 235% stimulation from one anionic to another. At 30 ppm this pattern was still evident.

Fischer (1958) observed that oxygen uptake by *Mycobacterium phlei* and *Staphylococcus aureus* was inhibited by TBS, noticeably at 10 ppm and completely (death of cells also) at 1000 ppm. The corresponding figures for *Proteus vulgaris* were 1000 and 10,000 ppm (0.1% and 1%). SDS (sodium dodecyl sulfate) gave similar results except that its bio-degradation complicated the picture somewhat. Fischer cited the survival and even the growth of bacteria at much higher concentrations on occasion, as high as 250,000 ppm (25%). He suggested that the presence of other nutrients and micelle formation might be involved. He also pointed out that the concentration of bacterial cells in the medium was quite important. At 5000 ppm of suspended solids (*M. phlei*), 1500 ppm of surfactant inhibited respiration almost completely, while at 15,000 ppm suspended solids inhibition was negligible.

Soaps also can affect bacterial respiration. Oleate at 30 ppm stimulated the oxygen uptake of *Mycobacterium avium* to almost double the endogenous level, but at 300 ppm reduced it to below endogenous. Under other conditions these organisms were about 100% killed by 300 ppm oleate, 95% killed by 30 ppm, and unaffected by 3 ppm (Minami, 1956).

Knöpp (1961) found thresholds for detectable inhibition of the bacterial degradation of protein at around 10 to 20 ppm of TBS, 5 to 10 ppm KBS, and 1 to 2 ppm LAS, while Meinck (1961) determined that inhibition of glucose degradation by *Escherichia* species was first evident at TBS concentrations around 60 to 120 ppm.

Activity of bacterial systems can also be estimated from their content of reductase enzymes, detectable by a color reaction with triphenyltetrazolium chloride. By that method Schönborn (1962a) found some 20% to 25% loss of activity in the presence of 30 to 70 ppm of KBS, TBS, or LAS.

TABLE 4.5

Influence of Surfactants and Phenol on Dehydrogenase Activity of
Activated Sludge Fed Glucose[a]

Additive, 50 ppm	Dehydrogenase activity, % of control
CPB (cetylpyridinium bromide)	5
CTAB (cetyltrimethylammonium bromide)	25
TBS	30
LAS (C_{11-15})	60
SDS (sodium dodecyl sulfate)	70
SAS (secondary alkane sulfonate)	70
Diisobutylnaphthalenesulfonate	90
OCE_{20} (oleyl–cetyl alcohols E_{20})	90
NPE_{20} (isononylphenol E_{20})	95
Phenol	100
No additive	100

[a] Pitter (1968d).

Pitter (1968d) studied a wider range of surfactants by the same method
(Table 4.5). At 50 ppm two cationics, CPB and CTAB, were the most
inhibitory, closely followed by TBS, while two nonionics and an alkyl-
naphthalenesulfonate had hardly any effect, nor did phenol, and SAS,
SDS, and LAS ranged in between.

The greater resistance of Gram-negative than Gram-positive bacteria
to the action of anionic surfactants remarked upon in the earliest reviews
has been repeatedly observed in later work (Fischer, 1958; Hartmann,
1963b, 1966b; Ōba, 1965a; Lambin, 1966). Concentrations on the order
of 10 to 20 ppm may cause noticeable effects in Gram-positives (including
Bacillus sp., *Sarcina* sp., *Staphylococcus aureus*) whereas several thousand
parts per million may have little effect on Gram-negatives (*Aerobacter
aerogenes*, *Escherichia coli*, *Pseudomonas* sp., *Serratia marcescens*). In
disagreement, Wayman (1963c) found *E. coli* to be considerably more
sensitive than indicated above, perhaps because he studied very dilute
systems containing only a few thousand cells per milliliter. Under his
conditions at 20°C and pH 7, 1000 ppm and even 100 ppm of TBS caused a
significantly reduced survival time (1.5 days) compared to that in distilled
water (6 days); 10 ppm had little effect. Survival time was greatly
influenced by factors of temperature and pH also.

According to Nakamura (1968) resistance of *E. coli* varies considerably
from strain to strain when exposed to 2000 to 50,000 ppm of SDS for 1 hr

in broth medium. At 10,000 ppm, for example, 10% to 100% mortality was observed, depending on the strain.

Taber (1962), using TBS, further confirmed that various bacterial species have widely differing sensitivity, and found similar differences among species of fungi. The human pathogen *Candida albicans* was killed by 600 ppm of TBS within 11 hr exposure, and 20 ppm gave a noticeable inhibition of growth. LAS was inhibitory at lower concentrations than TBS.

The bacteria responsible for anaerobic digestion are affected only slightly, if at all, by addition of TBS at 140 ppm (Hartmann, 1966b) or by LAS at 200 ppm (Bringmann, 1963). But these levels are very low compared with the 25,000 or 50,000 ppm or more of suspended solids typically present in such systems. At much higher surfactant levels inhibition does become readily apparent, for example, by decrease or cessation of gas production. Bruce (1966) concludes that the threshold for this effect lies somewhere around 15 to 20 mg of surfactant per gram of solids (1.5% to 2%), which would amount to perhaps 500 to 1000 ppm total surfactant in the system, dissolved and adsorbed. Both sulfate and sulfonate anionics cause this inhibition, but nonionics are inactive even at levels around 20% of the sludge solids. In addition to presenting his own data, Bruce also summarizes the extensive earlier literature on effects of surfactants on anaerobic digestion.

A.2. *Focal Points*

The mechanisms by which surfactants inhibit or stop bacterial activities have not been well explored as yet. Undoubtedly interaction with the bacterial proteins and enzymes is an important factor. Such interactions do occur in vitro, as discussed in the following sections, and if those essential cell components are altered in that manner within the cell itself, or at the cell wall, alteration of the life processes thereby would not be surprising.

Research on the "swarming" of *Proteus* species provides an example. These organisms are exceedingly motile, and even on solid culture media do not remain in fixed colonies but overrun the entire surface. Inhibition of the swarming is necessary if the usual bacteriological methods of isolation and identification of other organisms are to be used in the presence of *Proteus*. Lominsky (1942) found that soaps and surfactants had the desired inhibiting effect on *Proteus* without affecting its growth properties (aside from the disappearance of its characteristic flagella), or even those of much more delicate organisms. Later studies on *Proteus mirabilis* by Kopp (1965) showed that the linear primary alkyl sulfates inhibited more effectively with increased chain length. Fifty percent inhibition of motility in a liquid medium (5000 ppm casein–peptone) was achieved with about 20 ppm of the C_{14} homolog but required 4000 ppm of the C_6.

Absence of flagella in the inhibited organisms was reported in both studies. Whether this was the result of (i) direct attack on the flagella or (ii) inhibition of flagella formation, could not be established. SDS at levels of 100 or 1000 ppm may cause considerable breakdown or partial disintegration of the flagella of other bacterial species (Kopp, 1965), and at 1000 to 2500 ppm SDS can solubilize about twice its weight of protein from the isolated cytoplasmic membranes of *Bacillus subtilis* (Bishop, 1967). Likewise, Minami (1958) had observed cell wall damage after exposure to oleate soap at 3000 ppm.

SDS and other surfactants have been used in many other studies of the structure of cells and cell components, this by virtue of their reactivity toward proteins, often reversible, summarized in the next section.

B. Surfactant–Protein Complexes

One of the underlying factors in surfactant–bacteria interactions must be the reaction with proteins; such reactions may occur with the vital proteins of the intact bacterial cells, and they certainly do occur with purified proteins in aqueous solution. Study of these simpler aqueous systems has provided some fundamental information.

B.1. *Possible Modes of Interaction*

The backbone of a protein is its long polypeptide chain, derived from amino acids linked together by carboxamide groups (–CONH–) as shown schematically in Eq. (4.1).

$$\cdots CO_2H + H \overset{R}{N}\text{—CH—}CO_2H + H_2N\overset{R'}{\text{—CH—}}CO_2H + H_2N \cdots$$
$$\downarrow$$
$$\cdots CONH\overset{R}{\text{—CH—}}CONH\overset{R'}{\text{—CH—}}CONH \cdots \qquad (4.1)$$

The carboxamide group is moderately hydrophilic, sufficient to balance the various R groups of the amino acids, which are generally moderately hydrophobic, and to confer water solubility on the entire protein structure, at least in the absence of other complicating features in its chemical structure or in its aqueous environment.

Various secondary forces, attractive and repulsive, can act upon the R groups, between each other and the carboxamide groups; for example, secondary valences, Coulombic forces, steric effects, and the like. In solution the protein molecule may assume some characteristic preferred conformation, perhaps folded or coiled, in response to those forces, since the backbone is innately flexible. But the forces are relatively small and the conformation may be changed relatively easily by minor changes in the aqueous environment—in the pH, the temperature, or the presence of other solutes, for example.

Some of the Rs carry free carboxylic acid groups, and others free amino or other basic groups. Attractions and repulsions between these play an important part in determining the conformation of the protein molecule in solution. They also provide focal points for interaction with surfactant molecules, the positively charged amino groups attracting the negatively charged ions of anionic surfactants, and likewise with the carboxylate groups and cationics. Other, non-Coulombic, attractions may come into play also, capable of acting upon all three surfactant types: anionic, cationic, and nonionic. Once a surfactant molecule is bonded to the protein, it may itself provide a further center of attraction for others, by virtue of the same hydrophobe–hydrophobe forces responsible for micelle formation.

The surfactant–protein interactions may have various results—a dissolved protein may be precipitated or an insoluble one may be dissolved. Complex proteins may be dissociated into subunits just as a simple protein may be unfolded or otherwise changed in conformation if the secondary valence bonds of the protein are displaced by the surfactant. These same forces bind the surfactant to the protein, and the resulting complex is in mobile equilibrium with the free molecules. Under suitable circumstances, for example, by diffusion through a membrane permeable to the surfactant and not to the protein, the two can be separated. In some cases the original protein is recovered intact, in others it may have undergone some irreversible change.

B.2. *Serum Albumins and Anionics*

The earliest literature on surfactant–protein interactions was reviewed by Glassman (1948) and Putnam (1948). Much of the work before and since has been concerned with the relatively simple serum (or plasma) albumins and SDS (sodium dodecyl sulfate) or certain ABSs. The results are reasonably consistent in overall pattern, although they may differ in detail and in interpretation. Decker (1966, 1967) has summarized the literature in conjunction with presentation of his own results and a somewhat revised interpretation.

In aqueous solution the albumin (A) molecules exist in a folded state, each molecule having on its outer surface a number (m) of sites which can exert a strong bonding force upon anionic surfactant molecules (D). Depending on the particular type of albumin and surfactant involved, and on the specific conditions, m may range around 10 to 15. Thus addition of D to a solution of A results in formation of a series of complexes AD, AD$_2$, AD$_3$, . . . , AD$_m$, leaving very little free, uncomplexed D in solution until m mols have been added per mole of protein.

Further addition of D to the solution then results in partial unfolding of the protein molecule, exposing new, weaker bonding sites to give a total of n, ranging from around 38 to 55 depending on circumstances and the uncertainties of measurement. Each protein molecule evidently fills these new sites all at once; as the amount of D added is increased beyond m mols, the AD$_m$ is converted directly to AD$_n$ with no sign of complexes of intermediate composition. When all of the A is in the form AD$_n$, further addition of D initiates a second unfolding, exposing another n sites. Each AD$_n$ converts directly to AD$_{2n}$ and conversion is complete when $2n$ mols of D have been added per mol of A.

Further amounts of D can be bound by the complex if more D is added to the solution, up to a saturation point perhaps around AD$_{6n}$, but this appears to occur in a continuous manner rather than in discrete stages as before.

This general picture may differ in detail in individual cases. Depending on the exact chemical nature of A and D, or on the pH, one of the complexes AD$_n$ or AD$_{2n}$ may not be formed, or precipitation may occur. For instance, Putnam (1948) found that with horse serum albumin and SDS at pH 4.5, precipitation began when about $\frac{1}{2}n$ mols of D had been added and was complete at D$_n$. On continued addition of D, the precipitate began to redissolve when $2n$ mols had been added, and was all back in solution with only a few more mols of D.

It was very early evident that $2n$ was numerically about equal to the number of cationic spots (i.e., basic amino acids) in the protein molecule as determined by acid titration (Putnam, 1948). It seemed obvious that the bonding in the complex was between these positively charged groups and the negatively charged sulfate or sulfonate group of the surfactant anion. Yet both Ray (1966) and Decker (1967) conclude that the binding is not primarily electrostatic, but is probably mainly due to hydrophobic interactions. Beyond the AD$_{2n}$ complex it seems very likely that the further molecules of D which enter are held by hydrophobic forces of the same sort as are responsible for micelle formation in simple aqueous solutions of D.

Comparing the action of linear primary alkyl sulfates of 8, 10, 12, and 14 carbons and the corresponding alkane sulfonates of 8, 10, and 12 on bovine serum albumin, Reynolds (1967) concluded that both ionic and hydrophobic bonding are important. This was suggested principally by the fact that under the conditions used, the C_{12} sulfate unfolds the albumin molecule (as does the C_{14}), but the C_{12} sulfonate does not appear to (up to mol ratios around 100, anyway). Since the hydrophobes of the two surfactants are identical, the difference would seem to be more explainable if the linkage were through the ionic group. Although both have the same charge, they differ in geometry and presumably in charge distribution because of the additional oxygen atom in the sulfate group.

B.3. *Other Proteins and Anionics*

The general picture outlined above also fits reasonably well the results in the interaction of anionic surfactants with many other proteins of lower or higher molecular weight, for example, legumin (Brand, 1956), β-lactoglobulin (Hill, 1956), egg albumin (Yang, 1953, Aoki, 1959), α-amylase and Bence-Jones protein (Imanishi, 1965), myosin and light and heavy meromyosin (McCubbin, 1966), paramyosin and tropomyosin (Cowgill, 1968), and casein (Cheeseman, 1968).

A series of 14 proteins was examined by Pitt-Rivers (1968) to determine the maximum amount of SDS bound. The SDS:protein ratio by weight (which eliminates the variable of protein molecular weight) was remarkably constant at around 1.0 or a little less for the 10 (including bovine serum albumin) which contained cystine S—S cross-links. The four without such groups bound about 1.4 gm of SDS per gram of protein, and reached equilibrium in about half the time required by the cross-linked ones. This indicates that the S—S cross-linking interferes with the readjustment of the protein molecule's conformation to accept the maximum number of SDS molecules. When the S—S bonds were broken by reduction, the reduced form took up 1.4 gm/gm.

Bovine serum albumin has a molecular weight around 70,000 and took up 0.93 gm/gm, corresponding to a complex averaging around AD_{225}. This is well above the AD_{100} corresponding to the $2n$ complex and well into the range where SDS bonding is presumably accomplished by micellar forces. Pitt-Rivers pictures the complex as a series of micelles strung like pearls along the polypeptide backbone. However, the ordinary SDS micelles contain around 100 molecules each, and the entire serum albumin complex can only supply enough for two or three such structures per molecule. The micelle beads must be very small if they are distributed evenly among the many attractive centers along the backbone.

B.4. *Cationics*

The information on cationic surfactant–protein interactions is scantier than for anionics. Indications are that the same general pattern is followed, but the results are not as clear-cut.

Glassman (1950) found that bovine plasma albumin was precipitated by cetyldimethylbenzylammonium chloride in alkaline aqueous medium in much the same manner as by SDS in acid medium, redissolving upon addition of an excess of the cationic. Jerchel (1953) verified precipitation with a series of cationics but found that under some conditions precipitation did not occur with nonaromatic derivatives such as dodecyltrimethylammonium bromide (DTAB). In those cases he showed the existence of protein–cationic complexes in solution. Later Jerchel (1954) identified two complexes between bovine serum albumin and DTAB with approximate compositions AD_{40} and AD_{450}.

Timasheff (1951) found indications that the nonquatenary cationic dodecylamine hydrochloride gave complexes with egg albumin, around AD_{30} and AD_{100}.

Working with bovine plasma albumin, horse serum albumin, and egg albumin Foster (1954) and Aoki (1958, 1959) deduced that complexes were formed with dodecylpyridinium bromide in a manner similar to that with SDS, but various operational difficulties prevented estimation of the composition. Likewise, Imanishi (1965) found similarities in the action of cationics and anionics in changing the conformation of proteins, as also did McCubbin (1966).

Cowgill (1968) determined saturation values for tropomyosin with respect to cetyldimethylethylammonium bromide and SDS, getting values around AD_{30-50} and $AD_{200-180}$, respectively, at pH 2.0 and 8.5. These correspond to 0.2 and 0.75 gm/gm of protein, well into the micelle-type bonding range for the SDS but not for the cationic. Nevertheless, the relatively slight difference in the number of D molecules bound at widely differing pH suggests that hydrophobic forces are mainly involved rather than ionic. Cowgill found that neither of the surfactants seemed to have much effect on the helical conformation of the protein.

B.5. *Nonionics*

It now seems to be generally accepted that ionic forces are of only minor significance in forming the protein–anionic and protein–cationic surfactant complexes, and that hydrophobe interactions play a much more important role. One might expect, then, that nonionic surfactants would form protein complexes almost as readily as do the ionics. Unfortunately the methods which have been developed for studying the complexes with the ionics are not generally applicable for the nonionics. They have given little indication of any interaction of proteins with nonylphenol E_{20} (Imanishi, 1965) or with Tween 80 (oleoyl sorbitan E_{20}) (Cowgill, 1968).

Even so, evidence for such interaction is accumulating. Dowben (1961) found small differences in physical properties of bovine plasma albumin upon adding OPE_{10}, which suggested a change in the degree of hydration of the protein. Koehler (1961) used the same system except that a dimethylaminoazo dye had been added at certain locations along the albumin molecule. Changes in the spectrum of the azoprotein occurred when OPE_{10} was added; comparison with the spectra of simpler, known systems led to the conclusion that molecules of OPE_{10} did indeed become closely associated with the protein. Evidence of a more direct sort was obtained by Cheeseman (1968), whose ultracentrifuge studies indicated that OPE_{10} formed two complexes with casein.

C. Surfactants and Enzymes

C.1. *Weight Ratios in Complexes*

We have seen that surfactants interact with proteins in general; accordingly they will also interact with those special proteins which exhibit spectacular activity of one sort or another. Often loss of the activity results, reversibly or irreversibly. Thus Glassman (1950) found that type A or B botulinum toxin was inactivated by treatment with a number of anionic surfactants. In acidic solution less than 0.3 gm/gm toxin was required and inactivation was more effective with greater chain length of the surfactant. Similar results were obtained with cationics in neutral solution. In acid solution the toxin bears a net positive charge, in neutral solution a negative charge, and the inactivation would seem to involve interaction of the surfactant ions with oppositely charged groups on the toxin molecule. The anionics were much less effective in neutral solution and the cationics quite ineffective in acid. Nonionics had no effect at any pH.

Later Glassman (1951) studied an enzyme, lysozyme, in a similar manner, finding that much higher surfactant concentrations were required for inactivation. With enzyme solutions ranging from 20 to 1000 ppm, effects were first noticeable at a weight ratio around 5 gm/gm enzyme and inactivation was complete at around 50, for both anionics and cationics. Again, nonionics had no effect, at least up to 100 gm/gm.

In the surfactant–protein complexes discussed in the preceding sections the weight ratio of the components is around 1, order of magnitude. Enzyme inactivation occurs in this general range also, say weight ratios from 0.1 to 10, which is consistent with the view that the inactivation is the consequence of blocking of certain sites by surfactant molecules, and/or the resultant unfolding or dissociation.

Perhaps the best evidence for this has been presented by Gorin (1967), indicating that SDS dissociated purified urease (mol. wt. 480,000) into subunits with molecular weight around 60,000, which were associated with about an equal weight of SDS. The observations were made at a urease concentration around 5000 ppm with SDS ranging from 2000 to 20,000 ppm. The urease lost its enzymic activity, irreversibly.

C.2. *Threshold Concentrations for Inhibition of Enzymes*

Most of the work in this field simply reports the effects of the surfactant on the enzyme activity, as determined by measuring the velocity of a reaction catalyzed by the enzyme. Many of the enzyme preparations used have been of unknown or unstated purity, and ordinarily the impurities are themselves largely of a proteinaceous nature. In such a situation the surfactant:enzyme weight ratios and limiting concentrations as observed may not be generally applicable to all other specimens of the enzyme in all other circumstances.

Numerous enzymes are hindered in their catalytic activity by the presence of anionic surfactants at concentrations above 100 to 1000 ppm, and several workers have observed increased inhibition with longer hydrophobe groups. Thus Mathews (1954) found that inactivation of hyaluronidase at 2 ppm could be accomplished by 260 ppm of sodium decyl sulfate, but required only 90, 10, and 3.5 ppm of the C_{12}, C_{14}, and C_{16} homologs. Czok (1968) determined the inhibiting action of secondary octanesulfonate and its C_{17} homolog on acid phosphatase, alkaline phosphatase, papain, and invertase. The C_8 caused 50% inhibition at concentrations of 25,000 to 50,000 ppm, compared to 25 to 600 ppm for the C_{17}. He mentions that the linear primary alkyl sulfates show a similar increase of inhibitory action with increasing chain length up to about C_{17}, but that the trend reverses above that, perhaps because of decreasing solubility.

Wills (1954) studied the action of SDS on urease, finding complete inhibition by 300 ppm at pH 5.0 and zero inhibition at pH 5.4. Back at pH 5.0, 60 ppm gave zero inhibition and 150 ppm about 30% inhibition. At 150 ppm the inhibition was removed by raising the pH to 5.7, while at

300 ppm the inhibition was irreversible by such change in pH. He suggested that this might correlate with the micelle state of the SDS, said to be essentially 100% micellar at 300 ppm and 100% dissociated at 150 ppm. He examined 20 other enzymes, finding that 300 ppm of SDS had no effect on 6 of them at the specific pHs tested. Teepol (linear secondary alkyl sulfate) at 1000 ppm (perhaps 100 to 300 ppm of actual surfactant) was found to inactivate urease completely at pH 4.8. Hydrolysis of horse serum albumin by trypsin was some 20 times faster in the presence of 3000 ppm SDS than without SDS. This was attributed to action of the SDS on the albumin, making it more easily attacked by the trypsin, rather than to any promoting action on the trypsin itself. No effect was noticeable at 300 ppm SDS.

Viswanatha (1955) reports what might be a similar situation. Activity of trypsin (60 ppm) was completely destroyed by 10 min treatment with C_{10} linear primary ABS at 180 ppm and also at 3200 ppm, while intermediate concentrations were much less effective. At 600 ppm, for example, the tryptic activity was only about 30% below normal. The C_8 homolog gave a similar pattern, but at higher concentrations. In contrast, C_8, C_{10}, and C_{12} linear primary alkyl sulfate all showed simple, linear patterns with about 5%, 20%, and 45% inhibition resulting from treatment with 600 ppm.

Trypsin is also inhibited by soap. Peck (1942) investigated the action of several saturated and unsaturated soaps, finding that the activity of 6 ppm trypsin is completely inhibited by around 500 ppm of soap.

Wills (1955) investigated the effects of 11 anionic, 5 cationic, and 6 nonionic surfactants (mostly identified by trade names rather than by chemical structure) on the hydrolysis of fats and lower triglycerides by a lipase. In some cases inhibition occurred, but in others the hydrolytic reaction was accelerated, presumably because of emulsification of the fat and absence of enzyme inhibition. The anionics were generally inhibiting above 500 ppm. The cationics generally increased the hydrolysis rate (in one case by a factor of 11 at 1500 ppm) but also gave indications of enzyme inhibition at the higher concentrations, or when the cationic was added to the enzyme before the fat. The nonionics were somewhat inhibitory at 1000 to 10,000 ppm, but generally inactive at 200 ppm.

The relative lack of inhibition by some cationics is further indicated by Wilmsmann (1959), who found laurylpyridinium chloride to be inactive against phosphatase at weight ratios up to 2 gm/gm (although the cationic amide from stearic acid and diethylenetriamine accomplished 60% inhibition at that ratio). Wilmsmann (1959, 1963) also studied the inhibition of invertase by a series of anionics. Under the conditions used, it was more sensitive than phosphatase, and he too found noticeable differences from one anionic to another. Hexadecanesulfonate, 1-phenyldodecane sulfonate and TBS all gave 60% to 90% inactivation at around 100 ppm. SLS was relatively inactive at 100 ppm, but gave 90% inhibition at 350 ppm, as did its ethoxylate sulfate $C_{12}OE_2O\Sigma$. Igepon T was rather less active as an inhibitor, and Igepon A still less. A fatty acid–protein condensate was only 5% inhibitory at around 1 millimolar concentration; in the absence of molecular weight data its concentration by weight cannot be calculated, but would be very much higher than 300 ppm.

The SDS-hydrolyzing enzyme isolated by Hsu (1963) is an interesting case in that it is inactivated by its own substrate. Activity is lost at SDS concentrations above about 2000 ppm, except that this effect is neutralized by the presence of twice its weight of bovine serum albumin, which preferentially complexes the SDS.

C.3. *Possible Mechanisms of Inhibition*

The experiments of Mosolov (1966) perhaps give the beginnings of a picture of the mechanism of enzyme inhibition by surfactants. Noting that the number of SDS molecules complexed by trypsin was about equal to the number of basic amino acid residues in the trypsin molecule, he investigated the effect of masking the basic groups. Acetylation of trypsin converted the positively charged (cationic) ϵ-amino groups of the lysine residues **(4.1)** into essentially neutral acetylamino groups **(4.2)**; alternatively, reaction with succinic anhydride gave the succinylamino derivative **(4.3)**, wherein the ϵ-amino groups were not only blocked by linkage to the

$$
\begin{array}{ll}
\cdots\text{CONH—CH—CONH}\cdots & \cdots\text{CONH—CH—CONH}\cdots \\
\qquad\quad\text{C} & \qquad\quad\text{C} \\
\qquad\quad\text{C} & \qquad\quad\text{C} \\
\qquad\quad\text{C} & \qquad\quad\text{C} \\
\qquad\quad\text{CNH}_2^{(+)} & \qquad\quad\text{CNHCOCH}_3 \\
\qquad\quad\textbf{(4.1)} & \qquad\quad\textbf{(4.2)}
\end{array}
$$

$$
\begin{array}{l}
\cdots\text{CONH—CH—CONH}\cdots \\
\qquad\quad\text{C} \\
\qquad\quad\text{C} \\
\qquad\quad\text{C} \\
\qquad\quad\text{CNHCOCCCO}_2\text{H}^{(-)} \\
\qquad\quad\textbf{(4.3)}
\end{array}
$$

succinyl, but an opposite charge was provided in the near vicinity by the new carboxyl group. The acetyl derivative retained 88% of the original esterase activity of the trypsin, the succinyl 73%. But their pattern of inactivation by surfactants was markedly different.

SDS at 300 ppm reduced the esterase activity of 50 ppm trypsin to only

12% of its initial value, but had no effect on the other two derivatives. Presumably the negative dodecyl sulfate ions were bonding to the positive ϵ-amino groups in the trypsin and bonding did not occur when these were converted to neutral or negatively charged groups.

The picture was reversed with a cationic surfactant, CTAB. It had no effect on the original trypsin, inhibited the acetyltrypsin and, even more so, the succinyltrypsin. The CTAB was presumably bonding to the many negative centers in the succinyltrypsin molecules, and to the fewer ones present at other locations in the acetyltrypsin. These latter originate in the acidic amino acid residues of the trypsin (aspartic and glutamic acid). These same groups are also present in the original trypsin, but evidently overbalanced by the excess of ϵ-amino and other basic amino acid groups.

The nonionic emulsifier Tween 60 (stearoyl sorbitan E_{20}) had no inhibiting action on any of the three enzymes.

Thus it would appear that ionic interaction is somehow involved in the inactivation of these enzymes, even though (according to Section IV.B.2) it may not be the prime factor in bonding the surfactant to the enzyme.

In later studies Vas'kova (1968) obtained similar results with chymotrypsin. Using the gel filtration technique, SDS was shown to cause a change (possibly dimerization) in the chymotrypsin molecule, whereas succinylchymotrypsin was apparently unaffected. The esterase activity of the chymotrypsin was cut to about 30% in 300 ppm SDS and to about 5% in 30,000 ppm.

V. SURFACTANT ADSORPTION BY BACTERIAL MATTER

If ABS is added, say at 20 or 30 ppm, to a concentrated bacterial system, say an activated sludge containing 2000 or 3000 ppm of mixed liquor suspended solids, immediate analysis of the liquid phase may show only 2 or 3 ppm of ABS. This does not necessarily mean that 90% degradation has taken place within this first minute or so; upon closer examination the missing ABS can usually be found associated with the sludge, adsorbed on it. If the ABS is biodegradable, if it is LAS for example, the adsorbed fraction soon disappears, along with that in the liquid phase. If it is poorly degradable, for example TBS, the undegraded portion persists both on the sludge and in the solution.

Thus adsorption is an important factor which must be considered in the interpretation of surfactant biodegradation experiments (Chapter 5, Section VII.F.7). Accordingly, the basic principles governing adsorption will be covered in some detail and quantitative (more correctly, semi-quantitative) data will be presented to facilitate estimation of the resultant effects under various conditions.

A. The Freundlich Equation

Consider a liquid solvent containing a dissolved solute, to which an insoluble, particulate material, the *adsorbent*, is added. If the concentration of dissolved solute diminishes thereby, this is evidence that adsorption has occurred. In other words, some of the solute molecules have become bonded to the surface of the adsorbent and are no longer to be found dissolved in the solvent. In general a mobile equilibrium is established between the dissolved molecules and the adsorbed molecules (*adsorbate*). The mobility indicates that the bonding forces responsible are considerably weaker than valence bonds. If the solute concentration is lowered, as by adding more solvent, some of the adsorbed molecules redissolve. If the concentration of solute is increased, the concentration of adsorbed molecules on the surface of the adsorbent increases also.

The Freundlich equation [Eq. (4.2)] is an empirical relation between the concentration of adsorbed molecules, A, and the concentration of dissolved molecules, D, in the liquid phase,

$$A = kD^n \tag{4.2}$$

$$\log A = \log k + n \log D \tag{4.3}$$

where A is the concentration of adsorbate on adsorbent, mg/gm; D, the concentration of solute in solution, mg/liter; and k and n are constants (ideally). Equation (4.2) often gives a very good approximation of the observed values over wide ranges of concentration, provided the region where the adsorbent becomes saturated with adsorbate is not approached too closely.

The Freundlich equation may also be conveniently expressed in logarithmic notation [Eq. (4.3)], which is in the form $y = b + mx$. Thus a log–log plot of A and D will give a straight line, to the extent that the Freundlich equation is applicable to the system, with slope equal to n and the y intercept at $\log k$.

From Eq. (4.2) it is evident that the constant k is numerically equal to the value of A corresponding to $D = 1$. If the constant n equals 1, the equation says that the concentration of adsorbate is proportional to the concentration of solute remaining in solution. If $A = 0.5$ mg/gm is in equilibrium with $D = 1$ ppm, then A should be 5 and 50 mg/gm, respectively, when D is 10 and 100 ppm. In actual practice the constant n is often a little more or less than 1, in which case a tenfold increase in D results in somewhat more or less than a tenfold increase in A.

When working with activated sludge systems Sweeney (1964a) and his colleagues have found it convenient to speak in terms of a *sludge adsorption ratio*, the ratio of adsorbed to dissolved surfactant in a system containing 2000 ppm of mixed liquor suspended solids. They define it mathematically according to Eq. (4.4), which may be rearranged into Eq. (4.5)

$$R = \frac{2000}{1000\,G} \times \frac{AG}{D} = \frac{2A}{D} \tag{4.4}$$

$$A = \frac{R}{2}\,D \tag{4.5}$$

where A is the surfactant adsorbed on solids, mg/gm; D, the surfactant dissolved in mixed liquor, mg/liter (ppm); G, the mixed liquor-suspended solids dry weight, gm/liter; and R, the sludge adsorption ratio. If the sludge adsorption ratio, R, is constant, Eq. (4.5) is simply the Freundlich equation with $k = R/2$ and $n = 1$.

B. Equilibrations

In an ideal system where the Freundlich equation applies, the equation holds true regardless of the concentration of adsorbent present. Some illustrations are presented, with the aid of Table 4.6. System P is at

TABLE 4.6

Hypothetical Adsorption Systems

	System P		System Q	
	Amount	Concentration	Amount	Concentration
Water	1 liter	—	1 liter	—
Adsorbent	100 mg	100 ppm	1000 mg	1000 ppm
Dissolved ABS	10 mg	10 ppm (D)	10 mg	10 ppm (D)
Adsorbed ABS	0.5 mg	5 mg/gm (A)	5 mg	5 mg/gm (A)
Total ABS	10.5 mg	—	15 mg	—

equilibrium, containing 1 liter of water, 100 ppm of adsorbent, 10 ppm of dissolved ABS, and 5 mg of adsorbed ABS per gram of adsorbent. A second system, Q, containing 1000 ppm of adsorbent, must also contain 5 mg/gm adsorbed ABS at equilibrium. But it is important to note that Q cannot be obtained from P simply by adding another 900 mg of adsorbent, for this will adsorb some of the dissolved ABS. D will thereby drop below the original 10 ppm and consequently A will be lower than 5 mg/gm as well. The exact values can be calculated from the Freundlich equation if the constants k and n are known (or one of them in the present case, since we know that $k = 5/10^n$). System Q is obtained from P by adding another 4.5 mg of ABS as well, enough to supply the added adsorbent with its quota without depleting the dissolved ABS.

Similar considerations apply if Q is diluted back to 100 ppm adsorbent by adding 9 liters of water. The final value of D will not be 1 ppm but somewhat higher, while A will decrease somewhat from its initial value of 5 mg/gm. In other words, a certain amount of desorption has occurred, or a certain amount of ABS has been "washed" off from the adsorbent. Again, the final values of D and A can be calculated if k and n are known, to the extent that the Freundlich equation correctly describes the system. Likewise the amount of dilution or washing required for any desired degree of desorption can be calculated.

C. Surfactant Adsorption Data

Adsorption of anionic surfactants, particularly of ABS types, has been studied extensively with both inorganic and organic adsorbents. Minerals, for example sand, clay, or soil, are generally weak adsorbents. They hold as little as a few micrograms of ABS per gram, insufficient even to form a monomolecular layer on the mineral surface, very loosely held and easily washed off (McGauhey, 1957, 1959a, b; Renn, 1959; Ewing, 1962; Klein, 1963a, b; Wayman, 1963b; Suess, 1964; KrishnaMurti, 1966; Barbaro, 1967). In contrast, activated carbon may adsorb a thousand times as much ABS, milligrams per gram of carbon, from dilute solutions (Renn, 1959; Wayman, 1963b; Weber, 1963, 1964a, b, 1965; Hartmann, 1966a).

Biological matter such as primary sewage sludge, activated sludge, anaerobic digester sludge, and bacteria themselves can likewise adsorb large quantities of ABS from dilute solutions. Ewing (1962) passed natural and synthetic sewage through sand-filled columns, developing a biological slime growth on the sand grains. Addition of 10 ppm of TBS to the feed resulted in its adsorption on the slime, eventually reaching 10 mg/gm, or about 1%, dry weight basis. In the absence of the growth the sand alone adsorbed only about 5 μg/gm, or 0.0005%. Ewing calculated that a complete monomolecular layer of TBS on the sand grains he was using (0.838 mm average diameter) should amount to about 10 μg/gm, reasonably consistent with his observed value.

The intense adsorption of anionic surfactants onto biological systems very likely relates in some way to the surfactant–protein interactions covered in Section IV.B, in which complexes are formed with dissolved proteins. It is reasonable to suppose that the proteins of particulate biological matter form similar complexes, thus accounting for the strong adsorption. At variance with this concept is the report of Ewing (1962) that dead slime (killed by autoclaving the sand column described above) did not adsorb TBS to any greater extent than the sand itself. A similar idea is sometimes invoked to explain an anomalous rise in surfactant concentration occasionally observed after an initial drop during a biodegradation: the bacteria are said to have died and released surfactant back into solution. In any case, Schönborn (1962a) observed the opposite— that the adsorption of TBS by activated sludge was not changed when the sludge was killed by treatment with formalin. Perhaps this disagreement with Ewing's result arises from the different means used for killing the growth.

However there is no great disagreement in the general findings of the many workers who have examined surfactant adsorption. Unfortunately many of the data in the literature cannot be put into the universal framework of the Freundlich equation. It is expressed simply in terms of "percent adsorbed" without the information necessary to calculate the pertinent parameters—the concentrations on the adsorbent and in the dissolved state. Those listed in Table 4.7 have given this pertinent information in one form or another. Their data are plotted in Figs. 4.1 to 4.4 on a log–log scale showing the concentration of adsorbed surfactant (mg/gm of adsorbent) in equilibrium with dissolved surfactant (mg/liter of solution).

The approximate validity of the Freundlich equation for describing these systems is evident in the approximately linear relationships of the sets of data in the four figures. True, there are discrepancies, but they are to be expected. They could arise from differences in the nature of the sludges, in the chemical structures and molecular weights of the surfactants, in the experimental conditions, and in the analytical methods. Furthermore, the surfactant is in many cases a mixture of components with different chemical structures and hence with a range of different values for k and n; i.e., some of its components may be more tightly adsorbed than others. Thus Sweeney (1964a) reports that adsorption of TBS onto activated sludge is twice as great for the C_{15} homolog as for the ordinary C_{12}. And beyond that, the individual particles of sludge may differ from each other in adsorptive properties. In such circumstances deviations from the Freundlich equation are not surprising, nor are

TABLE 4.7

Biological Adsorption of Surfactants[a]

No.[b]	Adsorbent[c]	Surfactant[d]	k	n	Reference
1	AS	TBS	—	—	Degens, 1955
2	PS	*MBAS	—	—	Fairing, 1956
3	AS(A)	TBS[e]	—	—	Lockett, 1956
4	AS(B)	TBS[e]	—	—	Lockett, 1956
5	AS(A)	*MBAS	—	—	Lockett, 1956
6	AS(B)	*MBAS	—	—	Lockett, 1956
7	AS(A)	*MBAS	—	—	Lockett, 1956
8	AS(B)	*MBAS	—	—	Lockett, 1956
9	AS	KBS[f]	—	—	Manganelli, 1956
10	PS	TBS	—	—	McGauhey, 1959a
11	AS	TBS	—	—	McKinney, 1959b
12	AS(A)	TBS	—	—	Burgess, 1962
13	AS(B)	TBS	—	—	Burgess, 1962
14	DS(A)	TBS	—	—	Burgess, 1962
15	DS(B)	TBS	—	—	Burgess, 1962
16	AS	*TBS	—	—	Offhaus, 1962
17	AS	*TBS	—	—	Offhaus, 1962
18	AS	TBS	0.04	1.85	Schönborn, 1962a
19	AS	*TBS	—	—	Schönborn, 1962a
20	AS	*LAS(A)	—	—	Schönborn, 1962a
21	AS	*LAS(B)	—	—	Schönborn, 1962a
22	AS	*KBS	—	—	Schönborn, 1962a
23	AS	TBS	—	—	Hartmann, 1963a
24	DS	TBS	—	—	Hartmann, 1963a
25	S. aureus	TBS	—	—	Hartmann, 1963a
26	B. subtilis	TBS	—	—	Hartmann, 1963a
27	AS	TBS	—	—	Phillips, 1963, p. 67
28	AS (static)	TBS	—	—	Pitter, 1964b
29	AS (dynamic)	TBS	—	—	Pitter, 1964b
30	DS (static)	TBS	—	—	Pitter, 1964b
31	DS (dynamic)	TBS	—	—	Pitter, 1964b
32	AS	*TBS	1.8	1	Sweeney, 1964a
33	AS	*LAS[g]	5.2	1	Sweeney, 1964a
34	AS	LAS[h]	3.8	0.52	WPRL, 1965, p. 10
35	PS(A)	*MBAS	0.59	1.04	WPRL, 1965, p. 75[j]
36	PS(A)	MBAS[i]	1.44	0.93	WPRL, 1965, p. 75[j]
37	PS(B)	MBAS[i]	1.35	0.96	WPRL, 1965, p. 75[j]
38	PS(C)	MBAS[i]	0.71	1.09	WPRL, 1965, p. 75[j]
39	PS(B)	TBS	1.32	0.83	WPRL, 1965, p. 75
40	PS(B)	LAS[h]	1.64	0.77	WPRL, 1965, p. 75
41	DS(A)	*MBAS	—	—	WPRL, 1965, p. 76
42	DS(B)	*MBAS	—	—	WPRL, 1965, p. 76
43	DS(C)	*MBAS	—	—	WPRL, 1965, p. 76

TABLE 4.7 continued

No.[b]	Adsorbent[c]	Surfactant[d]	k	n	Reference
44	*E. coli*	ABS(F)	1.28	0.61	Hartmann, 1966a
45	*Sarcina* sp.	ABS(F)	0.21	1.39	Hartmann, 1966a
46	*B. subtilis*	ABS(F)	8.9	0.68	Hartmann, 1966a
47	*S. aureus*	ABS(F)	0.8	0.77	Hartmann, 1966a
48	TF	ABS(F)	0.89	0.96	Hartmann, 1966a
	TF	ABS(B)	0.7	1.01	Hartmann, 1966a
	TF	TBS(G)	0.46	0.92	Hartmann, 1966a
	TF	TBS(H)	0.43	1.0	Hartmann, 1966a
	TF	ABS(C)	5.1	1.1	Hartmann, 1966a
	TF	$C_{12-18}O\Sigma$	16.0	0.85	Hartmann, 1966a
	TF	$C_{12-18}OE_3O\Sigma$	4.9	0.91	Hartmann, 1966a
49	PS	LAS	0.45	1	House, 1966a
50	AS	LAS	2.1	1	House, 1966a
51	*A. faecalis*	TBS[k]	—	—	Marion, 1966
52	*A. faecalis*	LAS[l]	—	—	Marion, 1966
53	*A. faecalis*	$C_{12}O\Sigma$	—	—	Marion, 1966
	AS	*LAS[g]	3–6	1	Sweeney, 1966
	AS	*LAS[m]	5–35	1	Sweeney, 1966
	AS (unaccl)	LAS[h]	0.6–4.2	0.5–1.3	WPRL, 1967, p. 176
	AS (accl)	LAS[h]	1.0–4.0	0.4–1.1	WPRL, 1967, p. 176
54	Bacteria[n]	LAS	—	—	Tomiyama, 1968
55	Bacteria[n]	AOS	—	—	Tomiyama, 1968
56	*Pseudomonas* K5	LAS	0.13	0.94	Tomiyama, 1968
57	*E. coli*	LAS	0.28	0.84	Tomiyama, 1968
58	*B. subtilis*	LAS	0.34	0.83	Tomiyama, 1968
59	*E. coli*	CTAB	—	—	Salt, 1968
60	AS	$C_9APE_{9.5}{}^o$	—	—	Schönborn, 1966, p. 107

[a] Temperature 15° to 25°C; pH near neutral. Where necessary the original data have been cast into the Freundlich form $A = kD^n$.

[b] Numbered entries are plotted in Figs. 4.1–4.4.

[c] AS, activated sludge; DS, anaerobic digester sludge; PS, primary sewage sludge; TF, trickling filter sludge.

[d] Asterisk indicates residual after biodegradation and/or contained in sewage.

[e] Santomerse.

[f] Nacconol NR.

[g] European.

[h] Dobane JNX.

[i] Mixture of packaged detergents.

[j] Also reported in Bruce (1966).

[k] Ultrawet K.

[l] Nalkylene 500.

[m] U.S.

[n] *Pseudomonas* K5, *E. coli*, or *B. subtilis*.

[o] Merpoxen NO 95.

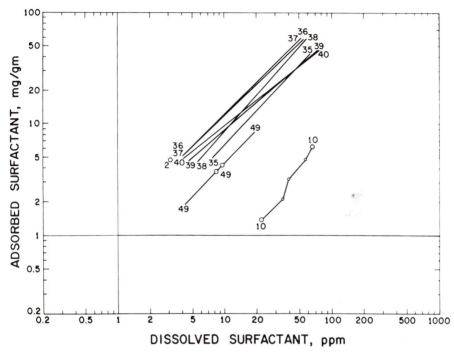

Fig. 4.1 Adsorption of surfactants by primary sewage sludge. (Numbers refer to Table 4.7.)

variations in the "constants" k and n from one system to another, or even within a single system.

Visual extrapolation of the lines in Figs. 4.1–4.4 to the y axis ($D = 1$; log $D = 0$) indicates that the intercepts (i.e., the values of the constant k) lie in the range from 0.1 to 10 mg/gm, while the values of the constant n (i.e., the slopes of the lines) range from somewhat under to somewhat over unity.

Hartmann (1966a) studied the adsorption of eight anionic surfactants by four inorganic adsorbents, six bacterial species, and trickling filter sludge. Bacterial adsorption was significantly higher at pH 3 and 4.7 than at 7, 9.2, and 11.4. He concluded that Gram-positive bacteria adsorbed more strongly than Gram-negative, but the effects of that difference in bacterial type, as well as differences in surfactant structure, are quite difficult to generalize among the rather wide variations in the results. Ōba (1965a) also noted that bacterial adsorption of TBS was greater at

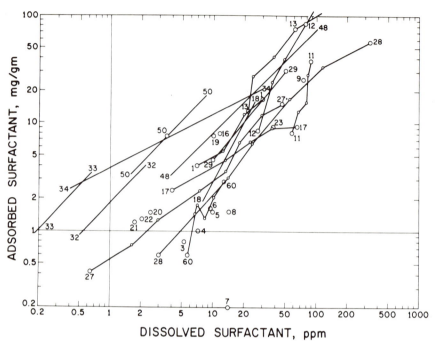

Fig. 4.2   Adsorption of surfactants by activated sludge. (Numbers refer to Table 4.7.)

low pH, and found an opposite trend with the cationic surfactant CTAB. Tomiyama (1968) compared LAS with α-olefin sulfonate (AOS) on several species of bacteria; again adsorption was greater at low pH. The AOS was slightly less adsorbed than the LAS, on the average, but the range of values for each surfactant was considerably greater than the difference in the averages.

Voss (1963) found that adsorption of TBS by *Staphylococcus aureus* and *Escherichia coli* was increased severalfold by presence of about 1000 ppm of calcium chloride in the medium, as well as by other bivalent metal ions. The suggested mode of action was that the negatively charged cell adsorbed the metal cation, and that in turn promoted the adsorption of the surfactant anion.

The single cationic entry in Table 4.7 (No. 59) seems to fit into the general pattern of the anionics, but the deviation from linearity is pronounced (Fig. 4.4).

Data on nonionic surfactants are quite scanty, and only one entry is

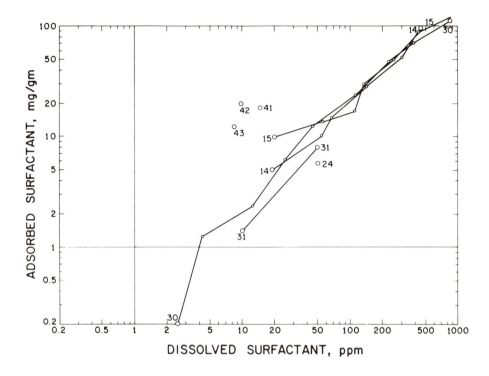

Fig. 4.3 Adsorption of surfactants by anaerobic digester sludge. (Numbers refer to Table 4.7.)

shown in Table 4.7 and the figures, that of Schönborn (1966) for nonyl-phenol $E_{9.5}$. The adsorption appears to be rather less than for the anionics, but this may be due in part to a difference in Schönborn's measurement procedure. He analyzed the sludge directly for adsorbed nonionic instead of calculating it by difference, and anywhere from 15% to 50% of his added nonionic was unaccounted for. He used this same procedure on polyethylene glycol 400 (E_9) and found only about 0.8 mg/gm adsorbed on the sludge in the presence of 10–15 ppm in solution.

As a whole the nonionic adsorption data in the literature are somewhat inconsistent with each other. Thus Oldham (1958) presents qualitative evidence for rather extensive adsorption of Lissapol N by the biological film of trickling filters, while WPRL (1965, p. 192) reports that adsorption of nonylphenol E_8 onto primary sewage solids is negligible. Lashen (1967b) found 1 to 3 mg/gm of OPE_{10} adsorbed on activated sludge when added at

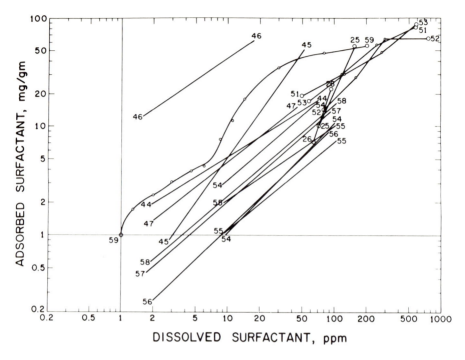

Fig. 4.4 Adsorption of surfactants by bacteria. (Numbers refer to Table 4.7.)

levels of 5 to 10 ppm; this is within the range of the anionics shown in Fig. 4.2. To the contrary, in a biodegradation field test Conway (1965) found nonionics (CTAS) adsorbed on the activated sludge to the extent of 2.8 mg/gm in equilibrium with 1.5 ppm of residual nonionic in solution (Table 4.8). This corresponds to a rather higher adsorption than the average of the anionics shown in Fig. 4.2. The MBAS results obtained at the same time fall closer to the average.

D. Adsorption or Biodegradation?

Figures 4.1–4.4 indicate that a biodegradation system containing 5 to 10 ppm of ABS dissolved in the liquid phase will also contain adsorbed ABS, perhaps 1 to 10 mg/gm of biological solids. If the solids content of the test system is high, the amount adsorbed may be significant. For instance, an activated sludge system may contain 2000 to 4000 ppm of mixed liquor suspended solids, and the adsorbed ABS may range from 2 to 40 mg/liter of mixed liquor, using the range given above for adsorption.

TABLE 4.8

Adsorption of Nonionics and Anionics[a]

		Test period	Adsorbed, mg/gm	Dissolved, ppm
Nonionics	Tergitol 15-S-9[b]	A	2.87	1.6
	"Hard" (CTAS)	B	2.74	1.5
Anionics	Tergitol 15-S-3[c]	A	1.1	1.9
	TBS (MBAS)	B	13.3	8.8

[a] Surfactant adsorbed on activated sludge, in equilibrium with dissolved, resistant surfactant remaining in effluent after degradation (Conway, 1965).

[b] Linear $C_{10\text{-}15}$ secondary alcohol E_9.

[c] Linear $C_{10\text{-}15}$ secondary alcohol $E_3O\Sigma$.

This is the same order of magnitude as the amount of ABS in solution, and must be taken into account when biodegradation is being measured.

Indeed, Hartmann (1963a) found that in his own experiments TBS removal was predominantly by adsorption rather than by biodegradation, and van Beneden (1964, 1965) also has expressed doubts. On the other hand, most workers who have examined the question have found that adsorption plays a rather insignificant part compared to biodegradation. In particular, Sweeney (1966) presents extensive data on LAS in activated sludge systems showing that 97% to 98% of the removal is by biodegradation, 2% to 3% by adsorption. Heinz (1967) states that adsorption accounts for only about 1% of the total LAS removal in the official German continuous-flow activated sludge test, while Lashen (1967c) showed adsorption to play a negligible part in OPE_{10} removal by activated sludge treatment.

This question is considered in further detail in Chapter 5, Section VII.F.7.

BIODEGRADATION TEST METHODS

I. BASIC PRINCIPLES

The three components essential to a surfactant biodegradation test system are the surfactant, the analytical method, and the biological agent. These have been viewed separately in Chapters 2, 3, and 4, and we are now ready to consider the varied ways in which they have been combined into test procedures.

The pertinent biological systems are predominantly bacterial. They may be chosen to simulate the conditions in a natural environment or in a sewage treatment process in attempts to get data more or less relevant thereto. [But not completely equivalent, as pointed out by Ettinger (1965) and Heinz (1966).] Or they may be designed with more fundamental considerations in mind so as to provide insight into the biochemical or other aspects involved.

A. Substrates in General

With increasing recognition of the possibility of environmental responses upon widespread use of new products, the need for standard test procedures for determining their biodegradability (and for estimating their potential impact upon the environment as well) is becoming self-evident, not only for surfactants, but for substrates in general. Many have been proposed— for example, see the extensive review of Ludzack (1963a). A few have achieved a certain tentative status in rather narrow fields, in that of surfactant biodegradation, for instance, by virtue of fairly extensive use within the field under fairly standardized conditions. But the adoption of uniform, detailed procedures across the whole wide range of wastes and wastewaters must await demonstration of their suitability in the hands of a wide range of workers and with a wide range of substrates; this has not yet been accomplished.

The Water Pollution Control Federation Subcommittee on Biodegradability has enunciated general principles which should apply in the testing of any substances (WPCF, 1967). Between the two extremes of primary biodegradation and ultimate biodegradation they offer the concept of "environmentally acceptable biodegradation . . . defined as susceptibility to biodegradation yielding end-products which are totally acceptable in the receiving environment which includes air, soil and water although principal interest may be in treatability in waste disposal facilities." Pertinent factors to be considered in establishing test conditions and measurements are enumerated, including adequate provision for acclimation; the necessity for correlation with results in the world at large is emphasized.

Two specific proposals for standardized procedures are representative of these which could eventually be chosen for the necessary exhaustive and diversified trials. Symons (1960) recommends 1.5 liter fill-and-draw activated sludge units, and presents a set of varied feeding schedules designed to meet the considerable difficulties in bacterial acclimation which may arise. On the other hand Young (1968) prefers a large scale 20 day BOD test and points out the necessity for demonstrating environmental acceptability of the degradation products by bioassay techniques.

B. Surfactants in Particular

Among the limitless variety of exposure conditions which might be used, certain ones have been the choice, for one reason or another, of workers in the surfactant field; these will be discussed in detail in this chapter. They differ from each other first of all in the numbers or concentration of bacteria present in the system, and this factor in turn may dictate such other conditions as surfactant concentration, presence and amount of other foods, time of exposure, and so forth.

The question of surfactant biodegradation test methods has been reviewed by Schönborn (1962b). Heinz (1964, 1966, 1967) considered in detail the general principles which should be observed, and how well the many test methods meet the basic requirements of (i) correctness of the biological principle, (ii) compatibility with the physicochemical properties of surfactants, (iii) pertinence to practical field conditions in sewage and rivers, and (iv) suitability for laboratory operation with clearly defined conditions and reproducible results. Heinz concluded that no single test can be satisfactory and suggested a combination of three: (1) a simple screening test (inoculated BOD dilution water), (2) the optimal test (closed bottle), and (3) a close-to-practice test (official German continuous activated sludge test).

C. Mixed Bacterial Species

Occasionally it is recommended that the long-sought standard bio-degradation test should involve a standard microorganism as well, to ensure the maximum in reproducibility. However, this appears to be self-contradictory in principle, since biodegradation generally involves acclimation, and once acclimated to the test compound the organism is no longer in its standard state. For the further reasons outlined in Chapter 1, Section II.B a mixture of bacterial species is to be preferred in a test procedure.

Development of fundamental information on and understanding of the sequences of biochemical processes occurring is quite difficult in the case of a pure bacterial species and a single food. With mixed species the situation is very much more complex, and if mixed substrates are involved also, the complexity is compounded once again. Thus much of the information available is descriptive and specific rather than generalized. Nevertheless, as indicated in a review by Gaudy (1966), beginnings have been made.

Cibulka (1967) describes a program to compare the biodegradation capabilities of pure bacterial species with mixtures thereof and with activated sludge. No results appear to be available as yet, beyond selection of the operating parameters to be used in the Warburg test procedures. Early experiments along this line by McKinney (1952) indicated that activated sludges made up of pure bacterial species (originally isolated from activated sludge) were less effective than the mixed sludge in removal of BOD (Table 5.1).

TABLE 5.1

BOD Removal by Pure Bacterial Species[a]

Species	BOD removal, %
Aerobacter aerogenes	66
Escherichia coli	76
Paracolobactrum aerogenoides	76
Flavobacterium sp.	78
Nocardia actinomorpha	88
Mixed culture from sewage	94–99

[a] Floc-forming species isolated from activated sludge, fed Weinberger synthetic sewage in 24-hr semicontinuous activated sludge operation (McKinney, 1952).

Stumm-Zollinger (1968) is studying the mixed bacterial species of river water. The bacterial concentrations in the original water are too low (by 3 to 4 orders of magnitude) for use of the desired techniques. Accordingly, she grows individually some 50 isolates from the original sample and then remixes them to give the higher concentrations necessary. Growth of the original sample as a mixture to the desired bacterial concentration was not suitable because the final species distribution would have been entirely different as a result of differing growth rates. The shifting population distribution in the growth of mixed bacterial species is discussed further in Section VII.F.3.

Even though mixed cultures are preferable for general biodegradation testing, pure cultures have been used with success in many studies. Table 5.2 lists some of those mentioned in the literature. Results are generally in accord with those obtained using mixed cultures, including a certain amount of disagreement, presumably due to differences in acclimation or in test conditions. Thus a negative response indicated in Table 5.2 for a given species and surfactant is not necessarily of universal validity. For example, *E. coli* degraded LAS in Huddleston's experiments, but not in Tomiyama's or Cook's, and similarly for *S. marcescens*.

Such inconsistency is not solely found in surfactant experiments. McKinney (1952) mentions that a floc-forming *Bacillus cereus* strain isolated from activated sludge "was not able to carry out the oxidation of organic matter to carbon dioxide and water." The organic matter referred to was the natural foods contained in Weinberger's synthetic sewage (see Table 5.3). This medium was readily utilized by several other floc-forming species isolated at that same time (Table 5.1).

In an extensive examination of LAS-degrading mixed cultures, Cook (1968) isolated numerous pure cultures, usually Gram-negative rods (*Klebsiella* sp., *Achromobacter* sp., *Flavobacterium* sp.) but on one occasion a Gram-positive *Micrococcus* species. In contrast with some others listed in Table 5.2, these isolates were much poorer than the original mixture in their ability to degrade LAS. Many were unable to degrade it at all in aerated cultures, and even with the powerful recycle trickling filter technique the degradation was very slow and erratic. She concluded from her experimental results that a key factor in those results was the flocculation or clumping of the bacteria during the tests, such as occurs on the surface of the packing medium in the trickling filter. When this was promoted by addition of stones or asbestos in shake culture runs, degradation was promoted, whereas parallel runs without such additive, or without shaking, did not flocculate, nor did they degrade the LAS. Nor was degradation accomplished by remixing the isolated bacteria.

TABLE 5.2

Biodegradation and Growth Studies with Pure Cultures

Organism	Surfactant[a]	Action[b]	Reference
Achromobacter thalassius	SLS, etc.	−	Williams, 1949
Achromobacter aquamarinus	SLS, etc.	−	Williams, 1949
Achromobacter sp.	LAS	+	Cook, 1968
Aerobacter aerogenes	SLS, etc.	+	Williams, 1949
	LPAS, etc.	+	von Riesen, 1955
Aerobacter cloacae (?)	SDS	+	Payne, 1967
Alcaligenes faecalis	SLS, etc.	−	Williams, 1949
	SLS, TBS, LAS	+	Marion, 1966
Bacillus subtilis	KBS, TBS	−	Kimura, 1962
	LAS, AOS	−	Tomiyama, 1968
Escherichia coli	SLS, etc.	+	Williams, 1949
	LPAS, etc.	+	von Riesen, 1955
	KBS, TBS	−	Kimura, 1962
	SLS, LAS, TBS	+	Huddleston, 1963
	ABS, TBS, etc.	+	Kölbel, 1964, 1967
	LAS, AOS	−	Cook, 1968
	LAS	−	Tomiyama, 1968
Escherichia freundii	SLS, etc.	−	Williams, 1949
Flavobacterium sp.	LAS	+	Cook, 1968
Klebsiella sp.	LAS	+	Cook, 1968
Micrococcus tweenis	Tween 80	+	Minami, 1958
Micrococcus sp.	LAS	+	Cook, 1968
Paracolobactrum aerogenoides	LPAS, etc.	+	von Riesen, 1955
Proteus morganii	SLS, etc.	+	Williams, 1949
Proteus vulgaris	SLS, etc.	+	Williams, 1949
	SLS, LAS, TBS	+	Huddleston, 1963
Pseudomonas aeruginosa	SLS, etc.	+	Williams, 1949
	LPAS, etc.	+	von Riesen, 1955
	SLS, LAS, TBS	+	Lambin, 1966
Pseudomonas fluorescens	SLS, etc.	+	Williams, 1949
	SLS, LAS, TBS	+	Huddleston, 1963
Pseudomonas C12, C12B	SLS, KBS, TBS	+	Payne, 1963a, b
Pseudomonas C12B	LAS, LPABS	+	Heyman, 1967, 68
Pseudomonas K5	LAS, AOS	+	Tomiyama, 1968
Pseudomonas 473	LPAS, etc.	+	Huyser, 1960
Pseudomonas sp.	SLS, etc.	+	Williams, 1949
Pseudomonas sp.	LPAS, etc.	+	Skinner, 1959
Pseudomonas sp.	LPAS	+	Hsu, 1965
Salmonella enteritidis	LPAS, etc.	+	von Riesen, 1955
Serratia marcescens	SLS, etc.	+	Williams, 1949
	LPAS, etc.	+	von Riesen, 1955
	KBS, TBS	−	Kimura, 1962
	SLS, LAS, TBS	+	Huddleston, 1963
	LAS	−	Cook, 1968
Sphaerotilus sp.	SDS, TBS	+	Pipes, 1963b
Staphylococcus aureus	KBS, TBS	−	Kimura, 1962

[a] AOS, α-olefin sulfonate; LPAS, linear primary alkyl sulfate; LPABS, linear primary ABS; SDS, sodium dodecyl sulfate; SLS, sodium lauryl sulfate.

[b] + denotes growth on or degradation of one or more of the surfactants tested; − denotes no growth or degradation.

The results of Nyns (1969a) on 40 pure bacterial strains likewise indicated less than satisfactory biodegradation capabilities. They had been isolated from detergent-polluted soil by the enrichment culture technique, 20 with lauryl sulfate and 20 with C_{13}LAS as the enriching agent. Using a modified shake culture method they could degrade C_{13}LAS only to the extent of 10–80%, mostly 30–60%. These poor results may perhaps have been due as much to the high initial concentration of LAS used, 100 ppm, as to any inherent inability of the individual bacteria. Degradation of lauryl sulfate (initial concentration 30 ppm) was 95% or better with 17 of the lauryl sulfate strains and 7 of the LAS strains. The other 13 LAS strains could only accomplish 25% to 50% degradation of lauryl sulfate, no better than they could do on the LAS.

II. REFERENCE STANDARDS AND NORMALIZATION

A. Reference Surfactants

The biodegradability of a particular surfactant sample may be expressed as the percent degraded, or percent remaining, under a specified set of biological conditions as detected by a specified analytical method. A time factor must also be specified, implicitly or explicitly, to provide a further basis for comparison and for extrapolation to natural conditions or to those existing in a sewage treatment plant.

With so many parameters involved, the biodegradability of a given sample obviously cannot be a unique number which turns up each time a test is run. However, a series of samples tested side-by-side will usually show a consistent order of merit from one time to another and often from one test method to another. Thus some degree of continuity can be established by including standard samples in each run to serve as reference points. In work on ABS biodegradation two convenient standards have been TBS and the C_{12}LAS made from pure α-dodecene. If the latter does not degrade to the extent of 98% to 100% the whole run is to be viewed with suspicion, since its complete degradability has already been well established. The TBS might degrade anywhere from 0% to 80% (or even higher), depending on the test method used and other factors; a test sample would have to do considerably better than the TBS to be considered satisfactorily biodegradable.

The need for a third standard to mark the limit of acceptability was conveniently met by the LAS from an early commercial variety of linear alkylate, Dobane JN. This product was tested very extensively in England both in the laboratory and in the field in the late 1950s, with the verdict that while it was a considerable improvement over TBS, it was at or below the borderline of acceptability as a solution to the waste detergent problem (STCSD, 1961, 1962). Accordingly, a sample of Dobane JN LAS included in a test series provided a reference point marking a lower limit of acceptability.

Dobane JN is no longer available, however. Its less than perfect performance was due to the presence of non-LAS impurities characteristic of the manufacturing process and raw materials, and it was made obsolete by subsequent improvements. Currently available commercial LAS can be used as a comparison standard, but with the disadvantage that it is perhaps too close to the pure α-dodecene LAS in quality. Standards with lower biodegradability can be made by mixing the α-dodecene LAS with TBS in some desired ratio, say 80:20, but their suitability has apparently not been checked very extensively.

B. Normalization

A step beyond the simple comparison of test sample with standard is the normalization of results as described by Sweeney (1964a). Here the standard sample is assigned a semiarbitrary biodegradation value, say 85% degradable for Dobane JN LAS. If the standard only shows say 80% in a given run, all the results in that run are revised upward by a normalization factor sufficient to raise the standard sample from 80% to 85%. Normalized results can be compared from one run to another with some degree of confidence because compensation for some of the variability in bacterial performance has been provided, to a certain extent, by this calculation.

The experimenter may use standard samples of his own choosing in his work, for the extra degree of reliability and precision they may afford. National or international standards have not yet been established performance standards being expressed simply as percent biodegradation under a specified treatment.

C. The Biodegradability Index

The biodegradability index is a useful device for comparing results from one set of conditions to another using the nonsurfactant organic components of sewage as an internal standard, measurable in the aggregate by determination of the BOD or COD. The efficiency of a sewage treatment is often expressed as the percent removal of the BOD, derived by dividing the BOD content of the effluent by that of the influent, or less often in terms of COD removal. Dividing the percent removal of the surfactant by the percent removal of the BOD or COD gives the biodegradability index of the surfactant.

If the surfactant is removed as readily as the "natural" components of the sewage, the percent removals are the same and the index is 1. An index below 1 results when the surfactant is removed less readily than the other components, and an index above 1 when it is more readily removed. Since the COD is a measure of the total oxidizable organics present, while BOD includes only the biologically oxidizable portion, the percent removal of BOD always exceeds that of COD. Thus a biodegradability index based on BOD will be numerically lower than one based on COD, and will represent a more stringent performance standard for the surfactant.

Ideally the biodegradability index should be a constant; in a biologically inefficient system, for example, the percent removal of the surfactant and of the natural components would both be poorer than in a good system, but hopefully they would both be affected to the same degree, leaving their quotient unchanged. In actual practice the situation is much too complex for this ideal to be generally realized. The sewage is a mixture of organic components which may vary widely from one sample to another, and the individual components may differ widely in their rates of degradation as well. It is unrealistic to expect that at all stages of the degradation process these will necessarily exhibit a constant ratio to the amount of surfactant (itself often a complex mixture) remaining. On the other hand, qualitative agreement is to be expected, and some degree of precision to within definable confidence limits should be achievable by statistical treatment of large numbers of observations.

The following examples give some idea of the range of variability which may be encountered. Schönborn (1962a) presents biodegradability index values for several surfactant samples under varying treatment conditions in batch activated sludge and trickling filter units. They are calculated on the basis of permanganate value, a form of COD determination. Sample A ranged from 0.86 to 1.01 compared to 0.66–1.05 for sample K and 0.48–0.79 for sample T (TBS). Renn (1964a), calculating on the basis of BOD removal in laboratory continuous activated sludge units, shows TBS data which scatter from 0.35 to 0.57. Three samples of LAS had ranges of 0.65–1.0, 0.85–1.05, and 0.95–1.05, compared to 1.03 in a field trial. He also gives data on laboratory trickling filters: 0.35–0.55 for TBS, 0.2–0.4 for a cracked wax LAS, and 0.85–1.15 for a paraffin derived LAS (Renn, 1965a).

Barden (1957) used a rather similar concept in calculating his Warburg respirometer data. He defined the "relative stabilization factor," F, as the ratio between the oxygen uptake due to surfactant (as percent of its total COD) and that due to the synthetic sewage (as percent of its total BOD). Data were derived from three parallel 6 hr Warburg runs: (i) unfed control, (ii) fed synthetic sewage, and (iii) fed synthetic sewage plus surfactant. Thus an F greater than 1 indicates that the surfactant underwent more rapid degradation than the synthetic sewage. Even with $F = 1$ the surfactant is actually somewhat the better, because the comparison has been made with respect to the BOD content of the sewage instead of its somewhat higher COD content.

D. Limits on Limits

With the development of test methods for estimating the biodegradability of surfactants it became possible to establish standards to be met by surfactants intended for widespread use. Thus the German government (1962) fixed by law a lower limit of 80% primary biodegradation on ABS-type surfactants, and the U.S. limit set by the members of the SDA (1965) is 90%, using a different test method.

These limits are realistic, high enough to accomplish the environmental objectives, yet not so high as to be unattainable commercially. Surfactants have been designed with greatly improved biodegradability compared to TBS, and such improvements are continuing. But there is an upper limit on what can reasonably be expected; that limit is what bacteria are able to accomplish on their natural foods.

In the uptake and biodegradation of natural foods, bacteria cannot attain complete oxidation to carbon dioxide and water, even with so readily assimilated a food as glucose. The closest approach is probably on the order of 90% to 95%, the incompletely oxidized remainder being nonbiodegradable cell matter or by-products synthesized by the bacteria (Chapter 4, Section II). Thus our definition of ultimate biodegradation (Chapter 1, Section II.A) does not call for complete oxidation, but must allow as well for the formation of these same normal metabolic products. This must be remembered likewise in assessing the biodegradation results, particularly as to ultimate biodegradability, on any surfactant, and in setting up further quality standards.

III. FEEDS AND MEDIA

Biodegradation tests are ordinarily made in an aqueous medium containing the surfactant under test and the organisms. Usually it also

TABLE

Feeds and Media—Concentrations in ppm (mg/liter); Salts

Component	BOD Water[b]	Butterfield, 1937	Cordon, 1968b[c]	Emschergenossenschaft[h]	German[k]	Gray-Thornton, 1928	Huddleston, 1964a, b	Husmann, 1963a, p. 32	Lashen, 1966
Water[a]	D	—	—	D	T	—	S	T	—
Ammonium	—	—	—	500[i]	—	—	—	50[q]	—
NH_4Cl	1.7	—	—	—	—	—	—	—	—
$(NH_4)_2HPO_4$	—	—	24.8	—	—	—	—	—	—
$(NH_4)_2SO_4$	—	—	—	—	—	500[m]	—	—	—
Calcium	—	—	2.9[d]	—	—	—	—	—	—
$CaCl_2$	27.5	7	—	—	3.2	100	—	—	3.8
Iron	—	—	0.4[e]	—	—	—	—	—	8[s]
$FeCl_3$	0.2	—	—	10	—	20	—	—	—
Magnesium	—	—	5.4[f]	—	—	—	—	—	—
$MgSO_4$	11	5	—	98	1	200	10	—	12.2
Potassium	—	—	—	—	—	1000[m]	—	—	—
KCl	—	7	—	—	—	—	—	—	—
KH_2PO_4	8.5	—	156	—	—	—	—	—	537
K_2HPO_4	21.8	—	183	500	—	1000	50	50	—
Sodium	—	—	—	j	—	—	—	—	—
$NaHCO_3$	—	—	—	—	—	—	—	100	—
NaCl	—	15	—	100	6.3	100	300[n]	—	—
Na_2HPO_4	17.7	50	—	—	—	—	—	—	1084
Other inorganics	—	—	1.0[g]	—	—	—	—	—	—
Infusorial earth	—	—	—	—	—	—	—	—	—
Acetate, NH_4	—	—	—	—	—	—	—	—	—
Benzoate, Na	—	—	—	—	—	—	—	—	—
Glucose	—	—	—	—	—	—	—	—	—
Meat extract	—	200	—	—	104	—	—	—	—
Nutrient broth	—	—	—	—	—	—	—	—	260
Peptone	—	300	—	—	156[l]	—	350[p]	100	—
Soap, castile	—	—	—	—	—	—	—	—	—
Starch, soluble	—	—	—	—	—	—	—	—	—
Urea	—	50	—	—	27	—	—	—	27
Yeast extract	—	—	—	—	—	—	—	—	20
Other organics	—	—	—	—	—	—	—	100[r]	—
Surfactant	—	—	40	10	20	—	—	10–300	20

[a] D, distilled; T, tap; S, 90T:10 sewage. [b] APHA (1965, p. 417). [c] SO_4-free medium. (1963a, p. 104). [i] NH_4NO_3. [j] NaOH to bring to pH 7.5. [k] German Government (1962). ductivity for level controller. [p] Soy peptone. [q] "Ammonium phosphate." [r] Sucrose. fied by Ludzack (1965) for pH control. [x] $CaCO_3$, introduced to unit directly as slurry.

5.3

Calculated to Anhydrous Form Except as Indicated

McKinney, 1959a	McKinney, 1959b	McKinney, 1959b	Nelson, 1961	Pitter, 1963a	Pitter, 1968a	Schönborn, 1962a	SDA, 1965[t]	SDA, 1965[v]	SDA, 1969a, b[w]	Urban, 1965	Urban, 1965	Weinberger, 1949a[ab]
—	—	—	T	T	T	T	D	T	T	—	—	T
—	—	—	—	—	—	—	3000	—	—	—	—	—
—	—	—	—	—	—	—	—	—	—	—	—	—
—	—	10	—	—	—	—	—	25	—	—	660	—
—	—	—	—	—	—	—	—	—	—	75[x]	150[x]	—
—	—	—	—	—	—	—	—	—	—	2	—	7
—	—	—	—	—	—	—	2[u]	—	—	2[v]	30[aa]	—
—	—	—	—	—	—	—	—	—	—	—	—	—
—	—	—	—	—	—	—	—	—	—	—	—	—
—	—	—	—	—	—	—	122	—	—	2	140	10
—	—	—	—	—	—	—	250	—	—	3	—	7
—	—	—	—	44	100	—	—	—	—	—	140	—
300	300	50	125	—	—	50	1000	130	130	—	—	—
—	—	500	—	—	100	100	—	—	—	—	—	168
—	—	—	—	—	—	—	—	—	—	—	300	30
—	—	—	—	—	—	—	—	—	—	20	—	25
—	—	—	—	—	—	—	—	—	—	—	—	5[ac]
—	—	—	—	—	—	—	—	—	—	—	—	25
—	—	—	—	—	—	—	—	—	—	140	—	—
300	300	—	125	—	—	—	—	—	—	—	—	—
—	300	—	—	—	300	—	—	130	300	—	—	—
—	—	—	—	—	—	—	—	130	—	—	—	—
300	300	200	125	—	—	—	—	130	200	—	—	100
—	—	—	—	240	200	100	—	—	—	140	—	—
300	—	—	125	—	—	—	—	—	—	—	—	50
—	—	—	—	—	—	—	—	—	—	80	—	100
—	—	—	—	—	—	—	—	—	—	40	—	30
—	—	—	—	—	—	—	300	—	—	—	—	—
—	—	—	—	—	—	100[r]	—	—	—	140[z]	—	—
—	—	—	50	—	—	15	30	20	20	15	—	—

[d] $Ca(NO_3)_2 \cdot 4 H_2O$. [e] $Fe(NO_3)_3 \cdot 9 H_2O$. [f] $Mg(NO_3)_2 \cdot 6 H_2O$. [g] $CoCl_2 \cdot 6 H_2O$. [h] Husmann [i] Milk peptone. [m] Either 500 $(NH_4)_2SO_4$ or 1000 KNO_3, not both. [n] To provide con- [s] $Fe_2(SO_4)_3$. [t] Shake culture medium. [u] $FeSO_4 \cdot 7 H_2O$. [v] Acivated sludge feed. [w] Modi- [y] $Fe_2(SO_4)_3 \cdot xH_2O$. [z] Skim milk powder. [aa] $FeSO_4$. [ab] Fuhs (1968). [ac] $Al_2(SO_4)_3 \cdot 18 H_2O$.

contains other materials added to aid in the normal functioning of the organisms or to simulate conditions in sewage treatment processes or in receiving waters. Some of the compositions frequently used are given in Table 5.3; from the magnitude of this very incomplete list it is easy to infer that each researcher tends to prefer his own recipe.

The dilution water which was developed for use in the BOD test is probably close to the minimum medium that can be used satisfactorily. It contains essential elements Na, K, Ca, Mg, Fe, N, S, P, Cl required for bacterial growth, while its phosphate components provide a substantial buffering capacity for pH control. When possible, organic nutrients are often added to the medium also, serving as food for the organisms, in addition to the material under test. This is generally avoided in BOD or respirometric work for obvious reasons.

A. Natural Sewage

Some workers prefer natural sewage as their medium. Deterioration during storage should be avoided. Countermeasures may be taken by refrigeration (McGauhey, 1957; Sweeney, 1964a; Loehr, 1967; see Chapter 3, Section I.A), or by sterilization. In the latter case Jones (1968c) shows that irradiation is less likely to cause changes in the sewage than is heat. However, Butterfield (1937) had shown earlier that there was no significant difference in results between fresh sewage and heat-sterilized sewage when used in semicontinuous activated sludge systems, at least during a 48 day period of operation.

For comparison with the various laboratory recipes of Table 5.3, the approximate composition of natural sewage and the primary sludge are given in Table 5.4, as well as of the treated secondary effluent. Earlier work on this up to 1957 has been comprehensively reviewed by Vallentyne (1957). Table 5.4 indicates that although about three-fourths of the organic carbon in the raw and settled sewage was identified as to general chemical type, only about one-third was so identified in the secondary treated effluent. Later Hunter (1965) subdivided the various fractions still further and concluded that the unidentified one-fourth of the primary sewage might be principally ether-soluble substances, neutral, amphoteric and basic.

In further investigations of the soluble components of primary sewage, secondary treated effluent and river water, Murtaugh (1965) showed that some 40% of the initial COD could be isolated as organic material by ether extraction after passage through a cation exchange resin. The organics then extracted were primarily acidic in nature, and a minor

TABLE 5.4

Organic Components of Sewage

	In sewage,[a] ppm carbon			In treated sewage,[b] ppm	In raw sewage sludge,[c] % of sludge
	Raw	Settled	Treated		
Protein[d]	36	27	3.1	1.6–7.4	0.6
Lipids	99	70	0.2	0.8–5.2[e]	1.1
Carbohydrates	55	46	1.6	0.8–2.4	0.4
Soluble acids	21	20	1.8	0.8–5.2[e]	0.1
Anionic surfactant	14	13	1.5	1.5–12.5	—
Crude fiber[f]	—	—	—	—	1.4
Total carbon	311	228	27	—	—
BOD	—	380[g]	12[g]	—	—
COD	—	—	—	—	5.8
Volatiles	—	—	—	—	3.6
Fraction unidentified	$\frac{1}{4}$	$\frac{1}{4}$	$\frac{2}{3}$	$\frac{2}{3}$	$\frac{1}{3}$

[a] Painter (1961).

[b] Bunch (1961).

[c] Teletzke (1967).

[d] Including amino acids.

[e] Other extractables; includes both lipids and acids.

[f] Cellulosic.

[g] Oxygen uptake, ppm.

fraction was identified as lower fatty acids from C_1 to C_6. The unidentified fraction ranged from about 50% to 95% in primary sewage, 80% to 98% in secondary, and 90% to 98% in river water.

B. Organic Foods

Most workers prefer the greater reproducibility and convenience of a feed mixture prepared in the laboratory from chemically defined ingredients. This is usually called "synthetic sewage," but "artificial sewage" would seem to be a better term in most cases. On the other hand, Mann (1957) describes "artificial sewage" compounded from excreta, soil, soap flakes, and tea, which should at least approximate a "synthetic sewage," in a sense.

The components available for preparation of bacteriological culture media are convenient sources of organic foods for use in these mixtures. They generally include proteinaceous material such as meat extract, nutrient broth, or peptone, often supplemented by a sugar such as glucose and occasionally by a fatty material such as soap. These natural foods are usually used, but for some purposes the surfactant itself may be the only organic component in the medium. Bacteria can utilize some surfactants as their sole organic nutrient; for example, an activated sludge system can be operated for extended periods of time (probably unlimited periods) with a surfactant (such as $C_{12}LAS$) as the sole food.

Some organic components in the medium, glucose and others, may interfere with acclimation of the microorganisms toward the test compound and their biodegradation of it, by suppression of formation or activity of necessary enzymes (Chapter 4, Section III.C). In actual practice the biodegradation test systems do acclimate and do accomplish degradation of the surfactant despite the presence of these potential inhibitors, but little is known as to the exact degree of trouble which they cause. Pragmatically this is perhaps unimportant, since they are also present out in the world, causing comparable difficulties, if any.

Nevertheless, fundamental knowledge of the mode and extent of action of these factors could provide useful insight in specific situations. Information is scanty, but we do have the report of WPRL (1965, p. 122–3) showing that presence of 50 ppm of glucose in the STCSD aeration test method considerably delays degradation of Dobane JNX sulfonate. Without the glucose, degradation was complete within 4 to 5 days, compared to about 15 days with it. This observation was confirmed by LGC (1966, p. 78); yeast acted similarly, while gelatin had no significant effect.

Ciattoni (1968) also found an inhibiting action of glucose. In river water, 50 ppm of glucose delayed the onset of degradation of 10 ppm of $C_{11}LAS$ as long as the glucose was replenished to maintain its concentration above 30 ppm. When the replenishment was stopped after 28 days and the glucose level was allowed to drop to zero (biodegradation), degradation of the LAS ensued. In the absence of glucose the LAS degradation started in about 4 days and was complete at 10.

Surprisingly (or perhaps not), Mann (1968) found that addition of proteinaceous food to inoculated BOD dilution water had the opposite effect. A high molecular weight LAS (C_{11-15}) failed to show its true biodegradability in this medium, sometimes persisting unchanged for as long as 40 days. Addition of meat extract–peptone–urea to the system, at the same level as in the official German activated sludge feed, resulted in prompt biodegradation in the normal pattern of lower molecular weight LAS.

The extreme sensitivity to supposedly insignificant changes in feed composition sometimes encountered was strikingly demonstrated by Vaicum (1968) in studying the official German activated sludge feed. The meat extract used is available either as a paste or in dried form. LAS biodegradation was 95% when the feed solution was made up from the paste (Feed P), but only 25% when made up from the dried product (Feed D). The difficulty with Feed D occurred during the sludge development stage of the test—the growth rate was poor, it had an abnormal microscopic appearance, and gave a cloudy, bad-smelling effluent. Feed D was completely satisfactory with LAS if used after a sludge had first been developed with Feed P. Furthermore, it could be used successfully from the beginning in a test of linear primary alkyl sulfate or in a control unit. Dried meat extract from several manufacturers all showed the same difficulty, as did also a sample of good paste which had been dried in the laboratory.

C. Nitrogen, Phosphorus, and Other Essentials

Provision of sufficient nitrogen, phosphorus, and sulfur in the feed is important if continued growth of the bacteria is desired, as in activated sludge tests, particularly when pure compounds such as glucose are used in the absence of more balanced foods such as meat extract or nutrient broth. Hattingh (1963) concluded that the optimum feed should contain at least 5.3 parts of nitrogen and 1.2 parts of phosphorus for each 100 parts of BOD. This is a useful guideline to follow even though Busch (1966) calculates that glucose, for example, should require some 50% more nitrogen. Actually, prolonged operation well below these limits is possible. The requirements for nitrogen and phosphorus as well as for minor and trace elements may be fulfilled by the sludge itself over long periods without replenishment via the feed, because those elements tend to be held in the sludge. A new and growing cell will pick up its requirements from the components liberated by old and dying cells. Thus Washington (1962) and Simpson (1964) suggest that a BOD:N ratio of around 100:1.1 is sufficient in extended aeration activated sludge treatment. However, the BOD removal is noticeably slower at such a low nitrogen level (Simpson, 1965, p. 175).

D. Control of pH

Activated sludge and trickling filter effluents are normally neutral, with a pH range around 6.5 to 7.5. Laboratory units may tend to go acid. The 24 hr effluents from semicontinuous activated sludge units may sometimes get down to pH 5 or even lower, or may slowly cycle between pH 5 and 7 with a period of several weeks or months. Although no effects on biodegradation performance have been observed to result from the low pH, operation closer to neutral is to be desired since acid conditions imply, or could result in, an abnormal distribution of microorganisms.

One approach to pH control has been the addition of large amounts of buffer to the feed. Pitter (1963b) used 20 times the amount of phosphate buffer called for in BOD dilution water, while Lashen (1966) used 0.01 M phosphate buffer as his medium. The presence of such massive amounts of phosphate—1000 ppm or more—in the medium may in some circumstances be no great improvement over the 0.00001 M concentration of hydrogen ions present at pH 5. Others (for instance, McKinney, 1959b) have included large amounts of sodium bicarbonate as buffer in the feed, again at the expense of having unrealistically high levels of inorganic ions.

Urban (1965) used a somewhat different means, adding a slurry of calcium carbonate to the system from time to time. Concentration of $CaCO_3$ in solution could never exceed its solubility in water, about 15 ppm, unless further solubilized as bicarbonate by the presence of carbon dioxide. However, the concentration of dissolved calcium ions could become very much higher, depending upon the amount of acids neutralized.

A more satisfactory approach to pH control is to prevent the acid from being formed in the first place, rather than to neutralize it as formed. An important source of acid is the oxidation of ammonium ions to nitrate in the nitrification process [Eq. (5.1)], which produces two equivalents of

$$NH_4Cl + 2\,O_2 \rightarrow HNO_3 + HCl + H_2O \tag{5.1}$$

acid from each equivalent of ammonium. Amino nitrogen unassociated with inorganic anions, for example, that present in amino acids, would yield only one equivalent of acid upon oxidation. As pointed out by Urban (1965) the same is true of ammonium salts of organic acids, wherein the organic anion leaves the system as carbon dioxide formed by oxidation. Accordingly he used ammonium acetate as nitrogen supplement instead of an inorganic ammonium salt.

According to Eq. (5.1) a feed containing a large excess of nitrogen should be more prone to development of acidity than one containing a nutritionally balanced amount. The feed used in the SDA (1965) semicontinuous activated sludge test provides an example. Ludzack (1965) suggested that its tendency to develop acidity was due to excess nitrogen, and recommended 200 ppm of nutrient broth as the sole nitrogen source (Table 5.3) (also, incidentally, eliminating the redundant beef extract).

The subsequent experience of the SDA Biodegradation Subcommittee indicates that the revised feed does indeed give better pH control (SDA 1969a, b). The standard SDA procedure may perhaps be changed accordingly at some future date when sufficient operating experience has been accumulated and a suitable occasion presents itself.

E. Surfactant Concentration

The surfactant content of the feed or medium usually lies in the range from several parts per million to several hundred parts per million, somewhat higher than ordinarily met in the receiving waters, or even in raw sewage. The lower limit is often dictated by the sensitivity and precision of the analytical method used. If the method is at best only able to detect and distinguish between 0.1 and 0.2 ppm, then a precision of only about 10% can be expected if the feed level is 1 ppm; 80% biodegradation might be distinguished from 90% under such circumstances, but not much closer. With a feed of 10 ppm this same analytical method should be able to differentiate 98% from 99% biodegradation, and 100% could be estimated with some confidence.

An upper limit for the surfactant concentration in the feed may be imposed by any of several factors: (i) toxic or inhibitory action on the bacteria (Chapter 4, Section IV.A) is an important consideration with anionic and cationic surfactants, less so with nonionics; (ii) foaming may become a problem in aerated systems, interfering with mechanical operation and making it difficult to keep the bacteria and medium within the apparatus; (iii) closer simulation of field concentrations may be desired, to avoid interference by atypical effects which might possibly occur at higher levels.

These potential disadvantages of operation at high surfactant levels have greater bearing in systems containing low concentrations of bacteria, such as in the river water test. With high bacterial concentrations and continuous-flow conditions as met in the complete mixing activated sludge process, the surfactant concentration in the feed may be very much higher. It may run at several hundred parts per million or more, provided the surfactant is reasonably biodegradable and the system is acclimated. In such a case, degradation will occur rapidly as each increment enters, and the surfactant does not reach a critical concentration in the mixed liquor.

F. Formulated Detergents and Surfactant Mixtures

Sometimes, as in the monitoring of commercial detergent samples, the surfactant to be tested may not be available in pure form, but only as a component in a formulation. In many cases, if a suitable analytical method for the surfactant is at hand, the entire formulation may be used as such in the biodegradation test. For examples see Mann (1957), Schönborn (1962a), Husmann (1963a, p. 101), Jendreyko (1963), and Orgel (1964). In fact, Pitter (1964c) reported a C_{10-13}LAS which appeared to degrade better in a built formulation (98%) than in pure form (91%).

Yet in some cases difficulties may arise. Bleaching, antiseptic, bactericidal, or other such agents if present could interfere with the proper functioning of the bacteria in the test procedure. (This would not likely happen in the outside world because such specialty formulations comprise only a minor fraction of all detergents used and their concentration in sewage would thus be much lower.) Even in the absence of such agents troubles might be experienced with formulations containing only a small percentage of surfactant; excessively high concentrations of the other components might be present at the desired level of surfactant.

Fischer (1965) reports on a tentative procedure the German Hauptausschuss is investigating for such cases, an extraction technique. The surfactant is separated from the inorganic components of the formulation by dissolving in alcohol. Isopropyl alcohol is preferred over ethanol because it can be easily separated from an aqueous system as a separate liquid phase by adding excess salt. Fischer mentions the disadvantages of working with such extracts, because of their unknown composition and chemical nature. If at all possible, the testing of the individual surfactant before formulation is greatly to be preferred.

Little work has been reported on biodegradation of mixtures of different surfactant types such as might be isolated from a detergent formulation. To date there has been no indication of interaction, nor is there any particular reason to believe there might be. As an example, Jendreyko (1963) found lauryl sulfate to be degraded readily in the presence of TBS, and the TBS was not degraded. Similarly Lashen (1966) found that LAS and OPE_{10} did not affect each other in a continuous activated sludge test, both degrading about 90%. In admixture with TBS the OPE_{10} still degraded 90% and the TBS 15%. Later Lashen (1967c) verified simultaneous degradation of LAS and OPE_{10} in a field test.

IV. BOD AND RESPIROMETRIC TESTS

The techniques used in BOD and respiratory procedures have been described in Chapter 3, Section V.B. Their specific application to problems of surfactant biodegradation will now be considered, starting with a discussion of the difficulties and uncertainties of interpreting the results from such tests.

A. Questions

A.1. *Questions of Measurement*

As pointed out in Chapter 4, Section II, in the absence of any added food, living bacteria are continually using oxygen at their slow rate of endogenous respiration. This results from their continuing oxidation of constituents of their own protoplasm or of the protoplasm of those less fortunate neighboring cells who have died. In addition to the oxidation of these carbon compounds to carbon dioxide and water, processes of nitrification may also occur, oxidation of ammonia and amino groups (occurring, for example, in proteins and their amino acids) to nitrite and nitrate.

Thus the total oxygen taken up by the system includes not only that for the oxidation of the test compound but also whatever is utilized in these other processes as well. Customarily a blank is run in conjunction with the test sample and its oxygen consumption is subtracted, hoping that the difference will represent the oxygen used for biodegradation of the test compound. Inasmuch as the blank and fed samples constitute different environments for the bacteria, the endogenous processes may not be parallel from the one to the other and if so the net result will be in error.

In extensive reviews Dawes (1962, 1964) has organized the current knowledge of the various processes of endogenous metabolism and the many factors which may exert altering influences. He indicates that most of the information is qualitative and that quantitative studies are needed on the metabolic reactions and interrelations. He later contributed such data for *Escherichia coli* (Dawes, 1965).

The difficulties of correcting for endogenous respiration are specifically explored in several researches. Ludzack (1964), by using ^{14}C-labeled activated sludge, found that the sludge evolved 100% more $^{14}CO_2$ when fed fish meal, glucose, or phenol than did the unfed control, but 20% less than the control when fed potassium stearate. Since the $^{14}CO_2$ evolved is a measure of the endogenous respiration of the sludge in each case, the futility of attempting to correct by simply subtracting the control value is readily apparent.

Dietrich (1967) used this same technique with pure cultures of *Escherichia coli*, *Aerobacter aerogenes*, *Bacillus cereus*, and *Rhizobium leguminosarum*; glucose feeding enhanced the endogenous respiration rate of the first two, decreased it for the other two. Allred (1964b) found that surfactants too could cause considerable change in the endogenous respiration rate of bacteria; to the contrary, Brink (1966) found no such effect in his experiments.

The error introduced by such alterations in the endogenous rate of the fed culture may be quite large, since the endogenous uptake may amount to a considerable fraction of the total. Thus Marion (1966), experiencing excessive oxygen uptake (100% to 150% of theoretical) in biodegradation experiments on sodium lauryl sulfate, attributed it to such stimulation of endogenous respiration.

Correction for nitrification may be approached somewhat more directly by calculation from the nitrogen balance of the system. Analysis for ammonia, amine, nitrite, and nitrate at the beginning and end of the run are required. Unfortunately cumulative analytical errors may make the corrections so uncertain as to hardly warrant the considerable efforts involved (Nelson, 1961). Symons (1963) has reviewed the whole question of nitrification correction. Others have favored addition of substances which inhibit nitrification such as thiourea or allylthiourea (Montgomery, 1967).

A.2. *Questions of Interpretation*

Assuming that the net oxygen used in actual biodegradation of the test compound can be determined adequately, there remains a second major obstacle to the accurate interpretation of the result. Complete degradation of the test substrate does not necessarily require complete oxidation. Some of it is used in the synthesis of new protoplasm, which is an oxidation state considerably short of that represented by carbon dioxide and water. For instance, Busch (1961) points out that in the bacterial degradation of glucose (and presumably with any other substrate as well) a plateau is reached when only about 40% of the theoretical oxygen has been absorbed. At this stage all of the glucose has disappeared, part of it having been oxidized to carbon dioxide and water, and the rest converted to new cells or protoplasm. Under the conditions of the standard BOD test this stage is reached in about 36 hr. Upon further aging with no further addition of food the system will resume oxygen uptake, usually reaching about 70% of theoretical at 5 days in the case of glucose. This results from the further oxidation of protoplasmic components, as discussed in more detail in Chapter 4, Section II.

On the basis of such observations as the foregoing, it has been proposed that any product showing oxygen uptake over 40% to 45% of theoretical should be considered as completely degraded (Nelson, 1961; Vath, 1964). However, such evidence is at best only qualitative, and a verdict based solely on the 45% criterion must be viewed with great caution, particularly if the product being tested is a mixture rather than a single, pure chemical compound. For many substances do approach their complete theoretical oxygen uptake if given sufficient time, and some of them do it rather rapidly; indeed, one of Vath's surfactant samples was in the latter category. A 50:50 mixture of such a substance with a completely non-degradable one might well show a 40% oxygen uptake and thus be erroneously classed as completely biodegradable.

Because of these uncertainties, an oxygen uptake test method would be quite incapable of differentiating between, say, 80%, 90%, and 100% biodegradation, in the absence of independent data of some other sort. Vath (1964) undertook to provide such supplementary information by measuring the increase in weight of the bacterial cells. Extreme precision in the filtration, drying, and weighing is, of course, necessary if the results are to be meaningful. Beyond that, some knowledge of the chemical composition of the increment of cellular growth and associated products might also be required, and knowledge of any changes in the dissolved organics in the system.

A further restriction on the use of BOD and respirometric methods in biodegradation studies is that they cannot be applied as such to mixtures of test compounds with ordinary foods such as occurs in sewage, where the oxygen consumption for the food would mask that for the test compound. Study of such mixed systems is important since degradation upon admixture could conceivably be either better or poorer than in pure form.

B. The Closed Bottle Test

Extensively used in England and Germany, the closed bottle test is simply a BOD test (Chapter 3, Section V.B.2) in which the analysis for dissolved oxygen content may be supplemented or replaced by analysis for surfactant content as well. Hammerton (1955, 1956) early used this technique in his studies differentiating primary from ultimate biodegradation. Fischer (1963) has described it in detail, pointing out that use of the food-free BOD dilution water medium and a very small inoculum of bacterial cells minimizes interferences in the analysis of surfactant and in the interpretation of the oxygen uptake. Since all oxygen required for the biodegradation is present in dissolved form at the beginning, problems of aeration and of oxygen transfer are avoided. However, this does impose a limitation: because of the limited solubility of oxygen in water (about

9 ppm in equilibrium with air at 20°C) the initial surfactant concentration should not exceed about 2 to 3 ppm, the aim being to have 4 to 5 ppm of oxygen remaining at the end of biodegradation. Fischer recommends analyses at 0, 5, 10, 15, and 30 days; MBAS removal had stabilized within 15 days in his experiments, but oxygen consumption continued beyond that time. His sources of microorganisms were sewage treatment plant effluent, laboratory activated sludge cultures, or garden soil suspensions; inoculation was with one drop per liter of medium.

C. The Warburg Test

Like the closed bottle test, the Warburg test derives its results from the amount of oxygen utilized in the biological oxidation of the surfactant. Since its early use by Bogan (1955) it has been applied frequently; Hunter (1964) has reviewed the techniques and shows typical results. The surfactant is exposed to the organisms in a closed, agitated system containing an excess of atmospheric oxygen. The amount of oxygen taken up is measured from time to time by noting the change in gas volume or pressure (Chapter 3, Section V.B.3).

In a comprehensive review of respirometry Montgomery (1967) discusses factors affecting oxygen demand: type of inoculum, concentration of inoculum and micronutrient elements, temperature, storage of samples before analysis, light, and protein synthesis inhibitors. He emphasizes the desirability of using ethanolamine instead of the more efficient KOH as carbon dioxide absorbent, thereby allowing 1% to 2% of CO_2 to remain in the gas phase. This is desirable in view of Gaffney's (1965) report that complete absence of CO_2 inhibits assimilation of food (i.e., biodegradation) by bacteria.

The higher the surfactant concentration, the greater the relative accuracy of the oxygen uptake measurements will be. Levels above 50 ppm are to be preferred, but care must be taken that bacteriostatic limits are not approached. This can be determined by making runs at several surfactant concentrations. Oxygen uptake should be proportionately greater at the increasing levels up to the bacteriostatic limit (Barden, 1957).

Acclimation of the organisms to the surfactant may not occur during the relatively short time of the run itself, but may be accomplished beforehand by appropriate means (Barden, 1957). Activated sludge from a unit acclimated to the surfactant is often used as the inoculum (Nelson, 1961; Brink, 1966). Inoculation is usually at a relatively high level, several hundred parts per million or more of activated sludge solids, preferably simulating the conditions of acclimation. However, a large inoculum means a correspondingly high rate of endogenous oxygen uptake, tending to overshadow that of the test compound.

The extent of endogenous oxygen uptake is estimated from a control run identical with the test run in all respects except that the surfactant is not added. The difference between the two is usually assumed to be the amount utilized in the oxidation of the surfactant. (The dangers of such an assumption are discussed in Section IV.A.1.) Ideally this difference will increase during the period in which the surfactant is being degraded; when degradation has proceeded to its fullest extent the difference becomes constant. The two cultures should thereafter absorb oxygen at approximately the same rate, the endogenous respiration rate. With a readily biodegradable substrate this state may be reached in a few hours if a suitably large inoculum is used, and if the organisms are already acclimated. In other circumstances runs may last for several days.

Brink (1966) used a nitrogen-free medium to minimize cell growth and nitrification. Concentration of cells used in the test was stated to be 100,000 to 130,000 ppm dry weight, but in a personal communication Brink advises that this is in error and that the actual range was 1000 to 2000 ppm. Theoretical oxygen uptake was approached in about 12 hr with n-dodecyl sulfate and 24 hr with LAS. Radiocarbon studies showed that there was no significant difference in endogenous respiration of the control and the fed culture.

For the reasons outlined in Section IV.A, the oxygen uptake results may be difficult to interpret quantitatively in terms of surfactant biodegradation. Supplementary analyses for such things as undegraded surfactant, intermediate degradation products, or COD may be called upon to provide further information. Ordinarily these would be made at the end of a run since withdrawal of samples at intermediate times would introduce complications.

Blankenship (1963) reported wide variability in replicate Warburg results on 1-dodecanol E_8. Out of 30 runs, 6 did not level off at a definite oxygen value. The other 24 ranged from 0.1 to 0.9 $\mu l/\mu g$ of surfactant oxygen pickup, with 11 in the vicinity of 0.75 $\mu l/\mu g$. Instability of bacterial species distribution during preliminary propagation of the mixed cultures was mentioned as a possible reason for the wide scatter.

V. DIE-AWAYS

The term *die-away* has long been associated with tests wherein the surfactant is exposed to the medium in an isolated system, the progress of biodegradation being observed by analysis from time to time as the surfactant "dies away."

Sometimes the die-away rate of the surfactant is expressed as a *half-life*, meaning the time taken for the concentration to drop to half its initial value, in analogy with the use of the term as a measure of radioactive decay rate. In surfactant biodegradation, however, it is generally far from constant. Thus the time to drop from one-half to one-fourth of the original concentration may be much longer or shorter than from the original concentration to one-half of it, and the time from one-fourth to one-eighth quite different again. In fact if the test surfactant is only 85% degradable under the given test conditions it never will reach the one-eighth point. Accordingly we shall try to avoid using the term "half-life," with its implications of constancy, and instead will simply call it the *halfgone* time if the occasion arises.

The die-away test system is usually considered as contrasting with a continuous-flow system, wherein it is most convenient to express the degradation as an overall figure calculated from the surfactant content of the influent and effluent. Yet these too are die-away systems in a sense, and die-away curves for the degradation of the surfactant can in principle be determined from the decreasing surfactant content of an increment of feed during its passage through the system. Likewise, the BOD and Warburg can be considered as die-away tests—in fact, just about any biodegradation system in which the surfactant disappears. But in common usage the term is usually applied only to the sort of tests discussed in this section, and in particular to:

A. The River Water Test

A.1. *Conditions*

This most simple of biodegradation test methods was early described by Degens (1950), Hammerton (1955), and Sawyer (1956), and later, with specified standard conditions, by Weaver (1964). The surfactant is simply dissolved in a sample of river water and the solution is analyzed at intervals thereafter to determine what degradation has been accomplished by the river bacteria. Figure 1.1 depicts the results in such a test.

The sampling site should be free of industrial pollution if possible. Suspended solids such as mud should be removed, for example by settling for a day or so and then decanting. The bacterial count in the sample will be relatively low—depending on the source and the occasion it may range from a few hundred to a few hundred thousand cells per milliliter.

The food content, i.e., the BOD, of the water is ordinarily quite low, on the order of a few parts per million.

The amount of surfactant usually added is between 1 and 10 ppm. This is rather higher than might be expected in a river; lower concentrations are preferable when sufficiently sensitive analytical methods are at hand. Surfactant levels above 10 ppm are to be avoided if simulation of natural conditions is desired and, more important, if possible bacteriostatic action of the surfactant is to be minimized. For example, LAS at 5 ppm will usually be degraded within a week or less, but in the same water at 20 ppm will sometimes require a month or more. Figure 5.1 illustrates such

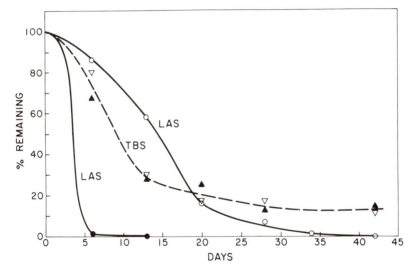

Fig. 5.1 Concentration effects in river water biodegradation of ABS (Swisher, 1962). Circles, C_{12}LAS; triangles, TBS; open points, 20 ppm; black points, 7 ppm (initial concentration).

a case and also shows that TBS is much less critical in that regard, suggesting that LAS has a lower bacteriostatic threshold than TBS.

Size of the river water sample is dictated by analytical requirements; 1 liter is usually sufficient. If the container is twice the volume of the sample, the equal volume of air will provide ample oxygen for the bacterial action even if the container is closed between analyses. Storage in the dark or semidarkness is preferable, to avoid algal growth, but this actually has little if any effect on the results. Ordinary laboratory temperatures in the region of 20° to 25°C are satisfactory.

A.2. *Speed and Acclimation*

The duration of the test will depend on the activity of the river water sample and on the nature of the surfactant. It may range from a few days to several weeks, rarely longer than 6 to 8. This time includes the induction period, the degradation itself, and, if degradation is not complete, a sufficient additional time at a constant final level of undegraded surfactant to allow its fraction to be estimated with reasonable confidence. From one river water to another these phases are qualitatively reproducible. Quantitatively, differences in bacterial activity are discernible, related to differences in the rates of acclimation and degradation. Such differences do not necessarily parallel the initial bacterial count; sometimes a high count sample is sluggish, while a low count sample may be very active. Nevertheless, the fraction of surfactant remaining undegraded is surprisingly independent of the source of the river water, ranging from about 0.5 to 1.5 times the average value.

To the extent that it represents an acclimation phenomenon, the induction period can be eliminated in subsequent runs on the same surfactant. The new test may be inoculated with, say, one-fourth its volume of liquor from a previously completed run (Huyser, 1960), or a fresh increment of surfactant may be added to the previous run after acclimation has occurred (Swisher, 1963b). In a system requiring a week or two for the first degradation of 5 ppm of LAS, subsequent 5 ppm increments disappeared within a day or two. This effect has been used by the SDA Biodegradation Subcommittee in a revised river water procedure, starting with a 5 ppm charge and adding a second 5 ppm when the first has dropped to 1 ppm. The disappearance of the second increment provides a surer estimate of the degradation rate than does the first, and the fraction of undegraded surfactant at the end is calculated somewhat more accurately on the broader baseline of 10 ppm.

A.3. *Effects of Variables*

Two detailed studies of variables in the river water test have been reported. Fuhrmann (1964) observed slower LAS degradation with (i) increased initial concentration of LAS, (ii) decreased bacterial count (by dilution), (iii) increased nutrient, (iv) increased storage time (beyond 5 days) of water before use, (v) aeration with oxygen, (vi) decreased temperature, and (vii) added phenol; aeration by bubbling with air had little effect on the degradation.

In the second study Setzkorn (1964) examined the degradation rates of LAS (i) in waters from five different rivers, (ii) in successive samples from the same river, and (iii) in replicate portions of the same sample. Standard deviations calculated for the MBAS analyses in the several experiments showed only small differences at beginning and end but considerable divergence during the period of active degradation. Retarding effects of increased initial concentration were evident, and acclimation effects were demonstrated.

Sweeney (1964b) has also pointed out the influence on the course of river water tests exerted by water variation, bacterial acclimation, and inhibition by higher molecular weight LAS. Blankenship (1963) reported on reproducibility of river water tests on biodegradability of nonionics, detected by surface tension.

B. Fortified and Inoculated Waters

Many modifications based on the general principles of the river water test have been devised, usually aimed at shortening the time required or at exerting somewhat more control over the biodegradation medium than is possible in simply selecting a river at random. While such modifications should in principle lead to greater reproducibility of biodegradation results than in the simple river water test, in practice the improvement achieved has been considerably less than spectacular.

B.1. *BOD Water*

BOD dilution water, developed long ago as the medium for the standard BOD test, is a dilute solution of calcium chloride, magnesium sulfate, ammonium chloride, ferric chloride, and sodium–potassium phosphate buffer in distilled water (Table 5.3) (APHA, 1965, p. 417). It would thus appear to be an ideal synthetic medium, and has been used in several methods.

Roberts (1960) inoculated BOD water containing 15 to 20 ppm of surfactant with sewage effluent or garden soil and incubated at 37°C with gentle aeration. In about 4 days he achieved biodegradation down to a fairly constant level of residual undegraded surfactant.

Borstlap (1963) accomplished equally rapid results by aeration of 10 ppm surfactant in BOD water at room temperature using an inoculum of washed activated sludge, amounting to about 500 ppm of sludge-suspended solids in the final medium. This is a very heavy inoculum, approaching the working concentration used in some activated sludge processes, and this procedure might equally well be classed as an activated sludge type. The sludge is obtained from a sewage treatment plant, washed 10 times with distilled water to remove substances interfering with analysis, resuspended in BOD water and aerated for 16 hr before use. Analyses are run on homogenous samples of the entire mixture so that any surfactant adsorbed on the sludge will not be missed. The procedure recommends that for a good comparison the surfactants in question be tested side-by-side with the same sludge, since variability in activity occurs from one sludge sample to another.

Heinz (1967) proposes the "open shake flask test" for the screening phase in a triple test procedure, and gives several examples of comparative results. BOD water containing 5 ppm of surfactant and a 1% inoculum of settled sewage, treated sewage, or soil extract, is shaken for 10 days, with analyses on days 0, 9, and 10. The latter two should not differ by more than 2% of the initial MBAS value. A control test with a standard LAS is run alongside, and if that does not reach its standard value by day 10 the test is prolonged to 14 days. This same test, with slightly modified provision in the running time (but not to exceed 20 days) and with further control of the operating variables, is the "provisional OECD open flask test of biodegradability of synthetic anionic detergents," under cooperative examination as a possible international standard test procedure (OECD, 1968).

B.2. *Other Salt Media*

Konecky (1963) developed a synthetic medium of undisclosed composition, made up of deionized water and inorganic salts, containing 40 ppm of surfactant and inoculated with 10 ppm of activated sludge solids. The mixture is stored in gallon bottles, agitated with magnetic stirrers, and sampled over a period of 21 days for analysis by COD and other appropriate methods. The activated sludge used for the inoculum is derived from a sewage treatment plant and propagated in the laboratory on a glucose–nutrient broth–sodium oleate medium for at least 2 weeks prior to use. This ensures that the organisms will not be acclimated to the surfactant being tested (unless it is sodium oleate).

The German Emschergenossenschaft also used BOD dilution water as the medium, held in a 2 liter tall-form glass beaker and aerated by a paddle stirrer turned at 50 rpm (Husmann, 1963a, p. 102). About 10 ppm of surfactant is introduced, along with a 20 ml inoculum of the bacterial suspension described below. The MBAS content of the system is determined daily for 9 to 10 days; degradation proceeds to almost its full extent within 4 to 5 days, after which the curve levels off. The bacterial suspension is prepared by inoculating 100 ml of a special mineral salts medium containing 10 ppm of LAS, with 5 ml of household sewage and stirring magnetically in a 750 ml Erlenmeyer flask for 6 days. Then 1 ml is transferred into fresh salts–surfactant medium. The bacteria can be propagated indefinitely by further transfers in the same manner.

B.3. *Super River Waters—the French IRChA Test*

Hitzman (1964) describes a test procedure using river water as the medium, but augmenting the bacteria naturally present by adding a bacterial concentrate. It was obtained by centrifuging two volumes of effluent from a municipal sewage treatment plant for each volume of river water used. Starting with 20 ppm of surfactant, biodegradation down to a constant level was achieved in 1 to 2 weeks. Unfortified river water reached the same level but required twice the time.

The French IRChA (Institut National de Recherche Chimique Appliquée) (Brebion, 1966) likewise uses a river water medium fortified by additional bacteria, in the form of a laboratory culture developed as follows. Sewage-polluted river water containing 20 to 60 million microorganisms/ml is used as the source of microorganisms. Food is added in the form of 300 ppm of meat extract and 10,000 ppm of peptone, and the water is aerated for 2 to 3 days during which time the bacteria multiply a hundredfold. Some 10% to 20% of this culture is then added to water from an unpolluted river to give the final medium with a bacterial count between 200 and 400 million/ml. Test solutions are 10 liters in volume, stored in 20 liter Pyrex flasks at 25°C with constant aeration. Initial surfactant concentration is 10 ppm (except for the blank control), and thoroughly mixed samples are taken at 0, 1, 2, 3, and 7 days for surfactant analysis and bacterial count. At 7 days a second 10 ppm increment of surfactant is added and the mixture is then further sampled at 7, 8, 9, and 10 days. An increase in bacterial count during this second period is further indication of biodegradation, supplementing the evidence obtained by analysis for surfactant content.

C. The Shake Culture Test

Adaptation of the widely used microbiological technique of shake culture to the study of surfactant biodegradation was described by Huddleston (1963); a more detailed description has been given by Renn (1964a). This procedure has been adopted with minor changes by the SDA (1965) as the presumptive part of its standardized test procedure for TBS and LAS. A very similar one serves as Japan Industrial Standard JIS K3363 (Tomiyama, 1969).

C.1. *Conditions*

The medium is distilled water containing 300 ppm of yeast extract as food for the bacteria, along with certain mineral salts (Table 5.3). Initial surfactant concentration is usually 30 ppm, and an inoculum of 1% of the volume of medium usually used. Depending on the amount necessary for analytical samples, the volume ranges from 100 ml to 1 liter, contained in an Erlenmeyer flask of 500 ml to 4 liter capacity. Aeration is accomplished on a reciprocating or rotary shaker operating in a horizontal plane. Free access to atmospheric air without excessive evaporation is provided by use of a loose aluminum foil cover or cotton plug at the mouth of the flask.

Bacterial growth is great enough to cloud the medium within a day or two, and biodegradation proceeds to its fullest extent within 1 or 2 weeks if acclimated organisms are used for inoculation. Acclimation is usually accomplished by making a series of runs on a given test compound, adding to each new flask a 1% inoculum from the preceding flask after it has been shaken for a few days.

The initial inoculum may come from any convenient source such as sewage, soil, or laboratory cultures. Or, indeed, no initial inoculum need be used at all; suitable bacteria are present in the general environment and develop in the medium without deliberate inoculation if sterile techniques are avoided (Swisher, 1966a). Ordinarily mixed bacterial species are used, although Huddleston (1963) reported success with several pure cultures as well.

C.2. *Nonaseptic Techniques*

So far as biodegradation results are concerned, sterilization of the medium and use of aseptic microbiological techniques are not necessary unless some special research objectives require it. This has been demonstrated experimentally in the work of the SDA Biodegradation Subcommittee, and indeed is almost a prerequisite if results are to be used in predicting the performance of the test surfactant in the field. Sterilization and aseptic techniques maintain the purity of a pure culture, or perhaps, rarely, constancy of species distribution in a mixed culture. But if a given surfactant can be degraded only under those conditions and not when the culture has been contaminated from outside sources, this would have little relevance to field results (although they would be of considerable scientific interest).

C.3. *Food Levels*

The concentration of food present in the shake culture medium is fairly high—300 ppm of yeast extract. This is sufficient for the development of a visible bacterial floc amounting to perhaps 100 ppm of suspended solids, somewhat lower than, but approaching, that met in some activated sludge processes. The presence of such numbers of bacteria and their associated normal metabolic products may well introduce interferences if study of the metabolic products from the surfactant is to be attempted.

Such interferences are minimized by reduction of the yeast level in the medium. With a biodegradable surfactant the shake culture operations can be conducted successfully without any yeast extract at all, using only the mineral salts portion of the usual medium, along with 30 ppm of surfactant serving as the sole food source for the bacteria. Such cultures have been maintained upon LAS with weekly transfer for over a year, with every indication that operation could be prolonged to any extent desired. The medium remains clear; the bacteria agglomerate to form a few discrete curds rather than remaining dispersed in the medium (Swisher, 1966c). Maintenance on the Bunch-Chambers medium (BOD water + 50 ppm yeast extract) is also readily accomplished (Swisher, 1968a).

C.4. *Reproducibility; Establishment of Quality Standards*

In its extensive studies of the suitability of the shake culture test as a standard biodegradation test method for ABS-type surfactants, the SDA (1965) sponsored some 600 test runs in 17 cooperating laboratories. Statistics on the results showed biodegradation averaging 94% to 96% for commercial LAS, and that 95% of the individual results on a given sample of LAS could be expected to lie within 7% of its average.

The SDA standard for biodegradability of ABS-type surfactants based on this work calls for 90% minimum removal (average of days 7 and 8) with the proviso that a sample falling between 80% and 90% may still be acceptable if it exceeds the 90% minimum in the SDA semicontinuous activated sludge test. Extension of the SDA standards to other types of surfactants would require studies of similar magnitude, and has not yet been accomplished by the SDA.

D. The British STCSD Test

D.1. *Conditions, Reproducibility, Applicability*

The British STCSD (Standard Technical Committee on Synthetic Detergents) adopted a standardized biodegradation test in 1966, using

10 ppm of surfactant in BOD dilution water inoculated with 30 ppm of air-dried activated sludge, aerated at 20°C and analyzed by a methylene blue method over a period of up to 21 days (STCSD, 1966).

Considerations involved in the adoption of this method are also discussed by Eden (1968), along with a report on collaborative comparisons by 11 laboratories. They found a grand mean of 87% $\pm$ 2% degradation for Dobane JNX·LAS, with a range from 83% to 92% in 69 runs.

In its discussion the STCSD (1966) stated that degradability over 90% in this test assured comparable removal at any normal, efficient sewage works, but that certain materials, a few, giving results below 90%, might do substantially better upon further acclimation. Need for a second test providing opportunity for acclimation was recognized.

D.2. *Varying Inoculum and LAS Type*

During the development of this procedure (WPRL, 1965, p. 122–124) the sludge was obtained from an acclimated fill-and-draw batch unit receiving 90% of its organic carbon as LAS. Decreasing the amount of undried inoculum from 30 to 0.3 ppm (dry basis) did not affect the 12% of undegradable found in a standard Dobane JNX LAS sample, but the time required to reach that level increased from 4 days to 20. An inoculum of 30 ppm of air-dried sludge required about 6 days, the same as 10 ppm of undried solids (dry basis). Dried sludge from several sewage treatment plants gave similar results, although some were several days slower; unacclimated sludge was slower also.

The potential influence of the inoculum in the test was dramatically demonstrated by Truesdale (1968) by using two stones from a recycle trickling filter acclimated to TBS. The TBS was degraded to about 80% compared to about 15% with the usual inoculum of dried activated sludge. Likewise ABS-4, a "slow to acclimatize" LAS, was increased to over 95% from 40% in a parallel experiment.

Difficulty in degrading LAS known to be degradable was also reported by Mann (1968) for a similar system. Using a 0.5% inoculum of treated sewage effluent in BOD water containing 10 ppm of LAS aerated at 20°C, he experienced widely varying results (0–90% degradation in 40 days) on a high molecular weight LAS (C_{11-15}) known to be at least 95% degradable. Increasing the inoculum to 5% or the temperature to 30° gave no improvement, but the trouble was corrected by adding 104 ppm meat extract plus 156 ppm peptone plus 27 ppm urea as in the official German activated sludge feed (Table 5.3). This gave 93% degradation in 19 days, 95% in

38 days. A medium molecular weight LAS required no such assistance—
in fact its degradation was slightly poorer in the presence of the foods.

Increasing the initial concentration of LAS in the STCSD test also
slows the degradation (Eden, 1968). At 10 ppm, Dobane JNX LAS was
half gone in 5 days, at 27 ppm in 5 days, at 48 ppm in 10 days, and at
87 ppm over 25 days, if ever.

The "slope culture technique" appears to differ from the STCSD test
only in that the inoculum is preacclimated by 7 days growth on an agar
slant containing 10 ppm of the test surfactant. Cook (1968) found that
this preacclimation was in fact ineffective, since cultures grown on
surfactant-free agar were equal or superior to the "acclimated" ones in
degrading LAS, 78–82% for unacclimated compared to 50–82% for the
others, on Dobane JNX LAS.

E. The Swiss EAWAG Test

A test under development by EAWAG (Eidgenössische Anstalt für
Wasser und Abwasserforschung und Gewässerschutz) is described and
critically appraised by Heinz (1967). It uses a salt medium containing
175 ppm of nitrogen and 89 ppm phosphorus, with 10 ppm of the surfactant as the sole carbon source. It is inoculated with washed activated
sludge fresh from a sewage treatment plant and aerated through a ceramic
filter for 5 days. Heinz says that the results are 5% to 10% lower on LAS
than with the official German continuous activated sludge test.

The EAWAG later simplified the procedure to use a BOD water
medium, and provided for optional use of shake flasks instead of bubble
aeration (EAWAG, 1968), thus evolving into the provisional OECD test
mentioned in Section V.B.1.

F. The Bunch-Chambers Test

This is a simple but very useful screening test (Bunch, 1967a). It
utilizes BOD water fortified with 50 ppm yeast extract, containing up to
20 ppm of the test compound and initially inoculated with 10% of settled
sewage. The mixture, 100 ml in an open 250 ml Erlenmeyer flask, stands
at room temperature for 7 days, at which time it is analyzed and used as
inoculum (10%) for a fresh batch of medium for a second 7 day run. This
is repeated twice more, for a total of four 7 day runs, thus providing
opportunity for bacterial acclimation.

If degradation is not evident or not satisfactory with the test compound
at 20 ppm, lower concentrations should be tested, since toxicity may have
been a factor rather than inherent nondegradability. Parallel runs with

compounds of known degradability are recommended, serving as reference standards and ensuring that conditions suitable for biodegradation are provided. Three replicates of each test are recommended, since there is sometimes considerable variation among replicates.

VI. TRICKLING FILTERS

A. The Topography and the Inhabitants

The trickling filter, alternatively termed percolating filter or bacterial bed, is usually a once-through type of treatment. The influent is cascaded intermittently over the packing medium and then leaves the system as effluent. The nutrients in the influent cause biological growth, which adheres to the surface of the packing as a film made up primarily of bacteria and their insoluble metabolic products. Although exposure of the liquid to the biological film may be of only a few minutes duration (Eden, 1964), solutes may be adsorbed from the liquid onto the film to varying degrees and thus longer effective exposure times are provided during which biodegradation may proceed at leisure.

Development, equilibration, and acclimation of the biological film may require weeks or months before steady state is reached. McKinney (1962b), Hawkes (1963), and Bruce (1969) have presented detailed analyses of the principles and practices involved in the development and operation, while Ludzack (1963a) has reviewed the various types of laboratory apparatus.

The biological film supports a diverse population of higher invertebrates, principally worms and insects—larvae, pupae, and adults. Although they probably do little direct biodegradation of the wastewater components, they do feed on the microorganisms and other film components. This is considered to be helpful in prolonged operation of the filter, which might otherwise become plugged with excessive film growth (WPRL, 1968, p. 91–8).

The invertebrate populations of trickling filters have been widely studied and catalogued, as summarized by Hawkes (1963). Solbe (1967) has observed and measured the development of film and the succession of species in a 12 cubic meter filter from the beginning of its operation. He presents data on the changes in species distribution with depth in the filter bed and with time. A census of 53 species of ciliated protozoa found in trickling filters is given by WPRL (1967, p. 171–2; 1968, p. 156); most of them have also been observed in activated sludge.

B. Laboratory Systems—Operation and Acclimation

Laboratory studies on trickling filters can be made without any great worry about scale-up factors. The most important dimension is the depth of the bed, which determines the retention time; this is around 6 feet, full scale. Horizontal dimensions are dictated by the volume of wastewater to be treated. Thus Goldthorpe (1950) was able to make a quite reasonable facsimile of a full scale plant using a vertical tube 3 inches by 6 feet filled with $\frac{1}{2}$ inch pieces of furnace slag. (He also describes construction of an elegant feeding device, continuous or intermittent, from parts of a discarded grandfather clock.)

The British Water Pollution Research Laboratory (WPRL) at Stevenage has made extensive use of laboratory and pilot plant size trickling filters in its studies on surfactants. Earlier work reported by Mann (1957) was done in units 18 inches in diameter by 6 feet, filled with 1 to 2 inch medium and requiring some 100 liters of feed per day. The rather large size was necessary to produce sufficient effluent for fish toxicity tests. Truesdale (1959) describes smaller equipment of a type used earlier by Lumb (1953), 6 inches by 6 feet. Less than 8 gallons of feed per day was required, applied at the overall rate of 100 gallons per cubic yard of packing per day, intermittently at 4 min intervals. This corresponds to 0.6 volumes of liquid per volume of filter per day. The feed was a composited detergent-free natural sewage to which about 13 ppm of the desired surfactant was added. These filters required about 14 weeks for development of a mature film, and an additional 4 to 8 weeks after starting LAS or TBS before acclimation to the surfactant was accomplished and steady state reached.

When feed was applied to the filter at a changing rate simulating the typical diurnal variation of sewage flow to a treatment plant ("square wave loading"), the degradation was somewhat poorer than with operation at a constant feed rate, particularly as regards surfactant removal (WPRL, 1962, p. 70–74). Klein (1965a, b) found that with a less drastic square wave pattern (1.33:0.67 instead of 1.5:0.5) the deterioration in performance was insignificant.

Eden (1965) showed a parallel decrease in BOD removal and LAS removal by a laboratory trickling filter as the feed rate (sewage plus Dobane JN036 LAS) was increased (Table 5.5).

Deacclimation of trickling filters occurred to different extents when different surfactants were omitted from the feeds for 2 weeks and then resumed again (WPRL, 1967, p. 154). There was a noticeable loss of acclimation with TBS, but not with ordinary LAS. A "difficult to acclimatize ABS RD1213," presumably a higher molecular weight LAS, experienced a slight loss in acclimation but it was regained within 3 weeks compared to the 12 to 13 weeks required for the initial acclimation.

Larger scale tests of 1 and $2\frac{1}{2}$ inch filter packing are described by Truesdale (1962), using $9 \times 8\frac{1}{2} \times 6$ foot beds and feeding municipal

TABLE 5.5

Deterioration in Trickling Filter Performance with Increased
Loading[a]

Filter loading, gal/yd³/day	Effluent BOD, ppm	Effluent MBAS, %[b]
60	3	4
90	5	8
120	16	10

[a] Eden (1965). Feed: sewage containing Dobane JN036 LAS.
[b] Percent of initial content.

sewage containing a 65:35 ratio of LAS:TBS. Depending on the packing, 10 to 16 weeks were required to reach steady state removal of surfactant compared to 3 to 5 weeks for BOD and 6 to 11 for ammonia. Removal of surfactant increased with increasing surface area of the packing.

The German Emschergenossenschaft (Husmann, 1963a, p. 32) used a relatively short tube, 15.5×55 cm, packed with 4 cm Lavalith slag. A layer of Raschig rings at the top acted as a distributor. The filter volume was 7 liters and the feed rate 7 liters/day, using natural sewage or a peptone–glucose–salts synthetic sewage and 10 to 200 ppm of surfactant.

Patterning after the unit earlier described by Degens (1955), Schönborn (1962a) used longer and narrower columns—10×110 cm. The volume was about 8 liters and the feed rate 8 liters/day. After acclimation on domestic sewage for 4 weeks the feed was changed to a synthetic (Table 5.3) containing about 15 ppm of surfactant. Schönborn gives data showing variation in filter performance with (i) increased feed rate, (ii) increased sewage concentration, (iii) recycling effluent back through the filter, (iv) decreased phosphate content, and (v) decreased sewage concentration. The resulting differences were for the most part small, and perhaps not statistically significant.*

C. The Rotating Tube

Gloyna (1952) developed a laboratory trickling filter comprising simlpy an empty hollow tube, $2\frac{1}{2} \times 24$ inches, open at each end, mounted slightly off horizontal and rotated slowly about its axis. Feed is introduced at the high end and trickles down, and the biological film builds up on the entire inner surface of the tube.

* The WPRL (1968, p. 134–5) standard trickling filter (6in × 4ft) is said to give reproducible and relevant data on surfactant biodegradation in a synthetic sewage. Acclimation to LAS (90–95% degradation) took 2–4 weeks, to TBS (65–75%) 12–13 weeks.

Weaver (1962) reports the application of Gloyna's apparatus to study of surfactant biodegradation using a glucose–peptone–beef extract–salts feed with 10 ppm of the surfactant, at a rate of 3 liters/day. Typically about 90% or more of the initial 400 to 450 ppm of BOD was removed, and about 20% to 30% of TBS.

Renn (1965a) used a feed containing meat and milk peptones and urea, corresponding to about 150 ppm of BOD, along with 10 ppm of the test surfactant. A feed rate of 5 gallons/day simulated a high rate trickling filter treatment, while one-third that rate corresponded to a standard rate treatment plant. Recycle studies were also made. Results showed that the removal of surfactant under the varying conditions was proportional to the removal of BOD. Kumke (1966) reported close agreement between results from the tubular unit and those from field tests on a full size filter.

D. Recycle Trickling Filter

Even a small trickling filter requires a relatively large amount of feed and of test surfactant when operated in the usual manner over the several weeks or months that may be necessary for acclimation and attainment of steady state. Burnop (1960) modified the system to batchwise operation by recycling the effluent back to the influent end of the filter, running a small fixed volume of liquor around and around. His filter was a 2 inch by 3 foot glass tube, mounted vertically and packed with chips of granite or limestone. The bacterial film was developed by several weeks of operation in the ordinary manner. For a test the feed contains the surfactant at 10 to 50 ppm along with an unspecified amount of malto-peptone; an unspecified volume of this is recirculated over the filter at a rate of seven times per hour and is analyzed at intervals up to 24 hr. Berger (1964) used a similar system.

This procedure has been further standardized by Edeline (1965), who used a 14 × 124 cm tube filled to 110 cm with 1 to 3 cm pieces of stone. The biological film is developed beforehand, and is used again and again. Prior to each test the bed is washed free of any surfactant by recirculating 0.9% sodium chloride solution for 2 days. Five liters of feed (the official German recipe, Table 5.3, including 20 ppm of surfactant) is then started through the filter. The pumping rate does not seem to be stated explicitly in Edeline's paper, but is apparently in the range of $\frac{1}{2}$ to 1 liter/hr.

During the first 2 to 3 hr the effluent stream is diverted from recycling so that the influent stream remains at a constant surfactant concentration, permitting measurement of the amount adsorbed by the biological film. The first effluent to leave the filter is depleted in surfactant because of both biodegradation and adsorption, analyzing close to zero. During the first hour or so the adsorptive capacity of the film becomes saturated and the surfactant content of the effluent rises to a constant level. This is lower than the input level by the amount of biodegradation occurring in a single pass through the filter. The amount of adsorbed surfactant is calculated from the area between this rising curve and the subsequent steady state concentration, and the initial biodegradation rate from the difference between the steady state and influent concentrations.

When the surfactant level in the effluent has reached its steady state in this single pass operation, recycling is begun. Operation is continued for a week, evaporation losses being made up automatically from a reservoir of distilled water. The recycling liquor is sampled for surfactant analysis at $\frac{1}{4}$, $\frac{1}{2}$, 1, 2, 4, and 7 days, providing data for a biodegradation curve and a value for the percent of residual undegraded.

A biodegradation rate constant k is also calculated, from the surfactant concentrations at the top and bottom of the column and the residence time, assuming a monomolecular reaction. This assumption is not valid; the calculated k is not constant but drops progressively. It is only about one-tenth its initial value by the time biodegradation is near completion. Nevertheless, a comparison of the k values for two different samples at corresponding stages in their degradation does reflect their relative rates. The changes in the values of k may result from the presence of several components in the surfactant, k decreasing as the more readily degradable ones disappear, leaving the more resistant ones with lower k's.

Edeline gives no information permitting judgment as to what effects bacterial acclimation might have on the biodegradation curve. It is conceivable that the preliminary washing may deacclimate the bacteria, and that if further successive runs were made on a given surfactant without intermediate washing, the successive curves might show progressive improvement over the first. If this is indeed the case, and if some surfactants show greater acclimation effects than others, then arises the question as to which mode of operation might give the more valid basis for comparison—a question for which there may be no satisfactory answer.

Jenkins (1967) has reported extensive studies on 3.75 × 75 cm recycle filters filled with 5 to 8 mm washed gravel. The feed solutions contained 3 to 20 ppm of surfactant as the sole organic carbon source, as well as BOD salts and 10 ppm of ammoniacal nitrogen, which appeared to be necessary for smooth degradation. The feed was recycled 7 to 18 times per day, and degradation of surfactant was finished in a few days. Degradation was more rapid upon replenishment of the surfactant, and after a few replenishments the biological growth was stabilized and reasonably consistent results could be obtained thenceforth.

A further variant has been developed by Alexandre (1967), using a 10 × 200 cm tube containing 150 cm of 1 cm pozzolana granules. Under standard conditions a 24 hr acclimation run is followed by a 24 hr final run. Several replicate curves made on a soft detergent of unspecified nature at different seasons of the year indicate good reproducibility of extent and rate of biodegradation.

VII. ACTIVATED SLUDGE

Activated sludge is a major tool for the study of surfactant biodegradation. Not only is it one of the most important agents for treatment of sewage, there is also an immense volume of accumulated knowledge (sometimes apparently contradictory, it must be admitted) resulting from years of study on it.

A. The Science and Technology of Activated Sludge

A.1. *Flocculation—Physical, Chemical, and Biological Factors*

Activated sludge is simply a bacterial floc in which the cells adhere to each other by virtue of the cementing action of metabolic polymers, perhaps polysaccharides (Brown, 1969). Crabtree (1966) implicated poly-β-hydroxybutyric acid also, but under other circumstances only minor amounts were found (Painter, 1968). Nishikawa (1968) found that nucleic acids were present and planned to investigate their contribution to the cementing action.

Embedded in the matrix along with the living cells are dead cells, cell wall fragments, and inert particulate matter, organic and inorganic, either existing originally in the feed or subsequently formed by chemical precipitation. Varying amounts of water-soluble components may be present in the sludge phase also, adsorbed from the surrounding aqueous phase.

At one time it was believed that particular specialized bacterial species, for example, *Zooglea ramigera*, were responsible for floc formation. Thus Butterfield (1937) isolated several pure strains of bacteria from activated sludge, capable of themselves forming activated sludge in pure culture. He considered them all to be related to *Z. ramigera*, which he had isolated earlier.

Later evidence indicates instead that the property of flocculation is exhibited by many of the common bacterial species and that flocculation occurs instead of dispersed growth when conditions of aeration, agitation, and food supply are suitable. This was demonstrated by McKinney (1952, 1953, 1956a) with *Escherichia, Pseudomonas, Alcaligenes,* and *Bacillus* species isolated from activated sludge. He considered that the most important factor was diminished food supply, leading to reduced motility of the bacteria and inability to break free against attractive (van der Waals) and cohesive (slime matrix) forces. However, he did find also that floc-forming ability could vary widely from one species to another. Polyvalent metal ions such as Ca^{2+} and Fe^{3+} have been implicated as well by van Gils (1964). Floc formation mechanisms have been reviewed and discussed in detail by Hartmann (1963b), Crabtree (1966), Busch (1968), Calaway (1968), and Coackley (1969).

Studies of the bacterial makeup of activated sludge are indeed laborious if the classical methods of isolation and identification of individual species are used. Some of the results are mentioned in Chapter 4, Section I. Prakasam (1967a) has reviewed the subject briefly in connection with his own approach to the problem by application of the replica plating technique.*

Further discussions on the biological, chemical, and physical factors entering into the establishment and operation of activated sludge systems, both in the laboratory and in the field, will be found in works by McKinney (1962b, 1968), Hawkes (1963), and van Gils (1964).

A.2. *Application to Sewage Treatment*

Flocculation and settleability are the essential features of activated sludge which suit it so well for sewage treatment; these properties make it possible to maintain the bacterial system at the very high concentrations necessary for degradation of the organics in the sewage within a reasonable exposure time.

Many variations of the basic activated sludge process have been developed for sewage treatment, but they all have some basic features in common. The process is typically a continuous one wherein the sewage is mixed with the activated sludge and air, to give what is known as mixed liquor. The vessel used is large enough to provide a retention time of several hours during which the mixed liquor suspended solids (MLSS) (the activated sludge floc and any other particulate matter present) remove the adsorbable and biologically oxidizable components from solution. After its sojourn in the aerator the mixed liquor flows to a settler from which the clear, treated sewage overflows at the top. The settled sludge is removed at the bottom of the settler for return to the aerator,

---

* WPRL (1968, p. 158–161) estimates bacterial counts of sewage, mixed liquor and effluent as 31, 180, and 5.5 million/ml respectively, and enumerates 13 bacterial genera therein.

perhaps with intermediate steps along the way depending on the exact process being used. Although most of the sludge is retained in the system by virtue of its settling characteristics, even in the best of circumstances the effluent does contain small amounts of colloidal materials which escape settling. From electron microscope studies Dean (1967) concludes that these colloids are predominantly fragments of bacterial cell walls, with smaller proportions of flagella and other cellular debris, viruses, and phages.

Classically the aerator is long and baffled, designed so that, ideally, each increment of feed plus sludge makes only a single orderly passage from inlet to outlet, air being added all the way along. At the other extreme is the complete mixing system, a later development wherein every effort is made to maintain the aerator contents in a steady state condition with uniform composition throughout its volume, the conditions being chosen to optimize biological or operational efficiency. The principle of contact stabilization may be called upon to reduce the residence time (and hence the volume) required in the equipment. Here much of the organic material is removed by adsorption or other processes upon first mixing with the sludge, and provision for its subsequent biooxidation is made by aeration of the settled sludge before its return to the mixing vessel.

Or the design may include extended aeration, wherein retention time in the aerator is prolonged 1 or 2 days or more, to hold the net formation of new sludge as close as possible to the irreducible minimum; this is economically attractive in smaller plants because it decreases the problems of sludge disposal (Simpson, 1964). The oxidation ditch, used extensively in Europe for many years (Pasveer, 1959, 1960a, b; Smith, 1968; Briscoe, 1969), likewise provides a retention time of several days, using instead of a tank an endless ditch in which the mixed liquor is circulated around and around. Batch operation is also feasible.

Details of activated sludge sewage treatment operations are beyond the scope of this present work—they have been outlined simply to provide a background helpful in dealing with the laboratory units used in surfactant biodegradation work.

B. Continuous-Flow Systems—Equipment and Conditions

Laboratory continuous-flow units may range in size from several gallons to a few hundred milliliters, depending upon such things as the objectives of the work and the philosophy of the experimenter. Ludzack (1963a) has critically reviewed laboratory-scale systems, several of which had been developed or used for surfactant biodegradation research.

An apparatus developed at the Water Pollution Research Laboratory (Truesdale, 1959) comprises a $6\frac{1}{2}$ liter rectangular aerator and a $\frac{1}{2}$ liter settler. The aerator is divided into four sections by vertical baffles, the mixed liquor flowing from section to section with overall retention time of 6 to 8 hr, in simulation of the classical type of treatment plant. Air is introduced at the bottom of each section; it is diluted with about 20% of extra nitrogen so that the gas flow rate necessary for proper agitation (1.3 liters/min) gives a dissolved oxygen content of no more than 2 to 3 ppm in the mixed liquor in the final section, characteristic of a full scale plant. Natural sewage is used as feed, with surfactants added at levels up to 13 ppm, and MLSS concentration is maintained at 3000 ppm. Up to 6 weeks acclimation is required to achieve steady state biodegradation of TBS or LAS. Eden (1965) experienced reduced efficiency of both BOD and LAS removal in such a system as the flow rate was increased (Table 5.6).

TABLE 5.6

Deterioration in Activated Sludge Performance with Decreased Retention Time[a]

Retention time, hr	Effluent BOD, ppm	Effluent MBAS, %[b]
8	5	5
6	5	6
4	20	8
2	31	12

[a] Eden (1965). Feed: sewage containing Dobane JN036 LAS.
[b] Percent of initial content.

The South African National Institute for Water Research (Urban, 1965) used a rather similar unit comprising a 4.7 liter aerator subdivided into five compartments in series, followed by a conical settling chamber equipped with a rotary scraper. An air flow of 5 liters/min maintained the oxygen level at 1 to 2 ppm in the last compartment, and the MLSS was held at 3000 ppm. Four to 12 weeks was required for acclimation using a synthetic sewage feed containing about 500 ppm of nutrients and 15 ppm of surfactant (Table 5.3), with a retention time of 11 hr in the aerator. After smooth steady state operation was attained, Urban was able to switch to a mineral salts medium containing no food aside from the 15 ppm of TBS or LAS with continued smooth operation and steady surfactant degradation for at least 9 weeks (but resulting in gradual depletion of the activated sludge). A convenient means for maintaining a neutral pH in the systems was devised—addition of 1 to 3 gm/day of solid calcium carbonate directly to the aerator.

The German Emschergenossenschaft (Husmann, 1963a, p. 40; Jendreyko, 1963) used a 200 liter aerator (divided into four 50 liter compartments in series) and a 400 liter settler, fed 4800 liters/day of natural sewage giving a 1 hr retention in the aerator. This simulated the full scale equipment at Soest, used in subsequent field tests. Surfactants were added at concentrations up to 20 ppm. A laboratory size model with 2 liter complete mixing aerator and 4 liter settler is also described by Husmann (1963a, p. 37).

Degens (1955) describes a 15 liter complete mixing aerator, fed natural sewage at 5 liters/hr to give a 3 hr retention, and air flow at 75 liters/hr (dissolved oxygen 1 to 2 ppm). Retention time in the settler was $1\frac{1}{2}$ hr, sludge being pumped from the bottom at 1 liter/hr into a sludge reaeration tank, retention 8 hr, then returning to the aerator. Return sludge suspended solids averaged 12,000 to 15,000 ppm, corresponding to 2400 to 3000 in the mixed liquor. BOD of the sewage feed was around 325 ppm, reduced to 10 to 20 in the effluent. Surfactants were fed at levels up to 50 ppm, and 1 to 2 weeks was required for return to steady state after a change in the feed.

Huddleston (1964a) developed a system of comparable size with automatic controls, designed around a complete mixing aerator holding 10 liters of mixed liquor and receiving 8 liters/min of air. Feed solution containing 350 ppm soy peptone and 10% natural sewage (Table 5.3) is introduced at rates corresponding to the desired retention time (generally $1\frac{1}{4}$ to 5 liters/hr for retentions of 8 to 2 hr), first passing through a pre-aerator with retention time one-sixth that in the aerator. Surfactant is introduced as a separate stream of stock solution at a proportional rate to give the desired concentration in the aerator, usually 20 ppm. MLSS is maintained at 2000 to 3000 ppm and BOD removal is around 90%. Equilibration to steady state in degradation of LAS or TBS requires as much as a week following a change in feed conditions. In addition to permitting calculation of degradation rates from retention time and influent and effluent concentrations, the equipment is also adaptable to direct measurement of the rate by stopping the feed, adding the desired amount of surfactant, and analyzing the aerator contents at subsequent intervals as it dies away.

McGauhey (1957, 1959a, b) describes complete mixing continuous systems with 1.6 and 10 liter cylindrical aeration sections, as well as a simulation of a compartmented system made up of six 1.6 liter units in series. Studies on TBS fed at 5 to 10 ppm indicated that its degradation increased from around 25% in synthetic sewage to 50% in natural sewage. But unpredictable variations of similar magnitude occurred subsequently in the same systems under supposedly constant conditions with no change at all in the operating variables usually considered pertinent—aeration rate, MLSS level (2000 to 6000 ppm), and BOD reduction (90% to 95%), for instance.

McKinney (1959b) used a very simple complete mixing unit—a rectangular box with a working volume of 1.5 liters, separated by a sloping baffle into an aerator section of 1.25 liters and a settler of 0.25 liters. The feed stream is introduced into the aerator where the entering air stream provides agitation. The flow proceeds underneath the baffle and into the settler section from which clear effluent overflows. Settled sludge returns to the aerator section by gravity back under the baffle, countercurrent to the flow of liquid. Operation on natural or synthetic (200 ppm nutrient broth) sewage at 4 liters/day (7.5 hr retention) led to a steady state MLSS concentration around 1000 to 2000 ppm. At 6 liters/day the MLSS built up to 3000 to 6000 ppm. Renn (1964a) reports the use of similar units with about 3 times the working volume, while Ludzack (1960a) developed a 5 liter unit with cylindrical rather than rectangular design.

Pitter (1964a) describes a system with a 4 liter aerator and $\frac{1}{2}$ liter settler, retention time 8 hr, MLSS 2000 to 3000 ppm and feed 240 ppm bactopeptone plus 44 ppm KH_2PO_4. Acclimation to surfactant was by incremental increase in its level from 0 to 2, 10, 20, and finally 30 ppm. In earlier work he had used a 12 liter aerator (Pitter, 1961a, b, 1963a).

C. The Official German Test Method

The official surfactant biodegradation test method of the German Government (1962) was developed under the auspices of a committee, the Hauptausschuss Detergentien und Wasser. Husmann (1962, 1963a, p. 95–105) discusses the background and the test method itself, while Bock (1960, 1961) shows what appears to be an interim model.

The official apparatus comprises a complete mixing aerator and a settler, working volumes 3 and 2.2 liters, respectively, fed at a rate of 1 liter/hr with synthetic sewage containing about 250 ppm of nutrients and 20 ppm of surfactant (Table 5.3). The air flow to the aerator is adjusted to give good agitation and to maintain the dissolved oxygen level above 2 ppm, but not so high as to cause foaming difficulties. If necessary the upper surface of the aerator, above the liquid level, is coated with an antifoaming agent. The test is started in the absence of any activated sludge; the sludge develops spontaneously within a few days, arising from growth of bacteria entering from the general environment. Return of sludge from the bottom of the settler to the aerator is via an air-lift pump. The MLSS concentration is measured weekly and held at or below 3000 ppm by discarding sludge if necessary. Since the sludge is developed in the presence of the surfactant, it is presumed to be acclimated.

The effluent is collected in 24 hr composites. Each is analyzed by the methylene blue procedure and the percent degradation is calculated for each day by comparison with the MBAS content of the feed for that day. The degradation usually reaches a reasonably steady value after the first week or so, after which a total of 21 consecutive days of smooth operation with reasonably constant daily values is required for completion of the test. The degradability of the sample is calculated as the average of the 21 daily values.

Reproducibility is reasonably good although, as with other test methods, erratic results may be encountered, particularly with samples of poorer degradability. The activity of the sludge might be expected to vary from laboratory to laboratory, depending on the bacterial species distribution in each particular environment. Husmann (1962) shows results of five laboratories on the same sample. All gave the same result in that the 80% degradation requirement of the German law was met, but the induction period ranged from 3 to 17 days and the steady state degradation levels ranged from around 85% to 100%. Fischer (1965) states that the spread between independent determinations is about $\pm 5\%$, $\pm 2\%$, and $\pm 1\%$ for substances with degradation of 75–85%, 85–90%, and 90–95%, respectively.

So far as the official text of the method is concerned, maximum time allowable for the initial induction period is not specified and appears to be left to the discretion of the tester. Likewise for any decision as to the magnitude of irregularity allowable without invalidating a consecutive 21 day period.

As written, the procedure is confined to those surfactants which can be analyzed by the methylene blue method, and to soap. An official analytical procedure is included for determining the soap content of a mixed product. The accompanying mathematical formula for calculating the biodegradability of such a mixture assumes the soap content to be 100% degradable. According to this formula a product containing 4 parts or more of soap for each part of MBAS would meet the 80% biodegradability requirement regardless of whether the MBAS was degradable or not. However, in current practice it appears that each individual component must meet the 80% limit.

Both theoretical and practical aspects of the official German test method have been discussed by Fischer (1965), based on experience gained in over 200 such tests. A bank of 12 units is described, as well as the associated analytical, manpower, and economic loads. Truesdale (1968) considers the German method to be very laborious, and found difficulties in maintaining sludge circulation and settling. He mentions that "... the Germans have found it necessary to ... weight the sludge floc by addition of ferric hydroxide." *

D. Miniature Continuous-Flow Units

Preparation, storage, and handling of feeds and effluents for the larger scale continuous units entails much effort, particularly if many are to be operated simultaneously. Miniaturization saves not only space, but also considerable time and labor, but introduces the possible difficulties of (i) further departure from characteristics of full scale treatment plants, (ii) unsteady biological performance, and (iii) limitations on sizes of samples which may be withdrawn for analysis. These problems are not serious: (i) even the larger laboratory units are so far removed from full scale that another factor of 10 need be no great additional obstacle, (ii) the biological performance is no more unsteady than in larger prototypes, and (iii) ample effluent is available for the usual analyses required in surfactant biodegradation work (although the quantity of sludge available for supplementary analyses is quite limited, no great inconvenience is introduced thereby).

Sweeney (1964a) gives construction and operating details on a complete mixing unit comprising a 600 ml working volume aerator and a 100 ml settler. A turbine agitator (150 rpm) in the aerator provides both mixing and circulation of the mixed liquor to and from the settler. The operating conditions suggested as a standardized test method include use of a natural sewage feed at a rate of 2 ml/min to allow a 6 hr retention, aeration at 40 ml/min to maintain dissolved oxygen around 2 ppm.

By use of the natural sewage feed in conjunction with activated sludge from a plant treating that same sewage, long acclimation periods are not required, as they might well be in equilibration of a sludge to a new and unaccustomed feed. Thus the test requires only 8 to 9 days operation.

---

* In extensive studies Janicke (1969) experienced occasional wide variations in acclimation time, and upsets in degradation; he found it advantageous to develop the sludge before starting surfactant addition. See also page 137, Vaicum (1968).

Variations in biodegradation activity from one sludge to another are compensated in the calculation of results by use of a normalizing factor determined from a standard surfactant included in each set of tests (Section II.B). Initial MLSS levels are chosen to match operating conditions in the treatment plant at the time the sludge is obtained, ranging from around 500 to 2000 ppm. The test surfactant is introduced at 3 ppm;

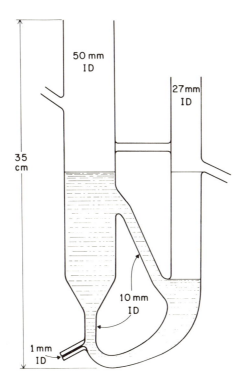

Fig. 5.2 Miniature complete mixing continuous activated sludge unit (Swisher, (1964a). Working volumes: aeration (shaded area), 300 ml; settling, 75 ml. Feed enters at upper left, air at lower left; effluent overflows at right.

radiotracer techniques (^{35}S) are used for the analysis to avoid interference from any surfactant already present in the sewage.

A still smaller complete mixing unit has been designed by Davis (1962), and used extensively by Swisher (1964a, 1967a, b) (Fig. 5.2). The 300 ml of mixed liquor in the aerator section is agitated by an air stream of 100 to 200 ml/min introduced into a constricted section at the bottom. This is designed to produce an airlift action as well, circulating the mixed liquor to and from the bottom of the adjacent 75 ml settler. Smooth operation is attained under a variety of operating conditions, feeding synthetic (150 ppm peptone) or natural sewage containing up to 200 ppm surfactant (LAS) at 3 to 7 hr retention times with MLSS at 2000 to 6000 ppm.

E. Batch and Semicontinuous Systems

A continuous activated sludge unit represents a considerable investment of space, time, and money for its installation, operation, and maintenance, and it is susceptible to the many difficulties inherent in keeping any continuous process in smooth operation. A batch unit is much more economical; although its simulation of full scale practice may on occasion be somewhat more remote, nevertheless it can give a wealth of pertinent information.

E.1. *Batch Sludge Die-Away*

In this simplest embodiment of a batch sludge system, settled activated sludge from a treatment plant is added to sewage containing the test compound, air is bubbled through the mixture and analyses are made over a period of hours or days. Sierp (1954) demonstrated applicability of the method using an alkyl sulfate and an alkylaryl sulfonate. With 3 hr aeration they were degraded 40–60% and 80–90%, respectively. Without aeration there was only 0% to 9% degradation. Without the activated sludge but with 3 hr aeration of the sewage, the alkyl sulfate was only 4% degraded compared to 70–80% for the alkylaryl sulfonate. (Their chemical structures were not specified further, but these results indicate that they were definitely not LPAS and TBS; in fact, the pattern resembles what might be expected if the alkyl sulfate were TBS and the alkylaryl sulfonate were LPAS.)

House (1956) used an 8 hr aeration and reported that the rate of TBS degradation was approximately proportional to the aeration rate. He also found that TBS degradation varied considerably from one sludge to another, ranging from about 25% to 85% in 8 hr of high rate aeration. BOD removal was much more constant, between 78% and 92%. The reason for this variation in TBS capability of the sludges was shown to be, in all probability, an acclimation effect dependent on the TBS level at the sewage plant prior to collection of the sludge. Laboratory experiments in which sludges were fed synthetic sewage containing 0 or 5 ppm of TBS for 7 days showed loss or gain, respectively, in TBS degrading ability of the sludge.

E.2. *The Romanian ISCH Test*

In the procedure of the Institutul de Studii și Cercetări Hidrotehnice described by Vaicum (1967) an unacclimated sludge is used so as to eliminate the possible variance in preacclimation observed by House, above. The test allows a total of 3 days aeration, giving some opportunity for development of acclimation to the test surfactant. The medium is the official German feed (Table 5.3) containing 1000 to 1500 ppm of MLSS. A 2 liter batch is aerated continuously and sampled daily for 3 days for analysis of surfactant content and permanganate-oxidizable organics.

Results on a series of alkyl sulfate and ABS samples closely paralleled those obtained in six widely used tests (river water, three continuous activated sludge types, BOD, Warburg) and had the advantage of being available in 3 days (see Table 5.15).

E.3. *From Batch to Semicontinuous*

The approach to acclimation that has been taken most commonly is just to run batch after batch. Thus in the semicontinuous process (also termed fill-and-draw) we simply have an aeration vessel containing activated sludge, to which the feed solution is added. After the desired amount of aeration (23 hr, for example) the air is turned off, the sludge is allowed to settle, the clear supernatant liquor (treated effluent) is drawn off, a like volume of fresh feed is added, and aeration is resumed to start a new cycle. This is often termed fill-and-draw operation.

Although the progress of biodegradation may be followed during an individual cycle by appropriate analyses and presented as a die-away curve, ordinarily analysis is only done at the end of the cycle. Each such point, then, represents the end of a complete die-away test carried out during the cycle, corresponding to the week or more which might be required in a river water test or several days in shake culture.

The semicontinuous system of operation dates back at least as early as Butterfield (1937). He used 8 liters of mixed liquor in a 10 liter bottle on a 24 hr cycle 5 days/week. After aeration the mixture was settled 30 min, 5 liters of supernatant was removed and replaced by fresh feed for the next cycle. Lumb (1953) used a modified schedule with 1100 ml of mixed liquor, aerating 10 to 12 hr, settling during the remainder of the day, and then removing 700 to 900 ml of supernatant.

E.4. *The SDA Procedure*

The standard procedure of the SDA (1965) resembles the standard treatability test proposed by Symons (1960). It is a direct descendant of that used by Helmers (1950) and Bogan (1955) in propagation of acclimated sludge as seed for Warburg respirometry. Chemical analysis of the sludge effluents themselves was later applied by Ryckman (1956, 1957) and McKinney (1959b) so that degradation therein could be calculated, and the respirometry has been used less and less in later years.

The procedure as standardized by the SDA calls for a 24 hr cycle (23 hr aeration, 1 hr for settling, drawing, and filling) in a 3×24 inch vertical cylinder containing 1500 ml of mixed liquor. The feed volume is 1 liter/day, containing 390 ppm organic nutrients and 20 ppm surfactant (Table 5.3), and the MLSS concentration is 2000 to 3000 ppm. The feed and effluent are analyzed by the methylene blue procedure and the biodegradability, or percent removal, is calculated as their quotient, averaged over 7 days of level operation. Results of a run are valid only if a parallel run made with standard pure $C_{12}LAS$ gives a removal over 97.5%.

The lower limit for adequate biodegradability of an ABS-type surfactant is set by the SDA at 90% by this procedure. Interlaboratory comparisons were made constituting 250 runs in 12 laboratories, forming the basis for a statistical treatment (SDA, 1965). It was calculated that 95% of the individual results for a given LAS sample would fall no more than about 5% below the average of all results on that sample. Variability was much greater in the case of TBS.

Inasmuch as the SDA study in this particular phase was limited to ABS-type surfactants analyzable by the methylene blue method, the performance standards which were established are applicable only to those types. Extension to other surfactant types would be dependent upon further studies of at least comparable magnitude. The SDA (1969a, b) has done this for nonionics, but their widely diverse chemical nature and the ambiguities of the analytical methods precluded establishment of clear-cut universal biodegradation performance standards and limits. During these studies several types of $6 + 18$ hr cycles were examined, but little was found to recommend them over the standard 24 hr. It was found feasible to go to a 5 day week, at some sacrifice of smoothness of results. Measurements made on the 3rd, 4th, and 5th days appeared to show little effect from the starvation over the previous weekend.

E.5. *Other Semicontinuous Procedures*

Much of the research on surfactant biodegradation has utilized the SDA procedure or slight variants thereof. In addition to those already mentioned, the work of McGauhey (1959a, b), Nelson (1961), and Schönborn (1962a) might also be cited.

An earlier, somewhat different system is described by Mann (1957). Designed to provide large volumes of effluent for fish toxicity studies, it utilized vessels with 16×12 inch cross section filled to about 36 inches with mixed liquor. Total working volume was 100 to 115 liters, aerated with 30 ml/sec of air and further agitated by a slowly rotating paddle at the bottom. At the end of each daily cycle, after $\frac{1}{2}$ hr settling, the volume of settled sludge was brought to 20 liters and a further 15 liters of supernatant was also retained in the system. Fresh feed (settled sewage) was added in amounts of either 65 or 80 liters. At those two levels of operation the MLSS averaged around 5000 or 3600 ppm, respectively. Surfactant, presumably TBS and LPAS, was introduced as a mixture of seven proprietary detergents, at levels from 20 to 80 ppm MBAS. Steady state degradation of the MBAS averaged around 80% and was attained about 7 weeks after start of addition; the units had been operated for 8 weeks previously without surfactant.

Weaver (1962, 1964) used alternating 7 and 17 hr cycles (6 and 16 hr aeration) in an effort to simulate more closely the exposure times in treatment plants (discussed in Section VII.E.6). Mixed liquor total volume was 4 liters, MLSS 2000 to 4000 ppm, effluent and feed volumes 3 liters. Feed was either natural sewage or a modified Butterfield (Table 5.3) containing 10 ppm of test surfactant.

E.6. *Relation to Continuous-Flow Systems*

Although each individual cycle in the semicontinuous operation constitutes a batch run on that particular increment of feed solution, the cycle is repeated again and again with fresh feed each time. Thus the overall aspect is a continuous—more precisely, semicontinuous—process with fresh feed going in and effluent coming out in very large increments instead of infinitesimal ones. This allows opportunity for acclimation and attainment of steady state operation.

Further, there is a close resemblance between the sequence of conditions during one batch cycle and those in a conventional-type activated sludge plant. That is characterized by plug flow conditions wherein feed and recycle sludge are mixed at the entrance to the aerator. Ideally, each increment of the mixture then flows undisturbed through each successive section of the aerator. The bacteria in the increment metabolize the foods present in that increment, they multiply, the food becomes depleted and, if there is sufficient retention time, the bacteria enter the endogenous respiration state. This is precisely the sequence of events occurring in the batch unit during each cycle.

Semicontinuous sludge units are usually operated on a 24 hr cycle. A closer simulation of plant operation would be given by a 6 or 4 hr cycle. Truesdale (1968) investigated such operation and found the expected poorer efficiency (Table 5.7) felt to be typical of continuous-flow

TABLE 5.7

Decreased Efficiency of Semicontinuous Activated Sludge with Decreased Cycle Time[a]

Cycle time, hr	ABS removal, %	
	TBS (Dobane PT)	LAS (Dobane JNX)
24	75.4	96.3
6	41.5	94.8
4	23.6	94.7

[a] Truesdale (1968).

operation at corresponding retention times. (Unexpectedly, a continuous unit with 3–6 hr retention may be equal or superior to a 24 hr semicontinuous, in biodegradation; see Swisher (1967a, b), for NTA and LAS benzene rings.) However, such short cycles are impractical for semicontinuous operation in the laboratory because they require around-the-clock attention. This could be countered by automation, but continuous units are probably to be preferred instead.

A compromise semicontinuous schedule such as 6 + 18 or 7 + 17 hr operation can be encompassed in a normal working day (Lockett, 1956; Weaver, 1962, 1964). Theoretically this sort of 6 hr result is not precisely comparable with that from a true 6 hr cycle because the preceding 18 hr phase may give the sludge an opportunity to degrade adsorbed, resistant components to a greater extent. In actual practice, however, the difference is probably too small to be noticed.

E.7. *Semicontinuous Arithmetic*

The large increments of feed and effluent involved in the semicontinuous operation introduce arithmetical oddities which are sometimes not evident, or perhaps are confusing, at first glance. For instance, the retention time is not equal to the cycle time. Further, the calculation of the percent degradation may seem to involve a certain amount of legerdemain. These apparent anomalies are only apparent, fortunately, and disappear under closer scrutiny.

Let us consider the SDA procedure, wherein $\frac{2}{3}$ of the total volume is removed as effluent at the end of each cycle. The desired amount of sludge, which occupies a settled volume of perhaps $\frac{1}{2}$ or $\frac{1}{4}$ of the remaining $\frac{1}{3}$, is retained. With such a proportioning, $\frac{1}{3}$ of the treated liquor from a given cycle remains in the system for the next cycle, $\frac{1}{9}$ for the second, $\frac{1}{27}$ for the third, and so on. This means that physicochemical steady state is approached to within about 1% within 4 or 5 cycles, or perhaps a little longer if adsorptive factors are important. (In contrast, biological steady state involves factors beyond mere arithmetic—such things as development of adaptive enzymes and redistribution of bacterial populations—and may require several dozen cycles or more, as discussed in Section VII.F.)

Examination of the arithmetic shows that the average retention time is 1.5 days with a 1 day cycle time. Two-thirds of the daily liter of feed is removed after 1 day, but the remaining third gets one extra day and one-third of that gets another extra day and so on. The total is $1 + \frac{1}{3} + \frac{1}{9} + \frac{1}{27} + \cdots = 1.5$ days.

The dilution of the fresh feed by half its volume of residual liquor at the beginning of each cycle tends to cause confusion in the calculation of percent degradation. Assume, for instance, that our system is accomplishing 90% degradation in steady state. One liter of effluent containing 2 ppm undegraded is withdrawn and replaced by 1 liter of fresh feed containing 20 ppm. Next day that is withdrawn, again with 2 ppm remaining. One is tempted to say that dilution of the 1 liter of 20 ppm feed with 500 ml of 2 ppm residual means an initial concentration of 14 ppm, so that the 2 ppm at the end of the cycle actually represents 14.3% remaining, and not 10%.

This view is incorrect, because the 2 ppm at the end of the cycle comes only in part from the new surfactant added at the beginning of the cycle; the rest comes from the undegraded residue of earlier feedings present in the previous cycle's sludge liquor.

The percent remaining is properly calculated, once steady state has been attained, by consideration only of what goes into the system and what comes out, that is, by dividing effluent concentration by feed concentration. What happens inside is irrelevant to this calculation.

This is readily seen in considering the extreme case of a completely undegradable surfactant. Neglecting adsorption for the moment, the effluent from the first cycle would contain 13.3 ppm, rising to 17.8, 19.3, 19.8, 19.95 and so on in the following cycles, approaching the steady state value of 20 ppm. If adsorption does occur to the extent characteristic of

TBS and if we assume a MLSS of 3000 ppm, about 40 to 50 mg of surfactant would be adsorbed on the 4500 mg of sludge solids in equilibrium with a dissolved surfactant concentration of 20 ppm (Fig. 4.2). This much surfactant would be contained in 2 or 3 liters of feed, or another two or three cycles before steady state. And once steady state is reached both effluent and influent contain 20 ppm and their quotient shows 100% remaining.

In the example cited previously the proper calculation is 2 ppm/ 20 ppm, or 10% remaining, not 14.3%. Detailed consideration of two extreme cases illustrating this will disclose another apparent anomaly, readily explained. Looking at the 2 ppm of undegraded in the 500 ml of residual sludge liquor, this amounts to 1 mg, and it may be (i) a completely undegradable fraction of the surfactant under test, or (ii) at the other extreme it may have exactly the same composition as the initial intact surfactant, still remaining at the end of the cycle because of insufficient degradation time. All other possibilities lie in between.

In case (i) the 20 mg of fresh surfactant loses 18 mg by degradation, leaving the 2 mg of undegradable, or 10%. The 1 mg of undegradable originally present in the 500 ml of sludge liquor remains unchanged, for a total of 3 mg in the 1500 ml of mixed liquor, or 2 ppm as observed.

In case (ii) the fresh surfactant has the same composition as that already present, remaining undegraded from the previous cycle. During the new cycle the 20 mg drops to 2.857 while the old 1 mg is dropping to 0.143, again adding up to a total of 3 mg undegraded, or 2 ppm, as observed. But this corresponds to 14.3% remaining, not the 10% we have been saying. The degradation is indeed only 85.7% during the 24 hr cycle, but in this case the residual surfactant is the same as that originally added and will continue degrading with further exposure. The total average retention time is 36 hr, not 24, and during the extra 12 hr the degradation proceeds on from 85.7% to 90%. As, of course, it must if we are to continue with the feed entering at 20 ppm and the effluent leaving at 2 ppm, as initially specified.

This discrepancy between cycle time and retention time can be minimized by increasing the ratio of effluent removed to sludge liquor remaining, requiring filtration or centrifuging if the natural settling limits of the sludge are to be exceeded. In actual practice, however, such measures are hardly necessary if a 24 hr cycle is used, because activated sludge will degrade most surfactants to their fullest extent within that time, if acclimated. Little further degradation is evident upon prolonging the treatment another 24 hr or more. In such a situation everything reduces to case (i), where the degradation during a single cycle is the same as the overall degradation.

F. Steady (and Unsteady) States

The continuous-flow activated sludge system is a complex one, and as McGauhey (1959a, b) has remarked, even when the mechanical operation is held in good control and no gross biological differences are evident, the biodegradation of a surfactant may vary greatly with no apparent cause, particularly if the surfactant is a resistant one such as TBS. The semi-continuous system is inherently rather more stable, but it too may have its ups and downs. Variations in the degradation of the more natural components of the feed, not only the surfactant, could undoubtedly be detected also if analytical methods of comparable selectivity and sensitivity were applied as assiduously. The cause for such variation, of course, lies in the sludge itself.

Not necessarily in the amount present, for sludge concentrations may be changed within wide limits, say from 1000 to 5000 ppm MLSS, with little effect on biodegradation performance. Rather, in the bacterial makeup, which despite apparent similarity in gross appearance may vary widely from one sludge culture to another, and in a given sludge from one time to another.

F.1. *Bacterial Populations—Their Rise and Fall*

The bacterial population of an activated sludge may differ in the range of individual species and strains present, in the relative numbers of individuals of each type and in their state of acclimation. A given species or strain may rise to dominance or dwindle to insignificance as a result of changing balances in the interaction of many environmental factors. To name a few, nature and amount of food supply, oxygen concentration, degree of agitation, temperature, flow rate, settling characteristics, competition with other species, presence of predators, or presence of toxic or inhibitory substances.

As an example, Downing (1966) (see also WPRL, 1962, p. 4–10) has calculated and demonstrated such effects in the case of nitrifying bacteria in activated sludge, and has defined circumstances under which they may find themselves washed out of the system. The failure of pathogenic bacteria to thrive under the conditions of sewage treatment in competition with the soil bacteria provides another example.

F.2. *Stabilizing Influences*

A particular compound, surfactant or other, will require a certain time for degradation, depending among other things upon its own concentration and the concentration and acclimation of the bacterial species doing the degradation. If this time is short compared to the retention time in the activated sludge unit the system will give stable performance and good degradation in spite of considerable fluctuations in the population of the necessary bacteria. This may well be an important factor in the greater operating stability of the semicontinuous system compared to continuous flow—the retention time is usually several times as long in the former.

Conversely, if the time required for the biodegradation is much longer than the retention time, a stable situation also results (no degradation).

But if the two times are of a comparable magnitude we can expect the results to differ from one sludge to another and from one time to another in response to relatively small differences in the numbers of the critical bacterial species present.

The factors of sludge settling and recycle can also serve as a stabilizing or damping influence on the system, tending to level out fluctuations. By the same token, these factors act toward prolonging the time required to reach steady state.

F.3. *Population Successions and Drifts*

Cassell (1966) has reported that mixed cultures in continuous-flow systems without sludge settling and recycle do not necessarily reach steady state at all. He found wide fluctuations in bacterial population distribution and effluent quality over long periods of operation with no indication of leveling out. In contrast, Shindala (1965) was able to reach steady state readily and reproducibly in a two-organism system (*Proteus vulgaris* and *Saccharomyces cerevisiae*) under similar circumstances. Here the bacterial growth was dependent on and limited by niacin synthesis by the yeast; addition of excess niacin upset the steady state population distribution until the excess was removed from the system by the continuous flow.

Parker (1966) showed that when three bacterial species were grown together in a continuous-flow system their relative growth rates were considerably different from the values in pure cultures, and also that changes in the medium could cause further alteration.

Prakasam (1964, 1967b) propagated the mixed bacteria of an inoculum of sewage by successive transfers in a medium containing sorbitol as the food; the initially diverse populations became dominated by only a few of the original species. Using a similar technique with a glucose–butyrate medium and an original mixed population from a river water inoculum, Chian (1968) obtained populations containing only two species, a pseudomonad and a coliform, not further identified. When this mixture was then cultured under continuous-flow conditions with retention times of about 3 hr, the proportion of coliforms increased from an original 10–30% up to about 90% during the first 2 to 4 days and stayed at that level thereafter. Unless the feed flow rate was increased: at retention times of 1 hr or less, the pseudomonads regained their original dominance. It is perhaps significant that at about this same threshold the glucose supply exceeded the capacity of the organisms to utilize it, and undegraded free glucose began to appear in the system. Mateles (1969) pointed out that while five or six replacements of the culture volume were usually considered sufficient to reestablish steady state in pure culture systems, in these experiments with two organisms together, 10 to 25 replacements were needed before population distribution had steadied at its new ratio. Thabaraj (1969) presents detailed experimental data on 8 hr activated sludge systems leading to this same conclusion. When the glucose level in the feed was increased he found an "ecological response" after 30–40 hr, wherein the bacterial species predominance in the sludge shifted and undegraded glucose and metabolic intermediates temporarily appeared in the effluent. The new steady state was not attained until some 50–80 hr after the glucose increase.

Changes in bacterial population distribution also occur from week to week or month to month during semicontinuous-type operation, even with rigid control to maintain constant operating conditions. Rao (1966) recounts in detail the changes in several semicontinuous activated sludge units fed glucose on a 24 hr cycle. The sludge changed significantly and reversibly as time went on, in color, cell morphology, and speed of removal of the glucose. Table 5.8 shows this succession in one of his sludge cultures during the observations. Painter (1968) has also remarked upon the bacterial inconstancy of systems open to the environment.

Adamse (1968a) observed the succession of bacterial species in activated sludge in a laboratory continuous-flow unit and in a full scale oxidation ditch over a period of 2 months when fed dairy wastes. The same succession occurred in both systems, and reproducibly from one year to the next (Table 5.9). The decline in pseudomonads and increase in corynebacteria is quite pronounced, and the observed reproducibility tempts one to think that it might possibly be universally characteristic of activated sludge–dairy waste systems. It would probably be best to resist this temptation until further evidence is at hand.

TABLE 5.8

Succession of Properties of an Activated Sludge Culture during Semicontinuous Operation[a]

Day	Color	Bacteria	Protozoa	COD removal, mg/hr/mg sludge
40	Golden brown	Floc	Many free swimming	0.745
50	Dark brown	Floc + filamentous	—	0.470
60	Black	Floc + predominant chains	—	0.320
70	Golden brown	Floc	Free swimming	0.505

[a] Rao (1966).

In further investigations Adamse (1968b) found that *Sphaerotilus natans* grew better than corynebacteria when dissolved oxygen was low and C:N ratio was high. Those are the conditions under which activated sludges both in laboratory and field often become dominated by *Sphaerotilus*, undesirable because of its poor settling characteristics.

In view of the foregoing observations it can be expected that a laboratory activated sludge system receiving a simple synthetic feed will tend to develop an overbalanced population dominated by the one or two species which are most favored by the particular conditions. Such a system might well have inherent instability toward any variations in operation which

TABLE 5.9

Bacterial Successions in Activated Sludge Treating Dairy Wastes[a]

	Bacterial distribution, %		
	9 days	28 days	56 days
Pseudomonadaceae	30–50	5–30	5–15
Achromobacteriaceae	39–42	20–60	10–25
Corynebacteriaceae	5–20	5–70	55–85

[a] Adamse (1968a).

might occur. Although no conclusive experimental evidence is at hand it is likely that the potential diversity of the population can be increased considerably by periodic inoculation with fresh sewage or activated sludge from a diversified treatment plant—say 1% of the working volume per day.

F.4. *Acclimation*

To accomplish degradation of surfactants and other substances less commonly met than ordinary foods, development of the appropriate enzymes often must be induced in the bacteria (Chapter 4, Section III). When the supply of exotic food is exhausted these enzymes may disappear, requiring reelaboration if that particular food again enters that particular system. This process requires some time, perhaps a matter of hours or days. This may be a factor involved when irregular performance is encountered, as in the poorer degradation reported by the WPRL (1962 p. 70–74) when using irregular flow rates with varying surfactant concentration ("square wave loading," Section VI.B).

This effect diminishes with increased retention time, as demonstrated by Huber (1968) in the full scale oxidation ditch at Grosslappen (Section X.A). He found that square wave feeding had no effect with 2 day retention time, degradation of both LAS and TBS being 80% to 90%. With $1\frac{1}{2}$ day retention LAS removal was still around 80% with steady feeding but TBS removal dropped to 56%, while square wave removals were 33% and 38%, respectively. (The poor square wave removal of LAS in this case was attributed to an unusually low MLSS concentration during the test rather than to the feeding schedule.) At 1 day retention with uniform feeding the removals were 30% and 20%, respectively for LAS and TBS. Huber felt that the two most important factors in his system were retention time and ratio of surfactant to MLSS.

With a retention time of a few hours one might expect that no more than a few days should be required for attaining a new steady state after a change in operating conditions—enough time for 10 or so replacements of the system contents. While this is true for a purely physicochemical system, a considerably longer time—perhaps several weeks or months—may be needed for the redistribution of the biological population into the new steady state composition. Thus, 1 or 2 weeks after a change from one synthetic feed composition to another in a continuous-flow system, a temporary upset is often noticeable, presumably reflecting some stage in the readjustment of the sludge to the change. The upset might take the form of poor settling, or perhaps poorer degradation of the test surfactant even though it may have been at the same concentration in both old and new feeds.

Similarly, when starting a unit with fresh sludge from a treatment plant on a synthetic sewage in the laboratory the operation may proceed quite smoothly for a week or so, only to be followed by several weeks of upsets before a final smooth state is reached. If natural sewage is used in a parallel unit, no such readjustment occurs. This is, of course, the basic reason for the use of the less convenient natural sewage feed in the Sweeney (1964a) (Section VII.D) test procedure—the system is already acclimated, except perhaps to the surfactant itself. Ludzack (1961, 1964) presents extensive data further confirming that several weeks or more are required for continuous units to reattain steady state after change in operating conditions.

F.5. *Mathematical Models*

Mathematical derivations and models for calculating substrate degradation and/or bacterial growth in continuous-flow systems have been developed, discussed, and experimentally checked by Washington and co-workers (Martin, 1964, 1965; Hetling, 1964, 1965) and also by McKinney (1962a), Schulze (1964a, b), Reynolds (1966), Westberg (1967, 1969), McLellan (1968), and Ramanathan (1969). The earlier work has been put into perspective in a review by Gaudy (1966).

Performance of systems in steady state can be described in terms of certain constants characteristic of the system, such as the specific bacterial growth rate (a function of the substrate concentration) and the bacterial yield (weight of cells produced per unit of substrate metabolized). If the inflowing substrate concentration is changed, a new steady state is eventually attained, but a significant time is first required for readjustment of the bacterial enzyme systems. Storer (1969) demonstrated this experimentally, showing that about 8 hr was required after so simple a change as tripling the influent glucose level, before the new steady state was reached. During the interim transient stage the specific growth rate varied widely from its equilibrium values, while the cell yield dropped to half and then rose to twice its normal value before finally resuming it. All this in response to the temporary overload of substrate in the mixed liquor, which in turn presumably triggered the development of additional supplies of the necessary enzymes, some of which were later demobilized as the substrate concentration dropped again in response to their action.

With this physiological readjustment to the higher glucose level completed, the characteristic mathematical constants resume (hopefully) their original values and the original mathematical relationship describes the new steady state. But not for long. Within another day further upsets occur and unutilized glucose and intermediate metabolites again appear in the effluent (Thaboraj, 1969). This time it is the result of a shift in the bacterial species distribution in response to the higher glucose level in the feed, termed the "ecological response." Once again steady state is eventually attained (within a day or two in these experiments), but now the original mathematical constants are presumably no longer applicable because different bacterial species are now involved.

Obviously description and prediction of real activated sludge systems by mathematical models is no simple task.

Downing (1966) (see also WPRL, 1965, p. 8–18) has developed a model specifically for the biodegradation of LAS-type surfactants in a continuous activated sludge system. Using pertinent constants determined from static or batch degradation experiments, the model gave results in reasonable agreement with those observed in actual practice. The input information necessary to achieve this included:

(i) Effect of acclimation on the growth rate constant of the LAS-degrading organisms.

(ii) Growth rate constant after acclimation, $k_D = 0.48$ day^{-1}.

(iii) Fraction of these organisms initially in the activated sludge, $C_0 = 0.25$.

(iv) Adsorption of the surfactant onto the sludge, $A = 0.38D^{0.5}$.

(v) Michaelis saturation constant of the surfactant with respect to the organisms (concentration at which the growth rate constant is $\frac{1}{2}$ the maximum), $X = 7$ mg/liter.

(vi) Fraction of undegradable in the LAS, 0.12.

The mathematical, computational, and experimental methods used in developing such a model are described in detail by Downing (1964) and Knowles (1965) for the case of nitrification of ammonia to nitrite and nitrate. Further experimental verification of the nitrification model is mentioned by WPRL (1965, p. 18–22) and Downing (1966).

Andrews (1968) reports some early results in the development of mathematical models which include an *inhibition function*, covering substrates which at higher concentrations depress the growth rate of the bacteria; anionic surfactants fit into this category. The drastic effects of too rapid change of concentration in the feed, for example, are quite dramatic in some of the theoretical cases considered. Rapid failure of degradation may result in such a case, the organisms washing out of the system as their growth rate drops below the critical value corresponding to the hydraulic retention time in the system. Yano (1969) is also developing the mathematics of such situations. Application to real systems such as continuous activated sludge with such characteristics as mixed bacterial species, retention of bacteria, mixed substrates, and whatever other complicating factors there may be, has not yet been accomplished. But the present simple model seems able to clearly suggest the generally favorable or unfavorable tendency of a particular change in the operating parameters.

F.6. *Bacterial Inhibition by Surfactants*

Some surfactants are inhibitory or toxic to bacteria at concentrations as low as 25 to 50 ppm (Chapter 4, Section IV.A), but nevertheless can be successfully degraded in a continuous activated sludge system when present in the feed at very much higher concentrations. This is possible because the surfactant can be adsorbed by the sludge as quickly as fed, and then rapidly degraded, so that its concentration in the mixed liquor and on the sludge is kept well below the critical threshold.

But if some irregularity of operation happens to occur in such a system a chain reaction may ensue, with results out of all proportion to the magnitude of the initial upset. Momentary interruption of degradation interrupts the freeing of occupied adsorption sites, incoming surfactant is not removed from solution as rapidly, its concentration in the liquid phase of the mixed liquor rises, inhibitory action on the bacteria increases, degradation becomes still slower, the adsorption sites become further clogged, further increasing the dissolved surfactant concentration, and so on. Even if the surfactant is not toxic to the bacteria a similar vicious circle may occur if the foaming threshold is reached; the foam will remove sludge from the mixed liquor, thereby further lowering the degradation rate.

Catastrophic upsets of this sort can be avoided by operating with surfactant levels nearer or below the toxic or foaming limits, say at 5 to 10 ppm in the feed, such as met in actual sewage. Much higher levels, say 100 ppm, are often desired in the laboratory to achieve greater analytical accuracy if the nondegradable fraction is small, or in the study of intermediate degradation products. Smooth operation at the higher levels for long periods of time, months, is readily attainable at the price of a little extra attention to the daily details of operation to ensure constant conditions.

F.7. *Adsorption Effects*

Surfactant adsorbed onto the activated sludge is a factor to be considered in drawing up a complete material balance in assessing biodegradation in activated sludge. Ordinarily adsorption does not introduce a serious error into such calculations if neglected, but this should be verified in each new situation.

As an example, Table 5.10 gives measurements made on an SDA

TABLE 5.10

Adsorption and Biodegradation of LAS on
Activated Sludge[a]

	A 23 hr	B 2 min	C 23 hr
MBAS in solution, ppm	0.2	1.7	0.2
MBAS adsorbed, mg/gm	1.1	6.5	0.3
MLSS, ppm	3700	3700	3900
MLSS, gm	5.55	5.55	5.85
MBAS in solution, mg	0.3	2.6	0.3
MBAS adsorbed, mg	6.1	36.0	1.8
MBAS total, mg	6.4	38.6	2.1

[a] Swisher (1964c).

semicontinuous unit in steady state degrading LAS. Column A shows the situation just before feeding. The liquid phase had 0.2 ppm MBAS in solution, corresponding to 0.3 mg dissolved in the 1.5 liters. Analysis of a sludge sample showed adsorbed MBAS of 1.1 mg/gm, for a total of 6.1 mg on the 5.55 gm of sludge in the system. After settling and removal of effluent, 1 liter of fresh feed was added, mixed in, and the system was quickly sampled for the analyses shown in column B. The liquid phase did not contain the 20 mg which had just been added, but only 2.6 mg; 17.7 mg of LAS had disappeared from solution within 2 min.

The sludge analysis shows that the missing LAS has been adsorbed. We now find 6.5 mg/gm on the sludge, corresponding to 36 mg, an increase of 30 mg over column A. This is not a very precise material balance because of interferences and resulting poor precision in the methylene blue analyses, particularly on the extract from the sludge. But the trend is quite unmistakable, and well beyond any analytical uncertainties. The next day at the end of the cycle (column C) the adsorbed LAS has dropped back to 1.8 mg, not significantly different from the 6.1 mg at the end of the previous cycle (considering the rather poor analytical precision) and the dissolved LAS is down to 0.3 mg, same as before. The 20 mg of added LAS is now neither in the solution nor on the sludge. It has been degraded.

Such phenomena are further illustrated by Tomiyama (1968), not only for TBS and LAS but also for AOS, in highly concentrated (4000 to 5000 ppm) cultures of *Pseudomonas* K5, *Escherichia coli*, and *Bacillus subtilis*.

The sequence of adsorption and degradation shown in Table 5.10 is repeated each cycle. In steady state the main body of sludge has become saturated with surfactant in adsorption equilibrium with the amount of undegraded surfactant remaining in the liquid phase and is unable to adsorb any greater amount, at the end of each cycle. Thus adsorptive removal of surfactant from any new increment of incoming feed can occur only to the extent that new sludge is grown from the foods in that increment of feed, and ordinarily this is very small.

For example, a liter of feed containing 400 mg of foods would grow at most only about 200 mg of new sludge. If the feed contains 20 ppm of surfactant, 95% degradable, the effluent will contain 1 ppm, and the sludge at adsorptive equilibrium will contain perhaps 1 mg/gm (Fig. 4.2). This amounts to 0.2 mg on the entire 0.2 gm of newly grown sludge, or 1% of the surfactant fed, leading to a corrected degradation figure of 94%. The higher the ratio of surfactant to normal foods in the feed, the lower will be the correction for adsorption (if other factors remain unchanged), since relatively less new sludge will be grown.

Even in the extreme case of zero biodegradation and high adsorption the surfactant content of the sludge would not be likely to exceed 10 mg/gm in equilibrium with 20 ppm in solution (Fig. 4.2). Here the 0.2 gm of newly grown sludge would take up only 2 mg of the 20 mg of surfactant fed, or 10%. Even that much is not very important in many circumstances, since the question of whether a particular surfactant is degraded 10% or zero would usually be an academic one.

As indicated at the end of Chapter 4, Section V.D, most of the workers who have examined the question have concluded likewise that adsorption corrections are usually quite minor.

VIII. THE SOIL—DRAINAGE FIELDS AND LYSIMETERS

A large fraction of the biodegradation which occurs on the land area of our planet is accomplished in the soil. Such action is, of course, an essential part of the carbon, nitrogen, and other cycles. Dead organisms, animal, vegetable, or microbial, are converted thereby back to their component parts, making them available for reuse by the living. The soil, with associated organisms, constitutes a very powerful degrading agent indeed—if given time in which to work.

Soil is by far the most important component of a septic tank sewage treatment system, so far as biodegradation is concerned. The septic tank itself is predominantly anaerobic; biodegradation is much slower under anaerobic conditions than aerobic, and little is accomplished during the few days retention of the sewage, so far as the soluble components are concerned. The principal action of the tank is the settling of insolubles, which subsequently must be removed periodically by mechanical means. The supernatant flows from the tank into the underground drainage field or tile field where, providing it has been properly constructed, the actual degradation of the soluble and unsettled sewage components takes place under aerobic conditions.

The field is a series of horizontal channels or beds lying a few inches below the surface of the soil, filled with gravel or similar material and covered by a top layer of soil. Air diffuses down from the surface, and aerobic bacteria in the soil and in the sewage oxidize the organic content of the sewage as it flows intermittently from the tank, along the channels, and eventually into the surrounding soil.

The soil lysimeter is used for experimental studies of the physical and chemical processes which may take place in a drainage field. It constitutes a cross section of one of the channels and the surrounding soil, with provision for introduction of feed at the top and collection of effluent from the bottom. The lysimeters used for study of surfactant biodegradation by Robeck (1963, 1964), Klein (1964a, 1965b), and Kempf (1968) were cylindrical or rectangular vessels several feet across by several feet deep, filled with sand or sandy soil. A pocket or channel at the top with dimensions of about 6 to 12 inches is filled with gravel or broken stones, simulating the channel of the drainage field. Influent is introduced into this pocket in individual doses of 10 to 20 liters during 1 to 20 min, repeated up to 6 times per day.

Useful information on biodegradation in soil may also be obtained with much smaller equipment such as the soil columns 2.5×25 and 3.8×78 cm used by Klein (1961, 1962, 1963a, b, 1964b).

Aerobic conditions are essential in all of these systems if satisfactory biodegradation is to be achieved. This is accomplished by feeding the influent intermittently with intermediate rest periods of several hours, allowing air to diffuse through the soil along with the percolating liquid. This mode of operation is called *unsaturated* flow, as opposed to saturated flow conditions wherein the column is flooded (saturated) with liquid at all times and air has no opportunity to penetrate. With saturated flow conditions biodegradation, including biodegradation of surfactants, is impaired considerably.

Development of suitable biological growth in the soil column is also essential; several weeks or months of maturation may be required before full efficiency is reached, depending in part upon the nature of the soil used. Robeck (1963, 1964) has explored the effects of soil varieties and operating variables upon lysimeter performance, particularly with regard to TBS removal (in which steady state values ranged from 40% to over 95%).

IX. ANAEROBIC SYSTEMS

A. Anaerobic Metabolism

Biodegradation is very much slower under anaerobic conditions than when there is an abundance of air; the reactions are thermodynamically less favored and the processes are less efficient. Nevertheless, two important areas of sewage treatment involve anaerobic systems (septic tanks and anaerobic digesters) and study of surfactant behavior under related conditions is likewise important.

The effects of surfactants upon the functioning of such systems have been extensively investigated, principally by measuring concentration thresholds for interference with gas production, since methane and carbon dioxide are the normal metabolic end products of anaerobic biodegradation. However, surfactant biodegradation deals with the converse, the effects of the systems on the surfactants. This can be studied by use of suitable analytical techniques, just as in the aerobic systems.

B. Die-Away Procedures

The simplest anaerobic procedure is merely to prevent access of air to the medium used in a die-away system. If food is present in sufficient amount, the initial dissolved oxygen will soon be consumed in aerobic biooxidation processes and the system will become anaerobic. Wayman (1963a) used a river water medium—actually consisting principally of primary sewage effluent—containing 10 to 25 ppm of surfactant, stored in a Brewer anaerobic jar for periods up to a month or two. Vath (1964) exposed 20 to 100 ppm of surfactant to settled sewage in amber bottles for 2 weeks. Klein (1965a) also used settled sewage as the medium: 2 liters with 25 ppm of surfactant, standing quiescent in a graduated cylinder over a period of 40 days.

Manganelli (1960) used a medium more similar to the contents of an anaerobic digester—a mixture of activated sludge and digester sludge containing around 30,000 ppm suspended solids, with provision for measuring gas evolution over a period of 40 days. Surfactant was added at 100 to 750 ppm, corresponding to about 0.5% to 2% of the suspended solids. Surfactant degradation was determined by analysis at the end of the run. Meinck (1961) and Pitter (1964b, d) used similar systems.

C. Semicontinuous Operation

Closer simulation of field conditions in septic tanks or anaerobic digesters is achieved by operating on a semicontinuous basis. Periodically a certain fraction of the reactor contents is removed and a like amount of fresh feed is added. The size and frequency of these increments determine the average retention time in the reactor.

Simulated septic tanks usually have a working volume of 3 to 4 liters, although Klein's (1964a, b, 1965b) ranged from 35 liters to 35 gallons. They have been fed whole sewage (Straus, 1963; Klein, 1964a, b, 1965b), or Butterfield (Weaver, 1962) or official German (Lashen, 1966) feeds (Table 5.3). Retention times have ordinarily been 1 to 5 days, except for Weaver (1962) who used 33 days. Surfactant content of the feeds has ranged from 10 to 30 ppm.

Laboratory anaerobic digester operations have been reported by Hurley (1952, p. 330), Raybould (1956), Johnson (1958), Maurer (1965), and Bruce (1966). Working volumes ranged from $\frac{1}{2}$ to 3 liters with retention times usually about 30 days. The initial charge was principally sludge from an operating anaerobic digester to give a good start, with subsequent feed being primary sewage sludge with surfactant level at 200 to 1200 ppm.

D. Natural or Synthetic Feed?

No detailed studies have been made on these biological systems in conjunction with surfactant biodegradation work. Gaudy (1966) has briefly summarized the biochemistry of anaerobic digesters.

Hattingh (1967) has detailed the biological and chemical changes occurring in an anaerobic digester during and after a 9 week acclimation to a nonsurfactant synthetic substrate, compared to a parallel unit fed raw sewage sludge. The synthetic feed contained dextrin 2.5%, sucrose

TABLE 5.11

Surfactant Biodegradation Field Tests

Location	Treatment method[a]	Rate tgd[b]	Population	Duration	Surfactant[c]	Introduction[d]	Reference
Wolverhampton (Coven Heath)	TF	450	—	15 mo	Teepol	Spi	Hurley, 1950, 52
Wolverhampton (Barnhurst)	AS	—	—	7 mo	Teepol	Spi	Hurley, 1952
Kettleman, Calif.	AS	10	100	10 days	TBS	RA	House, 1956
Wolverhampton (Coven Heath)	TF	650	—	9 mo	KBS	Spi	Raybould, 1956
Wolverhampton (Barnhurst)	AS	350	—	18 mo	K, T	Spi	Raybould, 1956
Wolverhampton (Barnhurst)	TF	5	—	12 mo	K, T	Spi	Raybould, 1956
Kingswood, England	TF	3	—	4 mo	K, T	Sub	Raybould, 1956
Luton, England	AS + TF	9000	110,000	2 yr	LAS	Sub	Eden, 1961a[e]
Munich (Grosslappen)	AS	20	—	5 yr	L, T	Spi	Scherb, 1962; Huber, 1962, 68
Marl-Ost, Germany	TF	—	40,000	1 mo	LAS	Spi	Jendreyko, 1962
Essen (Steele-Haferfeld)	TF	—	3,000	7 days	LAS	Spi	Jendreyko, 1963
Soest, Germany	AS	1400	42,000	18 mo	LAS	Spi	Jendreyko, 1963
Vossenbos, Holland	AS	40	1,000	18 mo	A, L, T	Sub	de Jong, 1964, 65
Brookside Estates, Ohio	AS	28	300	1 yr	L, T	Sub	Hanna, 1964
Dissen, Germany	AS + TF	500	—	15 mo	A, L, T	Spi	Spohn, 1964a
Friedrichsheim-Luisenheim	TF	65	—	7 mo	LAS	Sub	Spohn, 1964a
Hamburg (Volksdorf)	TF	400	6,000	8 mo	A, L, T	Spi	Spohn, 1964b

Location	Process[a]	Flow[b]		Duration	Surfactant[c]	Method[d]	Reference
Hamburg (Wensembalken)	AS + TF	100	3,000	13 mo	L, T	Spi	Spohn, 1964b
Woodbridge, Virginia (Elm Farm Park)	AS	—	400	6 mo	LAS	Sub	Renn, 1964a, 65b
Woodbridge, Virginia (Elm Farm Park)	AS	—	—	4 mo	SE	Sub	Conway, 1965
New Lisbon, New Jersey	TF	140	1500	7 mo	A, AK, L, T	Spi	Kelly, 1965
Manassas AFB, Virginia	AS	6	200	3 mo	LAS	Sub	Knapp, 1965
Plymouth, Wisconsin (Kettle Moraine)	AS	40	500	3 mo	L, T	Sub	Knopp, 1965
Johannesburg	TF	2000	—	1 mo	L, T	Spi	Urban, 1965
Johannesburg	Pond	600	—	2 mo	L, T	Spi	Urban, 1965
New Lisbon, New Jersey	TF	125	1500	6 mo	LAS	Spi	Kumke, 1966
Grossraschütz, E. Germany	AS	24	—	6 mo	A, AK, L	Spi	Walther, 1966
Preston, Herts.	TF	—	100	2 yr	LAS	Sub	Mann, 1968f
Montgomery Co., Pa. (Gwynedd-Mercy)	AS	25	1000	3 mo	OPE$_{10}$	Spi	Lashen, 1967b, c
Sandbach, Germany	TF	20	400	3 mo	AK, L	Sub	Krone, 1968

[a] AS, activated sludge; TF, trickling filter.

[b] Sewage flow, thousand gallons/day; 1 tgd = 3.8 m^3/day.

[c] Surfactants used; A, linear primary alkyl sulfate; AK, linear secondary alkane sulfonate; K, KBS; L, LAS; SE, linear secondary alcohol ethoxylate and ethoxylate sulfate; T, TBS.

[d] Method of introducing surfactant; RA, radiotracer; Spi, spiking; Sub, substitution

[e] See also STCSD (1960, 61, 62).

[f] See also WPRL (1966a, p. 129; 1967, p. 154; 1968, pp. 133–134).

3.5%, nutrient broth 1%, casamino acids 0.65%, plus minor amounts of various lower fatty acids and mineral salts. The 100 liter digesters contained 60 liters of mixed liquor and were fed 3 liters/day (after removal of 3 liters), giving a retention time of 20 days. There was no significant difference between the two effluents, liquid phase, as determined by analysis for alkalinity, volatile fatty acids, nitrogen, phosphorus, COD, and carbohydrate. However, the sludge phase of the synthetic-fed unit increased markedly in content of living cells, presumably because the inert solids of the raw sewage sludge were not entering. The bacterial species distribution changed. The content of protease and amylase enzymes increased, possibly in response to the nutrient broth and dextrin content of the synthetic feed.

The $CH_4:CO_2$ mol ratio in the gas mixture from the raw sludge unit remained around 1.9, whereas in the synthetic unit it declined to around 1.6. At the same time the efficiency of conversion rose from around 25% to around 70% in the synthetic unit; since only some 2% of the influent COD appeared in the effluent liquid phase, the missing 28% would appear to have gone into cell synthesis. However, the method of calculating the efficiency of conversion is open to some question on theoretical grounds (although the inaccuracy introduced thereby is probably insignificant). It is taken as mols of $CH_4 + CO_2$ gas mixture per "mol" (32 gm) of COD, but 32 gm of COD can represent anywhere from 0.5 to 2 or more atoms of carbon, depending on its oxidation state in the compound under consideration.

In any case, this study supplies baseline information of the sort necessary if degradation of surfactants in such systems is to be studied.

X. MEGATESTS AND FIELD TESTS

A. Advantages and Uncertainties; Analytical Interferences

The uncertainties of laboratory evaluation of extent and rate of biodegradability being what they are, it is not surprising that verification has often been sought through larger scale experiments using full size sewage treatment facilities. Results from such tests do carry a certain impact and convincingness beyond that which can be achieved in the laboratory, but are nevertheless equally subject to error. The operations must be planned and executed with great care if the results are to have any general validity. Although the expense is great, many such tests have been run. Some 30 are listed in Table 5.11.

In the usual municipal or private sewage treatment plant the researcher is not free to select and maintain the full range of operating conditions that might be desired in a detailed study. Such control can be exercised if a research-oriented facility is available. As an example the installation of the Bayerische Biologische Versuchsanstalt at Grosslappen near Munich may be cited. Huber (1968) summarizes their surfactant biodegradation experiments extending over a period of 5 years in an oxidation ditch designed to treat the sewage of about 500 people. This type of activated sludge system is usually operated with retention times of several days.

In the more usual situation a major source of difficulty is the inherent variability of the operating parameters of the treatment plant, at best under only partial control. The influent flow rate and composition may vary widely from hour to hour, day to day, and month to month. The treatment efficiency may vary widely also, for reasons known or unknown, uncontrolled or uncontrollable. Under such circumstances removal of the test compound may likewise vary, in which case meaningful results can be obtained only by observation over long periods of operation. Further, it is necessary to compare surfactant removal efficiency with the plant's efficiency in removing the more normal components of the sewage, usually averaged together in terms of BOD.

Beyond the instabilities experienced in treatment plant operation, a second major difficulty is analytical interference by surfactants present in the incoming sewage. If the test surfactant is of a new type easily distinguished from those already present by a specific analytical method, there is no trouble. But in the days when LAS was being developed the sewage contained TBS at comparable concentrations, while the analytical method of choice, because of its speed and sensitivity, was methylene blue. It could not differentiate between TBS and LAS. Differentiation could be accomplished by the IR or desulfonation–GC methods, but they were not developed until fairly late in the game, and besides they were prohibitively slow and expensive.

House (1956) demonstrated that radiotracer techniques were the ideal answer to this problem; in fact, even TBS degradation could be studied without interference from the "natural" TBS in the influent. Yet this method has not been used in field tests by others, perhaps because of unavailability of suitably trained personnel or facilities, or perhaps from concern about introducing radioactivity into the environment. Instead, interference has been handled in two less satisfactory ways, by substitution or by spiking.

B. The Substitution Technique—The Luton Test

The substitution technique requires replacement of all detergents used within the service area of the sewage treatment plant, replacement by similar formulations containing the test surfactant. This procedure was used in the granddaddy of all field tests, at Luton, England. Without any publicity, all stocks of TBS detergents in all stores in a wide area were replaced and maintained with new formulations based on the early LAS from Dobane JN (STCSD, 1960, 1961, 1962; Bolton, 1961; Eden, 1961a, b; Hammerton, 1961; Key, 1961; Squire, 1961).

A marked reduction in the MBAS content of Luton's treated sewage was achieved, but the improvement fell short of what had been anticipated on the basis of laboratory and pilot plant tests. The discrepancy was traced to the diurnal variation in the MBAS level entering the plant, a factor which was later shown in the laboratory to be detrimental to the biodegradation of the JN LAS ("square wave loading," see Section VII.F.4). With the complete conversion of the British detergent industry a few years later from TBS to LAS (a modern grade with lower content of branched impurities) the final results met expectations very well (Waldmeyer, 1968; WPRL, 1968, pp. 132–133).

During the test period at Luton, 2 years or more, complete replacement of TBS by LAS was never attained. The maximum fraction of LAS observed in the Luton sewage, as judged by the IR analytical method, was around 75% to 80%, reached after about 2 years. Contributing factors in the failure to reach 100% replacement would include (i) purchase of TBS products at distant locations, (ii) failure to completely exclude TBS from Luton retail detergent supplies, (iii) slow release of adsorbed and trapped TBS from sewer lines, and (iv) partial degradation of LAS before arrival at the sewage treatment plant.

More complete substitution has been effected in many of the tests listed in Table 5.11 because they were carried out in smaller, more isolated communities, making possible a more direct control over the detergent products used. Even so, difficulties were sometimes encountered. Thus de Jong (1964, 1965) describes continued release of adsorbed TBS during a period of over 3 months, from sludge held up in the system even after as complete a cleanout as practicable. In the opposite direction, Knapp (1965) encountered extensive degradation of the test surfactant (LAS) in the sewage lines before reaching the treatment plant, and had to resort to spiking to get quantitative data on the extent of its degradation.

C. The Spiking Technique

In spiking, the test surfactant is added continuously to the sewage in known amounts just before entering the treatment operation. Analyses are run on the stream (i) before the addition, (ii) after the addition before treatment, and (iii) after treatment. In the old days the usual situation involved TBS in the incoming sewage, LAS or some other anionic under test, and methylene blue as the only economically feasible analytical method. In such a case, the total MBAS removal during the treatment includes that of the "natural" MBAS (mostly TBS) as well as of the added test surfactant. The former is assumed to be the same as determined in blank runs made immediately before and/or after the test run; the degradation of the test compound is calculated by difference. The degradation values for the "natural" MBAS measured in the blank runs are quite variable, and the blanks should be run for a week to a month to get representative values. The trouble is that the average may change considerably from one blank run to the next, and this uncertainty is carried over, magnified, to the final value calculated for the degradation of the test compound.

For instance, Spohn (1964b) reports that in the second Wensenbalken series the three blank runs gave average MBAS removals of 26.5%, 44.4%, and 34.0%, grand average 34.3%. Calculated on the two extreme values, the LAS (Marlon BW1033) biodegradation was either 95.3% or 80.3%. To make matters worse, the degradation of spiked TBS in a companion run was almost as great, 93.0% or 75.4%. Spohn could not explain this surprisingly high result, but it does exemplify the difficulties inherent in field studies.

If analytical methods are available whereby the test surfactant can be readily distinguished from those already present in the incoming sewage, life is simpler. One needs wonder about the condition of the treatment system only for the test runs, not for the blank runs as well.

D. Field Tests Further Afield

Field tests outside of sewage treatment plants are rather rare. Urban (1965) describes work in a 6.6 mile canal (part of the Johannesburg treatment plant system, incidentally) with a flow-through time of 6 hr. This might be considered to model a short segment of a river. The spiking technique was used, with attendant uncertainties of TBS baseline, but the results indicated 35% degradation of LAS compared to 15% for TBS during passage. An associated pond system with 10 day retention time gave about 100% LAS removal compared to 22–35% for TBS, using a similar technique.

At Santee, California a series of recreational lakes was established in 1961, using water from an activated sludge sewage treatment plant. From the treatment plant the water flows through an oxidation pond and is then spread on a dry river bed. After percolation along the river bed for about $\frac{1}{2}$ mile, below the surface of the ground, the water arrives at the first of the four recreational lakes. The system was monitored exhaustively for 2 years (1962–1963), both biologically and chemically, and the MBAS results (Merrell, 1967, p. 56–57) in a sense represent field test results on TBS biodegradation. The influent to the treatment plant averaged 16.5 ppm, the effluent 8.5, or 49% removal. It was down to 6.7 ppm (59% removal) after the oxidation pond and about 3 ppm (82% removal) after percolation along the river bed. Upon the change from TBS to LAS by the U.S. detergent industry the oxidation pond effluent dropped to about 1 ppm; no later figures are presented.

After conversion of the detergent industries in Germany and the UK from TBS to LAS, there was indirect evidence suggesting significant removal of LAS in the sewer between home and treatment plant. To confirm this, the STCSD (1967, p. 7) undertook direct measurement using $LA^{35}S$ as tracer and $K^{82}Br$ as reference. During passage through 4.17 miles of the Hogsmill Valley sewer about 15% of the LAS disappeared; retention time in the trip was about 170 min.

Cesspool-groundwater tests were made in Nassau and Suffolk counties, Long Island, New York in 1965–1966. The substitution technique was used, to compare TBS, LAS, alkyl sulfate, and soap. An official report has apparently not yet appeared, but Cohn (1968) has given an informal summary. The MBAS removal by the cesspools appeared to be far from complete in most cases regardless of which surfactant was used. This is not surprising in view of the very limited biodegradation capacity of a cesspool in the absence of associated provision for aerobic action.*

Earlier, Page (1963) had surveyed wellwaters from a 60 square mile area near Denver which was irrigated with water from the South Platte River via irrigation ditches. The MBAS content of this water ranged from about 1 to 3 ppm, predominantly TBS introduced by a primary sewage treatment plant upstream. The wellwater showed a uniform MBAS content of 0.1 to 0.2 ppm regardless of the distance from the nearest ditch (20 to 7300 feet). Coliform bacteria counts in the wellwater ranged from 0 to 24/ml compared to 1000 to 100,000 in river and ditch samples. Removal of both MBAS and bacteria thus seemed to be substantially complete in the percolation through the 10 to 50 feet of sandy soil down to the water table. This suggests that ample aeration was available during percolation,

---

* The subsequent final report (Flynn, 1969) shows no significant difference between the four surfactants, and a sucrose ester product as well, in the ground-water pollution by the coliform bacteria or the organic matter (COD) of the sewage.

since indications are that degradation is impaired under the conditions of saturated water flow in which access of air is cut off (Section VIII).

E. From Field to Real Life

Despite their difficulties and uncertainties, we can nevertheless have some degree of confidence in the validity and applicability of field tests, at least if carried out in sufficient variety. The results of those on LAS have been in general agreement with predictions based on laboratory tests, and, more important, have been in agreement with the results of the change to LAS from TBS by the detergent industry. Those results, although rather slow to become evident in some cases, have been very satisfying. They are summarized in some detail for the UK by Waldmeyer (1968), for Germany by Husmann (1968) and Heinz (1968), and for the U.S. by Brenner (1968). (See also Chapter 1, Section I.C.4.)

In Japan, on the other hand, the early results have been most un-spectacular, judging by the report of Ōba (1968a). Rivers receiving un-treated sewage, sampled near the sewage discharge points, showed increasing amounts of undegraded LAS as conversion by the Japan detergent industry proceeded. Early in 1968 the MBAS levels in samples from these rivers ranged from 1 to 9 ppm, of which 20% to 40% was LAS. This approached the concentration found in the municipal sewage influents where there were treatment facilities—5 to 25 ppm, 30% to 40% LAS. Thus very little dilution of the sewage had taken place in the river, and presumably there was also little exposure time for biological action to occur before sampling. The portion of municipal sewage which did un-dergo treatment lost its LAS content thereby; the treated effluents showed 1 to 4 ppm MBAS, of which 0–5% was LAS.

XI. BIODEGRADATION POTENTIAL—COMPARISON OF METHODS

Even a cursory examination of the tables in Chapter 8 shows that the measured biodegradability of a particular surfactant may vary widely from one test method to another (sometimes, unfortunately, from one test replicate to another also). Obviously some test methods entail conditions more favorable for biodegradation than do others, or, in other words, some test methods have a higher *biodegradation potential* than others. The method with highest potential will be that with the most active and severe biodegradation conditions, and thus it will be the most lenient in passing a given surfactant on the score of biodegradability.

A qualitative estimate of the relative biodegradation potential of the various test methods is easily made on the basis of their primary bio-degradation of TBS. In general we find that the methods with lowest potential are those which use a low concentration of bacteria in a synthetic medium, such as the closed bottle test and the shake culture test. These ordinarily accomplish hardly any degradation of TBS, usually only 5% to 10% at best, even with prolonged exposure.

With much higher bacterial concentration, as in the batch or continuous activated sludge test, the biodegradation potential is much higher also, even when a synthetic medium is used. Here degradation of TBS is usually on the order of 50%. The river water test is also a high potential one; since its bacterial concentration is low, the high activity must be ascribed to the natural species distribution and the natural medium. TBS will often degrade to the extent of 65% to 75% in the river water test.

TABLE 5.12

Comparison of Biodegradation Test Methods[a]

	Percent removal				
	WPRL trickling filter	British STCSD aeration	German continuous sludge	SDA semi-continuous sludge	Recycle trickling filter
Dobane JNX LAS	92	89–91	93–94	96–97	90–93
Dobane JNQ LAS	96	95–96	97–98	98–99	96–99
Dobane 055 LAS	96	96–97	95–97	98	94–98
ABS-4[b]	90	28–42	66–68	98–99	96–98
Dobane PT TBS	72	15–22	34–36	75–76	86–91
Empilan KM9[c]	—	98	—	99	98–99
Empilan KM20[d]	—	98	—	—	96–99

[a] Truesdale (1968).
[b] "Slow to acclimatize."
[c] Linear primary alcohol E_9.
[d] Linear primary alcohol E_{20}.

The highest biodegradation potential is exhibited by soil systems with unsaturated flow conditions, as, for example, in the soil lysimeters described by Robeck (1963) and Kempf (1968). There, primary biodegradation of TBS was as high as 98%, and ultimate biodegradation was quite extensive also, judging by formation of inorganic sulfate.

Direct comparison of five test methods on LAS, TBS, and nonionics was made by Truesdale (1968) as shown in Table 5.12. Pilot scale trickling filters (15 × 180 cm) were taken as standard, representative of results to be expected in the field, and compared with the other four methods. The

TBS removals were surprisingly high in three cases. Relatively low bio-degradation potential is indicated for the British STCSD aeration test, which uses an inoculated synthetic medium, and the official German continuous activated sludge test. Furthermore, neither of these two showed the true degradability of the "slow to acclimatize" ABS-4. The SDA semicontinuous method and the recycle trickling filter both showed high biodegradation potential as judged by their performance on TBS. Similar indications are apparent in the results of Cook (1968) (Table 5.13),

TABLE 5.13

Comparison of Biodegradation Test Methods[a]

	Percent removal, MBAS			
	Dobane JNX	Dobane JNQ	Dobane 055	Difficult ABS[e]
Continuous sludge[b]	61 ± 5.2	66 ± 2.9	75 ± 5.0	34 ± 5.5
Slope culture[c]	74 ± 8.8	89 ± 1.6	0–66	20 ± 7.3
River water	88 ± 0.9	93 ± 0.6	96 ± 0.3	29 ± 1.9
Shake culture[d]	88	96	91	34
Semicontinuous sludge[d]	89 ± 0.4	96 ± 0.3	98 ± 0.3	70 ± 4.0
Recycle trickling filter	92 ± 1.6	96 ± 0.7	97 ± 0.4	83 ± 1.5

[a] Cook (1968).
[b] Official German.
[c] British STCSD aeration test inoculated from agar slant culture.
[d] SDA standard method.
[e] "Difficult to degrade."

who compared six test methods on the three Dobane LAS and on a "difficult to degrade ABS" (perhaps referring to difficulty of acclimation rather than of the degradation itself).

An earlier report on the Truesdale work at a preliminary stage (WPRL, 1966a, p. 131) is summarized in Table 5.14. Here the lower potential of the STCSD aeration test is further indicated by the alkylphenol ethoxylate results. Another very interesting point is that this test accomplished 44–56% degradation of TBS at that time compared to 15–22% in the later work. Although this difference may be within the expected scatter of results on a poorly degradable material, it is tempting to speculate that it may result from the elimination of TBS from British detergents at about

TABLE 5.14

Comparison of Biodegradation Test Methods[a]

	Percent removal			
	WPRL trickling filter	British STCSD aeration	WPRL continuous sludge	SDA semi-continuous sludge
Dobane JNX LAS	91	89	93	96
Dobane JNQ LAS	—	97	—	—
Difficult ABS[b]	94	33	93	97
Dobane PT TBS	74	44–56	75	77–79
Lauryl sulfate	—	99–100	—	100
Octylphenol E_x	61	49–50	—	47
Nonylphenol E_x (no. 1)	66	36	—	—
Nonylphenol E_x (no. 2)	72	44–50	—	—

[a] WPRL (1966a, p. 131).
[b] "Difficult to acclimatize."

that time. This might cause loss of TBS acclimation in the sewage treatment plants, with decreased effectiveness of the test inoculum if prepared by drying activated sludge therefrom.

Vaicum (1967) compared seven test methods, as shown in Table 5.15. A

TABLE 5.15

Comparison of Biodegradation Test Methods[a]

		Percent removal		
	Time	ABS[b]	AS[c]	SLS[d]
Inoculated BOD water[e]	?	30	94	100
Continuous sludge, natural sewage	6 hr	36	90	100
Continuous sludge, synthetic sewage	6 hr	29	89	100
Continuous sludge, official German	3 hr	23	98	100
Batch sludge, Romanian ISCH	3 days	35	98	100
Closed bottle (BOD)	5 days	17	100	—
Warburg	?	13	92	97

[a] Vaicum (1967).
[b] Commercial ABS (TBS?).
[c] Commercial alkyl sulfate.
[d] Pure sodium lauryl sulfate.
[e] Borstlap (1963).

commercial ABS, presumably TBS, was significantly lower in the two oxygen demand methods (13–17% degradation) than in the aerated systems (23–36%). The lower result in the official German test (23%) was attributed to the 3 hr retention time, compared to the other two continuous activated sludge methods. The Romanian ISCH test was judged to be the most rapid and convenient, and to give meaningful results.

XII. NONBIOLOGICAL METHODS

Biological test procedures such as those reviewed in this chapter are characteristically slow and are often expensive. In some cases the substitution of more rapid physicochemical methods is feasible. With sufficient knowledge of the biodegradation properties of a particular surfactant type, and with sufficient background experience gained from examination of a wide range of specimens, correlations may be developed between biodegradability and other parameters which may be measured more easily. These may then substitute for biological measurements in later specimens, with a minor degree of risk.

The risk is minimized and a certain degree of confidence can be enjoyed when the correlation is backed up with sound fundamental knowledge of the biodegradation chemistry of the particular surfactant type in question, the likely impurities which might accompany it when manufactured by the various possible routes, the selectivity and sensitivity of the physicochemical method chosen, and the like. On the other hand, a correlation based blindly on experimental results with no understanding of the underlying reasons is much more vulnerable to subsequent breakdown.

LAS would appear to be suitable for such an approach. Satisfactory biodegradability has been demonstrated repeatedly in the laboratory and, more important, in the field. Thus a given sample of LAS, or of its linear alkylbenzene precursor, would meet a biodegradability standard of 95% with reasonable certainty if it could be shown by physicochemical means to contain less than 5% of nonlinear impurity. Ordinarily at least part of the nonlinear impurity would be degradable also, providing an added margin of safety.

Accurate measurement of the total nonlinear impurity at such low levels is not easy—perhaps not as easy as running a biodegradation test. Even so, other correlations have been developed, such as those involving branching index or its equivalent, the average number of methyl groups per molecule (Chapter 6, Section II.C). De Jong (1967) reviews these possibilities, and in addition points out the particular utility of mass spectrometric information. The chain length distribution can be determined from the masses of the parent ions, while differences in chain structure are reflected by the relative amounts of certain ion fragments, **5.1**. Comparison of pure linear, slightly branched, and tetrapropylene alkylbenzenes showed mass 119 to be a sensitive indicator, the ratio of masses 119:91 increasing from 0.18 to 0.35 to 2.0 for the three types.

$$
\begin{array}{ccc}
 & \text{C} & \text{C} \\
\phi\text{C}^+ & \phi\text{C}^+ & \phi\text{C}^+ \\
 & & \text{C} \\
\text{Mass 91} & \text{Mass 105} & \text{Mass 119} \\
\textbf{(5.1a)} & \textbf{(5.1b)} & \textbf{(5.1c)}
\end{array}
$$

Such correlations should be quite valid in routine examination of production samples for control purposes, assuming, of course, that the samples fall within the range of structures examined in establishing the correlation in the first place. But if entirely new chain structures enter the picture, as by a new manufacturing process or new raw materials, biodegradability can no longer be estimated with confidence until the correlation has been reverified, or a new correlation developed, by comparison with biological results.

Van Cauwenberghe (1969) reports that much the same information results from pyrolysis GC of the desulfonated alkylbenzene, a quicker, cheaper technique than mass spectrometry. The amount of α-branching (at the chain carbon linked to the ring) can be estimated from the styrene: α-methylstyrene ratio in the pyrolyzate. Linear, slightly branched and tetrapropylene alkylbenzenes are clearly differentiated.

But it must be reemphasized that neither the mass spectrometric nor the pyrolysis GC correlations can be extended beyond their experimental foundations. For example, 2-methyl-2-phenylundecane (**7.16**) is completely branched at the α-position, yet its sulfonate is readily biodegradable. And its terminal QBS isomer (**7.17**) has zero α-branching but is nevertheless quite resistant.

CHAPTER **6**

CHEMICAL STRUCTURE
AND PRIMARY BIODEGRADATION

I. MOLECULAR STRUCTURE, CHAIN BRANCHING,
AND CHAIN LENGTH

In the beginning, the first great (also erroneous) generalization on surfactant biodegradation stated that alkyl sulfate is biodegradable and alkylbenzenesulfonate is not. This arose (and was subsequently first discredited) in England, where study of surfactant biodegradation had begun earlier than elsewhere because of several factors: the high population densities in England resulted in highly concentrated sewage and treated effluents, with correspondingly less effective dilution by the receiving waters; its highly developed sewage treatment plants and technology provided focal points for observation and also provided trained observers with long experience in tracing the short- and long-term variations so characteristic of treatment plant operations; its highly industrialized economy made conversion from soap to surfactants possible at a relatively early date.

The two major surfactants commercially used in England at that time, besides soap, were Teepol and ABS. The early biodegradation research indicated that Teepol, a linear secondary alkyl sulfate, was readily biodegradable. The ABS was much more resistant, at least on some occasions, and the above-mentioned erroneous conclusion followed naturally. Later the recognition gradually spread that "alkyl sulfate" and "ABS" were not chemical individuals but broad classes of complex mixtures, and that the rather erratic behavior of the ABS could be traced in part to its origin and chemical nature, that is, whether manufactured from kerosene or from tetrapropylene.

Bogan (1955) and Sawyer (1956) attributed the biological resistance of TBS to the branched structure of its alkyl groups, pointing out that linear primary ABS was, in sharp contrast, readily biodegradable.

Branched t-butylbenzene and linear n-butylbenzene, both unsulfonated, served as models with known structures to support that conclusion. Hammerton (1955, 1956) pointed out the same facts at the same time, with the further observation that alkyl sulfates were not necessarily easily degradable. For example, he found that the primary C_9 alkyl sulfate

$$\begin{array}{l} \text{C\ \ C} \\ \text{CCCCCCO}\Sigma \\ \text{C} \end{array}$$

showed 96% remaining after 3 weeks exposure in river water, whereas linear primary C_{10}ABS

$$\text{CCCCCCCCCC}\phi\Sigma$$

disappeared in 4 days. Hammerton suggested explicitly the first successful generalization of surfactant biodegradation, that the important factor is the linearity of the hydrophobic group, and that the chemical nature and mode of attachment of the hydrophilic group were of only minor significance: linear surfactants are readily biodegradable, highly branched ones are not. Lacking model surfactants having a wide range of hydrophobe structures with which to further check this hypothesis, he turned to BOD determinations on various C_4 to C_9 alcohols and some other compounds for supporting evidence. The results clearly showed increased resistance to biodegradation with increased branching.

Subsequent research through the years, employing an ever widening variety of model surfactants synthesized with known structure, has continued to verify this generalization. Although the effect of a single methyl branch in an otherwise linear molecule is barely noticeable, increased resistance with increased branching is generally observed, and resistance becomes exceptionally great when quaternary branching occurs at all chain ends in the molecule.

A corollary of this hypothesis is that the chemical nature of the hydrophilic group is of only minor importance in affecting biodegradability, and this too has been borne out in large part. Such is generally evident in the biodegradation tables in Chapter 8, and more specifically so in those studies making side-by-side comparisons of the various surfactant types. Table 6.1 lists many of the latter, and their results are for the most part quite consistent with this thesis.

Nevertheless, two effects attributable to the hydrophile structure are apparent among the much wider variations contributed by hydrophobe structure and by differing biological conditions. First, the most rapidly degraded of the alkyl sulfates, the linear primary (LPAS), are significantly faster than any other anionics, including the best of the alkane or

TABLE 6.1

Comparative Biodegradation Studies on Different Surfactant Types

	Bogan, 1954, 55	Hammerton, 1955, 56	Sawyer, 1956	Barden, 1957	Ryckman, 1957	Huyser, 1960	Winter, 1962	Ruschenberg, 1963a, b	Pitter, 1963b, 64d	Berger, 1964	Cordon, 1964	Čuta, 1964	Crauland, 1964	Kölbel, 1964	Sweeney, 1964a	Weil, 1964	Maurer, 1965	Oba, 1967	Tomiyama, 1969	SDA, 1969a, b
ABS	×	×	×	×	×	×	×	×	×		×	×	×	×	×	×	×	×	×	×
Alkylbiphenylsulfonate														×						
Alkylnaphthalenesulfonate		×		×		×	×					×		×						
Alkane sulfonate		×					×	×					×	×		×		×	×	
AOS											×								×	
Sulfo fatty acids and derivs	×	×	×		×		×									×	×			
Acyl isethionate	×	×	×	×	×		×			×						×				
Acyl tauride										×		×				×				
Sulfosuccinate esters		×		×						×										
Acylanilinesulfonate		×		×								×								
Alkyl sulfate	×		×	×	×	×	×	×	×		×	×	×			×	×		×	
Alcohol-EO sulfate													×			×				
Alkylphenol-EO sulfate							×									×				
Alkylgluconamide sulfate												×				×				
Alkyl phosphate																				
Soap	×		×	×		×	×	×		×	×	×								×
Alcohol-EO	×		×	×	×	×	×			×		×								×
Alkylphenol-EO	×		×			×				×		×								
Acid-EO				×								×								
Amide-EO																				
Amine-EO							×					×								
Acylsorbitan-EO																				
Benzyl-β-naphthol-EO												×								
Acyl sucrose																				×
Sulfonyl polypeptide																				
Amine oxide							×													×
Quaternary ammonium			×	×			×													

205

alkylbenzenesulfonates. Thus in a river water test one day or less is usually sufficient for disappearance of methylene blue reaction (MBAS) in the case of LPAS, compared to several days for the others; Ruschenberg (1963a, b) reports similar differences in inoculated BOD water. This lability cannot be considered as the result solely of the primary sulfate ester linkage since certain primary alkyl sulfates with branched hydrophobes may resist degradation for very long times, as mentioned above.

The second case where hydrophile structure has a noticeable influence on biodegradation is in the nonionic polyethoxylate group. Bogan (1954, 1955) and Sawyer (1956) found that fatty acid and fatty amide derivatives with 50 mols of EO were relatively resistant in respirometric studies, whereas the E_5 derivatives were much more readily attacked. They attributed the difference to poorer diffusion of the larger molecules through the cell membranes. Whatever the reason, the increased speed of degradation with decreased EO chain length has been amply confirmed by subsequent workers (Section IX.A; Table 6.19). Here too the effects of unfavorable (for biodegradation) hydrophobe structure often can far outweigh these relatively minor differences in rate related to degree of ethoxylation.

Both hydrophobic and hydrophilic groups are involved in a generalization first enunciated by Huddleston (1963) and Swisher (1963a). It derives from studies of LAS biodegradation—the effects of chain length and phenyl position along the chain—but it can be stated in more general terms: the greater the distance between the sulfonate group and the far end of the alkyl group, the more rapid will be the biodegradation. Here again the differences are rather small in comparison with other structural factors such as linearity, or with uncertainties of biological nature which may enter in comparing one test run with another. Accordingly the effect is most convincingly demonstrated by comparison of the surfactants mixed together in a single medium where they share exactly the same environment.

This requires analytical methods specific for each component of the mixture, met in the case of ABS by the desulfonation–gas chromatography technique. Such studies have led to verification for a wide range of LAS and other ABS structures. However, applicability of the generalization to other anionics or to other surfactant types such as nonionics has not been demonstrated unequivocally as yet.

A suggested mechanism explaining the effect is fixation of the sulfonate group onto the oxidative enzyme at a spot distinct from the site at which oxidation is initiated at the chain end. Increased distance between sulfonate and chain end would facilitate proper orientation of both on the enzyme (Swisher, 1963a). No work has been reported (or done?) toward testing this hypothesis.

In summary, the following generalizations relating surfactant structure and biodegradation have been advanced and are reasonably well validated:

(i) Structure of the hydrophobic group is a very important determinant of biodegradability; biodegradation is promoted by increased hydrophobe linearity and deterred by hydrophobe branching, particularly by terminal quaternary branching.

(ii) The nature of the hydrophilic group has only a minor influence on the biodegradability. The clearest examples of such influence are seen in (a) LPAS, which undergoes primary biodegradation significantly faster than other anionics, and (b) the polyethoxylate nonionics, where biodegradation is promoted by shorter EO chain length.

(iii) Increased distance between the sulfonate group and the far end of the hydrophobe group increases the speed of primary biodegradation of ABS and possibly of other surfactant types. It will be convenient to call this the *distance principle*.

II. ALKYLBENZENESULFONATES

The profound influence of alkyl chain structure on biodegradability of the ABS surfactants was clearly evident in the early comparisons of TBS with various other derivatives having linear alkyl groups, reported for example by Bogan (1955), Hammerton (1955, 1956), Sawyer (1956), and Ryckman (1957). Subsequently many pure ABS compounds with alkyl groups of known structure, both linear and branched, were prepared by the classical methods of organic synthesis and their biodegradation properties were determined. Over 50 such compounds are listed in Table 8.1, and these have been a major source of information on the relation of surfactant structure to biodegradation.

A. Phenyl Position

Application of desulfonation–gas chromatographic analysis by Huddleston (1963) in the biodegradation of LAS mixtures immediately disclosed a relationship that has subsequently been verified many times: the isomers with the phenyl attached near the end of the chain disappear somewhat more rapidly than those with more central attachment. Figure 6.1 shows this for a mixture of C_{12} and $C_{14}LAS$, where the degradation

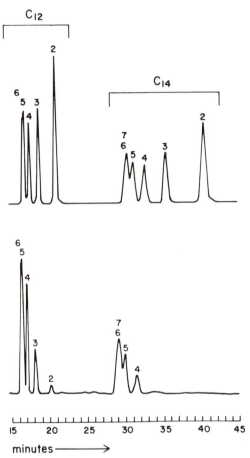

Fig. 6.1 Biodegradation of $C_{12} + C_{14}$ LAS mixture in river water, detected by desulfonation–GC (Swisher, 1963b). Upper: Day zero, 5 ppm MBAS. Lower: Day 15, 2.5 ppm MBAS.

rate is clearly dependent upon the position of the phenyl group along the alkyl chain.

Quantitative data calculated from the chromatographic peak areas in Fig. 6.1 are presented in Table 6.2. The original 5 ppm of MBAS had decreased to 2.5 ppm at 15 days, but at that time only 3% of the initial 2-ϕC_{12} remained, and 38% of the 3-ϕC_{12}, whereas essentially all of the original 5- and 6-ϕC_{12} was still present. (The value of 131% remaining indicated in Table 6.2 results from small analytical inaccuracies in the

MBAS and peak area determinations; it is not likely that these isomers could have been actually formed during the degradation.) A similar pattern is shown by the C_{14} isomers.

Behavior of the C_{12}LAS isomers in individual river water tests is shown in Fig. 6.2, manifesting the same pattern. Setzkorn (1964) obtained similar results with all six of the phenyldodecane sulfonate isomers. So too did Ruschenberg (1963a, b) with isomers of the even carbon number homologs

TABLE 6.2

River Water Biodegradation of C_{12} + C_{14}LAS Mixture[a]

Days	0	15	15	17
MBAS, ppm	5.0	2.5	2.5	1.6
Component	Composition, %		Fraction remaining	
2-ϕC$_{12}$	18.7	1.2	0.03	0
3-ϕC$_{12}$	9.6	7.2	0.38	0.04
4-ϕC$_{12}$	7.4	17.2	1.16	0.77
5,6-ϕC$_{12}$	12.7	33.4	1.31	1.31
Total C$_{12}$	48.4	59.0	0.61	0.47
2-ϕC$_{14}$	17.9	0	0	0
3-ϕC$_{14}$	9.0	0	0	0
4-ϕC$_{14}$	6.4	6.4	0.5	0
5,6,7-ϕC$_{14}$	18.3	34.6	0.95	0.51
Total C$_{14}$	51.6	41.0	0.40	0.18

[a] Swisher (1963b).

from C_8 through C_{16}; in each case the 2-phenyl isomer degraded more rapidly than the central isomer. It must be noted that comparisons of this sort, using the pure compounds separately, are not necessarily valid. The individual solutions do not constitute identical environments, and hence may not be truly comparable. However, in the present cases the results are all in agreement with those where all components are present together in a single solution and analyzed by desulfonation–gas chromatography.

Speedier degradation of each isomer than of the successively more internal ones occurs with all LAS homologs from C_6 to C_{16}, and probably beyond. This illustrates the validity of the distance principle, that speed of biodegradation increases with increased distance between sulfonate group and remote chain end. It appears that the sulfonate group itself is specifically involved rather than simply the position of attachment of the

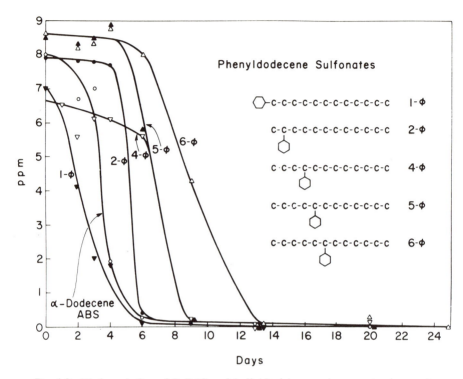

Fig. 6.2   Biodegradation of $C_{12}LAS$ and individual isomers in separate river water tests, detected by MBAS (Swisher, 1963b).

ring to the chain. This is indicated by the slower degradation of the 1-phenyldodecane and-tetradecane *ortho*-sulfonates compared to the corresponding para isomers (Huyser, 1960; Swisher, 1963b) and particularly by the linear diheptylbenzenesulfonate isomers (Swisher, 1963a), discussed in the following paragraphs.

These linear diheptylbenzenes were synthesized by alkylation of benzene with a 2 mol ratio of 1-heptene, giving a mixture of 6 meta and 6 para isomers. The 6 ortho isomers were not formed to any great extent, possibly because of steric hindrance. The 12 mixed isomers were converted to the corresponding mixed sulfonates, and these were found to degrade about 60% in the 24 hr semicontinuous activated sludge test as determined by MBAS loss. Relative degradation rates of the 12 individual isomers were determined by desulfonation–GC of the evaporated effluent. Figure 6.3 shows the gas chromatograms and Fig. 6.4 the percent of each

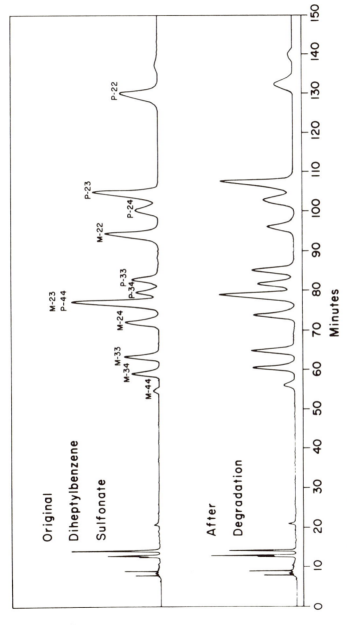

Fig. 6.3 Biodegradation of linear diheptylbenzenesulfonate mixture in 24 hr semicontinuous activated sludge, detected by desulfonation–GC (Swisher, 1963a). Upper: Time zero, 25 ppm MBAS. Lower: Time 23 hr, 10 ppm MBAS.

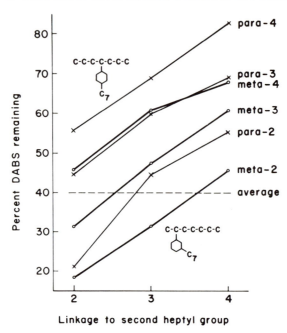

Fig. 6.4 Biodegradation of linear diheptylbenzenesulfonate mixture; amount of each isomer remaining in 24 hr activated sludge effluent expressed as percent of the amount in the influent (Swisher, 1963a).

isomer remaining after the 24 hr treatment. The data are in complete agreement with the distance principle. (Actually this is not overly surprising, since the principle was based in part upon these experimental results.)

The *meta*-2 family of isomers shown in Fig. 6.4 includes the three *meta*-diheptyls with the ring linked to the 2-carbon of one chain and to the 2-, 3-, or 4-carbon of the other chain. Lower speed of degradation when the latter link is closer to the center of the chain is quite evident, as with the other five families of isomers as well. Further, each meta family is somewhat more rapidly degraded than is the corresponding para family.

meta-2,3
(6.1)

para-2,3
(6.2)

This suggests again that the sulfonate distance is the underlying factor rather than the position of the ring–chain link; although the latter are in the same positions, the distance from the sulfonate to the most remote chain end will always average one carbon further in a meta isomer (6.1) than in the corresponding para isomer (6.2).

B. Chain Length

B.1. *LAS: Inhibition—Acclimation—Degradation*

Comparison of the C_{12} with the C_{14}LAS in Fig. 6.1 and Table 6.2 shows that the C_{14} is disappearing rather faster than the C_{12}, and that this is also true for each individual C_{14} isomer compared to the corresponding C_{12}, all in accord with the distance principle. Speedier degradation of longer homologs has been confirmed for all LAS chain lengths from C_6 through C_{16} (Huddleston, 1963; Swisher, 1963c; Allred, 1964b).

For LAS above C_{12}, where inhibition of bacterial action may be particularly noticeable in unacclimated systems, it is very important that comparisons be made with all components present in a single solution. Such interference is seen in Figs. 6.5, 6.6, and 6.7, showing river water

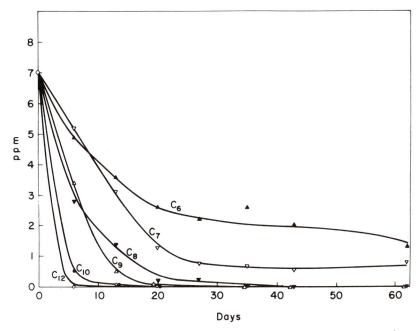

Fig. 6.5 Biodegradation of C_6 to C_{12}LAS individual homologs in separate river water tests, detected by MBAS (Swisher, 1963b).

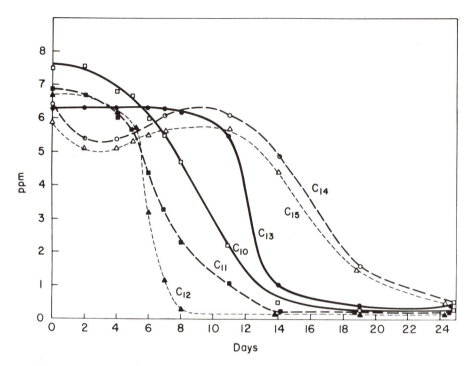

Fig. 6.6 Biodegradation of C_{10} to C_{15} LAS individual homologs in separate river water tests, detected by MBAS (Swisher, 1963b).

tests on individual LAS homologs in individual solutions. The degradation is increasingly rapid from the C_6 through the C_{12} homologs, but slower from C_{12} to C_{15} and then faster again to C_{18}. Behavior of the C_{12} plus C_{14} LAS mixture in Fig. 6.7 suggests strongly that bacterial inhibition is involved, since the C_{12} component of the mixture does not degrade at the same time as in the individual C_{12} sample. Instead, it is delayed until the bacteria overcome the inhibitory action of the C_{14} component. Once this is accomplished, both homologs degrade together, with the C_{14} leading slightly just as in Fig. 6.1 and Table 6.2.

This inhibiting action of the C_{14} LAS is overcome by acclimation, as illustrated in Fig. 6.8. Here the degradation did not start until after about 2 weeks, when the bacteria became inured to the C_{14} LAS. Once this acclimation was accomplished, subsequent increments of either the C_{12} or C_{14} homologs were readily degraded without delay. Huyser (1960) had earlier shown the importance of such acclimation in his experiments on

8-ϕC$_{15}$ sulfonate in water from the river Ij. In a fresh solution degradation started at day 4 and was complete at day 6. The experiment was then repeated using a mixture of 1 part of that water with 3 parts of fresh Ij water and the time required was only 1 to 3 days, and upon one further repetition it was down to 0 to 2 days. Setzkorn (1964) and Sweeney (1964b) report similar results. Sweeney further demonstrated that bacterial inhibition may still occur after acclimation, if the second increment is added at higher concentration than the first.

Ruschenberg (1963a, b) studied the degradation of a series of pure 2-phenyl and internal LAS compounds in inoculated media under rather mild conditions (Table 6.3). All of the 2-phenyls started degradation at 4 days and finished at 6 to 9 days, except for the C$_8$ homolog. Of the internals, the C$_8$ and C$_{16}$ did not degrade at all during the 30 day period, while the three intermediate ones all went at 9 to 21 days. These results suggest that phenyl position is a more important factor than chain length in determining the biodegradation (or acclimation plus biodegradation)

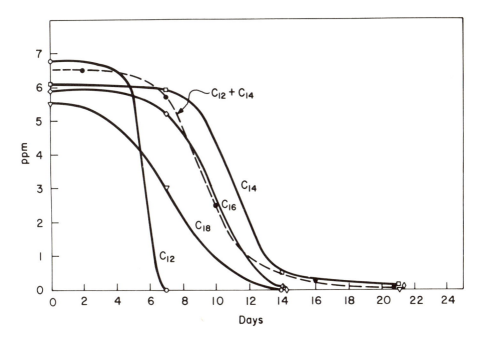

Fig. 6.7   Biodegradation of C$_{12}$, C$_{14}$, C$_{16}$, C$_{18}$LAS individual homologs, and one mixture, in separate river water tests, detected by MBAS (Swisher, 1963b).

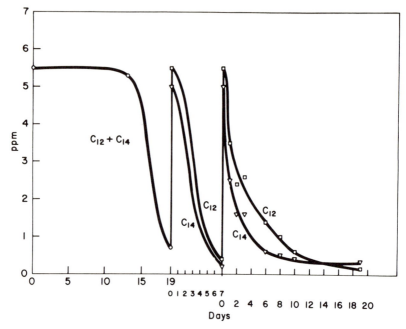

Fig. 6.8 Biodegradation of $C_{12} + C_{14}$LAS mixture in river water, detected by MBAS, with subsequent additions of the individual homologs to separate aliquots to demonstrate acclimation (Swisher, 1963b).

rate. This is in qualitative agreement with Fig. 6.1 and Table 6.2, which were obtained in a single system with all components present.

Huddleston (1963) investigated a single multicomponent system containing C_6, C_8, C_{10}, and C_{12}LAS, using desulfonation–GC analysis. The observed order of attack on the several components (Table 6.4) constitutes some of the experimental evidence upon which the distance principle is based.

In contrast with the above results, Simko (1965) reported little change in average phenyl position and homolog distribution between influent and effluent from a full scale activated sludge plant during a field test at Woodbridge, Virginia. He hypothesized that the distance principle, which had been developed mainly from die-away experiments, might not be applicable in a continuous-flow activated sludge system. However, the cause for the discrepancy must lie somewhat deeper than that, since Wickbold (1964) and Swisher (1963e) found laboratory continuous-flow

TABLE 6.3

Biodegradation of Pure LAS Components in Inoculated Media[a]

LAS	Start[b]	End[b]	LAS	Start[b]	End[b]
2-ϕC$_8$	9	13	4-ϕC$_8$	>30	—
2-ϕC$_{10}$	4	6	5-ϕC$_{10}$	9	21
2-ϕC$_{12}$	4	6	6-ϕC$_{12}$	9	21
2-ϕC$_{14}$	4	9	7-ϕC$_{14}$	9	21
2-ϕC$_{16}$	4	9	8-ϕC$_{16}$	>30	—

[a] Ruschenberg (1963a, b).
[b] Days on which the degradation started and was completed.

systems to behave as expected from the principle, and Hughes (1968b) verified it in a sewage treatment plant. In all of these cases the shorter chain and the internally phenylated components were slower to degrade.

Perhaps a contributing factor in Simko's disagreement is his use of mass spectroscopic analysis to establish the effluent composition. The effluent ABS contained up to 60–80% of nonlinear components of unknown and probably widely divergent side chain structure which contribute to the final mass spectroscopic result along with the LAS. It is perhaps not entirely coincidental that Simko's effluent mass spectra, which seemingly show 3-phenyl LAS homologs as the largest component, closely resemble his pattern for the influent control in this respect. That sample was taken before the introduction of LAS and presumably represents the pattern of TBS, which of course has no linear components at all.

TABLE 6.4

Relative Speed of LAS Biodegradation as Influenced by
Phenyl Position and Chain Length[a]

	C$_6$	C$_8$	C$_{10}$	C$_{12}$
First to go				2-ϕ
				3-ϕ
			2-ϕ	4-ϕ
		2-ϕ	3,4-ϕ	5,6-ϕ
		3-ϕ	5-ϕ	
	2-ϕ	4-ϕ		
Last to go	3-ϕ			

[a] Huddleston (1963).

Tarring (1965) shows biodegradation rates for a series of commercial-type LAS homologs derived from single carbon cuts of commercial-type olefins. The halfgone time decreases fairly uniformly from around 3 days to around 1 day from the C_8 to the C_{16} homolog. Tarring attributes this to a decreasing ratio of impurities with increasing chain length, but it seems likely that the distance principle is deeply involved as well.

The distance principle appears to hold for chain lengths up to about C_{16} or C_{18}, but de Jong (1967) presents data indicating that it may break down somewhere beyond:

LAS homolog	C_8	C_{10}	C_{12}	C_{14}	C_{16}	C_{21}
Halfgone, hr	108	37	17	16	17	108

This reversal at extreme chain length should be further verified, but need not be surprising. The higher homologs are so preponderantly hydrophobic as to considerably alter their water solubility and surfactancy, and biological differences would not be altogether unexpected.

B.2. *Limiting Concentrations for LAS Self-Inhibition*

Ciattoni (1968) made a detailed study of the inhibiting effect of LAS upon its own degradation. All homologs examined, from C_{10} to C_{15}, showed increased effect with increased chain length not only in river water but also in "system MT," prepared by inoculation of a nutrient medium (official German feed, Table 5.3). Each homolog had a limiting concentration as shown in Table 6.5, above which its degradation did not

TABLE 6.5

Limiting Concentrations for LAS Biodegradation[a,b]

Homolog	Brentelle canal, 20°C	Ticino river, 20°C	System MT[c], 25°C
C_{10}	90	140	—
C_{11}	38	55	300
C_{12}	12.7	22	—
C_{13}	4.2	8	—
C_{14}	1.7	2.8	14
C_{15}	0.6	1.2	—

[a] Concentration, ppm, above which degradation did not occur for at least 50 days.

[b] Ciattoni (1968).

[c] Official German activated sludge feed (Table 5.3), inoculated.

occur for at least 50 days, even though the bacteria retained their ability to degrade more normal substrates such as glucose and meat extract. Below the limiting concentration degradation of the LAS did occur, with induction periods which were shorter the lower the initial concentration.

Throughout the range from C_{10} to C_{15}, increasing the chain length by one carbon cut the limiting concentration to about one-third. This ratio held fairly constant even though the absolute values of the critical concentration varied from medium to medium as shown in Table 6.5. The limiting concentration was lower at lower temperature and at lower pH:

	20°C	10°C	5°C	pH 9	pH 7.3	pH 5.8
C_{13}LAS, ppm	4.2	3.5	1.8	5	4.2	3.2

The limiting concentration could be increased by acclimation to the LAS in question or to a higher homolog, and deacclimation could occur in the absence of LAS. With acclimated organisms in a urea–inorganic salts medium a given LAS degraded at a uniform rate which increased with increased concentration of organisms.

Inasmuch as the organisms retained their ability to degrade such things as nutrient broth and glucose in the presence of LAS above the limiting concentration, Ciattoni attributed the inhibition to interaction of the LAS with specific bacterial enzyme centers otherwise capable of attack on the alkyl chain. He pointed out that adsorption of LAS onto bacterial cells increases with increased chain length from C_{10} to C_{15}, consistent with the observed lowering of the limiting concentration throughout that same range. Since mixtures of homologs appeared to be less inhibitory than calculated from the individual components, he planned to seek further correlation by investigating adsorption from mixed homologs.

Ciattoni's results with System MT (Table 6.5) correlate well with those of Mann (1968) on a higher molecular weight variety of commercial LAS (C_{10-15}, average $C_{12.6}$). This material was degraded to around 95% in the official German activated sludge procedure and also in a field trial using a trickling filter, but the British STCSD inoculation test gave very erratic results with it, anywhere from zero to 90% degradation. A medium molecular weight sample (C_{10-13}, average $C_{11.8}$) degraded readily in all cases. Mann found that the high molecular weight LAS degraded easily in the STCSD test when meat extract, peptone, and urea were added to the medium at the same level as used in the German test, just as Ciattoni found that the limiting concentration was markedly higher in the German medium. Perhaps the proteins therein complex enough of the LAS (Chapter 4, Section IV.B) to reduce its inhibitory action.

B.3. *Linear Primary ABS*

The linear primary alkylbenzenesulfonates may also show tendencies in accord with the distance principle (Table 6.6), but the picture is not altogether clear. Perhaps the slower degradation of the C_{14} and C_{16} homologs indicated in the table simply reflects a greater inhibitory action on the bacteria as in the case of LAS, since these tests were run separately and in unacclimated systems.

TABLE 6.6

Biodegradation of Linear Primary ABS

| | | Halfgone, days | |
| | | Kölbel (1964) (*E. coli*) | |
Alkyl group	Swisher (1963b) River water	Series I	Series II
C_5	8	—	—
C_8	—	—	10
C_{10}	—	5.3	—
C_{12}	3	5.0	6
C_{14}	4	5.7	—
C_{16}	—	—	11

C. Chain Branching—Methyl Groups

Presence of a methyl group along the chain shortens the effective length by one carbon, and from this fact alone it would be expected to result in a noticeable but minor retarding of biodegradation speed in accord with the distance principle. If in addition the normal metabolic mechanisms of the bacteria have difficulty in handling the branched molecule, a much more drastic interference with biodegradation would be expected.

Such interference does not occur in the case of a single methyl group, judging by results reported by Swisher (1963b). Three model compounds were compared, 1-phenylundecane sulfonate, 2-phenyldodecane sulfonate and 1-phenyl-10-methylundecane sulfonate. As shown in Fig. 6.9, each of these has an effective chain length of 11 carbons, and they were found to degrade at substantially the same rate when mixed in river water. After 70% biodegradation had occurred, 77% of the first compound had disappeared compared to 64% each for the other two. The conclusion is

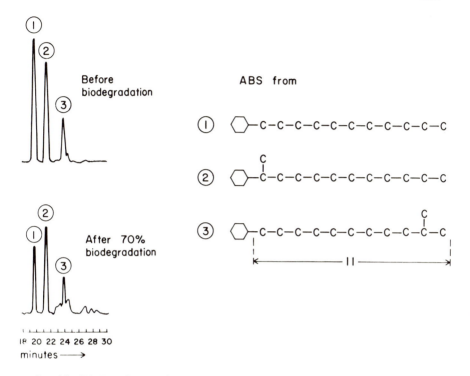

Fig. 6.9 Biodegradation of ABS mixture, effective chain length C_{11}, in river water, detected by desulfonation–GC (Swisher, 1963b). Upper: Day 0, 7.4 ppm MBAS. Lower: Day 4, 2.2 ppm MBAS.

that the extra terminal methyl group in the 1-phenyl-10-methyl compound has no adverse effects on the biodegradation. Sweeney (1964a) likewise reported little difference in the biodegradation rates of $C_{12}LAS$, its isomer

$$\overset{\text{C}}{\text{CCCCCCCCCCCC}\phi\Sigma}$$

and the next higher homolog

$$\overset{\text{C}}{\text{CCCCCCCCCCCCC}\phi\Sigma}$$

Presence of two isolated methyl groups on the side chain (as distinct from two on the same carbon atom when in a quaternary structure, discussed in the next section) also appears to have little adverse effect, judging by the observations of Cohen (1965) on

$$\overset{\text{C}\quad\text{C}\quad\text{C}}{\text{CCCCCCCCC}\phi\Sigma} \quad \text{and} \quad \overset{\text{C}\quad\text{C}\quad\text{C}}{\text{CCCCCCCCC}\phi\Sigma}$$

Nevertheless, it is conceivable that a multiplicity of methyl groups on adjacent carbons might be unsuitable for existing bacterial metabolic processes and hence might cause difficulty in biodegradation.

Such a situation does not appear to have been demonstrated as yet with compounds of known structure, although it may be that the resistance of TBS is due at least in part to such a circumstance. Tarring (1965) points out that TBS has 4 to 5 methyl groups per molecule compared to 2 for a completely pure LAS (i.e., the two ends of the chain), and presents data on impure LAS obtained from commercial-type raw materials during product development. As the number of methyl groups was increased from 2.3 to 3.0 per molecule the undegradable content increased, with considerable scatter, from about 8% to about 16%. Ruschenberg (1963a) found that 5% of methyl branches in $C_{11-13}LAS$ noticeably retards the completion of biodegradation, while 10% results in 3% undegradable as well. In similar vein, Pitter (1968c) presents data for KBS showing the biodegradability increasing from 5–23% to 48–63% to 80–95% as the linear alkane content of the kerosene was increased from 2–5% to 30–50% to 94–98%.

In all such cases, however, the key factor is not simply the percent of linear or branched alkyl groups, but rather the detailed structure of the branched fraction. Thus a singly branched ABS is readily biodegradable, even though it contains no linear alkyl groups at all.

Fujiwara (1968) developed a branching index formulated as $(H_{Me} - 6)/(H_t - 5)$, where H_{Me} is the number of methyl hydrogens and H_t is the total number of hydrogens in the molecule, determined by NMR and GC on the unsulfonated alkylbenzene. The index is thus the ratio of methyl branch hydrogens (not counting the two ends of the chain) to the total alkyl hydrogens. Eleven commercial alkylbenzenes, nine linear and two branched (presumably from tetrapropylene) were examined, with the results shown in Table 6.7. In addition to the obvious difference between the LAS and TBS samples, there might be a slight trend toward lower degradability with higher index within the two groups also. Unfortunately neither the physical nor the biological methods are precise enough to show such minor differences clearly. Two mixtures of samples H and K showed values for index and degradability reasonably close to those calculated from the proportions in the mixtures, indicating that both properties are additive.

A single methyl group on the ring instead of on the chain gives us an alkyltoluenesulfonate. These have been studied little, but do not appear to differ much from the corresponding ABS in biodegradability; for

TABLE 6.7

Effect of Branching Index on ABS
Biodegradation[a]

Sample	Index[b]	% Degraded[c]
F	0	97.3
B	0.8	96.7
E	0.8	95.0
H	0.84	96.9
G	1.2	96.4
A	1.6	93.9
C	2.1	95.4
D	3.0	89.2
I	3.4	93.8
K	23.1	16.8
J	23.6	13.3

[a] Fujiwara (1968).

[b] Methyl branch hydrogens $\times$ 100/total alkyl hydrogens.

[c] Official Japanese test method (shake culture).

examples see Pitter (1964c, d) on linear and Sweeney (1964a) on tetra-propylene. For the linear alkylxylenesulfonates (two methyls on the ring) Borstlap (1967b) found wide differences: the *para*-xylene derivative was 92% degradable, *meta*- 16%, and *ortho*- 61%. These are discussed further in Section III.

D. Quaternary Branching

The biological resistance of TBS was attributed by McKinney (1959a) to the presence of a quaternary carbon atom in the alkyl group in the typical structure **(6.3)** which had been assigned. (It must be remembered

$$
\begin{array}{c}
\text{C} \quad \text{C} \quad \text{C} \\
\text{CCCCCCC}\phi\Sigma \\
\text{C}
\end{array}
$$

(6.3)

that TBS is a mixture of dozens or hundreds of components, all minor, and that no single structure can represent it satisfactorily.) The premise was that the normal biological oxidation process involves formation of a double bond as one of the intermediate steps (Chapter 7, Section I.B), and this could not occur at a quaternary carbon atom, already bonded to four other carbon atoms. A chicken–egg situation results: one of the existing C—C bonds must first be broken in order to form the double bond, but the double bond must first be formed in order to break the C—C bond.

Actually the situation is not quite so hopeless. Bacteria have more than one way to skin a molecule, and they are able to handle quaternary carbons by other means (Chapter 7, Section IV). Even if they did not have this capability, primary biodegradation of **6.3** would not be blocked on that account. As shown by Nelson (1961), presence of a quaternary carbon does not interfere provided there is also present a sufficient length of open end chain. Indeed, Table 8.1 shows that many workers have found 2-phenyl-2-methylundecane sulfonate

$$\begin{array}{c} \text{C} \\ \text{CCCCCCCCCC}\phi\Sigma \\ \text{C} \end{array}$$

to be readily degradable, as well as others with the quaternary further out on the chain. Attack can proceed at the chain end quite sufficient to destroy the surfactancy and methylene blue response regardless of the presence of the quaternary carbon further down the chain.

Huddleston (1963) and Allred (1964a) show several internal quaternarys with varying degrees of biodegradability, but with these it is surely more a consequence of the distance principle than of the presence of the quaternary. For example, **6.4** is degraded 100% and **6.5** 0% under

$$\begin{array}{cc} & \text{C} \\ & \text{C} \\ \text{C} & \text{C} \\ \text{CCCCCCCCCCC} & \text{CCCCCCCCC} \\ \phi & \phi \\ \Sigma & \Sigma \\ \textbf{(6.4)} & \textbf{(6.5)} \end{array}$$

the relatively mild conditions of the shake culture test.

On the other hand, Nelson (1961) found that ABS with a terminal quaternary carbon atom may be quite resistant, and here the quaternary carbon itself is undoubtedly the prime factor. Such structures, typified in their simplest form by **6.6**, are generally quite difficult to degrade. Even

$$\begin{array}{c} \text{C} \\ \text{CCCCCCCCCC}\phi\Sigma \\ \text{C} \\ \textbf{(6.6)} \end{array}$$

so, acclimation can sometimes be accomplished, and Sweeney (1964a) achieved 84% degradation of the next lower homolog in continuous-flow activated sludge, with liberation of inorganic sulfate to the extent of 50%. On some occasions Swisher (1963b) observed 100% degradation of both the 8,8-C_9 and 9,9-C_{10} homologs in river water after a month or more of acclimation. Desulfonation–GC of partially degraded mixtures of the two showed that the longer one degraded more rapidly than the shorter, in accord with the distance principle. Similarly Kölbel (1967) found only a few percent degradation of the $9,9\text{-}C_{10}$, but up to 66% for the $11,11\text{-}C_{12}$.

Several other terminal quaternaries are listed in Table 8.1, having one or two additional methyl groups along the chain. There is some possibility that the high removals recorded in some cases represent adsorption or precipitation rather than biodegradation, since many of these derivatives are rather insoluble (Swisher, 1969). However, biodegradation is the likely mechanism in cases where oxygen absorption or sulfate release is observed, or where analysis of the entire sample (including solids and bacteria) shows disappearance, as in the desulfonation–GC of river waters.

The mere presence of the terminal quaternary group in the molecule does not interfere with biodegradation if an open end chain is also present. Thus the compound **6.7** degraded readily in river water, although it was marginally slower than its isomer **6.8**, which has two open end chains (Swisher, 1963b).

$$
\begin{array}{cc}
\begin{array}{l}
\text{CC} \\
\text{CCCCCCCCC} \\
\text{C } \phi \\
\;\;\Sigma
\end{array}
&
\begin{array}{l}
\text{CCCCCCCCCCC} \\
\;\;\;\;\;\phi \\
\;\;\;\;\;\Sigma
\end{array}
\\[6pt]
\textbf{(6.7)} & \textbf{(6.8)}
\end{array}
$$

Data presented by Tarring (1965) for impure LAS derived from commercial-type raw materials show that dimethyl groups have more influence than single methyls on the biodegradability. Residual unde-gradable increased from about 8% to 16% as the dimethyl groups went from 0.11 to 0.34 per molecule. Each dimethyl group corresponds to a quaternary carbon, but not necessarily vice versa.

E. Cyclic Groups

Aliphatic ring structures (naphthenes) are often present in petroleum fractions and also may be formed in petroleum processing. Hence the alkyl group of a commercially derived ABS may include some amount of cyclic structures. Although such rings can be biodegraded without particular difficulty, their presence does make the alkyl group more compact and this would be expected to slow down the degradation to some extent depending upon the exact structure.

Thus Huyser (1960) reports that 6-cyclohexylhexylbenzenesulfonate

$$
\begin{array}{c}
\text{C}\!-\!\text{C} \\
\diagup \qquad \diagdown \\
\text{C} \qquad\qquad \text{C CCCCCC}\phi\Sigma \\
\diagdown \qquad \diagup \\
\text{C}\!-\!\text{C}
\end{array}
$$

is degraded in river water within 15 days, compared to 5–11 days for various LAS isomers. Similarly Hammerton (1962) found that 2-phenyl-8-cyclohexyloctane sulfonate

was readily degraded in the BOD test, though somewhat more slowly than several LAS compounds.

With more compact cycloalkyl groups such as those in **6.9** and **6.10**,

(6.9)   (6.10)

Nelson (1961) found considerably more resistance, in accord with the distance principle. They showed no oxygen pickup in Warburg tests, compared to around 40% of theoretical for several linear ABS compounds under the same conditions.

If diolefins or dichloroalkanes are present in the alkylation step in the manufacture of alkylbenzenes, they give products with the chain linked to the benzene at two spots, forming condensed cyclic systems such as **6.11** and **6.12**. These products distil at about the same temperatures as the

(6.11)   (6.12)

corresponding alkylbenzenes, a little higher, and upon sulfonation give products with comparable surfactancy properties. Likewise their bio-degradability—the family of linear dialkylindanes and dialkyltetralins (four of each) derived from linear dichlorododecane gave a mixed sulfonate which was readily biodegradable (Swisher, 1959).

Tarring (1965) reports on a series of commercial-type LAS samples containing cycloaliphatic impurities in the alkyl groups. The general trend showed undegradable content increasing from around 8% to 16% as ring content increased from 20% to 27%, but one sample showed only 12% undegraded in spite of 32% rings. Interpretation of the results is somewhat hindered by the presence of methyl and dimethyl groups as well as rings.

F. Multiple Branching—TBS

Aside from the few quaternary derivatives mentioned in Section II.D, there is little information in the literature on highly branched ABS of known structure. However, it seems very probable that for a given carbon number the degradation will be slower with increased branching, if only because of the shorter effective chain length and the distance principle. Terminal quaternary branching introduces a characteristically higher order of resistance as discussed above. It is conceivable that even in the absence of a quaternary carbon the bunching of several branches on adjacent carbons of the chain would result in slower biodegradation than would be predicted simply on the basis of effective chain length alone (though, contrariwise, it might not). Or perhaps the resulting effects might vary widely, depending on the exact details of the structural linkages.

The exact side chain structures of the hundred or more "major" components which make up TBS have not been worked out. Nevertheless the biodegradability of TBS, with components ranging from moderately to extremely resistant, does appear to be in accord with the results on pure compounds discussed earlier in this chapter.

The generally poor biodegradability of TBS was early attributed to the considerable branching of the alkyl group (Sawyer, 1956), and later (McKinney, 1959a) more specifically to the presence of a quaternary carbon atom in the alkyl chain; this would make the normal pathway of the β-oxidation process impossible. But, as we have seen in Section II.D, the mere presence of a quaternary carbon atom in the chain does not necessarily impede primary biodegradation at all—the critical question is whether the quaternary carbon is at the end of an alkyl chain. If it is, it will certainly interfere with the bacterial attack at that point, but if there is another alkyl chain in the molecule, sufficiently long and open-ended, biodegradation will still proceed readily.

A terminal quaternary group is a t-butyl group,

$$
\begin{array}{c}
\text{C} \\
\text{—CC} \\
\text{C}
\end{array}
$$

Tetrapropylene-derived alkylbenzene shows a rather weak but definite

IR band near 1250 cm^{-1} (8.0 μ), in the region attributed by Hawkes (1960) to such groups. There is little doubt but that they are indeed present in TBS, but quantitative calculations of their amount are uncertain. From the intensity of the 8.0 μ band Hawkes (1958) has estimated that perhaps as much as 30% of the TBS molecules may contain such structures. If so, this would be enough to account for much of its biological resistance.

Cycloaliphatic groups may also impart biological stability if present in sufficiently compact configurations, but mass spectroscopy indicates that no significant amounts are present. Such components have mass numbers corresponding to C_nH_{2n-8}, and these are barely detectable, if at all, in tetrapropylene alkylbenzene.

If we wish to look beyond terminal quaternary groups for further structural factors to explain the resistance of TBS to primary biodegradation, a likely prospect is multiple branching and the distance principle. The TBS alkyl groups are certainly highly branched, and a 12-carbon group with four branches would only have an effective length of 8 carbons at best—it could be as low as 6. On this factor alone, the degradability would correspond to C_8 or C_6LAS. Add to this the further stability conferred by whatever terminal quaternaries are present, and possibly by multiplicity of branching, and the resulting resistance should be well into the range actually observed for TBS.

In two German patent applications Ruschenberg (1964) reports substantial improvement in biodegradability of TBS by modification of the sulfonation process (minimizing the excess of sulfuric acid used, increasing temperature to the range 60–90°C, and intensive mixing) or by use of a narrower cut of alkylbenzene (boiling range 285–295°C instead of 280–300°). Degradation in the closed bottle test, measured by MBAS reduction, was brought into the range of 90–95% compared to 70–80% without the improvements. Deep-seated changes in the average side chain structure through such minor changes in processing do not seem very likely, and perhaps this better biodegradability may instead reflect the uncertainties of biodegradation testing. In any case, the improvement achievable was not great enough to halt the commercial move away from TBS toward LAS.

Reference to Table 8.3 indicates that although TBS may be almost completely resistant to biodegradation in some biological systems (in the shake culture test or the closed bottle test, for example) it is substantially more susceptible in more active systems. In river water or activated sludge its degradation is often in the range of 50% to 75%. Results from the WPRL standard trickling filter (WPRL, 1968, p. 135) show steady 25–30% degradation during the first 8 weeks, then a rise to 65–75% by the 13th week. This suggests the presence of three groups of components with rather sharply differentiated susceptibilities. Heinz (1966) cites several instances of 80% degradation averaged over long periods of operation. With prolonged exposure, values approaching 100% are possible.

Thus Sharman (1964a) observed 97.8% degradation of TBS, still trending upward, after 196 days in a balanced aquarium. Even in a continuous-flow activated sludge system, 6 hr retention time, Sharman (1964b) and House (1965b) achieved a steady state removal of 90% (86% by biodegradation) in a year-long operation. Here the foam recycle principle was used: undegraded TBS was continuously removed from the effluent by foam stripping and returned to the aerator for further degradation, thus prolonging the effective retention time of the TBS fraction far beyond the nominal 6 hr.

As a further example, by percolation of TBS through a well-matured soil lysimeter Robeck (1963) found 95% to 98% primary biodegradation as well as 75% to 80% degradation to inorganic sulfate. Adsorption of the TBS onto the biological growth in the lysimeter gave prolonged retention time and was undoubtedly an important factor in promoting the degradation. Kempf (1968) too found about 98% degradation of TBS when fed at 20 ppm to lysimeters with a varied range of soils. Upon raising the influent level to 200 ppm TBS the degradation remained at 98% in some soils but fell as low as 75% in others.

Using bacteria isolated from soil by enrichment culture techniques Benarde (1965) was able to degrade TBS considerably further and faster than usual, provided that a medium containing 1500 ppm of glucose was used. He achieved 70% degradation in a shake culture system within 7 days, and 100% degradation in inoculated lake water in 20 days.

The above examples show that TBS is almost completely biodegradable under certain conditions. Obviously these special conditions were not met sufficiently often in the world at large to ensure satisfactory degradation in our general environment, and so the switch to LAS by the detergent industry resulted.

Exhaustive comparisons of the biodegradation of TBS, LAS, and LPAS under a wide variety of laboratory and pilot plant conditions have been made at the University of California Sanitary Engineering Research Laboratory by Klein (1964a, 1965a, b, 1966) and McGauhey (1964, 1966). They have summarized their results as shown in Table 6.8, providing the most realistic picture available of the probable performance of these major surfactants under sewage treatment conditions used in the United States in the era 1960–1965. The TBS was an industry composite prepared by the SDA, representative of the average U.S. tetrapropylene alkylbenzene. The LAS was likewise an industry composite, made from experimental

TABLE 6.8

Removal of Surfactants in Waste Treatment Processes[a]

Process	Average percent removal		
	TBS	LAS	LPAS
Primary sedimentation	2–3	2–3	—
Septic tank	9.2	11.8	62.1
Septic tank + percolation field (normal)	73.9	97.4	99.6
Septic tank + percolation field (ponded)	54.5	97.1	99.7
Standard rate oxidation pond	<40	93.1	98.0
High rate oxidation pond	<15	56.2	95.2
Standard rate trickling filter	35.0	84.7	—
High rate trickling filter	19.1	71.0	—
Activated sludge	45–50	95	100

[a] Klein (1966); McGauhey (1966).

products being developed by the alkylbenzene producers at that time; there is good indication that the products eventually commercialized were distinctly superior to the earlier experimental ones in biodegradability. The LPAS was a 50:50 mixture of those derived from coco and from tallow fatty alcohols, the two in most common industrial use.

III. OTHER ALKYLARYL SULFONATES

Table 8.5 shows that the less common alkyl aromatic sulfonates which have been examined may range from low to high in biodegradability. In general the results are reasonably consistent with those outlined for alkylbenzenesulfonates in the preceding section, but there are a few surprises as well.

Perhaps the most surprising is the extreme difference between the three linear dodecylxylenesulfonates reported by Borstlap (1967b) (Table 6.9). Although these are probably complex mixtures of isomers, as Borstlap points out, we can still guess at what might be the more likely structures, as set forth in the table, resulting from the sulfuric acid-catalyzed alkylation and the subsequent sulfonation. Although one would expect **6.13** (see Table 6.9) to be rather faster than the other two by the distance principle, the extreme differences which Borstlap found would seem to be quite unexpected. His die-away curve for the meta product shows an 18 day lag and is headed sharply downward at 21 days, so perhaps this is simply a case of marked inhibitory action by the initial 10 ppm of surfactant. But the ortho product lagged only about 14 days, dropped sharply to about 40% remaining by 16 days and held at that level thereafter. Dodecylethylbenzenesulfonate showed a similar behavior. Reasons for this wide diversity in behavior are not immediately apparent and further studies will probably be required for understanding it.

Borstlap found that linear dodecylphenol sulfonate required about 16 to 20 days for primary biodegradation compared to only 6–8 days for the corresponding alkylbenzene derivative ($C_{12}LAS$), while the dodecylthiophene analog substantially disappeared in 2–3 days. Here too the full meaning of these differences may be revealed by further study.

The dihexyl and diheptylbenzenesulfonates are rather resistant to biodegradation even with linear structure (Section II.A). This might be anticipated in view of their relatively short chains, and similar behavior of the propyl- and butylnaphthalenesulfonates is not surprising. Increasing

TABLE 6.9

Biodegradation of Linear Secondary Dodecylxylenesulfonates[a]

Xylene isomer	Possible structure[b]	% Degraded[c]
para	**(6.13)**	92
meta		16
ortho		61

[a] Borstlap (1967b).
[b] Assuming no rearrangement and o,p orientation in alkylation, and steric hindrance in sulfonation.
[c] River water, 21 days, MBAS.

the linearity and chain length, Kölbel (1964) found results consistent with the distance principle. Using a culture of *Escherichia coli*, the *n*-butylnaphthalenesulfonate was undegraded in 30 days, the *n*-hexyl disappeared during days 24–30, and the *n*-octyl during days 5–15. He further found that the *n*-butyl and *n*-hexylbiphenylsulfonates were undegraded in 30 days while the *n*-octyl and *n*-decyl derivatives disappeared during days 6–16, again consistent with the distance principle.

TABLE 6.10

Biodegradation of Thiaalkylbenzenesulfonates

| | | MBAS disappearance, days | | |
| | | Shake culture | | River water |
Compound	Structure	Lang (1965)	Long (1966)	Lang (1965)
LAS		8	3	7
1-Thia	CCCCCCCCCCCCS$\phi\Sigma$	5	2	5
2-Thia	CCCCCCCCCCCSC$\phi\Sigma$	9	2	7
3-Thia	CCCCCCCCCCSCC$\phi\Sigma$	15	2	14
4-Thia	CCCCCCCCCSCCC$\phi\Sigma$	14	2	14
5-Thia				
to 11-thia		>15	—	>14

An interesting series of thiaalkylbenzenesulfonates has been described by Lang (1965, 1967) and Long (1966), wherein various carbons in the chain of 1-phenyldodecane sulfonate have been replaced (figuratively) by sulfur. It should be noted that the reference compound in Table 6.10 is not the parent 1-phenyl, but LAS. The 1-thia degraded faster than the LAS, but as the sulfur atom was moved further out the chain the degradation generally became slower. Since the tests were run individually, not as mixtures, this may simply represent differing inhibition-acclimation properties. If such factors are not involved, the data suggest that different biodegradation mechanisms may be coming into play in this series. In any case, initial oxidation at the chain sulfur atom does not seem to be involved since the corresponding sulfoxides, for example,

$$O$$
$$CCCCCCCCSCCC\phi\Sigma$$

were found to be less biodegradable (Long, 1966).

IV. SULFONATES FROM ALIPHATIC HYDROCARBONS

The linear alkane and alkene sulfonates, both primary and secondary, are readily biodegradable when pure. They are in general slightly slower than the linear primary alkyl sulfates and slightly faster than LAS; for example, see Huyser (1960) and Ōba (1968b).

A. Alkane Sulfonates

Commercial alkane sulfonates are predominantly linear and secondary, with the sulfonate group attached randomly along the chain. They may contain nonlinear or other impurities which could make their degradation slower or slightly less than 100%. Di- or polysulfonate impurities are usually present, but according to Winter (1962) the polysulfonate is almost as rapidly and completely degradable as the monosulfonate, in the Warburg test. Degradation data are summarized in Table 8.6.

TABLE 6.11

Biodegradation of Commercial-Type Linear
Secondary Alkane Sulfonates[a]

Average chain length	Die-away, days[b]	COD removal at 21 days, %
C_{13}	4	26
C_{15}	4	24
C_{16}	4	28
C_{17}	7	27
C_{19}	10	14
$2\text{-}\phi C_{12}$ (reference)	10	30

[a] McAteer (1964).
[b] Mineral salts medium, unacclimated inoculum, MBAS.

Kölbel (1964) found that the pure linear primary alkane sulfonates, C_{10}, C_{12}, C_{14}, and C_{16}, were completely degraded by *Escherichia coli* within 6 days, remaining unchanged (MBAS) for at least the first 3 days. This would allow room for only minor differences, if any, in the degradation rates of the several homologs.

McAteer (1964) investigated the biodegradation of a homologous series of commercial-type secondary alkane sulfonates prepared by sulfoxidation of linear hydrocarbon cuts (Table 6.11). His tests were run in comparison with 2-phenyldodecane sulfonate, individually in mineral salts medium inoculated with unacclimated organisms. Thus his die-away times do not necessarily mean that the longer homologs have intrinsically slower biodegradation rates; acclimation times and, possibly, inhibitory effects on the bacteria must also be involved.

B. Alkene and Hydroxyalkane Sulfonates

The commercial α-olefin sulfonate (AOS) surfactants are ordinarily mixtures of approximately equal amounts of alkene sulfonate and hydroxyalkane sulfonate (Weil, 1965; Marquis 1966), along with some di- and polysulfonates. Ōba (1968b) reports on the biodegradation of three commercial-type, disulfonate-containing AOSs, along with pure synthetic prototypes of the two major components (Table 6.12). The degradability

TABLE 6.12

Biodegradation of α-Olefin Sulfonates[a]

Sample	% Disulfonate	% Degraded[b]
2-Pentadecene-1-Σ	0	99–100
3-Hydroxytetradecane-1-Σ	0	99
C_{15-18}AOS (C)	4	98
C_{15-18}AOS (D)	15	98
C_{15-18}AOS (E)	50	96

[a] Ōba (1968b).
[b] Shake culture, MBAS.

becomes slightly lower, perhaps, with increased disulfonate content. Degradability of the disulfonate content itself cannot be estimated with any certainty from these data, since the disulfonate is quite unresponsive in the methylene blue analysis (Chapter 3, Section III.A.2). However, Tomiyama (1968) was able to separate the disulfonate from the monosulfonate by foam fractionation and found that the disulfonate was indeed biodegradable, though somewhat more slowly than the monosulfonate.

Tomiyama (1968) also determined the relative degradation rates of the alkene sulfonate and the hydroxyalkane sulfonate components of the AOS, using a GC method after conversion to the sulfonyl chlorides to confer sufficient volatility. With approximately 50% degraded samples of the four homologs C_{15}, C_{16}, C_{17}, C_{18}, he found that in each case some 65–75% of the alkene sulfonate had disappeared compared to 28–32% of the hydroxyalkane. New peaks showed in these chromatograms also, indicating the presence of intermediate degradation products, as yet unidentified at the time of the report.

V. SULFOSUCCINIC ACID ESTERS

The sulfosuccinates are very effective wetting agents which find use in the textile industry, although not in detergent formulations. They are diesters of sulfosuccinic acid as in **6.14**, wherein the size and structure of

$$
\begin{array}{c}
\quad\quad\quad\;\; O \\
\;\;\; H \;\; \| \\
\Sigma\text{---}C\text{---}C\text{---}OR \\
\;\;\; | \\
\;\; HC\text{---}C\text{---}OR \\
\;\; H \;\; \| \\
\quad\quad\; O
\end{array}
$$

(6.14)

the alcohol group R may influence the surfactancy to a greater or lesser extent. Hammerton (1956) found that the biodegradation was influenced as well (Table 6.13). He concluded that the ester linkage was not the point of bacterial attack, since two of the esters were very resistant. Nevertheless,

TABLE 6.13

Biodegradation of Sulfosuccinic Acid Esters[a]

R group[b]	R structure	Halfgone, days[c]
Benzyl	—Cφ	4
n-Octyl	—CCCCCCCC	4.5
2-Ethylhexyl	C C —CCCCC	7.5
3,5,5-Trimethylhexyl	C C —CCCCC C	9
Isobutyl	C —CCC	11
4-Methyl-2-pentyl	C C —CCCC	d
Cyclohexyl	CC —C C CC	d

[a] Hammerton (1956).
[b] In **6.14**.
[c] River water, MBAS.
[d] No degradation in 28 days.

the opposite conclusion seems more attractive, to explain the relatively rapid disappearance of the terminal quaternary 3,5,5-trimethylhexyl ester. The two resistant compounds are both secondary esters, and it is easy to imagine that the biological hydrolysis mechanism is applicable to primary esters but less readily to secondary.

VI. FATTY ACYLAMIDE SULFONATES

The Igepon T's **(6.15)** are the major representatives of this class. The major commercial products are made from linear fatty acids and are found to be readily degradable (Table 8.7). Sawyer (1956) and Ryckman (1957) attribute this to rapid hydrolysis by naturally occurring amidase enzymes, with subsequent oxidation of the liberated fatty acid, but no direct experimental evidence seems to have been brought forward as yet.

Likewise, apparently no studies have been reported on direct comparisons to relate fine-structure to biodegradability. However, the related N-methyl anilides **(6.16)** have been examined in a homologous

$$
\begin{array}{cc}
\overset{O}{\overset{\|}{R-C}}\overset{C}{\underset{}{-NCC\Sigma}} & \overset{O}{\overset{\|}{R-C}}\overset{C}{\underset{}{-N\phi\Sigma}} \\
\textbf{(6.15)} & \textbf{(6.16)}
\end{array}
$$

series by Kölbel (1964). Upon exposure to *E. coli* their biodegradation was in reasonable accord with the distance principle:

R (in **6.16**)	$n\text{-}C_9$	$n\text{-}C_{11}$	$n\text{-}C_{13}$	$n\text{-}C_{15}$
Days to halfgone	11	8	7	7

VII. ALKYL SULFATES

The linear primary alkyl sulfates (LPAS) have long been recognized as extremely rapid in primary biodegradation, often disappearing in less than a day in shake culture or river water. Table 8.8 provides many examples, and certain entries in Table 8.9 indicate that the presence of a double bond or of two chlorine atoms along the chain does not interfere. Linear secondary alkyl sulfates, which have been studied particularly in the form of the commercial mixture Teepol, are likewise readily degradable, although somewhat more slowly (Table 8.10). Direct comparisons of several alkyl sulfates with each other and with their ethoxylate sulfates are given in Tables 6.15, 6.16, and 6.17.

A. Branching

The facile degradation of the alkyl sulfates first studied was attributed at the time to the presence of the readily hydrolyzable sulfate ester linkage in the molecule. Hammerton (1955, 1956) proved otherwise when he found that certain branched primary and secondary alkyl sulfates were very resistant:

$$
\begin{matrix}
\text{C} \;\; \text{C} \\
\text{CCCCCCO}\Sigma \\
\text{C}
\end{matrix}
\qquad\qquad 4\% \text{ degraded, 21 day river water}
$$

$$
\begin{matrix}
\;\text{C} \qquad\;\; \text{C} \\
\;\text{C} \qquad\;\; \text{C} \\
\text{CCCCCCCCCCCC} \\
\qquad\;\; \text{O} \\
\qquad\;\; \Sigma
\end{matrix}
\qquad\qquad 37\% \text{ degraded, 79 day river water}
$$

Huyser (1960) subsequently provided further examples:

$$
\begin{matrix}
\text{C} \;\; \text{C} \\
\text{CCCCCO}\Sigma \\
\text{C}
\end{matrix}
\qquad\qquad 0\% \text{ degraded, 18 day river water}
$$

$$
\begin{matrix}
\text{C} \;\; \text{C} \\
\text{CCCCCCO}\Sigma \\
\text{C}
\end{matrix}
\qquad\qquad 0\% \text{ degraded, 18 day river water}
$$

$$
\begin{matrix}
\;\;\text{C} \;\; \text{C} \quad\;\; \text{C} \;\; \text{C} \\
\text{C C CCCCCCCCC} \\
\;\;\text{C} \qquad\; \text{C} \;\; \text{C} \\
\qquad\qquad \text{O} \\
\qquad\qquad \Sigma
\end{matrix}
\qquad\qquad 0\% \text{ degraded, 56 day river water}
$$

$$
\begin{matrix}
\text{CCCCC} \quad\; \text{C---C} \\
\qquad\quad \diagdown \quad\! \diagup \qquad\;\; \diagdown \\
\text{CCCCCCC} \qquad\qquad \text{COΣ} \\
\qquad\quad \diagup \quad\! \diagdown \qquad\;\; \diagup \\
\text{CCCCC} \quad\; \text{C---C}
\end{matrix}
\qquad\qquad 10\% \text{ degraded, 59 day river water}
$$

Furthermore, the highly branched alcohol mixture obtained by oxonation of tetrapropylene gives a characteristically resistant alkyl sulfate, as has been well documented by Ruschenberg (1963a), Pitter (1963b, 1964c, d), and Berger (1964) (Table 8.9). Even so, these workers, as well as Sweeney (1964a) and Ōba (1967), do indicate that some degradation of this product usually occurs—under some circumstances quite extensively. This is attributable to the fact that the product is a mixture of many different chemical structures, each with its own specific biodegradation behavior.

Thus quite evidently a branched hydrophobe may impart biological resistance in either primary or secondary alkyl sulfates, not only when the branching is of the terminal quaternary type, but in some other cases as well. Yet other branched and cyclic alkyl sulfates may be readily degradable, and again Huyser (1960) provides examples:

$$\cdots \overset{\displaystyle \overset{\textstyle \cdots C}{\underset{\displaystyle |}{C}}}{CCCCCO\Sigma}$$ 100% degraded, 14 day river water

CCCCCCCOΣ 100% degraded, 7 day river water

On the basis of present data it does not seem possible to draw up any detailed relationship between structure and biodegradability which would allow prediction of performance, with reasonable certainty, of a new alkyl sulfate with nonlinear structure. Not only do we lack a sufficient range of known model compounds; the situation is further complicated by the evident possibility of attack either at the hydrophobic or at the hydrophilic end of the molecule by two different mechanisms, oxidative and hydrolytic, respectively.

B. Chain Length

Likewise, there has been little study of biodegradation as affected by chain length in the alkyl sulfates. Judging by the scanty data available, the effects, if any, are quite small and definitive results will be dependent upon methods of comparison more sensitive than side-by-side runs in separate systems. Winter (1962) investigated oxygen uptake in a series of LPAS in 24 hr Warburg runs as follows:

Homolog	C_{10}	C_{12}	C_{14}	C_{16}	C_{18}
% of theoretical	54	74	72	68	51

The 10 and 18 homologs would appear to be significantly slower than the intermediate ones, but further verification under other conditions would be desirable.

Gebril (1966) prepared a series of linear secondary alkyl sulfates from the corresponding linear alcohols, derived originally via chlorination of kerosene fractions. Shake culture studies (Table 6.14) showed increased biodegradation rate with increased chain length, assuming that acclimation and inhibition factors are constant. The few percent of undegraded material remaining appears to be resistant impurities rather than linear secondary sulfate, since the die-away curves leveled off at those values.

VIII. ETHOXYLATE SULFATES

The ethoxylate sulfates all have the structure $RE_nOCCO\Sigma$, and thus all of them are primary alkyl sulfates. Even so, as shown in Tables 8.11 and 8.12, their biodegradation behavior varies widely, depending on the

TABLE 6.14

Biodegradation of Linear Secondary Alkyl
Sulfates[a]

Average chain length	Halfgone, hr[b]	% Degraded, 48 hr[b]
$C_{10.5}$	24	95
$C_{12.5}$	21	97
$C_{14.5}$	17	98

[a] Gebril (1966).
[b] Shake culture, MBAS.

structure of the hydrophobe group R. This further exemplifies that biological hydrolysis of a primary sulfate is not necessarily easy; its dependence on other structural features of the molecule remote from the sulfate linkage suggests that a high degree of steric compatibility with the sulfatase enzyme is necessary.

To some extent the structural effects parallel those found for the simple alkyl sulfates and the alkylbenzenesulfonates. Table 8.11 indicates that the linear primary and secondary alcohol EO sulfates are in general readily degraded, while highly branched derivatives, for example, from tetrapropylene primary oxo alcohol, are much more resistant. This is quite evident in Table 6.15.

Tables 6.15, 6.16, and 6.17 give comparisons of alcohol EO sulfates with the corresponding unethoxylated sulfates. There seems to be a

TABLE 6.15

Alkyl Sulfates vs. Ethoxylate Sulfates[a]

Alcohol		% Degradation at 1; 3–4 days[b]	
Carbons	Type	Sulfate	EO Sulfate[c]
C_{12}	Linear primary (natural)	95; —	97; —
C_{12+14}	Linear primary (Ziegler)	96; —	97; —
C_x	Linear primary oxo	95; —	90; —
C_x	Linear primary oxo	95; —	90; —
C_{13}	Tetrapropylene primary oxo	65; 95	50; 78
C_{13}	Tetrapropylene primary oxo	68; 95	40; 76

[a] Berger (1964).
[b] Recycle trickling filter, MBAS.
[c] Degree of ethoxylation unstated.

TABLE 6.16

Alkyl Sulfates vs. Ethoxylate Sulfates[a]

Alcohol		% Degradation at 7; 7 + 2 days[b]	
Carbons	Structure	Sulfate	E_2 Sulfate
C_{18}	$n\text{-}C_{18}OH^c$	98; 97	84; 90
C_{14}	CCCCCCCCCCCCC C OH	97; 98	98; 96
C_{17}	CCCCCCCCCCCCCCCCC O H	100; 99	100; 100
C_{18}	$sec\text{-}C_{18}OH^d$	100; 100	—

[a] Crauland (1964).
[b] Inoculated river water test, with acclimation (Brebion, 1966), MBAS.
[c] Stearyl alcohol.
[d] From n-α-olefin.

TABLE 6.17

Families of Linear Ethoxylate Sulfates[a]

	$E_0{}^b$	E_3	E_4	E_5	E_6
	Days for 90% degradation[c]				
Primary alcohol	3–6	4–6	4–7	—	—
Primary oxo alcohol	5–10	5–8	6	—	—
Secondary alcohol	7–8	6–7	—	6–7	—
Primary alkylphenol	—	—	4	—	10
Secondary alkylphenol	—	—	14	—	6–21
	Oxygen uptake, % of COD[d]				
Primary alcohol	50	90	65	—	—
Primary oxo alcohol	—	—	—	—	—
Secondary alcohol	85–100	50	—	55	—
Primary alkylphenol	—	—	35	—	—
Secondary alkylphenol	—	—	—	—	35

[a] Steinle (1964).
[b] E_0, alkyl sulfate.
[c] River water, MBAS or surface tension.
[d] Warburg.

tendency for the EO derivatives to be somewhat more resistant. The difference is hardly noticeable when degradation is easy, but is more pronounced when degradation is poorer, as in the tetrapropylene oxo alcohol derivatives (Table 6.15).

The alkylphenol ethoxylate sulfates (Table 8.12) show widely divergent biodegradability, depending on hydrophobe structure. In general they parallel the corresponding alkylbenzenesulfonates, except perhaps they are somewhat more resistant, and there also appears to be a tendency toward greater resistance with greater degree of ethoxylation.

In particular, the results reported on linear secondary alkylphenol derivatives are surprisingly poorer than for the structurally similar LAS. So far as can be judged without direct comparison, the primary bio-degradation of LAS is both faster and further, as measured by MBAS. Smithson (1966) relates this to a more pronounced effect of phenyl position along the chain, since the internal isomers appear to be much more resistant than the 2-phenyl. For instance, under conditions where the 2-phenyl C_9 ortho isomer **(6.17)** was 90% or more degraded, the corresponding 5-phenyl isomer **(6.18)** was untouched. The 5-phenyl para

C—C—C—C—C—C—C—C—C $\qquad$ C—C—C—C—C—C—C—C—C

$-OE_xO\Sigma$ $\qquad\qquad$ $-OE_xO\Sigma$

(6.17) $\qquad\qquad\qquad\qquad$ **(6.18)**

isomer was intermediate in biodegradability. The same tendency toward improved biodegradability was shown by a "nonrandom" linear alkyl-phenol having a preponderance of 2-phenyl substitution (90% removal) compared to the more even distribution in the usual commercial-type product (70% removal).

Steinle (1964) presents extensive data on relative degradability of alcohol and alkylphenol EO sulfates, as influenced by hydrophobe structure and degree of ethoxylation, summarized in Tables 6.17 and 6.18. The first of these indicates a slight superiority for the linear primary alcohol derivatives over linear oxo and linear secondary alcohols. And, perhaps, that the linear primary alkylphenol derivatives may be slightly superior also, at least as regards primary biodegradation. The linear secondary alkylphenol derivatives appear poorer in this comparison, the E_4 and E_6 derivatives requiring 14 and 21 days for 90% disappearance of MBAS. (However, 90% disappearance of surface tension lowering was apparently more rapid, requiring only 2 and 3 days in three cases in Table 6.18.)

TABLE 6.18

Linear-Branched Alkylphenol EO Sulfates[a]

Methyl number[b]	Days for 90% degradation[c]			Oxygen uptake, % of COD[d]			
	E_4	E_6	E_7	E_4	E_5	E_6	E_7
0.95	2	—	—	—	—	53	—
1.17	—	2	—	37	—	—	—
1.32	—	9	—	—	22	17	13
1.39	—	3	—	—	—	—	—
1.50	15	—	19	—	—	—	—
1.60	18	—	—	17	—	—	—
1.76	—	12	—	—	—	9	—
1.81	38	—	—	—	—	—	—
1.97	—	45	—	—	—	—	—
tp[e]	—	—	—	—	—	4	—

[a] Steinle (1964).
[b] See text.
[c] River water, surface tension.
[d] 2 day Warburg.
[e] Tetrapropylene alkylphenol

Table 6.18 summarizes Steinle's results on a collection of commercial-type alkylphenols ranging from C_9 to C_{12} alkyl groups and differing in the apparent degree of branching of the alkyl chain and extent of ethoxylation. The data clearly indicate that the higher the methyl number the slower the degradation. A tendency toward slower degradation with increased degree of ethoxylation also seems indicated, this trend showing in both cases where direct comparisons can be made.

Steinle's methyl numbers do not precisely give the number of methyl groups in the alkyl chain, since they were determined by nuclear magnetic resonance. The NMR response of a methyl group near the ring is disturbed by the ring, so as to be obscured by the other hydrogens in the molecule. Considering the two unbranched alkylphenol isomers from which **6.17** and **6.18** were derived, both have two methyl groups. By NMR the internal 5-substituted isomer should have a methyl number near 2.0, whereas that of the 2-isomer would be considerably lower. Thus Steinle's poorer degradation with higher methyl number may reflect a higher content of the poorly degradable internal isomers instead of, or in addition to, increased branching.

IX. ETHOXYLATE NONIONICS

The two major factors influencing the biodegradation of the polyethoxylates are (i) the number of ethylene oxide units, E_n, in the hydrophilic group and (ii) the structure of the hydrophobic group. As attested by the work of many investigators, summarized in Tables 8.16 to 8.20, biodegradation is enhanced by decreased E_n and by increased linearity of the hydrophobe.

A. Degree of Ethoxylation

Qualitatively the effect of added EO groups on biodegradability has been verified quite extensively for a range of hydrophobe types. Bogan (1954) and Sawyer (1956) first pointed it out with pairs of ethoxylated fatty acids (Ethofat C/15 and C/60) and fatty amides (Ethomid HT/15 and HT/60), in which the hydrophilic groups averaged (perhaps) E_5 and E_{50}, respectively. The 5 day BOD was around 40–50% of theoretical for the lower ethoxylates, indicating extensive biodegradation, compared to only 5–15% for the higher ones. Bogan (1955) attributed the difference to the reduced lipophilic properties and increased molecular size of the higher ethoxylates, leading to poorer transport through the cell membranes. Oldham (1958, p. 146) also noted increased resistance with increased EO content, but gave no data.

In the many comparisons made subsequently this pattern is everywhere evident. Table 6.19 presents much of this, and the few exceptions involve the branched tertiary octyl- and tripropylene nonylphenol ethoxylates. Those hydrophobes are more resistant, so the attack is probably largely on the EO chain; perhaps that is a factor in their frequent (3 out of 7) reverse behavior.

B. Alcohol Ethoxylates

Study of Tables 8.16 and 8.17 indicates that the major factor affecting the biodegradability of the alcohol ethoxylates is the structure of the hydrophobe group, and in particular the linearity of its carbon skeleton. This has a more pronounced influence than other factors such as hydrophobe chain length, mode of attachment of the polyglycol chain and length of the polyglycol chain (i.e., degree of ethoxylation).

B.1. Hydrophobe Linearity

The linear primary alcohol ethoxylates are characteristically readily biodegradable. Primary biodegradation, whether detected by specific

TABLE 6.19

Degree of Ethoxylation and Extent of Biodegradation

Mols EO:	3	4–5	6–9	10–12	13–14	15–19	20–29	30–50	Test[j]	Time	Analysis[k]	Reference
	Percent removal or percent of theoretical[a]											
A. *Linear primary alcohols*												
C₈	—	100	—	—	13	—	0	—	RW	7d	PM	Huyser, 1960
C₁₂	—	—	99	99	—	—	80	90	RW	6d	σ	Blankenship, 1963
C₁₂	66	97	56	—	—	62	41	3	Wa	10d	O₂	Hunter, 1964
C₁₂	—	—	79[b]	73	—	—	—	—	BOD	30d	O₂	Heinz, 1967
C₁₆	—	65	—	53	—	—	33	—	BOD	20d	O₂	Ruschenberg, 1963a
C₁₆	—	—	—	2[c]	—	—	25[c]	—	RW	—	F	Weil, 1964
C₁₈	—	76	39	—	0	—	—	—	RW	7d	PM	Huyser, 1960
C₁₀₋₁₆	90	75	65	50	—	33	8	—	In	20d	PW	Pitter, 1968a, b
C₁₀₋₁₆	93	84	73	64	48	41	27	—	In	20d	COD	Pitter, 1968a, b
C₁₀₋₁₆	66	62	52	48	34	32	18	—	In	20d	O₂	Pitter, 1968a, b
C₁₂₋₁₄	—	—	—	94	—	85[d]	—	64	In	28d	Wt	Borstlap, 1967a
C₁₄₋₁₆	—	—	100	—	—	86	—	97	BAS	1d	SMB	Han, 1967
C₁₆₋₁₈	—	—	100	—	—	100	98	—	In	7d	TLC	Patterson, 1967
C₁₆₋₁₈	—	91	100	74	—	36	4[e]	—	In	20d	PW	Pitter, 1968a
C₁₆₋₁₈	—	87	—	70	—	46	23[f]	—	In	20d	COD	Pitter, 1968a
B. *Linear oxo primary alcohols*												
C₁₁₋₁₅	100	—	—	95	92	—	—	—	BAS	1d	SMB	Han, 1967
C₁₂₋₁₅	—	—	98[g]	—	—	83[h]	—	68	In	28d	Wt	Borstlap, 1967a

Mols EO:	3	4-5	6-9	10-12	13-14	15-19	20-29	30-50	Test[j]	Time	Analysis[k]	Reference
C. Linear secondary alcohols												
C$_{11-15}$			95		70	40	—	—	RW	21d	F	Booman, 1967
C$_{11-15}$			92		—	57	—	—	BAS	1d	SMB	Han, 1967
C$_{11-15}$			100		80	—	—	—	In	7d	TLC	Patterson, 1967
D. Linear alkylphenols												
C$_8$			71		—	51	—	—	In	28d	Wt	Borstlap, 1967a
E. Branched alkylphenols												
br-C$_8$			46		4[i]	49	—	—	In	28d	Wt	Borstlap, 1967a
t-C$_8$		5-7[i]		4-5[i]			—	—	SF		CT	Booman, 1965
br-C$_9$			29	11		12	4	0	In	20d	PW	Pitter, 1968a
br-C$_9$			16	6		6	3	2	In	20d	COD	Pitter, 1968a
br-C$_9$			11	6		3	2	0	In	20d	O$_2$	Pitter, 1968a
"Nonyl"		58		83		—	—	—	RW	34d	IR	Frazee, 1964b
"Nonyl"			44	29		—	—	—	BAS	1d	SMB	Han, 1967
F. Fatty acids												
Coco			42				—	13	BOD	5d	O$_2$	Sawyer, 1956
Coco			29				—	9	Wa	6h	O$_2$	Sawyer, 1956
G. Fatty amides												
Tallow			42				—	16	BOD	5d	O$_2$	Sawyer, 1956
Tallow			20				—	8	Wa	6h	O$_2$	Sawyer, 1956
H. Fatty ethanolamides												
Lauric	49			23		—	97	—	Wa	5d	O$_2$	Hunter, 1964
Coco		100			100	—	—	—	BAS	1d	SMB	Han, 1967

[a] See Chapter 8, Section III. [b] $E_6 = 76$; $E_8 = 82$. [c] Days for foam disappearance. [d] $E_{15} = 89$; $E_{18} = 81$. [e] $E_{20} = 7$; $E_{25} = 0$ [f] $E_{20} = 32$; $E_{25} = 13$. [g] $E_6 = 100$. [h] $E_{15} = 89$; $E_{18} = 77$. [i] Days for CTAS disappearance. [j] Abbreviations: Chapter 8, Section IV. [k] Abbreviations: Chapter 8, Section VI.

analytical methods or by measurement of surfactancy, is usually rapid and complete. Ultimate biodegradation, as indicated by oxygen uptake or other appropriate procedures, is usually extensive.

Although no studies on synthetic nonlinear hydrophobes seem to have been reported, Oldham (1958, p. 146) states that branched ones are more resistant. The effects of branching can be judged qualitatively by reference to the two major classes of oxo alcohols, derived from linear olefins or from tetrapropylene. The linear oxo alcohols (Chapter 2, Section II.E) are mixtures of linear and singly branched primary alcohols, ranging from perhaps 20% to 50% of the latter if pure linear α-olefins are used as raw material. Commercial linear olefins may contain branched olefins as impurities, and these would increase the branch content of the product. Even so, the total branching in a commercial linear oxo alcohol would probably average much less than one branch per molecule. In contrast, the tetrapropylene-derived oxo alcohol, also a primary alcohol, probably contains three or four per molecule.

Biodegradation results in Table 8.17 for these two classes of nonionics conform quite well with those of the corresponding anionics: the slightly branched linear oxo derivatives are readily biodegradable while those from the highly branched tetrapropylene are quite resistant.

Direct comparisons of the two have been made in several instances. Huddleston (1964b), in river water and detecting by CTAS, found that the linear primary $C_{12}E_9$ and $C_{16}E_{10}$ disappeared in 3 and 5 days, respectively, at which time the branched $tp\text{-}C_{13}E_{14}$ was only 55% degraded; 14% of it was still there at 26 days. In the shake flask test the linear primaries were gone within 2 days, whereas two-thirds of the tetrapropylene derivative was still present at 4 days. These comparisons are perhaps not entirely fair, because of the disparity in degree of ethoxylation, but a great difference in biodegradation is also exhibited when the ethoxylates are more evenly matched. Thus Bunch (1967a) found complete degradation of $n\text{-}C_{12}E_9$ compared to 0–10% for $tp\text{-}C_{13}E_8$ in the Bunch-Chambers test at 7 days. And Patterson (1967) reported 100% for $n\text{-}C_{18}E_8$ against 36% for $tp\text{-}C_{13}E_8$ in the British STCSD inoculation test at 11 days; the latter was only 70% degraded at 49 days, and leveling off.

On the other hand, the slight branching in the linear oxo derivative imposes little hindrance. Borstlap (1967a) compared a C_{12-15} linear oxo with a C_{12+14} linear primary at degrees of ethoxylation ranging from E_6 to E_{30} (Table 6.19). Differences between the two were negligible, probably well within the accuracy limits of the biodegradation test methods.

In summary, a single branch has no noticeable effect, but a multiplicity of branches can greatly impair the biodegradation.

B.2. *Alkyl Chain Length*

The chain length effect, if any, in the alcohol ethoxylates is much less evident than in the alkylbenzenesulfonates. Blankenship (1963) compared the linear primary derivatives $C_{10}E_{6.7}$, $C_{12}E_8$, $C_{14}E_{9.5}$, and $C_{16}E_{10.4}$ in river water, detected by surface tension. He found no significant differences—all degraded to around 97% in 5 to 7 days.

Huddleston (1964b) looked at a wider range of chain lengths (Table 6.20) and found what might be a trend, although further verification is necessary before a final judgment can be made. The degradation in river water speeded up with increased chain length from C_8 to C_{12} and then slowed down again from C_{12} to C_{18}. In shake flask runs no such effect was noticeable.

Earlier, Huyser (1960) had compared linear primary C_8 and C_{18} derivatives in river water (Table 6.19) and found the following percent degradation by phosphomolybdate analysis:

$$n\text{-}C_8E_4 \quad 100\% \qquad E_{14} \quad 13\%$$
$$n\text{-}C_{18}E_4 \quad 76\% \qquad E_{14} \quad 0\%$$

The C_{18} seems to be significantly slower than the C_8, but the retarding effect is much less than that produced by increased degree of ethoxylation. The results of Pitter (1968a, b) on C_{10-16} and C_{16-18} ethoxylates (Table 6.19 again) show a similar pattern.

B.3. *Secondary Alcohols*

Considerable data on linear secondary alcohol ethoxylates are included in Table 8.17. As would be anticipated from the linear alkyl structure, these are readily degraded. In view of the attachment of the hydrophilic group to an internal carbon atom on the hydrophobe chain, it is also not surprising that the secondaries are a little slower to degrade than are the corresponding primaries in the few instances wherein direct comparisons have been made.

As an example, Blankenship (1963) found that in river water 4-dodecanol E_8 required 10 days for 99% disappearance compared to 5–6 days for the 1-dodecanol derivative, as detected by surface tension measurements. The 6-dodecanol isomer still showed 30% remaining when the run was ended at 13 days. Although Myerly (1964) found the commercial C_{11-15} linear secondary alcohol ethoxylate to be substantially identical with the primary alcohol derivative, Vath (1964) found it to lag 3 to 10 days behind the C_{12+14} primary as detected by CTAS, surface tension, or foamability. (But their Warburg oxygen uptakes followed

TABLE 6.20

Effect of Linear Primary Alcohol Chain Length on
Biodegradation of Ethoxylates[a]

Compound	River water, days[b]	Shake culture, days[b]
C_8E_5	8	2.5
$C_{10}E_7$	5	2.5
$C_{12}E_9$	3	2.5
$C_{16}E_{10}$	5	1.5
$C_{18}E_{10.5}$	13	2

[a] Huddleston (1964b).
[b] Days required for complete disappearance of CTAS.

substantially identical curves.) Patterson (1967) likewise observed the
secondaries to be slower than the primaries in the British STCSD inoculation test, detected by TLC, as did Han (1967), using the 24 hr semicontinuous activated sludge test and his sulfation–MBAS analytical
technique:

Linear primary C_{14+16} (Alfol) E_8 100% E_{16} 86%
Linear secondary C_{11-15} (Tergitol) E_8 92% E_{15} 57%

The linear secondary alcohol ethoxylates, like most other classes,
exhibit slower biodegradation with increased degree of ethoxylation
(Table 6.19).

C. Alkylphenol Ethoxylates

Assessment of the biodegradability of the alkylphenol ethoxylates
(APE) has given rise to more disagreement, contradiction, and controversy
than any other area of surfactant biodegradation. Without doubt the main
underlying causes for the discrepancies have been (i) failure to make
sufficient provision for bacterial acclimation, and (ii) failure of the CTAS
analytical method to respond to biodegradation intermediates which still
show substantial foaming and other surface activity. This controversy is
probably a thing of the past, since by now it has been documented quite
thoroughly that with proper acclimation the branched APEs, the main
center of contention, do undergo substantially complete primary biodegradation. This can occur both in the laboratory and in the field, and
with respect both to chemical analysis and to surfactancy.

Perhaps because of varying degrees of acclimation, the APEs sometimes
do not show a very clear-cut relation between degree of ethoxylation and

their biodegradability (Table 6.19). However, the entries from Borstlap (1967a) on linear APE and Han (1967) and Pitter (1968a) on nonylphenol ethoxylates (presumably derived from tripropylene) do show such a trend fairly distinctly.

C.1. *Linear Alkyl Group*

Extrapolating from the quite satisfactory performance of LAS, which comprises linear secondary alkylbenzenesulfonates, one might expect the same to be true of the linear APEs. In general this has not been borne out as yet. The mixed linear secondary derivatives such as might be obtained by orthodox commercial processes seem to be considerably less than outstanding in their biodegradability, although certain specific linear structures are quite good. These latter have usually been obtained in the form of pure compounds of known structure by laboratory synthesis, and their biodegradation properties have provided guidance toward the future commercialization of improved products.

Smithson (1966) has summarized the situation: it appears that the linear APEs undergo primary biodegradation readily when the phenol is linked to the chain at or near the end. In contrast, linkage at or near the center of the chain increases the resistance to a much greater degree than is the case with LAS. It is not yet clear whether the increased resistance of the internal isomers relates to the biodegradation process itself or whether, as appears to be the situation with the branched APEs, discussed below, it is simply a matter of more difficult acclimation.

In any case, the findings of Blankenship (1963) illustrate the pattern. In the river water test, detected by surface tension, the primary *para-n*-octylphenol E_9 (6.19) and the ortho isomer (6.20), degraded much more easily than the mixed, presumably random, linear secondary octylphenol E_{10} (6.21). The difference is much greater than would be expected simply

C—C—C—C—C—C—C—C—⟨ ⟩—OE$_9$ C—C—C—C—C—C—C—C—⟨ ⟩

$\qquad\qquad\qquad\qquad\qquad\qquad\qquad\qquad\qquad\qquad\qquad\qquad\qquad$ OE$_9$

$\qquad$ 98%, River Water, 8 days $\qquad\qquad$ 95%, River Water, 10 days

$\qquad\qquad\qquad$ (6.19) $\qquad\qquad\qquad\qquad\qquad\qquad\qquad$ (6.20)

C—C—C—C—C—C—C—C

⟨ ⟩—OE$_{10}$

50%, River Water, 17 days

(6.21)

TABLE 6.21

Effect of Linear Secondary Alkylphenol Chain Length on Ethoxylate
Biodegradation[a]

Alkyl carbons	Mols EO	% Biodegradation, CTAS		
		RW,[b] 15 days	SF,[b] 5 days	CAS,[b] 4 hr
8	6.7	36	40	75
9	9.3 (9)[c]	65 (95)[c]	65 (45)[c]	88
10	10.3	90	92	100
12	17 (12)[c]	91 (92)[c]	90 (55)[c]	100
14	15.4	93	—	—

[a] Huddleston (1964b, 1965b).

[b] RW, river water; SF, shake flask; CAS, continuous activated sludge.

[c] Figures in parentheses from Huddleston (1965b).

from the higher degree of ethoxylation, and must be attributed to a hindering effect of the secondary linkages in **6.21**. The results cited by Smithson (1966) indicate that the secondary linkage per se is not the important factor, but rather its position, i.e., whether it is near the middle or near the end of the chain. This tendency is illustrated in the percent removals of CTAS in 6 hr continuous-flow activated sludge tests reported in Smithson's original source, describing a nonrandom linear alkylphenol made by a process yielding a preponderance of the 2-phenyl isomer compared to the ordinary, random product. Whereas a random linear alkylphenol ethoxylate showed only 62.5% removal, the nonrandom linear decylphenol $E_{8.5}$ was 91.3% removed and the pure ortho-(2-decyl)phenol $E_{9.5}$, 99.6% removed.

Turning now to chain length, the data of Huddleston (1964b) summarized in Table 6.21 seem to show a trend in accord with the ABS distance principle: the longer the chain the easier the biodegradation. This was evident even though in this series the degree of ethoxylation was increased along with the chain length, which would tend to give an effect in the opposite direction. Unfortunately the trend disappeared in later data by Huddleston (1965b), also given in Table 6.21, which show no significant difference between C_9APE_9 and $C_{12}APE_{12}$. Convincing proof that the distance principle does or does not apply to these nonionics probably must await the use of analytical techniques capable of differentiating the individual homologs, so that they can be compared when mixed together in a single solution.

C.2. *Branched Alkylphenols*

The biodegradation of branched APEs is, on occasion, considerably better than might be anticipated. The prominent discrepancies between results of the various workers are most likely due to difficulties of acclimation. The existence of acclimated organisms at certain locations in the field can be attributed to the many years' use of such products for specialized industrial purposes at those locations. Given equivalent attention in the laboratory and equivalent exposure in the environment, perhaps the linear alkylphenols too would prove to be correspondingly better than present laboratory data suggest.

Little, if any, work has been reported in the literature on pure derivatives with a range of known branched structures such as those used in the fundamental research in the ABS series. The closest approach seems to be the results summarized by Steinle (1964) on a series of alkylphenol ethoxylates with various "methyl numbers" determined by nuclear magnetic resonance (Table 8.19). These products ranged from linear primary alkylphenols through linear secondaries to other derivatives of unstated origin and structure, presumed to be more highly branched because of higher methyl number. But the NMR spectrum is responsive not only to the number of methyl groups but also to their proximity to the ring. Thus the higher methyl numbers could reflect a higher proportion of centrally linked isomers (which have been shown in the preceding section to be less readily degraded) equally as well as a higher degree of branching. Steinle's data show products with methyl numbers 0.95, 1.17, 1.32, and 1.39 to be 90% degraded within 5 to 6 days in river water, detected by surface tension, while products with 1.50, 1.76, and 1.97 took 38 to 45 days. (None showed any significant oxygen uptake in 48 hr Warburg runs.) But because of the NMR ambiguity the actual amounts of branching would seem to be uncertain and the inferred relation of increased branching to poorer degradation inconclusive, at least on the basis of the information presented.

Most of the data on branched APEs come from products based on three commercial-type alkylphenols, the octyl, with alkyl group derived from diisobutylene, the nonyl from tripropylene, and the dodecyl from tetrapropylene. The former is said to be predominantly the tertiary octyl derivative

$$\begin{array}{c} \text{C C} \\ \text{CCCC}\phi\text{OH} \\ \text{C C} \end{array}$$

The two polypropylene derivatives are mixtures of many isomers with a range of alkyl groups having closely related fine structures, just as is the case with TBS. In the (frequent) absence of any specific information on structure it is fairly safe to assume that these are meant when an author refers simply to "octylphenol" or "nonylphenol" or "branched" derivatives.

Side-by-side comparison of these branched products with the corresponding linear secondary derivatives in laboratory biodegradation tests usually shows the linear to be superior, as in Table 6.22. There, both

TABLE 6.22

Comparative Biodegradation of Linear and Branched APEs

| APE | % Biodegradation | | Method[b] | Analysis[c] | Reference |
	Linear[a]	Branched			
C_8APE_9	71	46	In, 28 d	Wt	Borstlap, 1967a
	51	49	In, 20 d	Wt	Borstlap, 1967a
C_9APE_9	65	25	RW, 15 d	CT	Huddleston, 1964b
	65	30	SF, 5 d	CT	Huddleston, 1964b
	88	55	CAS, 4 h	CT	Huddleston, 1964b
	57	33	In, 9 d	CT	Garrison, 1964
	66	32	In, 9 d	σ	Garrison, 1964
	75	0	In, 9 d	F	Garrison, 1964
	62	10	SF, 7 d	CT	Garrison, 1964
	60	18	SF, 7 d	σ	Garrison, 1964
	0–50	0	SF, 7 d	F	Garrison, 1964
	0.1G[d]	0	Wa, 5 d	O_2	Garrison, 1964

[a] Linear secondary.
[b] Abbreviations: Chapter 8, Section IV; d, days; h, hours.
[c] Abbreviations: Chapter 8, Section VI.
[d] G, grams O_2 per gram sample.

linear and branched derivatives fall below the 80% to 90% figure usually taken as the minimum for an environmentally acceptable product, and this trend is generally evident in Tables 8.18 and 8.19 as well.

Vath (1964), Huddleston (1965b), and Osburn (1966) have all pointed out that the cobalt thiocyanate analytical method for nonionics in some instances does not respond to some intermediate biodegradation products which nevertheless may have pronounced foaming properties and other surface activity, and that CTAS reduction alone is thus insufficient evidence for acceptable biodegradability. There has been a resulting tendency to discount all favorable APE results on this account, but Table

8.19 still includes a significant number of entries indicating very extensive degradation even when judged by the more stringent criteria.

At least three workers have presented such evidence indicating extensive biodegradation of the nonylphenol ethoxylates. Sato (1963) reported 96% disappearance of the E_{10} derivative by a phosphomolybdate analytical method and, more significantly, 100% disappearance of the benzene ring as detected by its UV absorption at 275 nm, upon 3 days aeration in activated sludge. In river water tests on nonylphenol E_{10} Osburn (1966) used a foam-stripping prepurification procedure to isolate the intact original material plus any surface active intermediates. These disappeared progressively down to about 5% remaining at 34 days as analyzed both by IR and UV methods. Non-surface-active intermediates did develop during the degradation, but remained behind in the foam-stripping separation step.

Patterson (1968) examined nonylphenol E_9 using the British STCSD inoculation test in conjunction with TLC analysis. Throughout the biodegradation, which ordinarily leveled off at 60% to 80% completion, the remaining nonionic correlated closely with the foaming properties. In many of the runs sewage effluents from various locations were used instead of BOD dilution water, with very little difference in results. Except in a few cases: in a few of the sewage effluents the degradation was, reproducibly, much more rapid and complete, reaching 95–100% in 3 to 4 weeks. Although a differing degree of acclimation was undoubtedly one factor involved, it was evidently not the only one, for sewage generated under detergent-free conditions showed this property in one instance. Further, pH appeared to be important also. Whereas the usual sewage dropped in pH during a run, say from 7.5 to 5.5, the active ones rose, say from 7.5 to 8.5. When the initial pH of the synthetic medium was raised to 9.2 the subsequent degradation was significantly better than without such adjustment. In each case the more rapid alkaline degradation was accompanied by the appearance of intermediate degradation products not evident during the normal degradation, detected as more mobile components in the TLC.*

The biodegradation of the t-octylphenol ethoxylates, particularly OPE_{10}, has been extensively and thoroughly studied by Booman's group (Booman, 1965, 1967; Lashen, 1966, 1967a, b, c). With careful attention to acclimation they approximated complete degradation, as detected by CTAS or by foaming properties, under a variety of test conditions including a field test. Existence of acclimated organisms in the field was

* The WPRL (1968, p. 133–4) reports extensive removals of an APE (structure undisclosed) in trickling filters after sufficient acclimation—70% in the laboratory and 80% in a field test.

demonstrated by river water tests on OPE_{10} using water collected above and below locations of large industrial usage. Also by comparison of activated sludges from industrial waste treatment with others from domestic wastes. Where prolonged exposure to OPE_{10} (which is principally an industrial surfactant, not used in household detergents) had occurred, acclimation had also occurred, or was readily brought about in the laboratory.

D. Other Ethoxylates

Considerable work has been done on other members of the ethoxylate nonionic family, as attested by Table 8.20. It should not be necessary to discuss these in any detail since the results are satisfactorily in agreement with the general principles already discussed, and no new principles appear to have developed from them as yet.

For the most part these surfactants are ethoxylates of linear fatty acids and fatty amides. They are easily degraded even with ethoxylation as high as E_{20}, but two E_{50} derivatives appear to be more resistant.

X. MISCELLANEOUS SURFACTANTS AND SOAPS

Several of the categories of surfactants listed in the tables of Chapter 8 will not be discussed specifically. The results on them are mostly isolated or scattered; if cases appear to be contrary to the principles of surfactant biodegradation already developed, the data are too scanty to provide a useful basis for new generalizations. However, three interesting points will be cited here.

First, the use of a sugar as the hydrophilic group does not result in any spectacular improvement in biodegradability, but falls in line with the usual principles. If the hydrophobic group is resistant, if it is derived from tetrapropylene for instance, primary biodegradation may be quite incomplete. With a linear hydrophobe primary degradation may be rapid and complete, but, as it must, ultimate degradation lags behind and is more difficult to accomplish, just as with other surfactants.

Second, cationic surfactants are susceptible to biodegradation despite their strong bactericidal and bacteriostatic properties. Obviously this can happen only when the test conditions minimize the antibacterial effects, as by working at a low cationic concentration, or at a low ratio of cationic to organisms. Such conditions should exist in the field, since cationics make up only a very minor fraction of total surfactant usage.

Third, the natural soaps and fatty acids are not particularly outstanding in their biodegradability in comparison with the modern commercial surfactants. It is evident that their environmental acceptability in the past was largely due to precipitation as calcium or magnesium soaps rather than to any great superiority in biodegradability.

A. Soaps

Loehr (1968) has studied the biodegradation of a series of fatty acids and soaps using Warburg techniques. He was more concerned with degradation rate than with extent, but casual inspection of his oxygen uptake curves indicates that oxidation exceeding 50% of theoretical was often achieved within 6 to 24 hr. Lag periods of several hours to a day or more sometimes occurred, indicating a need for acclimation of the seed. His results may be briefly summarized as follows:

(i) The sodium soaps up through C_{18} could be metabolized by the bacteria. The corresponding calcium soaps could also, as long as the insoluble particles remained finely divided. If agglomeration occurred, degradation slowed or ceased.

(ii) Degradation of the sodium soaps was slower with greater chain length in the series C_8, C_{12}, C_{16}, C_{18}, C_{20}.

(iii) Unsaturated soaps were degraded more readily than the corresponding saturates, except in one case, sodium elaidate, a C_{18} homolog with a trans configuration at the double bond.

(iv) Concentrations of 300 to 400 ppm were used in most cases, but concentrations of several thousand parts per million did not appear to be toxic, at least in the case of the oleate.

The WPRL (1968, p. 77) reports on removal of stearic acid in batch activated sludge systems. Maximum rates were in the range 1–2 mg/gm per hour, comparable to the rates for LAS found earlier (WPRL, 1967, pp. 175–6). In an activated sludge plant, fatty acid removal occurred at a similar rate, about 60% of it being degraded, 10% appearing in the effluent suspended solids, and 30% leaving the system on wasted excess sludge.

CHAPTER 7

METABOLIC PATHWAYS
AND ULTIMATE BIODEGRADATION

I. BIOCHEMICAL OXIDATION

The chemical mechanisms used by bacteria for the biodegradation of surfactants are those which they already have, either fully developed or latent, for utilizing normal foods in their normal life processes. These reactions are catalyzed by enzymes. The mere presence of an exotic organic compound may often trigger the development of modified enzymes capable of accepting the new compound as a substrate, leading to its utilization as a food. So, what is an exotic chemical in one situation may be a quite normal food for some other bacterial community in a different environment.

Accordingly, a wide variety of organic compounds can serve as foods for bacteria, providing for their growth and energy requirements. The variety of biochemical mechanisms needed for doing this is much narrower, since a given mechanism can often be used by the organism for a whole series of related compounds, and since many of the intermediate degradation products are the same. Surfactants, to the extent that they are biodegradable, serve as foods also, and their degradation is brought about by these same reactions. Some minor modifications of some of the processes may be needed sometimes because of the peculiar molecular combination of hydrophobic and hydrophilic groups which makes a surfactant a surfactant.

The net overall reaction in bacteria is oxidation, as it is in animals. We will briefly discuss three of the general oxidative mechanisms which appear to be particularly pertinent in the bacterial utilization of surfactants. These deal with (i) terminal or ω-oxidation (which, in a semantic quirk, is often also initial oxidation), the first step in the degradation at the terminus of the hydrophobic group, (ii) β-oxidation, the process whereby the aliphatic portion of the hydrophobic group is degraded, and (iii) aromatic oxidation, which is applicable when the hydrophobic group contains a benzene ring.

A. ω-Oxidation—Aliphatic Hydrocarbons

Evidence presented in Section III indicates that the attack on LAS, and presumably on many other surfactants as well, begins with the oxidation of a terminal methyl group to a carboxyl group. In the absence of any indication to the contrary we can assume, at least as a first approximation, that the mechanisms used are adapted from those used in the initial attack on unsubstituted aliphatic hydrocarbons.

Bacteria can thrive on linear aliphatics of almost any chain length, so readily, indeed, that petroleum is a potential raw material for conversion to food protein, i.e., to bacterial protoplasm. Many common species of bacteria have this ability, with or without acclimation, some preferring shorter chain lengths, some longer. Entry into the literature of bacterial attack on hydrocarbons can be gained via reviews by Fuhs (1961), Foster (1962a, b), Johnson (1964), McKenna (1965), van der Linden (1965), and Humphrey (1967). These form the basis for the unreferenced statements following in this section.

Fredricks (1966) showed that in at least two cases, *Pseudomonas fluorescens* and a *Corynebacterium* species, the ability to utilize *n*-dodecane as a sole carbon source was not impaired even after 38 transfers on a 1% glucose medium in absence of dodecane. He suggested that the enzymes necessary for hydrocarbon oxidation were thus constitutive rather than adaptive. This was not proved conclusively, since his experimental procedure did allow 6 days exposure to the hydrocarbon medium before cell counts were made, but even if not constitutive, the necessary enzymes were certainly induced quite readily. Van Eyk (1968) found that induction of certain enzymes was necessary before his strain of *Pseudomonas aeruginosa* was able to oxidize *n*-hexane. Paraffins from *n*-butane to *n*-octane were good enzyme inducers, as were a number of other compounds which were not themselves metabolized, COCCOC and CCOCCOC, for example. On the other hand, the inducing action of the paraffins was repressed by the presence of many other materials such as yeast–peptone extract, glucose, and many simple carboxylic acids.

Initial attack on the longer *n*-paraffins (C_6 and higher) is predominantly terminal, at one end of the chain, yielding the corresponding fatty alcohol and fatty acid as the first identifiable products. Evidence on the exact mechanism and the intermediate steps is conflicting, probably because several different pathways may be used depending on (i) the particular

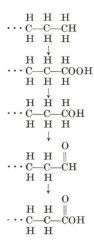

Fig. 7.1   ω-Oxidation by oxygenation.

microorganisms involved, (ii) the chain length, (iii) other structural features of the hydrocarbon, and (iv) the operating conditions.

There is strong indication that one of the pathways involves addition of molecular oxygen to the hydrocarbon, catalyzed by an oxygenase enzyme, to give the primary hydroperoxide which in turn is converted to the primary alcohol, to the aldehyde, and to the carboxylic acid (Fig. 7.1). A second possible pathway, more questionable, or perhaps only more rarely noticed, involves initial formation of a double bond at the chain end (Fig. 7.2).

After formation of the fatty acid, the next steps in the biodegradation

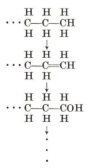

Fig. 7.2   ω-Oxidation by dehydrogenation.

are usually those of β-oxidation, shortening the chain two carbons at a time as discussed in the next section. But other reactions have also been observed in particular cases, for example, fatty ester formation, or diterminal oxidation wherein the ω-oxidation is repeated at the other end of the chain. Formally the ester can be viewed as the reaction product of the fatty alcohol and the fatty acid, but more likely it is formed directly from more reactive intermediates in the biooxidation. The diterminal oxidation occurs when the β-oxidation is for some reason slow enough or ω-oxidation fast enough to permit simultaneous attack at both ends of the chain. This results in α,ω-dicarboxylic acids, susceptible to subsequent β-oxidation from both ends.

Cyclohexane, which has no chain end, also undergoes bacterial attack. There is evidence indicating that the initial stage is again a hydroperoxide **(7.1)**, which converts to the corresponding secondary alcohol. Secondary

(7.1)

alcohols or their oxidation products, ketones, have also been found in bacterial oxidation of the shorter, gaseous, linear hydrocarbons. Yeasts have been observed to give secondary alcohols from the longer hydrocarbons also. For example, Klug (1967) found that *Candida lipolytica* formed both primary 1-alkanols and secondary 2-alkanols in about a 2:1 ratio from the C_{14}, C_{15}, C_{16}, C_{17}, and C_{18} alkanes. Jones (1968a) likewise reports extensive formation of both primary and secondary (2-hydroxy) derivatives in a variety of long chain compounds having substituents already present at the other end, by another yeast, *Torulopsis gropengiesseri*. He showed that this organism operates by direct introduction of the oxygen at or next to the chain end. Bacteria will probably be found to make similar conversions on occasion also, if they are watched closely enough.

Central attack on a hydrocarbon chain also occurs. Abbott (1968) fed *n*-hexadecane to glucose-grown cells of *Nocardia salmonicolor*, whereupon the dehydrogenation product 7-hexadecene accumulated in the system, accompanied by minor amounts of the 6- and 8-isomers, up to 20% or more. Acclimated cells did not accumulate the olefin, presumably because they were then able to degrade it further as soon as formed. A likely path would be a splitting of the molecule at or near the double bond to give two molecules of short fatty acids which would then be further oxidized. Internal oxidation of *n*-hexadecane to ketones via secondary alcohols has been observed by Klein (1969) in an *Arthrobacter* species.

Bacterial attack on branched hydrocarbons also occurs, the degree of difficulty apparently depending on the structural details of the molecule.

B. β-Oxidation

In living cells, fatty acids are degraded by the process of β-oxidation. In principle, they are also synthesized by the reverse reaction, although somewhat different synthesis pathways are commonly used. Fatty acids are a universal component of all living cellular organisms, and the β-oxidation mechanism is used in all types—animal, plant, or microbial. The literature on β-oxidation is voluminous; discussion of the basic principles is given in detail by Stumpf (1960) and in summary by Anderson (1967) and Overath (1969). Induction and repression of the necessary enzymes in *E. coli* has been studied by Weeks (1969).

Briefly, the reaction is an oxidation of the fatty acid chain two carbons at a time into a succession of acetyl groups which are used for energy or synthesis reactions by the cell. The reaction is actually a series of reactions, enzymically catalyzed. A coenzyme is also involved, called coenzyme A, a moderately complex mercaptan **(7.2)** (Moffatt, 1959) (containing, incidentally, a quaternary carbon atom).

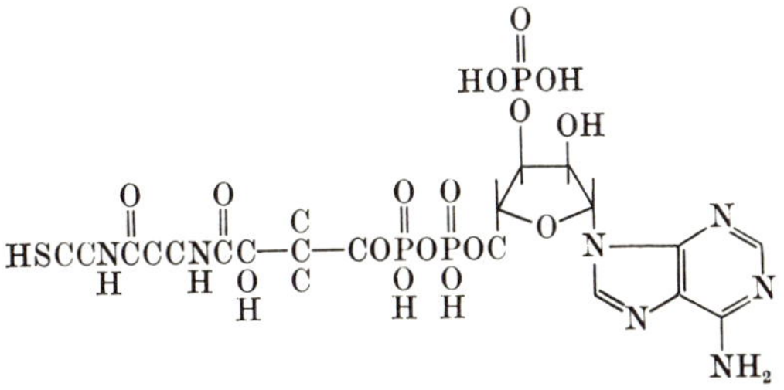

Coenzyme A (HSCoA)

(7.2)

The overall reaction is outlined, not in full detail, in Fig. 7.3. First the carboxyl group is esterified with coenzyme A; at least two intermediate steps are involved there. Next, two hydrogens are removed to give the α,β-unsaturated derivative, which is then hydrated to the β-hydroxy and then dehydrogenated to the β-keto derivative. Finally another molecule of coenzyme A adds between the α- and β-carbons, splitting off acetyl coenzyme A and leaving a fatty acid coenzyme A ester two carbons shorter than the original. It is ready to engage in the same sequence of reactions for still further degradation.

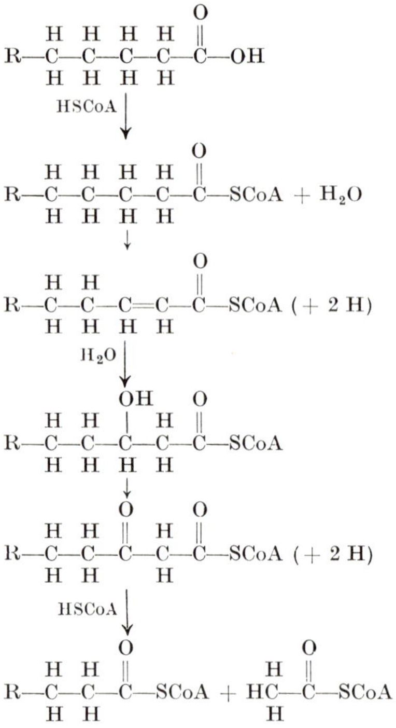

$$
\begin{array}{c}
\quad\quad\quad\quad\quad\quad \overset{O}{\underset{\|}{}} \\
\text{H H H H} \\
\text{R—C—C—C—C—C—OH} \\
\text{H H H H}
\end{array}
$$

HSCoA $\downarrow$

$$
\begin{array}{c}
\quad\quad\quad\quad\quad\quad \overset{O}{\underset{\|}{}} \\
\text{H H H H} \\
\text{R—C—C—C—C—C—SCoA} + H_2O \\
\text{H H H H}
\end{array}
$$

$\downarrow$

$$
\begin{array}{c}
\quad\quad\quad\quad\quad\quad \overset{O}{\underset{\|}{}} \\
\text{H H} \\
\text{R—C—C—C=C—C—SCoA } (+ 2 H) \\
\text{H H H H}
\end{array}
$$

H_2O $\downarrow$

$$
\begin{array}{c}
\quad\quad\quad \text{OH} \quad O \\
\text{H H} \quad | \quad \text{H} \quad \| \\
\text{R—C—C—C—C—C—SCoA} \\
\text{H H H H}
\end{array}
$$

$\downarrow$

$$
\begin{array}{c}
\quad\quad\quad \overset{O}{\|} \quad\quad \overset{O}{\|} \\
\text{H H} \quad\; \text{H} \\
\text{R—C—C—C—C—C—SCoA } (+ 2 H) \\
\text{H H} \quad\;\; \text{H}
\end{array}
$$

HSCoA $\downarrow$

$$
\begin{array}{c}
\quad\quad \overset{O}{\|} \quad\quad\quad\quad\quad \overset{O}{\|} \\
\text{H H} \quad\quad\quad\quad\quad \text{H} \\
\text{R—C—C—C—SCoA} + \text{HC—C—SCoA} \\
\text{H H} \quad\quad\quad\quad\quad \text{H}
\end{array}
$$

Fig. 7.3 β-Oxidation (HSCoA = coenzyme A).

Each of the reactions indicated in Fig. 7.3 is itself a sequence of reactions. Each is catalyzed by its own specific enzyme and activators, and is reversible under suitable conditions. The hydrogen indicated in Fig. 7.3 does not actually appear as free hydrogen atoms, but instead is plucked off by hydrogen transfer agents such as nicotinamide adenine dinucleotide (NAD) or flavin adenine dinucleotide (FAD), which pass it on to other labile components of the cell. In aerobic systems the hydrogen may ultimately be passed on to atmospheric oxygen, forming water. In anaerobic systems, too, the β-oxidation mechanism is used, with the ultimate hydrogen acceptor being perhaps a carbon compound (forming methane) or a sulfur compound (forming hydrogen sulfide) or the like, depending on the particular organism and circumstances.

McKenna (1966) advances evidence that branched hydrocarbons such as pristane

$$
\begin{array}{c}
\text{C} \quad\; \text{C} \quad\; \text{C} \quad\; \text{C} \\
\text{CCCCCCCCCCCCCCCC}
\end{array}
$$

may follow the same metabolic pathways as linear ones, i.e., ω-oxidation followed by β-oxidation. Stokke (1969) has proved that the latter can indeed be done, except that α-oxidation occurs when the branch is on the β-carbon. He found that guinea pig kidney tissue degraded 3,6-dimethyl-octanoic acid **(7.3)** by successive steps of α-, β-, α-, and β-oxidation. When

$$C{-}C\overset{C}{\underset{}{|}}C{-}C\overset{C}{\underset{}{|}}C{-}C{-}CO_2H$$

(7.3)

the methyl group is on the β-carbon, β-oxidation is avoided, presumably because formation of a β-keto group in such a case would involve a pentavalent carbon. Results from α,β-dimethyl and α,β,γ-trimethyl derivatives should be quite interesting. Or perhaps frustrating.

C. Aromatic Oxidation

The benzene ring occurs in all living systems, for example, in several of the amino acids, and it should not be surprising that metabolic mechanisms are available for the synthesis and degradation of aromatic compounds. McKinney (1956b) has explored some of the pathways used by the organisms of activated sludge, and has briefly reviewed some of the early milestones. Knox (1961), Evans (1963), Dagley (1965a), van der Linden (1965), and Gibson (1968) have reviewed the various ring degradation mechanisms, while Fuhs (1961), Foster (1962a), and McKenna (1965) have dealt more particularly with the attack on benzene itself and other aromatic hydrocarbons by bacteria and other organisms.

Two of the common routes for benzene ring biodegradation are shown in Fig. 7.4. Catechol is taken as the starting point since it is the common intermediate formed from benzene itself and many benzene derivatives— benzoic acid or phenol or salicylic acid or others. From any of these catechol is formed first in an enzyme-catalyzed oxidation with molecular oxygen, and the ring is then split between or adjacent to the two hydroxyl groups. In the first case a dicarboxylic acid is formed, which is converted by three successive molecular rearrangements into β-ketoadipic acid. This can then be split by the β-oxidation process to give acetate and succinate groups, both of which are ordinary cell components.

In the second pathway the initial rupture of the ring occurs adjacent to the two hydroxyls, leading to formic acid, acetaldehyde, and pyruvic acid, all of which are common cell metabolites.

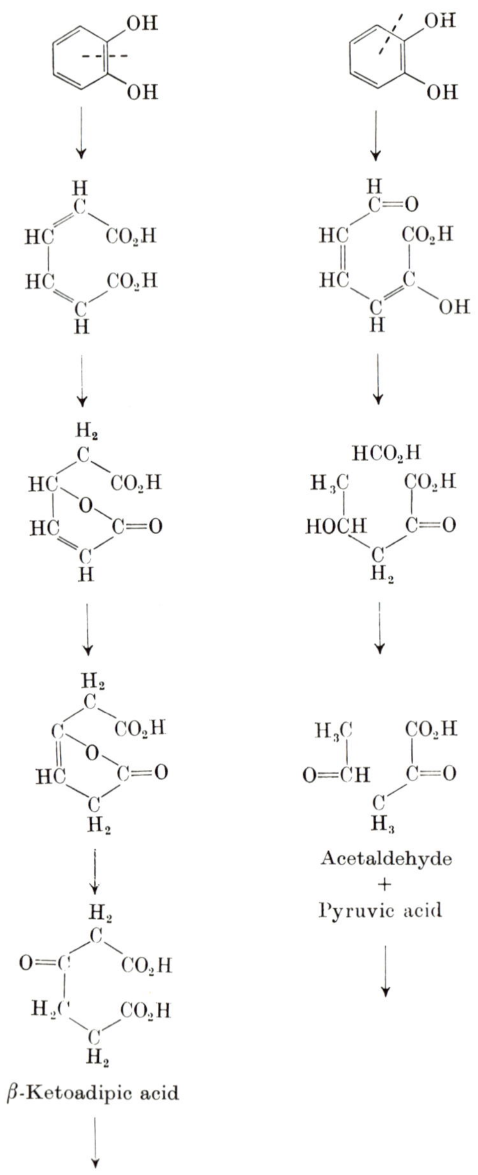

Fig. 7.4   Biodegradation of benzene ring via catechol.

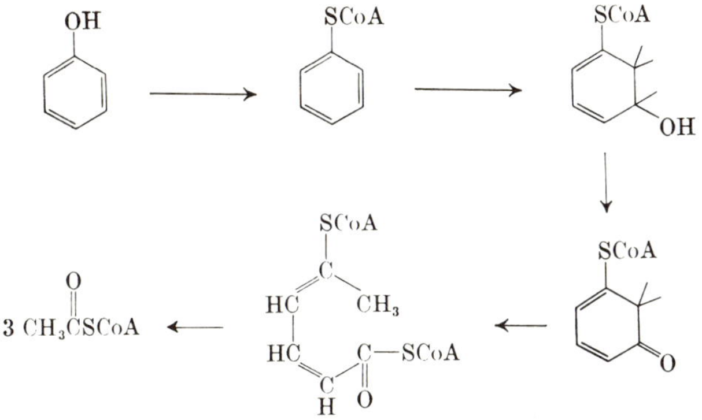

Fig. 7.5   Benzene ring degradation by β-oxidation only (McKinney, 1956b).

McKinney (1956b) observed that a phenol-acclimated sludge oxidized resorcinol somewhat more readily than catechol and suggested a third pathway (Fig. 7.5) via phenyl coenzyme A. The *meta*-hydroxylated intermediate should be readily formed from resorcinol directly, as well, and the subsequent degradation should be accomplished using the β-oxidation process only.

D. Unsulfonated Alkylbenzenes

Unsulfonated higher alkylbenzenes are not necessarily valid models for study of ABS biodegradation because (i) the absence of the sulfonate group may influence the applicability of various metabolic pathways, and (ii) the very limited water solubility of the higher alkylbenzenes introduces problems and uncertainties in making them physically available to the microorganisms in known and reproducible amounts. Even so, the results from these compounds are very interesting in their own right, and further, they do parallel those found with the sulfonates. Although they perhaps cannot be drawn upon for rigorous confirmation of the sulfonate results, it is comforting to see the similarity—attack begins at the end of the alkyl group and then the chain is degraded by the β-oxidation process.

Two general procedures have usually been used, either observation as to whether the alkylbenzene can support bacterial growth when it is the sole source of carbon, or isolation and identification of metabolic products. Even though the compound may not support growth when fed alone, in some cases it is attacked when fed along with another substrate which can serve as a food source, such as a linear alkane.* Ability to utilize

* Raymond (1969) gives many examples in a recent review.

alkylbenzenes differs from species to species of microorganism and from structure to structure of alkylbenzene, just as with other compounds. For a given alkylbenzene in a given set of conditions one species may not be able to attack it at all, a second may be able to degrade it to a certain extent, while a third may be able to carry the degradation to a more advanced stage, or not as far. ‿

Using a strain of *Nocardia* which was unable to degrade phenylacetic acid, Webley (1956) noted a buildup of that acid upon feeding 1-phenyl-decane or the C_{12} or C_{18} homologs. He concluded that the most probable course of reaction was ω-oxidation followed by β-oxidation [Eq. (7.1)]. In a similar manner 3-phenyleicosane yielded α-phenylbutyric acid [Eq. (7.2)]. The odd carbon 1-α-naphthylundecane yielded an analogous product [Eq. (7.3)].

$$\phi CCCCCCCCCC \longrightarrow \phi CC'CC'CC'CC'CC'CO_2H \longrightarrow \phi CCO_2H \qquad (7.1)$$

$$\underset{\phi}{CCCC'CC'CC'CC'CC'CC'CC'CC'CC} \longrightarrow \underset{\phi}{CCCCO_2H} \qquad (7.2)$$

$$(7.3)$$

Davis (1961), using another *Nocardia* species, obtained similar results. He was able to isolate 80% of the theoretical yield of phenylacetic acid starting with 1-phenyldodecane. However, the odd-chain homolog 1-phenylnonane gave only small amounts of the odd acids phenylpropionic **(7.4)** and cinnamic **(7.5)**. These accounted for only 5% of the original

$$\phi\!\!-\!\!C\!\!-\!\!C\!\!-\!\!CO_2H \qquad \phi\!\!-\!\!C\!\!=\!\!C\!\!-\!\!CO_2H$$
$$\textbf{(7.4)} \qquad\qquad \textbf{(7.5)}$$

nonylbenzene; the other 95%, including the benzene ring, was apparently oxidized to new cells and carbon dioxide. Experiments with the acids themselves showed that this particular *Nocardia* species was unable to oxidize phenylacetic acid but readily utilized benzoic, phenylpropionic, and cinnamic acids.

TABLE 7.I

Alkylbenzenes as Substrates for Bacterial Growth[a]

C$_{10}$	C$_{11}$	C$_{12}$	C$_{20}$
φCCCCCCCCCC[b]	φCCCCCCCCCCC[b]	φCCCCCCCCCCCC[b]	φCCCCCCCCCC · · ·[b]
	C φCCCCCCCCCC	C φCCCCCCCCCC C	CCCCCCCCCC · · ·[b] φ
C$_{13}$			
CCCCCCCCCCCCC φ	CCCCCCCCCC φ	CCCCCCCCCCCC φ	CCCCCCCCCC · · · φ
		C CCCCCCCCCC φ	CCCCCCCCCC · · · φ
C$_{14}$			
CCCCCCCCCCCCCC φ		CCCCCCCCCCCC φ	CCCCCCCCCC · · · φ
		CCCCCCCCCCCC φ	CCCCCCCCCC · · · φ
			CCCCCCCCCC · · · φ

[a] McKenna (1964, 1966).
[b] Supported growth of *Micrococcus* and *Pseudomonas*; all compounds supported growth of *Nocardia*.

Although resistant to the *Nocardia* under the conditions mentioned, the even-carbon phenylacetic acid can be utilized by other organisms. For instance, both Dagley (1965b) and Blakley (1967) report degradation by *Pseudomonas* species with splitting of the benzene ring. On the other hand, Douros (1967, 1968) found conditions under which half a dozen organisms, including a *Micrococcus* and several *Pseudomonas* species, were unable to further degrade the odd-carbon acids **7.4** and **7.5**, this time resulting from 1-phenylpentane and 1-phenylheptane.

Bacterial growth experiments are reported by McKenna (1964, 1966) in which 19 species of *Micrococcus, Pseudomonas, Mycobacterium*, and *Nocardia* were cultured with many linear and branched alkanes and alkylbenzenes; the latter are listed in Table 7.1. The *Nocardias* were the most versatile, exhibiting growth (although questionable in a few cases) on all hydrocarbons presented except for a highly branched Me$_5$-heptane;

they did grow on the still branchier homolog Me$_7$-nonane:

```
    C C C                           C C C C
    CCCCCCC    (no growth)          CCCCCCCCC    (growth)
    C    C                          C  C   C
```

The *Mycobacterium* species were almost as versatile, but the *Pseudomonas* and *Micrococcus* were not. The latter two grew on the linear 1-phenyl-alkanes, but not on the linear secondary isomers, nor on the branched derivatives.

Failure of the several *Pseudomonas* species to grow on most of the unsulfonated alkylbenzenes in Table 7.1 may reasonably be ascribed to other factors than resistance in the chemical structures. First, Warburg respirometry (McKenna, 1966) showed significant oxygen uptake when acclimated *Pseudomonas* were fed the secondary phenylalkanes, even though growth was not supported. (Evidently the oxidation products could not be further utilized by the organisms.) Even more significant, all of the C$_{10}$ to C$_{14}$ alkylbenzenes listed in Table 7.1 readily undergo primary biodegradation if sulfonated (Table 8.1), and representatives of all the structural types have been shown to undergo benzene ring degradation as well, thus at least approaching ultimate biodegradation. (The C$_{20}$ homologs apparently have not been investigated.) Although tests on sulfonates are usually run in mixed bacterial systems, it seems quite likely that *Pseudomonas* in pure culture could achieve at least primary biodegradation of all these sulfonates.

Nyns (1969b) investigated 21 unidentified bacterial strains isolated from soil by enrichment culture on tetrapropylene alkylbenzene, and another 22 on C$_{13}$ alkylbenzene mixed isomers, presumably linear. Upon cross-feeding, 17 of the C$_{13}$ strains could grow on the *tp*-alkylbenzene, but only one of the *tp* strains on the C$_{13}$. Only five of each grew on C$_{13}$LAS, and none on TBS, indicating that acclimation to the alkylbenzenes did not impart any special proficiency toward the corresponding sulfonates. Nor were another 22 strains, isolated using C$_{13}$LAS, especially suited for the alkylbenzenes: 6 of them showed growth on the C$_{13}$ and one on the *tp*-alkylbenzene.

II. BIODEGRADATION OF LOWER SULFONATES

Lower molecular weight sulfonates have been studied in hope of gaining, by extrapolation, further insight into the processes of surfactant bio-degradation, or into the reasons for biological inertness. Considerable caution must be exercised in drawing any inferences from such work because the presence of a hydrophobe group, or a degradation product thereof, may conceivably alter its biodegradation behavior compared to that of the unsubstituted hydrophile. But regardless of the possible difficulty of applying them to the problem at hand, the results are of considerable interest in any case.

A. Aromatic Sulfonates

Bogan (1955) and Hammerton (1955) both studied the separate components of surfactant molecules. Both agreed that sodium benzenesulfonate was readily degraded, as evidenced by extensive oxygen uptake

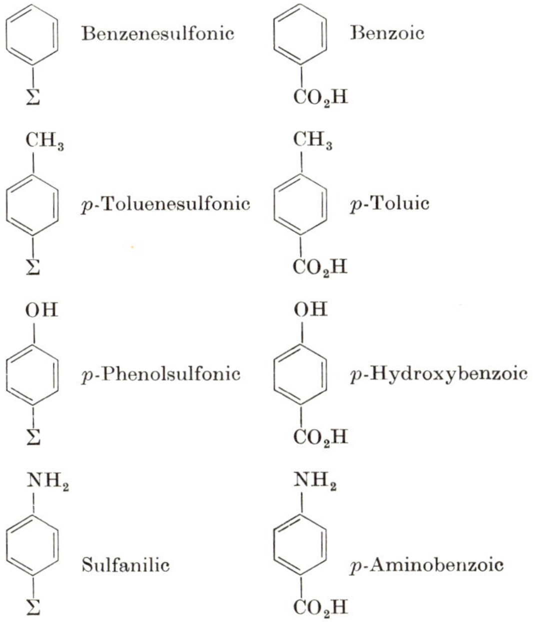

Benzenesulfonic

Benzoic

p-Toluenesulfonic

p-Toluic

p-Phenolsulfonic

p-Hydroxybenzoic

Sulfanilic

p-Aminobenzoic

Fig. 7.6 Sulfonates and carboxylates used in acclimation studies (Symons, 1961).

in Warburg or BOD systems. The listings in Table 8.27 show that others have confirmed this as well, using respirometry, UV spectroscopy, and formation of inorganic sulfate as criteria.

Symons (1961) acclimated eight batch activated sludge cultures to four benzenesulfonic acid derivatives and to the corresponding four benzoic acid derivatives as sole carbon sources. These are illustrated in Fig. 7.6— the plain, the p-methyl, the p-hydroxy, and the p-amino. Warburg measurements were then made for each sludge on each of the eight substrates, seeking evidence for interrelations between the various metabolic pathways through comparison of rate, extent, and profile of the oxygen uptakes.

The initial oxygen rates of Symons' acclimated systems are given in Table 7.2. The sulfonates were slower than the corresponding carboxylates,

TABLE 7.2

Biodegradation of Benzenesulfonates and Carboxylates[a]

Substituent	Sulfonate[b]	Carboxylate[b]
None	133	310
p-Methyl	70	156
p-Hydroxy	120	550
p-Amino	58	110

[a] Symons (1961).

[b] Initial oxygen uptake rate, ppm/hr, by acclimated sludges, each fed its own substrate, in Warburg respirometer.

and the presence of a substituent in the ring slowed the degradation in every case except for the p-hydroxybenzoate. The cross-feeding studies showed that of the sulfonate sludges, only two were fully acclimated to each other's substrate, namely the benzene- and p-toluenesulfonates. These two sludges also degraded the p-phenolsulfonate, but the uptake rate was only about half that of full acclimation. The sulfanilate sludge rapidly degraded both sulfanilate and p-phenolsulfonate, but the uptake curves were concave or flat instead of convex, taken to indicate incomplete acclimation. None of the carboxylate sludges was fully acclimated to any of the sulfonate substrates, although the p-aminobenzoate sludge did accomplish fairly extensive attack on the benzene- and toluenesulfonates.

Symons felt that both the benzene- and toluenesulfonates were degraded by the benzenesulfonate sludge through attack on the sulfonate group, believing that attack on the methyl group was difficult, and unlikely to occur rapidly in an unacclimated sludge. Further, he felt it likely that the toluenesulfonate sludge worked through the methyl group, and that its apparent ability to oxidize the benzenesulfonate equally readily was due to an irrelevant abnormality in those particular comparative runs, where the toluene oxidation rate was considerably slower than had been observed on other occasions.

The p-phenolsulfonate sludge showed no cross-acclimation to the other sulfonates, taken to indicate that it was working through the hydroxyl group. The sulfanilate presented a special difficulty in that it had some of the antibacterial properties of its relative, sulfanilamide. Although the sulfanilate sludge was able to oxidize sulfanilate rapidly and extensively, the concave shape of the curve indicated some deviation from normal acclimation. The other sludges did very poorly on sulfanilate. The comparative results were interpreted by Symons as probably indicating attack through the amino group by the acclimated sludge, and that the same mechanism could be brought to bear on the hydroxyl group of the p-phenolsulfonate, but not on the methyl group of the p-toluenesulfonate.

Symons concluded that if any other group besides sulfonate is present on the ring, this alternate group is the point chosen for attack, and that the possibility of degrading a higher alkylbenzenesulfonate from the sulfonate end is remote.

He further concluded that the presence of a sulfonate group makes a compound hard to degrade, but this last does not seem to be completely substantiated. True, the sulfonates he examined were not as rapid as the corresponding carboxylates, but other data suggest that a sulfonate group may be much better than no group at all, at least in some cases. For instance, compare Bogan's (1955) 38.8% of theoretical oxygen uptake for sodium benzenesulfonate with the 3.3% he found for benzene, both with acclimated sludges in 6 hr Warburgs. Or the 78% and 5% figures of Winter (1962). Perhaps this indicated improvement derives from physicochemical reasons such as increased solubility or reduced volatility, or from biochemical ones such as increased susceptibility to biooxidation or reduced toxicity, but whatever the reason, the presence of the sulfonate has here made the molecule easier to degrade.

Conclusive proof that benzenesulfonate and toluenesulfonate can be degraded through attack on the sulfonate group has been provided by Cain (1968). He isolated 17 *Pseudomonas* strains by enrichment culture on sulfonate medium from sewage, river water, or soil. These fell into three groups. All of them grew on benzene- and toluenesulfonates as the sole carbon source, but 12 were unable to do so on the higher sulfonates, from ethylbenzene up to octadecylbenzene. The other five strains could live on any of the benzenesulfonates from C_0 to C_{18}.

The first 12 strains further fell into two groups. Nine, typified by strain A, used metabolic pathway A for the ring degradation and 3, including strain B, used pathway B. The detailed steps in the two pathways were determined by isolation and identification of the intermediate degradation products, and it was further shown that the enzymes involved were adaptive, developed by exposure to the benzene- or toluenesulfonate.

Both pathways begin with oxidative scission of the sulfonate group yielding catechol and inorganic sulfite [Eq. (7.4)]. The sulfite is rapidly

$$\text{(benzene ring)}\ SO_3Na \longrightarrow \text{(catechol ring)}\ \substack{OH \\ OH} + NaHSO_3 \tag{7.4}$$

oxidized to sulfate, perhaps by simple inorganic oxidation by atmospheric oxygen. The catechol is rapidly oxidized via the two different pathways shown in Fig. 7.4. Strain A works via pyruvic acid, strain B via β-ketoadipic acid. When strain A is fed p-toluenesulfonate the methyl group is not oxidized but persists unchanged through all the steps shown in Fig. 7.4, from 4-methylcatechol to propionaldehyde.

In addition to working out the details of these benzenesulfonate and p-toluenesulfonate degradations, Cain (1968) also reports a few experiments with longer alkylbenzenesulfonates. Sulfite was detected in cultures fed the linear primary para C_4, C_6, and $C_{18}ABS$. The catechol 2,3-oxygenase enzyme was detected in cells grown on the C_2, C_4, C_{10}, C_{12}, and C_{16} homologs, as well as with 2-phenyloctane LAS. But the enzyme activity was very low in the case of the C_4 and higher homologs, amounting to only a few percent of that developed on the toluenesulfonate. Further investigaton will be required before it is clear whether the long chain products and their shorter chain carboxylated intermediate degradation products can follow this same sequence of sulfonate removal before ring degradation.

The possibility also exists that still other pathways may be utilized in other circumstances. Thus Heyman (1968) suggested that p-toluenesulfonate first underwent oxidation to p-sulfobenzoic acid when degraded by *Pseudomonas* C12B, because the acclimated cells were able to oxidize benzoic acid and catechol without delay whereas unacclimated cells showed a lag. Perhaps these data are consistent with other pathways as well.

B. Aliphatic Sulfonates

In any cases where ring splitting does occur before, removal of the sulfonate group, as well as in degradation of alkane sulfonates and olefin sulfonates, the intermediate products will be aliphatic sulfonates. Such products have been isolated as crude mixtures, not further characterized, in the biodegradation of LAS (Section III.A). Fundamental information on mechanisms for the further biodegradation of such compounds is scanty, but at least two pertinent cases have been studied, taurine and a sulfonated sugar.

Taurine, $H_2NCC\Sigma$, is a metabolic intermediate and end product widely distributed in animals. Ikeda (1963) isolated a taurine-degrading species of *Agrobacterium* by the enrichment culture technique from river mud. Complete oxidation of taurine, without nitrification, should require 3 mols of oxygen per mol and should give 1 mol of ammonia and 1 mol of sulfate [Eq. (7.5)]. Warburg tests resulted in the absorption of a little over half

$$H_2N—CH_2—CH_2—SO_3H + 3\ O_2 \rightarrow 2\ CO_2 + H_2O + NH_4HSO_4 \qquad (7.5)$$

the theoretical amount of oxygen and production of a little less than half the theoretical carbon dioxide. Search for other products to account for the missing, unoxidized carbon was unsuccessful, leading to the conclusion that it might have been converted to new cellular components. Production of sulfate and ammonia ranged up to the theoretical amount.

Cell-free extracts of the bacterium contained an enzyme which partially degraded taurine in vitro, with absorption of oxygen. Ammonia was formed in amounts from 60% to 100% of theory, but only 5% to 30% of the theoretical sulfate. Thus it appears that oxidative deamination precedes breaking of the carbon–sulfur bond. Two sulfonate-containing intermediates of unknown structure were isolated by elution from an ion exchange column. They were rather labile, liberating inorganic sulfate at room temperature. These may be key intermediates in sulfonate bio-degradation, since the carbon–sulfonate bond is usually quite stable.

Starkey (1964) reported a similar sequence in the bacterial degradation of taurine, and of cysteic acid **(7.5)** as well. Ammonia was first liberated,

$$\begin{array}{c} H_2NCC\Sigma \\ | \\ CO_2H \end{array}$$
(7.5)

after which the sulfur appeared as sulfite, which was then oxidized to sulfate.

The mechanism for formation of the carbon–sulfonate bond in nature is obscure. Perhaps the best documented case is the plant sulfolipid **7.6**, a surfactant widely distributed in significant amounts in all photosynthetic tissues of green plants (Benson, 1963). The sulfolipid molecule contains a hydrophobe consisting of two fatty acids linked to the α- and β-hydroxyls of glycerol. This diglyceride is linked in turn through its γ-hydroxyl to the α-pyranose form of 6-sulfo-D-quinovose. Structurally this is identical to the D-glucose molecule except that the 6-hydroxyl is replaced by a sulfonate group. The sulfonate is synthesized very rapidly in illuminated plant tissues, detectable amounts being formed within 1 min, traced by incorporation of ^{35}S and ^{14}C from inorganic sulfate and carbon dioxide.

Martelli (1964) studied the biodegradation of 6-sulfo-D-quinovose methyl pyranoside (the diglyceride of the sulfolipid **7.6** being replaced by a methyl group) by a *Flavobacterium* species isolated from soil by the

(7.6)

enrichment culture technique. Sulfoacetic acid, ΣCCO_2H, was found as a biodegradation intermediate, and this was degraded in turn to give inorganic sulfate. Thus the degradation of the sulfo sugar proceeded to a rather advanced stage with the carbon–sulfonate bond still intact, and the bond was broken only toward the end of the sequence of reactions.

No information seems to be at hand regarding the chemical mechanism by which these aliphatic carbon–sulfonate bonds are severed. At the present state of knowledge, judging by the taurine and sulfo sugar results, it seems plausible that in a situation where an aromatic sulfonate retains its sulfonate group intact through the ring opening, it could also survive subsequent degradation stages down pretty close to the last step before inorganic sulfate is released.

III. THE BIODEGRADATION PROCESS IN LAS

LAS has proved to be a satisfactory solution to the waste detergent problem posed by TBS. Primary biodegradation of LAS occurs readily, with loss of the two most noticeable and sensitive indicators of its presence in our environment, its foaming properties and its methylene blue response. Furthermore, direct tests on LAS biodegradation effluents have shown them to be nontoxic to mammals and to aquatic life—or at least, less sweepingly, nontoxic to rats and to bluegills (Borstlap, 1964; Swisher, 1964a).

Nevertheless, this accomplishment of environmental acceptability does not necessarily mean that ultimate biodegradation of the LAS to carbon dioxide and water has also been accomplished. It is conceivable that the biodegradation process might stop at some point short of the ultimate, leaving resistant intermediates undetectable by toxicity, foaming properties, or methylene blue response. If indeed formed, these products would join the much larger fraction of inert, resistant, anonymous organic-compounds of natural origin characteristically present in our waters and wastewaters (Chapter 5, Section III.A).

The chances of subtle, second order environmental effects arising from these hypothetical intermediates, if present, would seem to be quite small, and becomes smaller yet in proportion to the extent that ultimate biodegradation does occur. In the case of LAS, the evidence indicates that ultimate biodegradation is readily accomplished in the laboratory, given suitable acclimation, and there is no reason to believe that it is not being accomplished in nature. The metabolic pathway used begins with attack on the end of the alkyl chain to form a carboxyl group, followed by rapid β-oxidation of the chain, then more slowly by oxidation of the ring with simultaneous conversion of the sulfonate group to inorganic sulfate. Oxygen uptake during the process is in good agreement with the theoretical amount calculated for biological oxidation to completion. The evidence is discussed in detail in the following sections.

A. Oxygen Uptake and Oxidation Products

Even though oxygen uptake measurements cannot be interpreted precisely, because of the multitude of biochemical reactions involved, the data do indicate that LAS is extensively degraded under suitable conditions. Thus Ryckman (1956, 1957) found that the individual LAS components 2-sulfophenyloctane, -decane, and -dodecane all absorbed about 50% of the theoretical oxygen during 6 hr in the Warburg respirometer. He pointed out that this might well be consistent with complete degradation of the LAS, since protoplasm synthesis could account for the other 50%. He was able to support this view by other experiments on these sulfonates as well as on the corresponding linear 1-phenyl isomers:

(i) Methylene blue response dropped to zero.

(ii) Extended term (10–15 day) BOD measurements showed 60–75% of theoretical oxygen uptake.

(iii) Inorganic sulfate was formed to the extent of 95–100% of theoretical; sulfate release paralleled the oxygen uptake except for an initial delay, suggesting that sulfate release was one of the later steps.

(iv) IR spectra of the biodegradation residues were identical with those from the control systems; the original absorption bands, characteristic of alkyl chains, benzene rings, and sulfonate groups, disappeared.

Validity of these criteria was strengthened by the fact that none of them were met by TBS homologs under the same treatment.

Although it does not constitute additional proof for ultimate biodegradation, it is still interesting that Ryckman was able to operate his activated sludge units using the linear alkylbenzenesulfonates as the sole food. The sludge continued to grow and propagate with no particular indication of harm for at least 50 days on the primary derivatives and at least 10 days on the secondary.

Examination of Tables 8.1 and 8.2 reveals ample confirmation of Ryckman's findings both with pure LAS components and with commercial-type LAS mixtures. Oxygen uptake ranges from 50% to 100% of theoretical, depending on conditions, and so too do sulfate formation and COD removal.

Actual isolation of biodegradation intermediates is described by Krüger (1964) and Wickbold (1964). The official German activated sludge test method was used, 3 hr retention time, fed commercial LAS. The evaporation residue from the effluent was separated into fractions by extraction with various solvents and the fractions were characterized by elementary and functional group analysis as follows:

(A) 5–10% undegraded LAS
(B) 35–40% alkyl aromatic hydroxy carboxylate sulfonates
(C) 10–20% aliphatic carboxylate sulfonates
(D) 35–50% degraded to inorganic sulfate

Fraction (A) consisted of intact $5\text{-}\phi C_{10}$, $4\text{-}\phi C_{10}$, $6\text{-}\phi C_{11}$, and $5\text{-}\phi C_{11}$ sulfonates, in accord with the distance principle. (B) included subfractions averaging from C_{12} down to C_4 in side chain length, 0.2 to 1.2 hydroxyls and 0.8 to 2 carboxyls per molecule. Increasing the retention time in the sludge unit gave shorter side chains and more hydroxyl and carboxyl groups. Fraction (C) could not be obtained in reproducibly purified form, but one specimen averaged as the sulfonate of a $C_{5.5}$ acid containing 1.5 carboxyl groups. It was emphasized that none of these were end products; they were biologically and chemically labile and subject to further degradation upon further exposure.

Borstlap (1964) studied the intermediates obtained in batch runs

TABLE 7.3

Surfactant Biodegradation Intermediates[a]

Substrate	Time, weeks	% MBAS remaining	Intermeds., % found	Alkyl carbons	OH groups	CO$_2$H groups
LAS—DOBS C300	4–18	0	47–52	10	0.2	0.8
LAS—DOBS JN	2–18	11–6	59–65	10	0.4	0.7
TBS—DOBS PT	3–18	35–8	68–70	12	0.6	0.3
AS[b]—n-C$_{14}$	1	0	7–10	—	—	0.9
AS[b]—n-oxoC$_{14}$	3	0	13–18	—	—	0.3
AS[b]—tp-oxoC$_{13}$	3	18	51	—	—	0.2

[a] Borstlap (1964).
[b] AS, alkyl sulfate.

starting with 50 liters of 500 ppm surfactant, inoculated with 7.5 liters of washed, unacclimated activated sludge. The intermediate degradation products were isolated without fractionation, by isopropyl alcohol extraction of the evaporated, bacteria-free residue, and analyzed for elements and functional groups. The amount and nature of the intermediates showed no significant change after the first few weeks, up to 18 weeks. Borstlap's results on LAS, TBS, and several alkyl sulfates are given in Table 7.3. The pattern of chain shortening, hydroxylation, and carboxylation noted by Krüger and Wickbold, above, is evident, but under Borstlap's conditions the degradation of the LAS has stopped at an early stage, with only about two carbons gone.

B. Initiation of Oxidation

All experimental data on biodegradation of LAS are consistent with the hypothesis advanced by Ryckman (1957), that attack begins at the end of the alkyl chain via mechanisms also used in the degradation of aliphatic hydrocarbons. In particular, Huddleston (1963) isolated phenyldodecanoic acid from desulfonation of partially degraded 2-phenyldodecane sulfonate [Eq. (7.6)]. Although no work has been reported regarding the earliest stage intermediates between the original LAS and this first ω-carbokylate, it seems reasonable that the pathway followed might be one of those used for the linear hydrocarbons (Section I.A), perhaps via hydroperoxide, alcohol, and aldehyde (Fig. 7.1).

$$\underset{\underset{\Sigma}{\phi}}{\text{CCCCCCCCCCCC}} \rightarrow \cdots \rightarrow \underset{\underset{\Sigma}{\phi}}{\text{CCCCCCCCCCCCCO}_2\text{H}} \qquad (7.6)$$

The necessity for acclimation before LAS biodegradation begins indicates that adaptive rather than constitutive enzymes are involved in these early stages, most probably those catalyzing the initial reaction of the terminal methyl group. Any or all of several aspects might be involved: (i) formation of permease enzymes to transport the LAS molecules into the cells, (ii) formation of oxidases required for attack on the methyl group, or (iii) formation of new models of such enzymes, already present in "natural" form but not capable of accommodating the particular chain lengths or other structural features of the LAS molecules.

Using enrichment culture technique starting with soil from near a sewage treatment plant outfall, Payne (1963a) isolated two pure bacterial strains capable of growth and multiplication on sodium lauryl sulfate. One of these, a *Pseudomonas* species designated as C12B, could do likewise on KBS (kerylbenzenesulfonate), on individual isomers of phenyldodecane sulfonate, on sodium benzenesulfonate, and on various fatty acids and fatty alcohols. Heyman (1967) experienced difficulty in reacclimating C12B and other organisms to the oxidation of LAS and linear primary ABS following propagation in nonsurfactant media, but found that acclimation could be reestablished by prefeeding with short or long chain aliphatic alcohols, aldehydes, or carboxylic acids. He concluded that the most likely mechanism for this type of acclimation was induction of permease enzymes necessary for transport of the ABS across the cell wall. Failure of the ABS itself to induce the necessary enzyme was attributed to the presence of its benzene ring, but this may be held as speculative in the absence of any experimental evidence.

Marion (1966) exposed *Alcaligenes faecalis* cultures to 250 ppm concentrations of TBS, LAS, and SLS, prepared cell-free extracts from the cultures, and made Warburg runs on mixtures of the extracts with the respective surfactants. Significant oxygen uptake was observed with SLS, about 40% of that achieved with whole cells, indicating the presence of an oxidative enzyme. With TBS and LAS, however, oxygen uptake was only slightly higher than the controls with both cell-free extracts and whole cells. Marion surmised that any enzymes may have been denatured by the relatively high concentrations of TBS and LAS used, which he had already found to be somewhat more deleterious than SLS.

C. Alkyl Chain Oxidation

Huddleston (1963) has presented direct evidence for the formation of a terminal carboxylate and for its subsequent β-oxidation in the biodegradation of LAS. Starting with pure 2-sulfophenyldodecane, after 8 days exposure in shake flask culture he subjected the products to desulfonation and obtained two organic fractions, one acid and one neutral, which were examined by IR, GC, NMR, and mass spectroscopy. The acid fraction, amounting to about 13% of the original 2-ϕC_{12}, consisted of phenyl-hexanoic and phenyloctanoic acids along with smaller amounts of the C_{10} and C_{12} homologs. This clearly indicates a process of ω- followed by β-oxidation, removing two carbons at a time (Fig. 7.7).

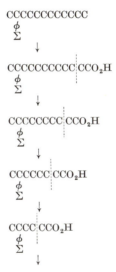

Fig. 7.7 ω,β-Oxidation of LAS (Huddleston, 1963).

The neutral fraction corresponded to 31% of the original 2-ϕC_{12} and was characterized as phenyl-substituted ketones or other aromatic carbonyl structures. Very likely this originated from the next product after the sulfophenylhexanoic acid, namely sulfophenylbutyric acid, which would undergo cyclodehydration during desulfonation to give 1-methyl-3-indanone as discussed below. Altogether, acid and neutral fractions totaled 44% of the original carbon. Other experiments had indicated that 50–55% of the 2-ϕC_{12} had been converted to CO_2, so most of the original compound was accounted for.

Further evidence for β-oxidation of the side chain was advanced by Swisher (1963c), who found by desulfonation–GC that during river water biodegradation of LAS, intermediates were formed and subsequently degraded. The transient chromatographic peaks, shown in Figs. 7.8 and 7.9, are actually from indanones and tetralones formed during the

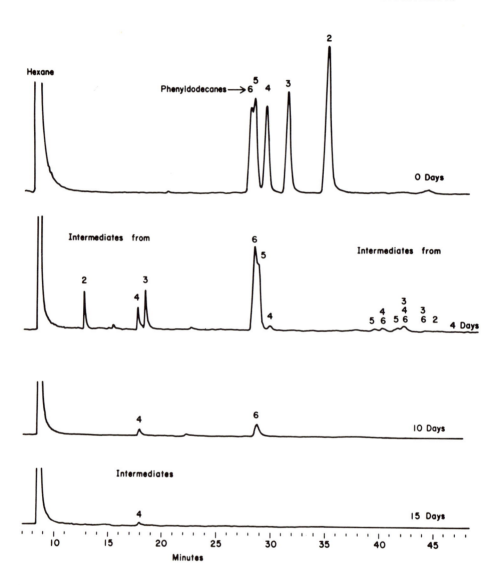

Fig. 7.8   Biodegradation of C_{12}LAS in river water, detected by desulfonation–GC (Swisher, 1963c).

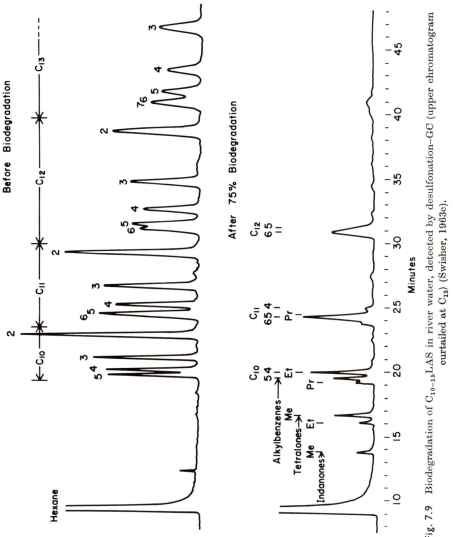

Fig. 7.9 Biodegradation of C_{10-15}LAS in river water, detected by desulfonation–GC (upper chromatogram curtailed at C_{13}) (Swisher, 1963c).

desulfonation by cyclodehydration of the true intermediates, the short chain sulfophenylalkanoic acids resulting from β-oxidation.

Individual river water runs on each of the pure C_{12}LAS isomers identified the intermediate peaks as numbered in Fig. 7.8. Thus intermediate peak 3 at around 15 min came from the 3-ϕC_{12} isomer and was actually 1-ethyl-4-tetralone, formed as indicated in Fig. 7.10. The next step beyond

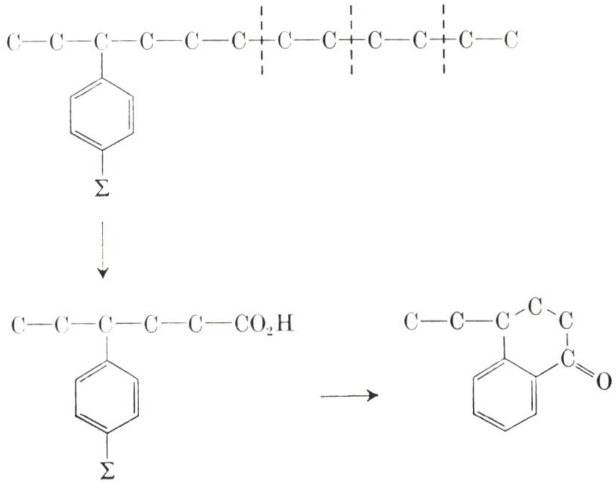

Fig. 7.10 β-Oxidation intermediate and its desulfonation product, from 3-ϕC_{12}LAS (Swisher, 1963c).

the C_6 acid, γ-sulfophenylcaproic acid, would be attack on carbon number 4 preparatory to scission of the 4,5 bond. Evidently this attack is more difficult than the previous steps, perhaps due to proximity to the ring or to the sulfonate group, and the C_6 acid builds up in the solution until the bacteria learn how to degrade it. Very likely the subsequent attack is through carbon 1 or 4, or possibly through the ring; in any case the intermediate then disappears.

This same intermediate, γ-sulfophenylcaproic acid, would of course be formed by β-oxidation of any 3-phenyl homolog of even chain length. The 3-phenyl homologs of odd chain length degrade to the intermediate with one less carbon, β-sulfophenylvaleric acid, before meeting the temporary resistance, and this cyclizes to 1-ethyl-3-indanone upon desulfonation. The 2- and 4-phenyl LAS components behave similarly, leading to a total of six of these desulfonation intermediates from odd plus even LAS. Their structures are shown in Fig. 7.11 and their chromatographic pattern in Fig. 7.9.

Failure of the 5- and 6-phenyl isomers to accumulate similar intermediates can be attributed to the longer unattacked chain end remaining after β-oxidation reaches a corresponding stage. For example, when $6\text{-}\phi C_{12}$ arrives at stage **7.7** the other end of the chain, being further away from the sulfonate group than in the 2-, 3-, and 4-phenyl isomers, is

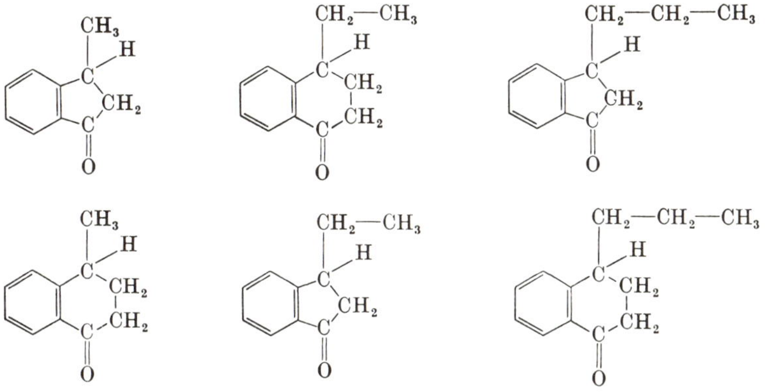

Fig. 7.11 Desulfonation products from LAS biodegradation intermediates (Swisher 1963c). Upper: From 2-, 3-, and 4-phenyl even-carbon LAS. Lower: From 2-, 3-, and 4-phenyl odd-carbon LAS.

presumably more susceptible to attack and hence the amount of the intermediate would not be expected to build up.

$$\underset{\underset{\Sigma}{\phi}}{CCCCCCCCO_2H}$$

(7.7)

In the river water experiments just described, contrary to the shake culture results of Huddleston (1963) discussed previously, there was no indication that any significant amount of the longer chain carboxylates was ever present, although they must have been formed as intermediates in the stepwise degradation. This was further verified by Swisher (1964b) by methylating the desulfonation products prior to gas chromatography. The carboxylate methyl esters are much more readily detected than are the free carboxylic acids, which ordinarily give broad and diffuse chromatographic peaks. Figure 7.12 pictures the course of degradation of a mixture of C_{11}LAS and the corresponding carboxylate, sulfophenylundecanoic acid (SϕU).

Before the LAS had even begun to degrade, the SϕU had all disappeared, forming the expected short chain intermediates indicated by peaks B and

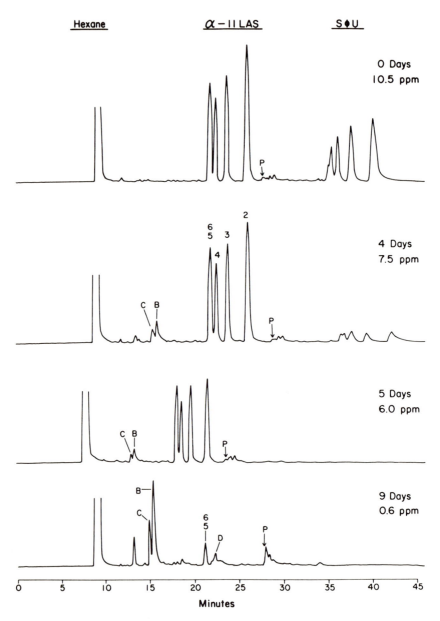

Fig. 7.12 Biodegradation of SφU + C$_{11}$LAS mixture in river water, detected by desulfonation–methylation–GC (Swisher, 1964b). Peak B: 1-Methyl-4-tetralone. Peak C: 1-Ethyl-3-indanone. Peak D: 1-Propyl-4-tetralone. Point P: Column pressure doubled to speed elution of SφU peaks.

C (the third one, D, is obscured by the 4-ϕC_{11}). Although the intact SϕU is easily detected, there is no indication of the presence of its C_9 homolog (its first β-oxidation product) at any stage. Nor is there any indication of formation of more SϕU during the subsequent degradation of the C_{11}LAS, which in this case occurred only after the original SϕU was all gone. Although this might be taken as evidence that the ω,β-oxidation mechanism for LAS is incorrect, it is much more reasonable to suppose that the longer carboxylates are very quickly degraded and disappear as soon as formed.

Apparently this same course of biodegradation occurs in a rat. Michael (1968) found that after oral feeding of LA^{35}S, the radioactivity was almost entirely excreted within 3 days. During that time most of the LAS had been converted to sulfophenylbutyric and -valeric acids, the β-oxidation products of the even and odd components of the original LAS.

Although β-oxidation is undoubtedly the most important mechanism for the chain degradation, other reactions may be involved also, to a minor extent. Close examination of the 4 day chromatogram in Fig. 7.8 shows two tiny peaks around 15 min and a third at $22\frac{1}{2}$ min. They are in exactly the positions and proportions to be expected for the three odd-carbon intermediates (Fig. 7.12, peaks C, B, and D). If one is persuaded that the resemblance of the patterns is too close to be a mere coincidence and that these are indeed the odd-carbon intermediates, then one is forced to conclude that some other mechanism besides β-oxidation must have been operating, to a minor extent, to form them from even-carbon LAS. Most likely this would be the occasional removal of a single carbon instead of a pair at one of the intermediate stages; removal of three at once would seem to be less likely.

Heyman (1968) found that when *Pseudomonas* C12B was acclimated, to C_{12}LAS it was also acclimated to benzoic acid: Warburg oxygen uptake on benzoic acid began almost immediately, whereas unacclimated cells showed an initial lag period of about an hour. Cells acclimated to C_{11}LAS also gave some evidence of benzoic acid acclimation, but considerably less marked. When exposed to homogentisic acid **(7.8)** (2,5-dihydroxyphenylacetic acid, a common cell metabolite) the C_{11}LAS cells began oxygen uptake immediately and the C_{12} cells lagged almost an hour.

(7.8)

Heyman pointed out that these results were consistent with the chain degradation pathways outlined in Fig. 7.13: ω-oxidation and β-oxidation down to the vicinity of the ring. The 2-, 4-, and 6-phenyl C_{11}LAS isomers would give sulfophenylmalonic acid **(7.9)**, which could be further decarboxylated to sulfophenylacetic **(7.10)**, which in turn could degrade further via homogentisic acid. The other two C_{11} isomers would go through 3-sulfophenylglutaric acid **(7.11)**, sulfophenylsuccinic **(7.12)**, β-sulfophenylpropionic **(7.13)**, and sulfobenzoic **(7.14)** acids. All five of the C_{12}LAS isomers would join this latter path at **7.12**.

Heyman's scheme (Fig. 7.13) seems reasonable in the main and the

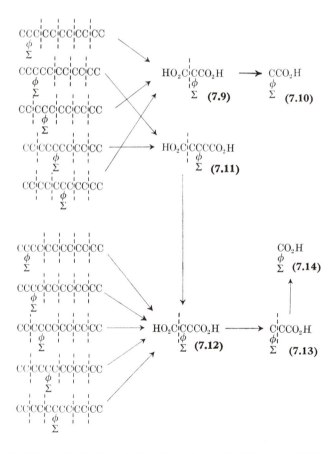

Fig. 7.13 Side chain biodegradation, C_{11} and C_{12}LAS (Heyman, 1968).

pathways chosen chemically justifiable, with minor reservations. It is convincing that **7.11** should not engage further in a complete β-oxidation, because a keto group cannot be formed on the central carbon atom, and that it would be open to the postulated α-oxidation to give **7.12**, which is in accord with the findings of Stokke (1969) on methyl-substituted chains (Section I.B). But reasons are needed to explain why this same α-oxidation should not also occur to convert **7.12** to **7.9** and **7.13** to **7.10**, thus opening the phenylacetic–homogentisic pathway to all LAS components. Perhaps the α-oxidation is much slower than the alternative, postulated reactions, decarboxylation in **7.12** and β-oxidation in **7.13**.

With linear primary ABS these complicating features are not necessary; the results are explained with no more than ω,β-oxidation for the chain degradation pathway. Heyman's experiments indicate that the odd homologs (toluene- and n-amylbenzenesulfonates) go via benzoic and the evens (ethyl- and n-dodecyl-) via phenylacetic, as would be expected.

Difficulty encountered in oxidation of preformed p-sulfobenzoate by intact cells of *Pseudomonas* C12B was attributed to an antagonistic action against p-aminobenzoate, a cell component with important metabolic functions. Nevertheless, Heyman found that an intracellular enzyme capable of degrading p-sulfobenzoate was formed upon acclimation to p-toluenesulfonate or n-amylbenzenesulfonate. Extracts prepared by disruption of the cells degraded the rings of p-sulfobenzoate, and of benzoate as well, as shown by UV analysis. Thus it would appear that biodegradation of appropriate alkylbenzenesulfonates can proceed via p-sulfobenzoate, presumably because it is degraded as soon as formed and toxic concentrations are not built up.

D. Ring Degradation and Beyond

The fact that LAS-acclimated cells are able to immediately degrade certain benzene ring compounds (Heyman, 1968; see above) shows that ring-degrading enzymes have been developed in the cells and implies that they are active in degrading the rings of the LAS. Other evidence for the ring degradation is available also. Elementary analysis of fractions isolated from later stages in LAS biodegradation indicate the presence of less than six carbon atoms and more than one carboxyl group per sulfonate group (Krüger, 1964; Wickbold, 1964; Section III.A). Obviously there is no room for a benzene ring in the molecules in such a mixture, at least not in all of them; the ring must have been disrupted in the biodegradation. And there is further evidence.

D.1. *Detection of Benzene Rings by UV Analysis*

A more specific proof of ring degradation is afforded by application of UV spectroscopy. Benzene rings show three characteristic absorption bands, two extremely intense and one rather weaker, at wavelengths which are somewhat dependent on the nature of any substituents on the ring. In the case of LAS the two strong peaks lie at 193 and 223 nm, with very little difference from one isomer and homolog to another. Oxidation of the LAS side chain to an ω-carboxylate (sulfophenylundecanoic acid) or completely (*p*-sulfobenzoic acid) has very little effect on this characteristic pattern (Fig. 7.14). The absorption is so intense that the method

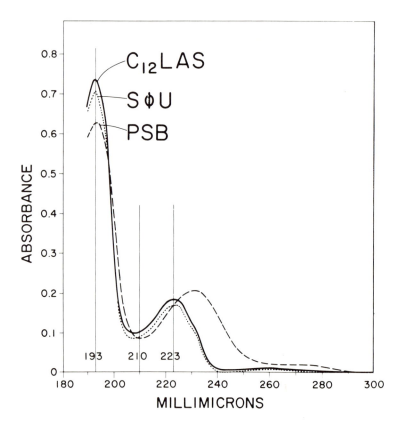

Fig. 7.14 UV spectra of benzene rings (1 cm light path) (Swisher, 1967b). C_{12}LAS: 5 ppm in aqueous solution. SϕU: 5.55 ppm sulfophenylundecanoic acid disodium salt. PSB: 3.45 ppm *p*-sulfobenzoic acid potassium salt.

is directly applicable at the concentrations ordinarily used in biodegradation systems, but is subject to interference by comparably strong absorption in the same general region which might also be present, for example, from nitrite, nitrate, or nucleic acids.

Disappearance of these bands, then, constitutes strong evidence for destruction of the benzene ring. Such evidence has been provided by Foster (1964) and Krüger (1964) for commercial-type LAS. Further, Swisher (1967b, c, 1968a) has investigated several LAS components in pure form in activated sludge, river water, and shake cultures with the results outlined in the following section.

D.2. *Acclimation, Cross-Acclimation, and Pathways*

Acclimation for LAS ring degradation takes somewhat longer than for primary biodegradation, probably because the appropriate intermediates are not present in the system until after primary degradation has occurred, with resulting shorter exposure time. Figure 7.15 shows this effect in two continuous-flow activated sludge cultures separately acclimated to 3-ϕ- and 6-ϕC_{12}LAS. Switching the feeds impaired ring degradation for several days until acclimation was reestablished. Primary biodegradation was undisturbed, as indicated by zero MBAS in the effluents throughout.

Acclimation to ring degradation in shake cultures is illustrated in Fig. 7.16. Here a 14 day transfer schedule was used because the usual 7 day period would have allowed only 2 to 4 days exposure to the pertinent ring-containing intermediates formed in the primary biodegradation. Even so, some eight to ten transfers were necessary before reasonably complete ring degradation could be accomplished within the 14 day total.

Pitter (1968c) did not achieve LAS ring degradation within 20 days in his test system—BOD water inoculated with acclimated activated sludge. This, like the shake culture system, has a low biodegradation potential. The failure was probably due to incomplete acclimation of the organisms to the conditions of the BOD water system, even though they were acclimated to the activated sludge conditions.

The exact sequence of reactions in LAS ring degradation has not yet been completely elucidated. There is no reason to suppose that the general pathway should be different from those used for other, nonsurfactant, benzene derivatives, as by oxidation to a catechol derivative and rupture between or adjacent to the two OH groups. In the case of LAS the ring would have two substituents, the sulfonate group and the remnants of the oxidized chain. Possibly ring opening might occur with the chain only partially oxidized, or possibly complete oxidation of the chain down to

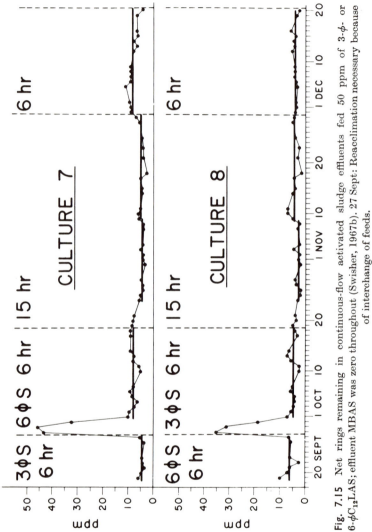

Fig. 7.15 Net rings remaining in continuous-flow activated sludge effluents fed 50 ppm of 3-ϕ- or 6-ϕC$_{12}$LAS; effluent MBAS was zero throughout (Swisher, 1967b). 27 Sept: Reacclimation necessary because of interchange of feeds.

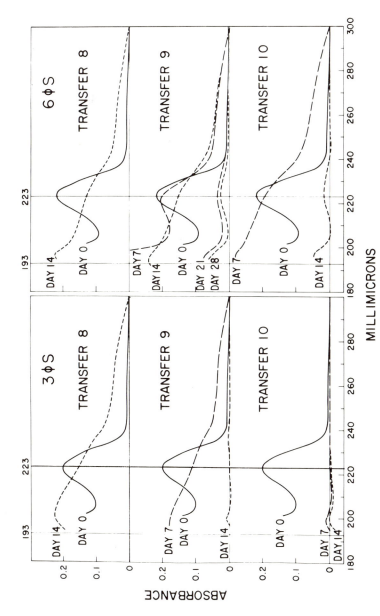

Fig. 7.16 Development of acclimation for benzene ring biodegradation in shake cultures as detected by UV spectra during 8th, 9th, and 10th transfers in SDA medium (Swisher, 1968a).

the ring might first be required, or perhaps prior removal of the sulfonate group, as Cain (1968) has proved for benzene- and toluenesulfonates.

Isolation of aliphatic sulfocarboxylates as reported by Krüger (1964) and Wickbold (1964) indicates that sulfonate removal need not precede ring opening. Swisher (1968c) has found that liberation of inorganic sulfate and disappearance of benzene ring absorption bands are approximately simultaneous in shake culture biodegradation of the individual C_{12}LAS isomers, but obtained no clear indication as to which occurred first. With either sequence the resulting product would be an unsulfonated aliphatic compound as in Eq. (7.7). The product would have a singly

$$\text{(structure with } R \text{ and } SO_3^-) \longrightarrow \text{(branched ring structure with } R) + SO_4^{2-} \qquad (7.7)$$

branched carbon skeleton if R is a carboxyl group, or doubly branched if R includes further remnants of the secondary alkyl chain of the original LAS. Or, R might be split off before or during ring opening. In any case, the product would contain no outlandish structural features and would degrade readily in the cell's metabolic system.

The fact that acclimation to ring degradation of one LAS compound does not necessarily include its isomers (Figs. 7.15, 7.17) suggests that the key structure involved comes prior to complete oxidation of the chain. Application of ω,β-oxidation to any even-carbon LAS component of any chain length leads to a single final structure, sulfophenylsuccinic acid (**7.12**). Referring to Fig. 7.15 we see that Culture 7 can degrade the rings of the 3-phenyl isomer, but requires acclimation to do the 6-phenyl. Conversely with Culture 8. Since both isomers should presumably reach structure **7.12** near the end of chain degradation, the structures involved in the acclimation to ring degradation must occur at some earlier stage, where they are still different. Thus the ring-adaptive enzymes involved may have nothing to do with ring oxidation, directly, but rather may be the ones necessary to convert the two earlier different intermediates to the later stage **7.12**. Or, on the other hand, the paths may not involve **7.12** at all, but may proceed to ring splitting directly from the earlier, differentiated, intermediates. Or perhaps both such pathways are used.

Existence of multiple pathways is suggested by the results in Fig. 7.17. Under the mild conditions of the shake culture test the degradation

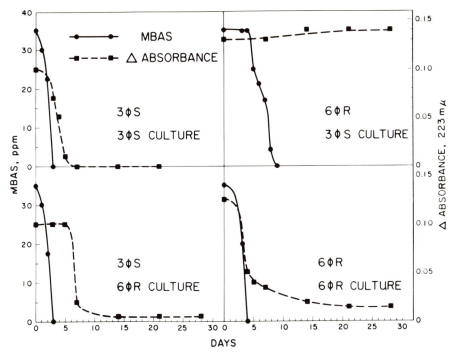

Fig. 7.17 Primary biodegradation (MBAS) and benzene ring biodegradation (Δ absorbance) in shake cultures (Swisher, 1968a). Left: 3-ϕC$_{12}$LAS. Right: 6-ϕC$_{12}$-LAS. Upper: 3-ϕ-acclimated culture. Lower: 6-ϕ-acclimated culture.

processes are slow enough to see that for the 3-phenyl isomer there is a lapse of about a day between primary and ring degradation in the 3-phenyl culture (upper left). In contrast, for the 6-phenyl isomer with its acclimated culture (lower right) over half of the ring degradation is substantially simultaneous with the primary degradation. Here the chain is undoubtedly attacked first also, but ring attack ensues immediately without the delay shown by the 3-phenyl.

Considering the degradation of the 6-phenyl isomer only, multiple routes seem to be involved. The initial 6-phenyl molecules are all identical, but ring degradation levels off with some 30% of the rings remaining (lower right). Evidently 30% of the 6-phenyl molecules follow a path via some intermediate which is resistant to further attack under the shake culture conditions (but not in activated sludge, where only 5–10% of the rings remain in the 24 hr cycle).

Strictly speaking, the 6-phenyl molecules are not all identical: the carbon atom to which the ring is attached is asymmetric, and half of the original molecules are thus mirror images of the other half. Even so, the above argument for multiple pathways still holds, since the resistant fraction does not amount to 50%. In fact, all LAS components have an asymmetric carbon except for the central isomers with odd carbon number (e.g., 6-ϕC$_{11}$, 7-ϕC$_{13}$), and those too will become asymmetric if biodegradation alters the two segments of chain unequally. No studies seem to have been reported on optical isomers in surfactant biodegradation, but the effects, if any, must be quite small—certainly no greater than the effects of differing phenyl positions and chain lengths in LAS.

The differences cited above in the biodegradation behavior of the 3-phenyl and 6-phenyl isomers in their respective acclimated shake cultures appear to be truly characteristic; they are repeatable, with minor differences, time after time. The cross-feeding results illustrated in Fig. 7.17 (lower left, upper right) are less reproducible, showing the influence of uncontrolled, unidentified factors.

IV. QUATERNARY ALKYLBENZENESULFONATES (QBS)

A. Degradation of QBS

Information on the metabolic pathways involved in the biodegradation of QBS was provided by Kölbel (1967), who studied benzene ring degradation in several examples in comparison with the corresponding linear primary ABS. Under the action of an *Escherichia coli* culture compounds **7.15** and **7.16** (as well as their C$_{14}$ homologs) were degraded (MBAS) in 9 and 16 days, respectively (incidentally, note the distance principle operating). However, the benzene ring did not degrade in either case, indicated by the continued presence of the 225 nm UV ring band even after 30 days. But in the case of the linear derivative **7.15** a new absorption

$$\Sigma\phi\text{CCCCCCCCCCCC} \qquad \Sigma\phi\overset{\text{C}}{\underset{\text{C}}{\text{CCCCCCCCCC}}} \qquad \Sigma\phi\text{CCCCCCCCCC}\overset{\text{C}}{\underset{\text{C}}{}}$$

$$(7.15) \qquad\qquad (7.16) \qquad\qquad (7.17)$$

band had appeared at 268 nm. Kölbel identified this as being characteristic of a double bond adjacent to the ring and to a carbonyl group, and attributed it to sulfocinnamic acid, $\Sigma\phi$C=CCO$_2$H. No such band developed with **7.16**, indicating that the quaternary group next to the ring was stable enough to prevent formation of a double bond there.

TABLE 7.4

Ring Biodegradation in QBS[a]

Compound	Ring degradation, %[b]
C Σφ CCCCCCCCCCC C	85–95
C Σφ CCCCCCCCCCC C	80
C Σφ CCCCCCCCCCC C	85–95
C Σφ CCCCCCCCCCC (**7.19**) C	?[c]

[a] Swisher (1969).
[b] 24 hr semicontinuous activated sludge, UV analysis.
[c] Results indeterminate because of extensive adsorption on sludge.

The terminal quaternary derivative **7.17** showed some 30% MBAS degradation within 27 days, and by 30 days the 268 nm band had appeared; the 225 nm ring band was still present also. This was interpreted as rupture of the chain at some intermediate point, followed by chain degradation as with **7.15**. The work of Tryding (1957) on biooxidation of 2,2,17,17-tetramethylstearic acid (**7.18**) was cited in support of mid-chain attack, and the subsequent work of Abbott (1968) (Section I.A) further confirms the possibility.

$$\begin{array}{ccc} \text{C} & & \text{C} \\ \text{CCCCCCCCCCCCCCCCCCCCCO}_2\text{H} \\ \text{C} & & \text{C} \end{array}$$

(**7.18**)

Parallel experiments using mixed bacteria from treated sewage instead of *E. coli* gave similar results, except that ring degradation did occur within 30 days for both **7.15** and **7.16**. With the former, disappearance of the ring band was preceded by the development of the sulfocinnamic band at 268 nm, but not in the case of **7.16**. Ring degradation was not achieved at all in the terminal quaternary **7.17**. The oxygen uptake in all experiments was reasonably consistent with the spectroscopic results.

Benzene ring biodegradation approaching 100% in three QBS isomers has also been observed by Swisher (1969) using the 24 hr cycle semicontinuous activated sludge system (Table 7.4). With the terminal QBS **7.19**, only about 10–20% of the input MBAS and rings appeared in the daily effluent, but extensive buildup of undegraded QBS occurred in the sludge, in part as the magnesium salt. When corrected for this accumulated QBS the actual primary biodegradation was estimated to be in the range of 0–40% for isomer **7.19** (see Table 7.4), depending on the particular assumptions made as to the amount of new sludge synthesized and its content of QBS. It seems likely that ring degradation would have occurred to the same extent as the primary degradation, but direct proof was lacking because of insufficient precision. Ring degradation observed with the other three QBS isomers was unequivocal, since no such adsorption then occurred.

Terminal QBS of the type **7.19** has been observed to undergo complete primary biodegradation in some circumstances (Chapter 6, Section II.D), although acclimation is difficult.

B. Mechanisms for Quaternary Biodegradation

Mechanisms do appear to be available for such degradation by attack almost anywhere on the molecule: (i) at the ring end (Cain, 1968; Section II.A), (ii) at an intermediate point on the chain (Abbott, 1968; Section I.A), or (iii) on the terminal quaternary group itself. Biodegradation of nonsurfactant quaternary compounds has been studied by several workers.

McKenna (1966) examined a series of fatty acids **(7.20)** for ability

$$\begin{array}{c} \text{C} \\ \text{C}_n\text{CCO}_2\text{H} \\ \text{C} \end{array} \qquad (n = 1, 3, 6, 9)$$

$$\textbf{(7.20)}$$

to support bacterial growth. None did. Nor did the quaternary alkyl-benzene

$$\begin{array}{c} \text{C} \\ \phi\text{CCCCCCCCCC} \\ \text{C} \end{array}$$

However, requirements for growth support are more stringent than for mere oxidation or biodegradation in some respects, and a full assessment of the biostability of these structures has not been made as yet.

Earlier, Tryding (1957) had shown that rats were able to degrade 2,2-dimethylstearic acid. It was largely converted to 2,2-dimethyladipic acid [Eq. (7.8)], obviously by ω,β-oxidation without involvement of the

$$\text{CC}|\text{CC}|\text{CO}|\text{CC}|\text{CC}|\text{CC}|\overset{\text{C}}{\underset{\text{C}}{\text{CCCCCCO}_2\text{H}}} \rightarrow \text{HO}_2\overset{\text{C}}{\underset{\text{C}}{\text{CCCCCO}_2\text{H}}} \qquad (7.8)$$

quaternary group. But the diquaternary 2,2,17,17-derivative was also attacked, though much less readily and less extensively [Eq. (7.9)]. Major

$$
\begin{array}{c}
\text{C} \quad \vdots \quad \vdots \quad \vdots \quad \vdots \quad \text{C} \quad\quad\quad \text{C} \quad\quad\quad \text{C} \\
\text{CC}\vert\text{CC}\vert\text{CC}\vert\text{CC}\vert\text{CC}\vert\text{CC}\vert\text{CCCCCCO}_2\text{H} \rightarrow \text{HO}_2\text{CCCCCCCCCCCCCCCCCCO}_2\text{H} \\
\text{C} \quad \vdots \quad \vdots \quad \vdots \quad \vdots \quad \text{C} \quad\quad\quad \text{C} \quad\quad\quad \text{C} \\
\downarrow \\
\text{C} \quad\quad\quad\quad\quad\quad \text{C} \quad\quad\quad\quad\quad \text{C} \\
\text{HO}_2\text{CCCCCCO}_2\text{H} \quad (+ \text{HO}_2\text{CCCCCCCO}_2\text{H} + \text{HO}_2\text{CCCCCO}_2\text{H}) \\
\text{C} \quad\quad\quad\quad\quad\quad \text{C} \quad\quad\quad\quad\quad \text{C}
\end{array}
$$

$$(7.9)$$

products identified were dicarboxylic acids with 22 and 8 carbons, with much smaller amounts of the 9- and 7-carbon homologs. These shorter products might have been formed by β-oxidation subsequent to chain splitting, which, if random, could give both even and odd fragments. Predominance of the 8-carbon product thus suggests that such attack is not random, or else that β-oxidation has proceeded from the end of the chain right through one of the quaternary groups. Central attack on the chain is also suggested by the presence of unidentified early oxidation products with 22 carbons, presumably hydroxylated derivatives.

Jones (1968a) observed that the yeast *Torulopsis gropengiesseri* was able to oxidize terminal quaternary groups in long chain compounds to the corresponding alcohol,

$$
\begin{array}{c}
\text{C} \quad\quad\quad\quad \text{C} \\
\cdots \text{CCCH}_3 \rightarrow \cdots \text{CCCH}_2\text{OH} \\
\text{C} \quad\quad\quad\quad \text{C}
\end{array}
$$

but found no indication of further oxidation to carboxylate.

Attack on the quaternary carbon structure itself was demonstrated by Mohanrao (1962): 3,3-dimethylglutaric **(7.21)** and 2,2-dimethylvaleric **(7.22)** acids and a number of others were degraded after suitable acclima-

$$
\begin{array}{cc}
 \quad \text{C} & \quad \text{C} \\
\text{HO}_2\text{CCCCCO}_2\text{H} & \quad \text{CCCCCO}_2\text{H} \\
\text{C} & \quad \text{C} \\
\textbf{(7.21)} & \quad \textbf{(7.22)}
\end{array}
$$

tion. Ease of the biodegradation varied from structure to structure. 3-Methyl-3-ethylglutaric acid **(7.23)** was readily metabolized while 3,3-diethylglutaric acid **(7.24)** was resistant. The quaternary alcohols

$$
\begin{array}{cc}
\quad \text{C} & \quad\quad \text{C} \\
\quad \text{C} & \quad\quad \text{C} \\
\text{HO}_2\text{CCCCCO}_2\text{H} & \text{HO}_2\text{CCCCCO}_2\text{H} \\
\quad \text{C} & \quad\quad \text{C} \\
\quad \text{C} & \quad\quad \text{C} \\
\textbf{(7.23)} & \quad\quad \textbf{(7.24)}
\end{array}
$$

studied, for example, 2,2-dimethylpropanediol,

$$\text{HOCCCOH} \quad \text{(with C above and C below the central C)}$$

were resistant under all conditions tried.

Detailed study of **7.21** revealed the probable biodegradation mechanism, the first step being disruption of the quaternary structure by conversion to acetic and dimethylacrylic acids [Eq. (7.10)], with subsequent steps

$$\text{HO}_2\text{CCCCCO}_2\text{H} \rightarrow \text{HO}_2\text{CC} + \text{C}{=}\text{CCO}_2\text{H} \qquad (7.10)$$

following more orthodox patterns. Several genera of common soil bacteria (*Bacillus, Flavobacterium, Pseudomonas, Nocardia*) were able to accomplish this, and it may well be that similar reactions are involved in the biodegradation of the quaternary ABS structures.

Pantothenic acid is one of the structural elements of coenzyme A **(7.2)** and contains a quaternary carbon atom. It is degraded in activated sludge

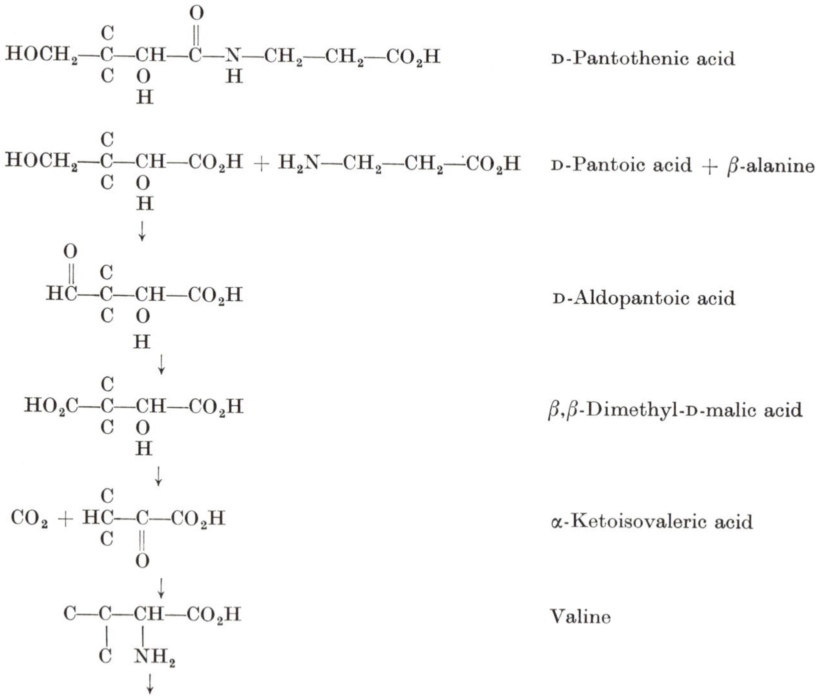

Fig. 7.18   Degradation of quaternary group in pantothenic acid (Goodhue, 1966a,b; Nurmikko, 1966; Magee, 1966).

(Mohanrao, 1962), and Snell's group has traced the metabolic pathway for its biodegradation, including disruption of the quaternary structure, by *Pseudomonas* K5 (Fig. 7.18).*

V. ULTIMATE BIODEGRADATION OF TBS

Although some have expressed doubts as to the biodegradability of TBS, attributing its removal to adsorption (Hartmann, 1963a), nevertheless there is overwhelming evidence that under the proper conditions biodegradation of TBS does occur. In addition to the many reports based on MBAS analysis, Table 8.3 includes an even more convincing body of radiotracer data. Here the initial TBS is marked by introduction of ^{35}S atoms into its sulfonate group, thus allowing the degradation products to be segregated into three fractions: (i) intact TBS, (ii) intermediate degradation products still containing the sulfur atom, and (iii) inorganic sulfate.

House (1956) pioneered in this radiotracer technique, finding as high as 80–90% TBS removal by activated sludge in field tests, and also reporting that most of the original radioactivity was found in the inorganic sulfate fraction. His laboratory studies with activated sludge showed somewhat lower TBS removals and also a great discrepancy in sulfate formation, which only ran from 0% to 7%. He later found that the field sulfate results were in error because hydrogen peroxide had been used in the sulfate analysis to oxidize any sulfite present. It turned out that hydrogen peroxide would decompose the TBS biodegradation intermediates, liberating their sulfate also. This trouble did not occur when bromine was used, as was done in analyzing the laboratory samples (House, 1966b).

Klein (1962, 1963a, b, 1964a, b, 1965a, b, 1966) and McGauhey (1959a, b, 1964) also used the radiosulfur method in their exhaustive studies on TBS, often observing significant inorganic sulfate formation, up to about 10% in activated sludge and up to 30% or more in soil.

A very close approach to ultimate biodegradation of TBS was reported by Robeck (1963), using a soil lysimeter. During a period of almost 2 years, 183 gm of TBS was fed, of which 8 gm appeared in the effluent and about 1 to 2 gm was found adsorbed on the soil–slime mixture in the lysimeter upon final dismantling. Radiotracer analyses of the effluent showed intact TBS amounting to 2–5% of that in the influent, and 75–80% of the initial TBS sulfur in the form of inorganic sulfate. Sulfur-containing intermediates accounted for the other 20%.

---

* Sorlini (1969) has found *Achromobacter* sp. which degrade 2,2-dimethylmalonate and -succinate, the former via decarboxylation.

In other lysimeter experiments Kempf (1968) observed 95% primary biodegradation of TBS. He examined carbon tetrachloride extracts of the effluent (pH 4) by IR spectroscopy and noted that the ring bands at 6.25 and 13.15 μ and the sulfonate bands at 8 to 8.8 μ were substantially absent. Aliphatic chain bands at 3.4 and 7.25 μ were still present, and new bands at 5.75 and 7.72 μ had appeared, attributed to ester, carboxyl, or hydroxyl groups.

Actual isolation of intermediate fractions from TBS biodegradation also indicates that the pathways are similar to those followed in LAS degradation, giving compounds containing hydroxyl and carboxyl groups. In the inoculated medium used by Borstlap (1964) the intermediate had undergone very little change. It still averaged about 12 carbons in chain length, with about 0.6 OH groups and 0.3 carboxyls per molecule (Table 7.3). In a continuous-flow activated sludge system reported by Krüger (1964) two intermediate fractions were isolated, (i) a $C_{8.5}$ABS with an average of 0.8 OH and 1.1 CO_2H groups per molecule and (ii) a $C_{4.6}$ aliphatic sulfonate averaging 0.4 OH and 1.4 CO_2H groups.

The pattern here is the same as with LAS. The primary biodegradation, as determined by disappearance of MBAS or of $TB^{35}S$, is well under way before inorganic sulfate starts to appear, and thus the attack on the sulfonate must occur, on the average, subsequent to the initial attack on the molecule. However, there is as yet no conclusive evidence that sulfate liberation necessarily represents the final step in the complete biodegradation of the molecule. Depending on its structure, such an organic residue might or might not present difficulties in further biodegradation.

In his rat studies Michael (1968) found that most of the radioactivity from feeding $TB^{35}S$ was excreted within 3 days. Most of it was in the form of carboxylated oxidation products with 11, 12, or 13 carbons in the alkyl group, and some with two double bonds also. Thus the rat's mechanism for ω-oxidation was quite effective, but little further could be accomplished before excretion, presumably because of the branching of the alkyl group. Michael theorized that the double bonds originated by oxidation of the carbon adjacent to the ring to an alcohol, with subsequent dehydration to a double bond, hydroxylation at the allylic position and dehydration again. Throughout the sojourn in the rat there was no attack on the sulfonate group, as shown by absence of inorganic sulfate-35.

The work of Benarde (1965) stands in some disagreement with the foregoing radiosulfur results. He reports that in certain cultures and in the presence of a food such as glucose, sulfate liberation from TBS is simultaneous with MBAS disappearance, indicating that attack on the sulfonate group is the initial step. The microorganisms were isolated originally by the enrichment culture technique from soils, using a TBS medium. The significance or validity may become more apparent in the light of future work. The existence of more than one pathway for degradation is not at all uncommon, and initial attack on ring sulfonate in benzene- and toluenesulfonates has been demonstrated by Cain (1968) (Section II.A). But possibly Benarde's use of hydrogen peroxide (for oxidation of any lower inorganic sulfur compounds to sulfate for analysis) may have formed inorganic sulfate from partially degraded organic intermediates as well, as experienced by House (1966b), above. If so, the amount of sulfate actually arising from bacterial action may have been much less, and formed at a late stage in the biodegradation.*

Earlier, Phillips (1963) had noted in Warburg studies that unacclimated activated sludge accomplished greater oxygen uptake on TBS than it did after several days' acclimation (1.2 vs. 0.8 gm O_2/gm TBS), whereas methylene blue analysis of the Warburg mixtures showed zero removal of MBAS by the unacclimated sludge versus 25–45% removal after acclimation. He hypothesized two modes of attack on the TBS: (i) oxidation of the side chain with resultant oxygen uptake, leaving a product still responsive to methylene blue, and (ii) removal of the sulfonate group only, destroying methylene blue response but leaving the unsulfonated alkylbenzene which exerts little or no oxygen demand. The first mechanism was presumed to predominate with the unacclimated culture, the second with the acclimated. In the light of the other work on TBS discussed above, neither mechanism is especially compelling. Perhaps Hartmann's (1967) hypothesis fits the observations better: some of the unacclimated cells are killed by the TBS, liberating cell components which the remaining cells can oxidize (Chapter 4, Section III.C).

VI. SULFONATED ESTERS AND AMIDES

To account for the rapid degradation of Igepons A and T (7.26), Sawyer (1956) and Ryckman (1957) suggest initial hydrolytic attack to liberate the fatty acid, which could then enter the normal metabolic reactions readily since these surfactants are made from the natural linear fatty acids. No direct experimental evidence for this mechanism seems to have been developed. Even though it may prove to be the actual pathway taken

---

* Ichikawa (1966) inferred that his own bacterium degraded TBS (up to 40%) via sulfonate attack also; it degraded LAS (Ucane) no more than TBS.

in this particular case, it would not necessarily be universal. For example, it is by no means certain that the ester or amide linkage would be broken as readily in the case of nonlinear derivatives, just as nonlinear alkysulfates may resist biological hydrolysis.

Sheers (1967) advanced a similar hypothesis for the biodegradation of *N*-secondary alkyl-3-sulfopropionamide (NASP) **(7.25)**, which has a

$$
\begin{array}{cc}
\quad\quad O & \quad\quad O \\
\quad\ C \ \| & \quad\quad \| \\
C_{9-18}CN{-}CCC\Sigma & C_{17}C{-}NCC\Sigma \\
\quad\ H & \quad\quad C \\
\text{NASP} & \text{Igepon T} \\
\text{(7.25)} & \text{(7.26)}
\end{array}
$$

structure resembling but somewhat different from Igepon T. Sheers found that NASP was as readily degraded in unacclimated as in acclimated systems, with rapid loss of foaming properties (4 days in river water). This he attributed to the presence of the –NHCO– group, also representative of the peptide link which holds proteins together and which is hydrolyzed by the proteolytic enzymes found in all living systems. Such attack on the NASP would split the molecule and destroy the surfactant properties even in the absence of any further biodegradation. Upon treatment of NASP with the proteolytic enzyme trypsin in aqueous solution (about 1% of each) he observed about 5% disappearance of NASP in 19 days, 10% in 32 days. Parallel experiments with Igepon T and an ABS (TBS?) showed no such disappearance of surfactant. These results may be irrelevant, since the concentrations were much higher and the times much longer than in the biodegradation studies, but they are at least consistent with the suggested mechanism. But then, Igepon T, which was not hydrolyzed by the trypsin, is readily degraded in unacclimated systems also (Bogan, 1955; Table 8.7), perhaps as readily as NASP.

VII. ALKYL SULFATES AND ETHOXYLATE SULFATES

A. Initial Hydrolysis or Initial Oxidation?

Ryckman (1957) suggested that the first step in the biodegradation of alkyl sulfates was most likely hydrolysis to the alcohol, with subsequent oxidation thereof. This is undoubtedly true for the linear primary alkyl sulfates (LPAS), where hydrolysis by bacterial enzymes is quite facile and has been accomplished in vitro with cell-free enzyme concentrates. But many other alkyl sulfates are much slower to degrade, and some are quite resistant (Chapter 6, Section VII.A).

Since hydrolysis would result in immediate loss of surfactancy and methylene blue response (in other words, in primary biodegradation) it is obvious that the resistant alkyl sulfates are not readily hydrolyzed by the bacterial action. It is not clear as yet whether the eventual attack in such cases is by hydrolysis or by oxidation at the other end of the molecule as with ABS.

B. Linear Primary Alkyl Sulfates

B.1. *Hydrolysis by Sulfatases*

The course of biodegradation of LPAS has been delineated by Payne and co-workers, using principally a *Pseudomonas* strain (C12B) originally isolated by enrichment culture from soil taken near a sewage outfall (Payne, 1963a, b). It was capable of growth in sodium dodecyl sulfate (SDS) medium with no other source of organic carbon, and in so doing, of course, degraded it. Such growth was made possible by adaptive enzymes which were not present when the cells were grown on normal food, but which developed upon exposure to the SDS. Biochemical comparisons of unacclimated with acclimated cells established the probable pathway: (i) hydrolysis to alcohol and sulfate, (ii) oxidation of alcohol to fatty acid, (iii) degradation of fatty acid by β-oxidation.

Williams (1964) demonstrated the presence of a sulfatase enzyme, which accomplishes the hydrolysis, in the acclimated cells but not in the unacclimated. The enzyme was isolated by disruption of the cells, extraction, centrifuging, and dialysis, and was detected in the resulting cell-free solution by liberation of inorganic sulfate from SDS. Payne (1965) recovered the other hydrolysis product, dodecanol, by extraction and identified it by GC.

Hsu (1963, 1965) independently reported isolation of a primary alkyl sulfatase. He separated six strains of SDS-splitting bacteria from sewage, all being *Pseudomonas* species, obtained the enzyme as a cell-free extract and purified it further by fractionation and column chromatography. It too was an adaptive enzyme, its formation by the cells being induced by the presence of SDS in the growth medium. Even though induced by the C_{12}LPAS, the free enzyme could split the C_8, C_{10}, and C_{14} homologs equally easily as well. Absence of oxygen uptake during the splitting indicated hydrolysis rather than oxidation. [Later Marion (1966) found evidence that an SLS oxidative enzyme was also present in cell-free extracts from an acclimated *Alcaligenes faecalis* culture (Section III.B).]

Hsu reported experiments with intact acclimated cells showing that on the average one cell could split 1.4×10^{-9} μmols of SDS per hour at 32°C. This is about 850 million molecules, so that the average cell is splitting some 250,000 molecules/sec. This impressively large number reflects not so much any unusual busyness on the part of the bacterium as, rather, the minute size of a molecule.

One milligram of Hsu's purest enzyme concentrate would split 200 μmols (about 60 mg) of SDS per hour. The proportion of pure enzyme present in such a concentrate is not known, but it would appear to be about 200 times as active as Payne's preparation. A cell splitting 1.4×10^{-9} μmols/hr would require 0.7×10^{-11} mg of Hsu's concentrate. This would amount to about 1% of the cell's mass, assuming the cell's volume to be about 1 cubic micron, mass 10^{-9} mg.

Upon feeding a selection of nonsurfactant aliphatic and aromatic sulfates to *Pseudomonas aeruginosa*, Harada (1964a) induced the formation of at least three different sulfatase enzymes. They were not induced by feeding inorganic sulfate. Sulfatase formation also occurs in *Aerobacter aerogenes* upon feeding a variety of organic sulfur compounds. Here too inorganic sulfate, as well as sulfite and certain organic sulfur compounds, repressed the enzyme formation (Harada, 1964b; Rammler, 1964).

The results reported by Nyns (1969b) are also consistent with initial hydrolysis of LPAS to the free alcohol during biodegradation, according to the principle of sequential induction (Chapter 4, Section III.A). He isolated 20 unidentified strains of soil bacteria by enrichment culture on SLS medium and found that 18 of them could grow readily on lauryl alcohol as well, whereas less than half of some 60 other strains, isolated on C_{13}LAS, C_{13} alkylbenzene, or *tp*-alkylbenzene media, were able to do so. Conversely, only 6 out of 20 strains isolated on lauryl alcohol medium could grow correspondingly on SLS.

B.2. Oxidation of the Alcohol

The next stage after hydrolysis of the LPAS appears to be oxidation of the alcohol, accomplished by dehydrogenation, a reaction catalyzed by dehydrogenase enzymes. Payne (1963b) and Feisal (1966) found that unacclimated *Pseudomonas* C12B cells readily oxidized C_6 and C_8 linear primary alcohols in Warburg experiments, but that the response decreased stepwise nearly to zero with the C_{10}, C_{12}, C_{14}, C_{16}, and C_{18} alcohols. Cells acclimated to SDS showed the same pattern except for the C_{12} alcohol, which surpassed even the C_6 and C_8 alcohols. Corresponding results were obtained when cell-free extracts were tested for dehydrogenase enzyme activity, although the increase in C_{12} response with SDS acclimation was less spectacular. Subsequent experiments (Williams, 1966) showed that the unacclimated cells contained constitutive dehydrogenases active toward all linear primary alcohols from C_2 to C_{11}, but not toward higher ones. (Incidentally, dehydrogenase activity toward the secondary alcohols 2-propanol and 2-butanol was also demonstrated, and toward the branched primary alcohol

$$
\begin{array}{c}
\text{C} \\
\text{CCCCOH}
\end{array}
$$

while a quaternary primary isomer

$$
\begin{array}{c}
\text{C} \\
\text{CCCOH} \\
\text{C}
\end{array}
$$

was inert.)

Thus it appears that upon hydrolysis of the LPAS the constitutive dehydrogenases of the C12B are capable of catalyzing the immediate oxidation of any liberated alcohols up to C_{11}, and that formation of dehydrogenases for the longer alcohols is readily induced.

Presumably the oxidation of these alcohols under the influence of the dehydrogenase enzymes first gives the corresponding aldehydes and then the corresponding fatty acids; there can be little question but that the fatty acids are then degraded by the usual β-oxidation process, and substantiation has been provided by Payne's group. Prochazka (1967) reported the transient accumulation and subsequent disappearance of dodecanoate in *Pseudomonas* C12B cultures growing on *n*-dodecanol (in place of the alkyl sulfate) during the disappearance of the dodecanol. And earlier, Williams (1964) had measured the amounts of isocitrate lyase and malate synthetase in the cell-free extracts of the C12B organisms. These enzymes are induced by growth on C_2 compounds such as acetate, and their production was increased severalfold upon growing the C12B in SDS medium (although not as much as in acetate medium). This is consistent with the idea that β-oxidation is involved, since that process produces such C_2 fragments.

The findings of Dronkers (1964) are quite consistent with this picture also. As shown in Table 7.5, Run A, he was able to isolate significant amounts of dodecanol and dodecanoic acid from an SDS solution which had undergone 80% primary degradation (MBAS). He obtained these by extraction with petroleum ether after the degradation had been interrupted, in amounts of 37% and 5%, respectively, of those calculated for the SDS which had disappeared. It is reasonable to assume that much of the other 58% had undergone further degradation toward carbon dioxide and water. In Run B, carried to 90% primary degradation, the amount of petroleum ether solubles was still lower, in accord with the greater exposure to the biodegradation conditions.

TABLE 7.5

Biodegradation of Sodium Dodecyl Sulfate[a]

	Run A		Run B,
	mg	mmol	mg
Initial SDS	500	1.735	500
Final SDS	100	0.347	50
Degraded SDS	400	1.388	450
Petroleum ether extracts	132	—	60
Neutral fraction	115	—	—
Dodecanol	95	0.510	—
Unknown A	20	—	—
Acid fraction	17	—	—
Decanoic acid	0.2	—	—
Dodecanoic acid	14	0.070	—
Unknown B	0.2	—	—
Unknown C	2.6	—	—
Foamate, alcohol soluble	—	—	52
Surfactant	—	—	45
NaCl	—	—	6

[a] Dronkers (1964).

B.3. *Oxidation before Hydrolysis*

In Run B, Table 7.5, Dronkers (1964) isolated the 10% remaining surfactant components from his biodegradation mixture by foam stripping the aqueous solution after the petroleum ether extraction. The dried foamate yielded 52 mg of alcohol-solubles, of which 45 mg was surfactant. After exchange of sodium for hydrogen by means of an ion exchange resin, potentiometric titration of the surfactant fraction showed a small amount of weak acidity along with the strong acidity, interpreted as carboxyl groups on some 5% of the alkyl sulfate molecules titrated. Dronkers advances this as evidence that the alkyl sulfate occasionally undergoes oxidative attack on the hydrophobe group prior to hydrolytic removal of the sulfate. Unfortunately he gives no evidence that the carboxyl groups are indeed on the alkyl sulfate molecules; the possibility thus remains that they may actually arise from simple carboxylates which escaped extraction by the petroleum ether.

For this same reason the lower equivalent weight of his surfactant fraction (268 vs. 288 for the original SDS) also cannot be considered as unequivocal proof of an initial oxidative attack if the figure was calculated as total acid equivalent weight or as strong acid equivalent weight. If, on the other hand, the 268 is a methylene blue equivalent weight, Dronkers' case is quite strong, provided such lower molecular weight components were not present in the original SDS.

Although not directly pertinent to biodegradation in wastewater, it is interesting to note in passing that oxidation before hydrolysis does occur when SDS is biodegraded by a rat following intravenous injection. Denner (1969) found that 95% of the injected sulfate was excreted in the urine within 12 hr. A fraction appeared as inorganic sulfate, but most of it was in the form of the sulfate ester of 4-hydroxybutyric acid. Thus the rat's preferred metabolic path appeared to be ω-oxidation of the hydrophobe chain followed by β-oxidation [Eq. (7.11)]. Hydrolysis of the sulfate ester group, as determined from inorganic sulfate formation, was

$$\Sigma\text{OCCCC}|\text{CC}|\text{CC}|\text{CC}|\text{CC} \rightarrow \Sigma\text{OCCCCCO}_2\text{H} \qquad (7.11)$$

surprisingly slow, only about 20% during the limited residence time in the rat before excretion.

Earlier, Knaak (1966) had found that the shorter, branched 2-ethylhexyl sulfate was excreted by the rat largely unchanged. About two-thirds of it was intact and one-third appeared as a dihydroxy acid as shown in Eq. (7.12). Here too oxidation, at least the first step, preceded hydrolysis, but oxidation of the terminal methyl group apparently could not be accomplished as it was with SDS.

$$
\begin{array}{ccccc}
\text{H} & \text{H} & & \text{H} \ \text{H} & \\
\text{O} & \text{O} & & \text{O} \ \text{O} & \\
\text{CCCCCCO}\Sigma \rightarrow \text{CCCCCCO}\Sigma \rightarrow \text{CCCCCCOH} \rightarrow \text{C--C--C--C--C--CO}_2\text{H} & (7.12) \\
\text{C} & \text{C} & \text{C} & \text{C} & \\
\text{C} & \text{C} & \text{C} & \text{C} & \\
\end{array}
$$

B.4. Anaerobic Oxidation

Unlike many organic compounds and most surfactants, LPAS has been reported to undergo extensive biodegradation in anaerobic systems (Pitter, 1964b, d; Klein, 1965a; Maurer, 1965). The work of Ōba (1967) suggests that this may be simple hydrolysis and that little if any actual degradation of the organic portion of the molecule occurs. From a 3 day anaerobic culture in which 257.5 mg of coco alcohol sulfate had degraded, as determined by MBAS analysis, he isolated a nonacidic ether-soluble fraction of 205.7 mg, compared to 37.6 mg of similar fraction from a control culture. The difference, 168.1 mg, checks closely with the amount of coco alcohol which would be formed by hydrolysis of the original sulfate, 166 mg. In contrast, a 15 hr aerobic culture (corrected by control) gave no significant amount of nonacidic fraction, but did show an acidic ether-soluble fraction amounting to some 38% of the original coco alcohol. Thus, (i) under aerobic conditions the alcohol liberated by hydrolysis of the alkyl sulfate is rapidly oxidized to fatty acid which is then degraded, while (ii) in the anaerobic system the liberated alcohol, like many other compounds, was inert.

In the course of this work Ōba also looked for any sulfide that might have been formed from the degradation of sulfate or sulfonate groups under the anaerobic conditions. Although the sulfide concentrations in his cultures approximately doubled during 7 days, from about 0.5 to 1 ppm, there were no significant differences between the control culture and those with 20–25 ppm of LAS, TBS, AOS, or primary alkyl sulfates (linear, branched, or ethoxylated).

C. Secondary and Branched Alkyl Sulfates

The enzyme concentrates which have been isolated in studies of LPAS biodegradation are quite specific for that particular class of surfactant. For example, Payne (1965) reported that his enzyme would not hydrolyze a C_{10-20} linear secondary alkyl sulfate mixture, nor aryl sulfates (p-nitrophenyl sulfate, phenolphthalein sulfate), nor even the substituted primary alkyl sulfate $Cl_2\phi OCCO\Sigma$. Likewise Hsu (1965) found his enzyme to be inactive toward the branched secondary alkyl sulfate **7.27**.

$$
\begin{array}{c}
\text{C} \\
\text{C} \quad \text{C} \\
\text{CCCCCCCCCCC} \\
\text{O} \\
\Sigma
\end{array}
$$

(7.27)

But these and other alkyl sulfates may have their own specific sulfatases, at least in some cases. Thus Payne (1967) isolated a strain of *Aerobacter cloacae* from soil by enrichment culture using a C_{10-20} linear secondary alkyl sulfate medium. Cell-free extracts prepared from this organism contained a secondary alkyl sulfatase as demonstrated by hydrolysis of 3-pentanol sulfate, giving the alcohol and inorganic sulfate. It was an induced enzyme rather than constitutive, since cells grown on broth in the absence of secondary sulfate did not contain the enzyme. This secondary sulfatase was more versatile than the LPAS ones, since it was able to hydrolyze the primary SDS as well, but it did not act on aryl sulfates.

Presumably this enzyme could hydrolyze the C_{10-20} secondary sulfate which induced its development in the first place, but this is not so stated.

The later stages in biodegradation of the nonlinear or nonprimary alkyl sulfates are almost unexplored. Virtually the only information available comes from Borstlap (1964) (Table 7.3) and Pitter (1964c), who reports oxygen uptake data from 20 day respirometer runs (Table 7.6). It is noteworthy that under Pitter's conditions he was able to achieve substantially theoretical oxygen uptake with glucose during the 20 day run.

TABLE 7.6

Biodegradation of Alkyl Sulfates[a]

			Oxygen demand[b]			BOD/COD,
Type	Structure	Homolog	Theor.	COD	BOD	%
Prim	Linear	C_{12}	2.00	1.96	1.92	98
Prim	Lin. oxo	C_{11-15}	2.07	2.50	2.15	86
Prim	Tp[c] oxo	C_{13}	2.07	2.17	0.65	30
Prim	2-Et-C_6	C_8	1.66	1.74	1.20	69
Sec	Linear	C_{10-13}	1.97	1.92	1.47	77
Glucose (reference)			1.07	1.055	1.05	99.5

[a] Pitter (1964c).

[b] 20 day respirometer, gm/gm.

[c] Oxo alcohol derived from tetrapropylene.

This occurred likewise with the LPAS, suggesting essentially complete degradation to carbon dioxide and water.

Pitter's data on the linear oxo alkyl sulfate are indeterminate, to a certain extent. His sample was presumably a commercial-type product, and evidently contained a considerable amount of oxidizable impurities, since the COD value was almost 25% higher than the theoretical figure for a C_{13} average product. The BOD was a little higher than theoretical, but was only 86% of the observed COD. At one extreme, if we assume that the impurity contributed nothing to the BOD, then the oxygen uptake of the linear oxo alkyl sulfate was in excess of theoretical. At the other extreme, if the impurity was completely oxidized in the respirometer, the BOD exerted by the actual alkyl sulfate must have been somewhat below theoretical.

The oxo product derived from tetrapropylene was quite resistant, as would be anticipated from its highly branched structure, and its oxidation reached only 30% of completion. Yet it was attacked to a significantly greater extent than TBS, which showed only about 10% in a parallel test.

The linear secondary alkyl sulfate, at 77%, approached completion fairly closely, but fell short of the linear primary. If this was a commercial product containing undefined quantities of nonlinear impurities, it would not be surprising if the performance of a pure product would come considerably closer to that of LPAS.

Partial oxidation of the branched primary alkyl sulfate of 2-ethyl-hexanol by a rat is described in Section VII.B.3 (Knaak, 1966).

D. Ethoxylate Sulfates

In the absence of direct experimental data on the compounds themselves, the metabolic pathways involved in the biodegradation of the ethoxylate sulfates can be discussed only on the less substantial basis of inference and conjecture, and to little profit.

Their most significant and striking feature is that despite the fact that they are all primary alkyl sulfates, $ROE_nOCCO\Sigma$, their primary biodegradation, as measured by methylene blue response, may range from rapid and complete to substantially none. This indicates once again the specificity of the primary alkyl sulfatase enzymes, since resistance of an ethoxylate sulfate to primary biodegradation necessarily includes resistance to enzymic hydrolysis of the sulfate group. Thus the geometry of the hydrophobe group R controls not only the possibility of direct oxidative attack upon itself, but also controls the possibility of hydrolytic attack on the sulfate group, even though the gross structure at the sulfate end of the molecule is identical in all ethoxylate sulfates.

The oxygen uptake data listed in Tables 8.11 and 8.12 indicate that extensive or possibly complete oxidation of the linear primary alcohol ethoxylate sulfates can occur during biodegradation. The extent appears to be somewhat less, and/or the rate somewhat slower, with secondary alcohol hydrophobes, still less for linear alkylphenols, and least of all for branched alkylphenols.

VIII. ETHOXYLATE NONIONICS

In contrast to the anionic surfactants, the nonionics are organic in both the hydrophobic and hydrophilic groups, and hence are in principle open to oxidative biodegradation at both ends of the molecule. The great majority of nonionic surfactants in use today are polyethoxylates, and the following discussion of biodegradation of the free polyglycols will serve as groundwork for the subsequent sections on the surfactants themselves.

A. Unsubstituted Glycols and Polyglycols

The unsubstituted polyglycols without hydrophobe attachments have received considerable study, as has the monomer, ethylene glycol. The resulting information, if used with appropriate caution, can be helpful in understanding the biodegradation process in the corresponding surfactants. As indicated in Table 8.26, many of these materials undergo extensive biodegradation. In the absence of specific and sensitive analytical methods for the intact starting materials, much of this work has made use of BOD or respirometric techniques and thus provides information on ultimate as well as primary biodegradation.

A.1. Oxygen Uptake

Ethylene glycol (E_1) itself was found by Lamb (1952) to be readily degradable in the BOD test when sufficient time was allowed for acclimation. Although only 12% of the theoretical amount of oxygen was absorbed in the usual 5 days, the rate accelerated rapidly after that time and reached 52% in 10 days, 71% in 15, and 78% in 20 days. The latter figure represents a close approach to ultimate biodegradation, assuming that some fraction of the substrate was converted to bacterial protoplasm rather than being completely oxidized to carbon dioxide and water.

In contrast, diethylene glycol (E_2) and triethylene glycol (E_3) were much more resistant in Lamb's experiments, absorbing a little less than 20% of the theoretical oxygen in 20 days. Mills (1953, 1954) obtained much the same results with E_2, E_3, and E_4 in the BOD test, and also with polyethylene glycol 400 (PEG 400), a higher polymer averaging around E_8 to E_9. Mills did find, however, that removals of 50% to 70%, as determined by measurement of residual COD, could be attained for these materials in a continuous-flow system under anaerobic conditions with 7 day retention time.

In the Warburg respirometer also, Gerhold (1966) found that the monomer was more readily oxidized than the dimer or trimer. With unacclimated seed (activated sludge from three different sewage treatment plants) E_1 picked up about 40% of the theoretical oxygen in 24 hr, while E_2, and E_3 only absorbed one-half to one-fourth as much. But Fincher (1962a, b) had earlier found to the contrary that although E_1 reached 67% of theoretical in 8 hr, E_2, E_3, and E_4 attained 114%, 92%, and 99%, respectively. His greater success was doubtless due to his use of a selected organism, called TEG-5 (a member of the *Pseudomonas-Achromobacter* group, isolated from soil by enrichment culture with E_3), which had been preacclimated to each individual substrate for the Warburg runs. Using the same TEG-5 strain in large volume Warburg runs Prochazka (1967) obtained $54 \pm 3\%$, $54 \pm 4\%$, and $54 \pm 1\%$ for E_2, E_3, and E_4.

Earlier, Bogan (1955) has shown the importance of acclimation in his work on the monoethyl ethers of E_1 (CCOCCOH) and E_2 (CCOCCOCCOH). These had shown 5 day BOD values around 8% of theoretical with un-acclimated seed, 54% and 34% with acclimated. His Warburg results were 52% and 40%.

Supplementing Warburg oxygen uptake measurements with simultaneous determination of the increase in bacterial cell mass, Vath (1964) concluded that E_9 was completely metabolized. The sum of the two corresponded to about 90–110% of the initial E_9. Borstlap (1967a) approached the problem by measuring the nonvolatile organic residue remaining after the biodegradation. He reports that E_9 left only 2%, compared to 10% for the higher polymer E_{23}. Pitter (1968a, b), with a synthetic medium inoculated with acclimated organisms, analyzed by oxygen uptake and COD removal, found essentially complete degradation of E_2, E_3, E_4, and E_6 but essentially zero for E_{13}, E_{22}, E_{34}, and E_{80}. He remarked upon the anomaly that although the E_{13} polyglycol was stable, several nonionics in the same range or higher showed significant degradation of the polyethoxylate chain under the same conditions.

Others have used more specific analytical methods for the polyglycols, as listed in Table 8.26. Their results thus relate more closely to primary than to ultimate biodegradation, depending on the specificity of the method, but they too suggest that many of the polyethylene glycols, at least up to E_{23}, can be degraded under suitable conditions, and that primary degradation can reach 100%.

A.2. *Oxidative Pathways*

The details of the biochemical mechanisms involved here have not been well elucidated as yet, but a beginning has been made. Fincher (1962a, b) and Payne (1963b) demonstrated the presence of a dehydrogenase enzyme in cell-free extracts of their *Pseudomonas* TEG-5 after culturing on E_4. The enzyme catalyzed the oxidation of polyglycols ranging from E_2 up to E_{14}, detected by decolorization of methylene blue or ferricyanide added to the system as hydrogen acceptor. The oxidation rate decreased with increasing degree of polymerization, and E_{23} and E_{135} were not attacked.

In later work Payne (1966) also observed dehydrogenation (but slower than with E_2 to E_{14}) of the polyglycol ethers nonylphenol E_{10-20}, dipropylene (glycol?) E_{10-20} and C_{11-15} linear secondary alcohol E_9. The corresponding linear secondary E_3 sulfate was not reactive, leading Payne to the conclusion that a free terminal OH is necessary if dehydrogenation is to occur. Although the possibility must be considered that this failure might have been due to inactivation of the enzyme by the ethoxylate sulfate, which is after all an anionic surfactant, nevertheless it seems reasonable to assume that the dehydrogenase is involved with the terminal OH, which would presumably proceed through the aldehyde to the carboxylic acid.

In an extensive review on oxidation of glycols and glycol derivatives by the acetic acid bacteria *Gluconobacter* and *Acetobacter*, de Ley (1964b) cites many examples of such reactions, particularly the conversion of ethylene glycol to glycolic acid, $HOCCO_2H$, and occasionally to oxalic acid, HO_2CCO_2H. Diethylene glycol, E_2, undergoes similar reactions, but less readily, to give $HOCCOCCO_2H$ and $HO_2CCOCCO_2H$, and likewise E_3, still less readily. De Ley found that at least four enzymes involved in the oxidation of glycols could be isolated from *Gluconobacter suboxidans*. Two are the soluble primary and secondary alcohol dehydrogenases, while the others, much less specific in their oxidation capabilities, are probably associated with the cytoplasmic membrane of the cell. The primary alcohol dehydrogenase was less than half as active (oxidation rate) toward ethylene glycol as toward ethanol, and was inactive on E_2 and E_3. The secondary alcohol dehydrogenase was inactive toward primary alcohols and diols but oxidized propylene glycol and other compounds containing a secondary OH group.

A beginning has likewise been made on the mechanism of anaerobic degradation of ethylene glycol. Abeles (1961) prepared cell-free extracts of an *Aerobacter aerogenes* strain after anaerobic growth on glycerol, which converted propylene glycol to propionaldehyde and ethylene glycol to acetaldehyde [Eq. (7.13)]. The overall result is that one carbon is oxidized one equivalent and the other is reduced by the same amount. The net change in energy is thus very small; the energy content of ethylene glycol and acetaldehyde are almost identical.

Later Abeles and colleagues isolated the enzyme involved, diol dehydrase. This, in the presence of a coenzyme (a cobamide, related to vitamin B_{12}), could accomplish the reaction in vitro. Using deuterium or tritium tracer techniques they found that the reaction involved removal of one of the terminal methylene hydrogens from the glycol by the coenzyme, which then reintroduced it at the adjacent carbon as shown in Eq. (7.13) (Frey, 1967). Studies by Rétey (1966) using ^{18}O tracer showed

$$CH_3-\underset{\underset{H}{|}}{\overset{\overset{H}{|}}{C}}-\underset{\underset{H}{|}}{\overset{\overset{T}{|}}{C}}-OH \rightarrow \cdots \rightarrow CH_3-\underset{\underset{H}{|}}{\overset{\overset{T}{|}}{C}}-\underset{\underset{H}{|}}{\overset{\overset{OH}{|}}{C}}-OH \rightarrow CH_3-\underset{\underset{H}{|}}{\overset{\overset{T}{|}}{C}}-C{=}O \qquad (7.13)$$

that the oxygen atom was transferred in the opposite direction. Gaston (1963) traced the possible mechanism a step further using the anaerobe *Clostridium glycolicum*, isolated from pond mud. During growth of this organism in a medium containing ethylene glycol, about $\frac{1}{2}$ mol of ethanol and $\frac{1}{2}$ mol of acetic acid were formed for each mol of glycol disappearing. Gaston theorized that the glycol was first converted to acetaldehyde, which then underwent self-oxidation and reduction. Even though no net oxidation takes place, this reaction would yield a small amount of energy, some 5% of that from complete oxidation of the glycol.

On the basis of the foregoing work, it is reasonable to suppose that the first step in the aerobic biodegradation of the polyglycols would be oxidation of the terminal CH_2OH group to aldehyde and then to carboxylate. The mechanism for the subsequent steps, particularly the details of breaking the ether linkage, still remain to be disclosed. (If an internal carbon could be oxidized in a manner similar to the formation of terminal aldehyde and carbolylate, the corresponding products would be hemiacetal and ester. Either of these would probably hydrolyze much more readily than the original ether.) Prochazka (1967) has taken a first step by discovering, through GC, a short chain intermediate in the degradation of E_2, but its nature was not determined at that time.

B. Alcohol Ethoxylates

B.1. *Formation of Polyglycol or Polyglycol Carboxylate*

It has been quite well established that biodegradation of a linear primary alcohol ethoxylate (LPAE) is characterized by rapid disappearance of the original nonionic accompanied by liberation of the ethoxylate portion of the molecule as free polyethylene glycol (or something closely resembling it), which subsequently disappears also. Although the chemical steps involved appear to be quite obvious, some of the details have not yet been worked out.

Thus it is not clear whether the initial step is hydrolytic or oxidative (or both) [Eq. (7.14)]. In both cases the initial reaction would probably

$$CCCCCCCCCCCCCOE_n \begin{cases} \nearrow & CCCCCCCCCCCCCOH + E_n \text{ (hydrolytic)} \\ \\ \searrow & HO_2CCCCCCCCCCCCCOE_n \text{ (oxidative)} \end{cases} \tag{7.14}$$

require acclimation. In the first case the alcohol liberated would then be readily oxidized to the fatty acid, which would in turn be equally readily degraded by β-oxidation. In the second, the terminal carboxylate should be readily degraded by β-oxidation down to HO_2CCOE_n (or an odd-carbon homolog), which might resemble the polyglycol fairly closely in most of its properties. Subsequent degradation of it, or of the free polyglycol, would again probably require acclimation.

Either of these pathways is consistent with the observations of Blankenship (1963), who reported rapid oxygen uptake by $C_{12}LPAE_8$ in 3 hr Warburg runs, usually extrapolating to asymptotic levels around 48% of theoretical. He calculated that this corresponded to substantially complete oxidation of the C_{12} chain (proposing a process of ω,β-oxidation), leaving the E_8 untouched. In view of the inherent ambiguities of respirometric data this agreement might be viewed as simply a coincidence, at least in the absence of other supporting analytical data on the system. Particularly so in view of numerous other respirometric data in Table 8.16 showing considerably higher oxygen uptakes for such materials.

Frazee (1964b) did provide other analytical data. He degraded $C_{14}LPAE_{8.3}$ in river water, isolating any remaining undegraded surfactant by foam stripping and analyzing it by IR. By the 5th day the surfactant content had dropped to zero and the foam isolation technique could no longer be used because the solution did not foam. Chloroform extraction did yield something—unalkylated PEG, E_x, of undetermined chain length, amounting to 5.7 ppm. The initial LPAE concentration was 19–20 ppm, corresponding to around 13 ppm of $E_{8.3}$. Thus the data suggest that the hydrophobe group is degraded first, leaving the polyglycol, which degrades somewhat more slowly.

Using her TLC analytical technique and the British STCSD inoculation-aeration test procedure, Patterson (1967) was able to demonstrate clearly the progressive disappearance of the original LPAE within 7 days, accompanied by the appearance and subsequent disappearance of PEG-type material. The lower ethoxylates, E_6 and E_9, gave only small amounts of the PEG, perhaps 1 ppm at maximum, about 10% of the original surfactant, and it was all gone again within a week or two. With higher ethoxylates, E_{15}, E_{20}, and E_{30}, the PEG maximum was correspondingly higher and its subsequent diminution slower, in good agreement with the increased resistance of the higher PEGs indicated in Section VIII.A.1.

These results still do not give sufficient information for a definitive picture of the biodegradation process. The TLC method used does not resolve the individual components of the liberated "polyglycol" and conceivably might not distinguish between an unsubstituted PEG and a carboxylated derivative HO_2CCOE_n. Isolation of further amounts of "polyglycol" by further chloroform extraction after acidification also suggests that carboxylated polyglycols may well be present. Although the data do not give a clear answer as to whether the material is formed as a result of oxidation or hydrolysis, in either case it seems quite likely that its subsequent disappearance should involve the same chemical mechanisms whether it is unsubstituted or carboxylated PEG.

Patterson's work is briefly summarized in an LGC (1966, p. 140–144) report along with allusion to further work on examination and properties of the biodegradation intermediates which led to the conclusion that LPAEs degrade via both pathways, initial hydrolysis and initial oxidation. However, the supporting experimental data have apparently not been published as yet. [Details are now available in Patterson (1970).]

Pitter (1968a, b) examined LPAE in inoculated BOD medium and found COD removals substantially in excess of the amount calculated for hydrophobe oxidation alone. He concluded that oxidation of the ethoxylate portion of the molecule must have occurred also. The difference was less for the higher ethoxylates and was negligible for the E_{15-20} derivatives, again confirming the greater resistance of the higher ethoxylates.

B.2. Secondary and Branched Alcohol Ethoxylates

The linear secondary alcohol ethoxylates (LSAE) appear to follow the same general path as the primaries. Vath (1964) made a Warburg study on an E_6 tagged with ^{14}C in the ethoxylate group. By the time 39% of theoretical oxygen had been absorbed, 26% of the ^{14}C had been evolved as carbon dioxide and caught in the KOH well. This suggested that oxidation of the EO chain did not lag very much behind the oxidation of the hydrophobe.

A much sharper demarcation between the two was found by Wickbold (1966) in the 3 hr exposure time allowed in the official German activated sludge test. Surprised that $C_{14}LSAE_9$ and E_{12} showed an apparent degradation of only about 35%, he investigated further and found that primary degradation was actually 95% and that the interference was caused by nonsurfactant intermediate biodegradation products of the type 7.28. These accompanied the intact surfactant in his prepurification step

$$HO_2CCCO_2H$$
$$O$$
$$E_n$$
$$(7.28)$$

prior to analysis (n-butanol extraction from HCl-acidified solution), in contrast to free PEG which was rejected some 98%, and responded equally well in the subsequent bismuth iodide analysis. The dicarboxylate intermediates were rejected when the preextraction was made from sodium bicarbonate solution, their sodium salts remaining in the aqueous phase.

Patterson (1967) found that $LSAE_9$ and $LSAE_{13}$ behaved much like the primary derivatives. TLC analysis showed liberation of PEG-type material which disappeared within the next 2 weeks. Primary degradation of the $LSAE_{13}$ slowed down considerably in its later stages, requiring about 20 days for complete disappearance, whereas its PEG-type intermediate peaked at about 6 days and was gone 15 days.

The highly branched tetrapropylene-oxo $C_{13}E_8$ was much more resistant in Patterson's experiments, with 30% remaining at 49 days. PEG liberation was only barely noticeable; presumably the PEG was degraded about as fast as it was formed in the primary degradation (which was slower than the linears) and so did not accumulate.

C. Alkylphenol Ethoxylates

C.1. Oxygen Uptake

In contrast to primary biodegradation of the APEs, which can proceed substantially to completion with suitable acclimation, Tables 8.18 and 8.19 show that oxygen uptake falls far short of theoretical in almost all of the studies. There are two main exceptions. Lashen (1966) found 7 day BOD values of 23–29% for OPE_{10} with acclimated seed compared to 0–15% without acclimation. Contrariwise, Hartmann (1967) found Warburg uptakes of 10% for acclimated and 58% for unacclimated activated sludge, which difference he attributed to factors other than oxidation of the APE (Chapter 4, Section III.C).

Failure of oxygen uptake to proceed beyond 15–20% (indeed, the results are in the range 0–5% more often than not) in the determinations made by other workers can conceivably be due to acclimation difficulties at least in part. Conditions for BOD and respirometric measurement are ordinarily rather different from those under which acclimation is usually developed in the laboratory and quite different from conditions in the field—oxygen uptake measurement has not been accomplished at all in the latter case. Since the acclimation might not carry over from the one environment to the other, it remains uncertain whether the good primary biodegradation occasionally observed for these compounds is accompanied by extensive oxygen uptake as well.

C.2. Weight Loss

Measurement of oxygen absorption allows estimation of extent of ultimate biodegradation, subject to the characteristic ambiguities in interpretation of the data, but throws little light on the chemical pathways involved. Measurement of the weight of material left can provide supporting information, also of a nonspecific sort, but perhaps less ambiguous. It is of course necessary to recover all of the biodegradation intermediates originating with the surfactant, and to exclude normal metabolic products.

Borstlap (1967a) undertook such determinations in systems inoculated with activated sludge, using n-butanol as solvent for the undegraded material, after evaporation of the aqueous test medium to dryness. The residual organics so determined were corrected by subtracting the amount

TABLE 7.7

Biodegradation of Ethoxylate Nonionics[a]

	% Remaining[b]	
Hydrophobe	E_9	E_{15}
Linear oxo primary alcohol, C_{12-15} (Dobanol 25)	5	11
Lauryl:myristyl alcohol 1:3	0	11
Linear C_8 alkylphenol	29	49
Branched C_8 alkylphenol	54	51

[a] Borstlap (1967a).

[b] Percent of initial weight recovered after 28 days in an inoculated system.

found in a control run, fed no surfactant. He found that APE left a considerable percentage of incompletely degraded materials compared to the corresponding alcohol ethoxylates (Table 7.7). Subject to the probably small inaccuracies arising from incomplete separation of the pertinent organics and from possible metabolic deviations between the activated sludges in the control and surfactant runs, it appears that under Borstlap's test conditions the APE progresses some 50–70% toward complete, ultimate biodegradation.

Earlier Huddleston (1965b) had measured the weight of chloroform-extractable (after acidulation) material remaining in river water and continuous-flow activated sludge systems. Any biodegradation intermediates not extractable with chloroform would be missed, and hence the results may not give the full story with regard to ultimate biodegradation. But it does give a minimum figure for the amount of incompletely degraded material remaining; it was about 50% or more. IR and NMR examination showed that the EO chains had been shortened significantly (Table 7.8). The activated sludge data on the $C_{12}APE_{12}$ are puzzling—it is not clear how the EO chain could have been shortened from E_{12} to E_7 and still leave 95% of the weight—but even so the general trend is quite evident: under Huddleston's conditions too, the APE did not approach ultimate biodegradation very closely.

TABLE 7.8

Biodegradation of Ethoxylate Nonionics[a]

	River water (20 days)		Semicontinuous AS (24 hr)	
	%[b]	EO[c]	%[b]	EO[c]
C_{10-12} LPAE$_6$	10	0	8	0
Isooctyl APE$_{9-10}$	47	E_4	60	E_4
Linear C_9APE$_9$	54	E_5	40	E_3
Linear C_{12}APE$_{12}$	75	E_{11}	95	E_7

[a] Huddleston (1965b).

[b] Percent chloroform-extractable remaining.

[c] Average EO chain length of chloroform-extractable.

C.3. *UV Spectroscopy*

Since APE contains a benzene ring, UV spectroscopy is applicable to detection of any biodegradation in that region of the hydrophobe. Sato (1963) gave evidence for deep-seated change in C_9APE$_{10}$ (presumably derived from tripropylene) during aeration of a 400 ppm solution in washed, unacclimated activated sludge, MLSS concentration 2250 ppm. The phenolic benzene ring band peaking at 275 nm was evident at the beginning of the aeration. After 7 hr the absorbance had increased about 50% and the maximum had shifted to 270 nm, suggesting formation of biodegradation intermediates. At 24 hr the absorbance in the 270–275 nm region was only half the original and at 48 hr was down to 10–15%, substantially flat, with no indication of any ring remaining. During this same time the phosphotungstate analysis value for nonionic content dropped to about 10% of the initial value. It is difficult to see how such analytical changes could have resulted from merely superficial changes in the original APE molecules.

Later Osburn (1966) likewise observed disappearance of the 275 nm ring band in river water medium, down to around 25–35% of its original value in 35 days. This was verified by IR spectroscopy, which showed parallel disappearance of the phenolic ether band arising from the link between the alkylphenol and the PEG group.

On the other hand, Pitter (1968a) did not see any change at all in the UV spectrum of branched C_9APE_8 (nor of the APE_{30}) during 20 days in his respirometer, while oxygen absorption and phosphotungstate and COD analysis indicated only small to moderate removals.

C.4. *The Course of Biodegradation*

Shortening of the EO chain during biodegradation is one of the more prominent APE reactions. In contrast, hydrolysis at the phenolic ether link to liberate a full size PEG group has not been observed, contrary to the course of reaction in LPAE. If the latter is truly a hydrolytic reaction, it would seem that steric factors could be responsible for the enhanced stability in APE, and not necessarily the aromatic character of the ether link, since the tetrapropylene-derived tridecyl alcohol ethoxylate is likewise resistant to this reaction. As pointed out in Section VIII.B.2, failure to find the PEG does not necessarily mean that it is not formed, but only that it does not accumulate, as when its rate of formation is slower than its specific degradation rate.

However, there is direct evidence that the EO chain is shortened, possibly one EO group at a time, while it is still attached to the hydrophobe. A terminal glycolic hydroxyl group is left at each step rather than a carboxyl group. The chain shortening may be purely hydrolytic, liberating ethylene glycol which would be subsequently oxidized [Eq. (7.15)]. A second possibility is oxidation of the terminal carbon to carboxyl as a preliminary to the hydrolysis [Eq. (7.16)]. If so, the hydrolysis must follow

$$R\phi OE_n + HOCCOH \longrightarrow HOCCO_2H \longrightarrow \cdots \quad (7.15)$$

$$R\phi OE_nOCCOH \longrightarrow R\phi OE_nOCCO_2H \longrightarrow R\phi OE_n + HOCCO_2H \longrightarrow \cdots$$
$$\textbf{(7.29)} \qquad\qquad\qquad (7.16)$$

$$R\phi OE_n\overset{O}{\overset{\|}{O}}CCOH \longrightarrow R\phi OE_n + HOCCO_2H \longrightarrow \cdots$$
$$(7.17)$$

very quickly, since the hypothetical intermediates **(7.29)**, carboxylated at the end of the EO chain, have never been observed, only the hydroxylates. Alternatively, the intermediate oxidation step, if it does occur, might be at the inside carbon atom to give a glycolic acid ester, presumably easily hydrolyzed either biologically or chemically, as in Eq. (7.17).

Meanwhile, at the other end of the molecule, oxidation of the chain can be occurring. Methyl groups are converted to carboxyls, and the further steps of β-oxidation then occur if the chain is structurally suitable, or of the less usual mechanisms if not, just as in the case of ABS. And finally, the benzene ring is also susceptible, as discussed in the preceding section, although the exact sequence of reactions leading to ring degradation has not been elucidated as yet, nor have the exact structural and biological requirements.

We now review in the following section the experimental work which has given us the above picture.

C.5. *Polypropylene APEs*

The biodegradation pathways for the alkylphenol ethoxylates were charted in some detail by Frazee (1964b) and further explored by Osburn (1966), using IR and UV analytical procedures. The IR absorption at 1250 cm^{-1} provided a measure of the phenolic ether content (one per molecule) and at 1120 cm^{-1} the aliphatic ether links ($n - 1$ in a molecule containing E$_n$). The ratio of these two leads to the number of EO units, n, and if the structure of the hydrophobe is known the molecular weight and hence the concentration by weight can be calculated. Changes in the hydrophobe structure can be estimated by examination of the CH bands at 2900 cm^{-1}. Carboxyl groups (plus aldehydes and ketones) can be estimated from the carbonyl band at 1700 cm^{-1} and the primary OH group at the end of the EO chain from its band at 1065. The benzene ring content is measured by the UV absorption at 275 nm.

These measurements could not be made on the river water solution as such; isolation and concentration of the compounds present was first necessary. This was accomplished by (i) foam stripping, which separated the surface active components and left carboxylated intermediates behind, or (ii) chloroform extraction from neutral or alkaline solution, which did likewise, or (iii) chloroform extraction from strongly acid solution, which pulled out carboxylated products as well (if not acidified they remained in the water phase as calcium or magnesium salts). A further procedure (iv) was also used on some occasions: treatment with ion exchange resin which removed ions and carboxylated intermediates, followed by passage through activated carbon to adsorb the nonionics, which were then eluted by methanol:chloroform.

Application of these techniques during 30–35 days of river water biodegradation of branched C$_9$APE$_{10}$ led Frazee and Osburn to the following picture of the biodegradation process:

(i) The APE disappeared to the extent of 70–90%, and the 10–30% remaining averaged around E_4 in EO chain length.

(ii) About half of the initial nonionic had been converted to carboxylated products, average EO chain length likewise around E_4.

(iii) Both uncarboxylated and carboxylated fractions showed the presence of one alcoholic OH group, the one at the end of the EO chain.

A sample of APE_4 prepared to match that remaining in the river water gave a similar pattern during its own degradation: the EO chain diminished to about $E_{2.5}$, and the alkyl group was carboxylated to a considerable extent. The corresponding APE_6 behaved in a similar manner, but the longer E_{11} and E_{15} derivatives did not degrade at all in 35 days.

The dramatic difference reported here—E_{10} extensively degraded, E_{11} untouched—would seem more likely to result from some factor other than the small difference in average EO chain length. After all, each of the two materials should be a mixture of the same series of individual ethoxylates ranging perhaps from E_5 to E_{15}, in slightly varied proportions. Other, more likely factors might be differences in bacteria, acclimation, or inhibitory impurities, but at the present stage this can be little more than idle speculation.

Like spectroscopy, TLC has also been a powerful tool in exploring the biodegradation of APE. Patterson (1966a) observed that the pattern of TLC ethoxylate spots obtained in the examination of sewage or river water was displaced slightly downward from that of Lissapol NX (a commercial branched C_9APE_9) supposed to be the origin of the ethoxylates therein. Later Patterson (1968) demonstrated that this same depressed pattern was obtained from Lissapol NX during biodegradation in the British STCSD inoculation-aeration test. She stated further that no acidic degradation products were detected, presumably deduced from absence of response to further extraction of the spent sample after acidification.

This depressed pattern presents several difficulties of interpretation. It appears to have the same range and distribution of ethoxylate components as the original Lissapol NX, and so can only represent some minor change in the hydrophobic group. This could be conversion to a slightly more hydrophilic structure, which would migrate a little more slowly in the developing solvent. It is tempting to speculate that we are dealing with carboxylated APE derivatives, but Osburn (1966) indicates that these are not extractable from neutral solution. Lower oxidation stages resulting from introduction of oxo or hydroxyl groups might be suitably hydrophilic, and extractable as well, but it is not clear why the oxidation should have gone so cleanly to the intermediate stage and then stopped so abruptly. Substantially complete conversion of the original structure to the depressed structure is indicated by the sharp TLC pattern with no suggestion of spots at the original locations. Perhaps analysis at earlier stages in the degradation would have given a suitably smeared pattern indicative of intermediate stages wherein conversion was incomplete.

A further difficult point is that these TLC patterns apparently give no evidence for the stepwise shortening of the EO chain noted in Osburn's work, and showing in Table 7.8 (Huddleston, 1965b) as well.* Shorter ethoxylates do show up in the less common cases when Patterson's abnormal, alkaline, rapid biodegradation of the APE occurs (Chapter 6, Section IX.C.2) as indicated by new spots closer to the advancing solvent front, and diminution of the spots of the higher ethoxylate components is then evident also. The new spots do not fall on a continuation of the pattern of the original ones, and they show some irregularity and multiplicity as well. Perhaps this means that further changes in the hydrophobe have occurred along with shortening of the EO chain.

Of these discrepancies between Patterson's and Osburn's results, perhaps the most troubling is the absence versus the presence of (i) the intact hydrophobe group and (ii) carboxylated intermediates, during the biodegradation. These may simply reflect differences in the biodegradation test systems used, but convincing interpretation probably must await simultaneous application of the several analytical methods during the biodegradation.

Patterson remarked upon the absence of detectable amounts of free PEG during the degradation of Lissapol NX in contrast to the linear alcohol ethoxylates. It could be that such liberation does occur, but at a rate slower than its own subsequent degradation so that it never accumulates in the medium. In a summary of Patterson's work, an LGC (1966, p. 144) report presents the conclusion, on the basis of experimental data not published, that the biodegradation pathway taken by the branched C_9APE_9 involves initial carboxylation of the alkyl group, oxidation to a shorter carboxylate of the type **7.30**, and finally "gradual disappearance by aromatic ring attack, ketonization and end group hydrolysis." [Details are now available in Patterson (1970).]

$$\overset{\text{C}}{\underset{\text{C}}{\text{HO}_2\text{CC}\phi\text{OE}_9}}$$

(7.30)

---

* Using other biodegradation conditions and his own TLC analytical method, Bürger (1967) did not find the lower ethoxylates either.

C.6. *Tertiary Octylphenol Ethoxylates*

Applying the IR analytical method to t-OPE_9 in river water, Osburn (1966) showed degradation of the EO chain to $E_{2.5}$. The hydrophobe portion of the molecule remained essentially unchanged, as was expected from the

$$
\begin{array}{l}
\text{C C} \\
\text{CCCC—} \\
\text{C C}
\end{array}
$$

structure of the alkyl group. The absence of a hydrogen on the carbon adjacent to the methyl groups was thought to preclude carboxylation, and what little carboxylation did occur was attributed to the possible presence of minor amounts of other isomers with more favorable alkyl structures. However, as we have seen in the discussion of QBS (Section IV) there is also some possibility of slow attack on the major component itself by such means.

Using radiotracer techniques Lashen (1966) too showed extensive degradation of the EO chain in t-OPE_{10}, using activated sludge in both continuous-flow and semicontinuous systems. The benzene ring was tagged with tritium and the EO chain with ^{14}C. After some 10–15 days of acclimation the loss of ^{14}C averaged around 65% (loss of CTAS was 90–100%) during the biodegradation, while the tritium content remained unchanged. This was interpreted as oxidation of the EO chain either to carbon dioxide which was removed from the system by the aeration, or to a lesser extent by conversion to bacterial protoplasm or other insoluble metabolites remaining in the sludge. (Depending on the alkali reserve of the feeds, some CO_2 may also have been retained in the effluent as carbonate or bicarbonate, in which case the oxidation of the EO chain would actually have been correspondingly higher than 65%.)

The complete recovery of tritium in the effluent proved that the 65% loss of ^{14}C could not have been due to adsorption of the original OPE_{10} onto the sludge and must have been indeed due to biodegradation. The tritium balance gives little information on the fate of the benzene ring, since even upon complete oxidation the tritium would have remained in the effluent as T_2O, except for minor evaporation loss and some incorporation into the bacterial protoplasm.

C.7. *Linear Alkylphenol Ethoxylates*

Osburn (1966), using the IR technique, found that linear secondary C_9APE_{10} was removed somewhat faster than the branched isomers during the earlier stages of a comparative river water test, but at 35 days their levels were about identical, 10% remaining. The EO chain length was down to E_5 in the linear product at 30 days. Carboxylated intermediates were present but their amount and alkyl chain length were not reported.

With one unit higher average degree of ethoxylation, linear C_9- and $C_{11}APE_{11}$ degraded to 7% and 40% remaining, respectively, but the EO chain of the residual material had not been shortened at all; the presence of residual carboxylated intermediates was not mentioned. Thus it would seem that the higher ethoxylates of the linear secondary alkylphenols, E_{11} or above, are degraded from the hydrophobic end of the molecule exclusively, in contrast to E_{10} or lower, where hydrophilic attack may parallel or precede.

Such an abrupt change in properties from the E_{10} to the E_{11} ethoxylates is as difficult to rationalize as it was in the case of the branched nonylphenol ethoxylates (Section VIII.C.5). Both the E_{10} and E_{11} derivatives actually are mixtures of the same E_5 to E_{15} individuals, and it seems probable that the explanation must lie elsewhere than in the slightly different average degree of ethoxylation.

D. Ethoxylated Sugar Derivatives

The experiments of Brebion (1964) give some information on the reactions involved in the biodegradation of a sucrose surfactant, ethoxylated tallow sucroglyceride. The chemical structure of this material is not specified beyond the statement that it is obtained by ethoxylation of the reaction product of tallow with sucrose. As a first approximation we may assume that it is a mixture of fatty acid mono- and diglycerides and fatty acid sucrose esters ethoxylated on the free hydroxyl groups (those unesterified by the fatty acids) of the glycerol and sucrose.

The degradation was conducted in an inoculated medium containing 100 ppm of the surfactant over a 7–10 day period (Brebion, 1966), monitored by several methods. Among these, GC indicated an instantaneous hydrolysis of the surfactant upon first introduction, giving the free fatty acids (detected after analytical conversion to methyl esters), which then slowly disappeared by biodegradation during the next several days.

The speed of the initial hydrolysis is surprising, but was confirmed by acidimetric titration. A solution of 1 gm in 50 ml of sewage water showed an acid number of 132 within $\frac{1}{2}$ hr, compared to acid numbers of 3 in distilled water and 8 in river water. Here there is a second surprise also, because the saponification number of the surfactant was found to be only 48. Thus over twice as much fatty acid was liberated by action of the sewage water as was liberated upon hydrolysis with boiling alcoholic potassium hydroxide.

Parallel analyses by surface tension indicated 90% disappearance of surfactancy within 24 hr, although the polyethoxylate content (CTAS analysis) was removed only slowly and incompletely, down to around 35% remaining in 7 days. Thus the surfactancy properties were rapidly destroyed by the initial hydrolysis, but the hydrolysis products persisted in the biodegradation medium for a considerably longer time.

The emulsifiers Tween 20, 40, 60, and 80 are the E_{20} derivatives of sorbitan monolaurate, -palmitate, -stearate, and -oleate, respectively. These structures should be rather analogous to the ethoxylated tallow sucroglyceride, without the glycerides and with the sucrose replaced by sorbitan (which is the anhydride of the sugar alcohol sorbitol).

Sierra (1957) used these Tweens as substrates in his method for detecting bacterial synthesis of lipase enzymes. The Tween was included at about a 1% level in peptone–agar medium. Upon growth of bacteria on plates of this medium, positive indication of lipase production was the formation of an opaque zone around the colony, caused by the precipitation of the calcium salt of the liberated fatty acid. Obviously, then, the fatty acid portion of the Tween molecule is readily split off by bacterial action. Fourteen species of *Pseudomonas*, *Micrococcus*, *Vibrio*, *Bacillus*, and *Streptococcus* gave positive results in this test, while three others did not. A negative result does not necessarily mean that the species in question was unable to carry out the hydrolysis within its cells, but simply that the lipase was not excreted into the medium.

Earlier, Archibald (1946) had developed a lipase assay method in which Tween 20 was the substrate and the hydrolyzed fatty acid was determined by extraction and titration with standard alkali. About one-tenth of the theoretical amount of fatty acid was liberated from a 40% Tween 20 solution in 1 hr in the presence of 2500 ppm of pancreatic lipase.

Using Tween 80 as an enrichment culture medium, Minami (1958, 1959) isolated from soil a hitherto unknown organism which he christened *Micrococcus tweenis*. It was capable of growth on Tween 80 as its sole source of organic carbon, with rapid oxygen uptake. In the presence of inhibitors such as sodium azide, oleate was found to accumulate in the medium, suggesting that the normal course of biodegradation involved hydrolysis to give oleate which was then (in the absence of inhibitor) oxidized. Fates of the EO and sorbitan portions of the molecule were not indicated.

CHAPTER 8

BIODEGRADATION DATA

I. ORGANIZATION AND SCOPE OF TABLES

The tables which make up most of this chapter show representative data selected from the literature. Coverage is reasonably complete through 1968 with regard to the publications which have appeared and the different surfactants treated in them. However, the amount of data presented has been cut to a minimum because its full volume would be far beyond any reasonable size in comparison with its additional value. The data given herein are intended to be indicative only, and the original reference must be consulted if the true applicability and limitations are to be judged.

Basically the tables show extent of biodegradation as related to the chemical nature of the surfactant. Many variables are involved, often uncontrolled and uncontrollable, attested by the frequent inconsistency of the results obtained by different workers, and sometimes by the same worker. In the latter case the best results (i.e., those indicating the most biodegradation) are usually entered, but often there is indication of the range also. In the absence of gross experimental errors the good result is more indicative of the biodegradation capability of a given surfactant; the bad result may simply reflect failure of bacterial growth or acclimation through no particular fault of the surfactant.

No attempt has been made to show the state of acclimation of the organisms or the initial concentration of the surfactant. Although both are very important factors, their inclusion would have increased the size prohibitively.

Six items are shown for each entry in the tables: (i) the chemical nature of the substrate, (ii) the extent of degradation, (iii) the general biological conditions, (iv) the exposure time, (v) the analytical method, and (vi) the reference. In the following sections, details of the organization, definitions, and abbreviations are presented individually for each column.

II. SUBSTRATE – CHEMICAL NATURE: COLUMN 1

There are four groups of tables, covering anionics, nonionics, cationics, and nonsurfactants. No attempt has been made toward exhaustive coverage of many of

327

the nonsurfactants. These latter include (i) compounds which have served as models of the various moieties in surfactant molecules, useful in fundamental studies on the surfactants themselves; (ii) naturally occurring bacterial foods such as glucose, which provide a meterstick for comparison with biodegradation of synthetic products such as surfactants; and (iii) compounds used in detergent formulations, such as hydrotropes and chelants.

Each of the four major groups is subdivided into individually numbered tables of chemically related materials, and these are further subdivided to the extent necessary for easy location of the desired compound without reference to an index, by simply consulting the detailed arrangement as listed in the Table of Contents. Within each subdivision the arrangement is usually by chain length or size of alkyl group, from smaller to larger, followed by products with mixed lengths, followed by those in which the size is unspecified or unknown. The ethoxylated derivatives are further arranged according to the number of glycol units, n, in the hydrophilic group E_n.

The detailed chemical nature of the compound or material tested is identified in the first column, "Substrate," to the extent that that may be possible. The structure, if known, is indicated in usual chemical shorthand. For example, the entry $1\phi2PrC_5$ in Table 8.1 means the sulfonate of 1-phenyl-2-propylpentane (8.1). Nonstandard or less familiar abbreviations such as Hx for n-hexyl and Cy

$$CCCCC\phi\Sigma$$
$$C$$
$$C$$
$$C$$

(8.1)

for cyclohexyl or cyclohexane are listed below. In a few cases the chemical structure has been deduced from the context in the original source, but usually has been left unresolved if the information is insufficient.

The term "linear" if often applied to commercial products which, although predominantly so, may still contain as much as 5% or 10% of branched or cyclic impurities. Incomplete biodegradation in commercial types of LAS, for instance, may only reflect the presence of such impurities rather than failure of the truly linear components.

Trade names have been used frequently in the literature, and sometimes the researcher gives little indication of the chemical nature of the material. It should be remembered that trade names often are retained unchanged even after the producer has made significant change in the chemical nature of the product. On other occasions the producer may indicate product varieties by letters or numbers in the trade name; these are sometimes omitted by the researcher, and furthermore their significance becomes increasingly difficult to track down as time goes on. In some cases the researcher himself has differentiated his samples by letters or numbers; such identification has often been retained in the tables.

Abbreviations used in the Substrate column:

AOS	α-Olefin sulfonate
APE_n	Alkylphenol ethoxylate with n mols of EO
Am	n-Amyl
BI	Branching index, variously defined
br-	Branched
CWO	Cracked wax olefins
Cy	Cyclohexyl or substituted cyclohexane
CyG	Cycloaliphatic groups per molecule
DEA	Diethanolamide
DMG	Dimethyl groups per molecule
E_n	Polyglycol or polyglycol derivative, averaging n EO units per molecule; percentage figure indicates weight percent of EO used in making the derivative.
EA	Ethanolamide, ethanolamine
ex	Derived from
Hx	n-Hexyl
i	Iso
MEA	Monoethanolamide
MG	Methyl groups per molecule
P_n	Polyglycol group derived from propylene oxide, averaging n PO units per molecule
PEG	Polyethylene glycol
s	Secondary
SAS	Secondary alkane sulfonate
So	Sorbitan
t	Tertiary
TEA	Triethanolamine
tp-	Alkyl group derived from tripropylene, tetrapropylene, or other polypropylene
x	As subscript (e.g., E_x, C_x) indicates number unstated or unknown
ϕ	Benzene ring with substituents in accord with context
Σ	Sulfonic or sulfonate group; $O\Sigma$ is a sulfate group

III. EXTENT OF BIODEGRADATION: COLUMN 2

The second column in the tables, "Extent," shows the percent biodegradation (or simply percent removal in many cases where biodegradation was not explicitly proved) found, using the indicated exposure conditions and analytical method. With nonspecific analytical methods the figure given is percent of the theoretical amount. Thus results of respirometric measurement of oxygen absorbed are usually the percent of that calculated for complete oxidation of the substrate, usually corrected for endogenous respiration determined in a parallel, unfed control. In some cases the percent is based on the experimentally determined COD of the substrate rather than the calculated theoretical.

Generally in die-away type tests the extent of degradation at first increases with increasing exposure time and then tends to level off, approaching some constant value. Where possible, that level value is the one entered in the table, and the approximate time to reach that level is entered in column 4. If no level value is reached, or if the extent of degradation is low, the extent and time given are those at the end of the run. Some workers have reported only the time at which some arbitrary extent of degradation was reached, 50% or 80% or 90%, with no indication of what extent might ultimately have been attained. Such cases are indicated with a dagger (†).

In some situations the extent figure is not in terms of percent; these are indicated by special abbreviations:

B indicates biodegradability index, Eq. (8.1),

$$B = \frac{\% \text{ removal of surfactant}}{\% \text{ removal of BOD}} \qquad (8.1)$$

F indicates that oxygen uptake is reported as the "relative stabilization factor" of Barden (1957), which compares the surfactant with the natural foods making up Barden's synthetic sewage, Eq. (8.2),

$$F = \frac{\text{surfactant } O_2 \text{ uptake/its COD}}{\text{synthetic sewage } O_2 \text{ uptake/its BOD}} \qquad (8.2)$$

G indicates oxygen uptake as weight of oxygen absorbed per unit weight of substrate, e.g., milligrams of oxygen per milligram of surfactant, usually with correction for endogenous respiration.

M indicates oxygen uptake expressed as mols of oxygen per mol of substrate.

R or Rate indicates study of relative degradation rates of individual components of a substrate mixture, for example by the desulfonation−GC technique.

+ indicates qualitative evidence for biodegradation without presentation of quantitative information.

† indicates arbitrary value chosen by the researcher for termination of his test (usually 50%, 80%, or 90%), not necessarily the full extent possible.

IV. CONDITIONS—TEST METHODS: COLUMN 3

The third column in the tables shows the test method used. They are discussed in detail in Chapter 5. The following entries and abbreviations have been used in this column:

Alg	Algae were a major component in the biological system; often their role in the biodegradation was not clearly demonstrated.
An	Anaerobic
AnSew	Anaerobic sewage die-away
Aqu	Aquarium used as medium for die-away test
ASEff	Activated sludge effluent die-away
BAnD	Batch anaerobic digester
BAS	Batch or semicontinuous activated sludge
BOD	Standard BOD procedure
CAnD	Continuous anaerobic digester, usually fed in small daily or weekly increments (actually semicontinuous)
CAS	Continuous-flow activated sludge
Chan	Biodegradation measured in sewage effluent during residence time while flowing through a long channel
Enz	Enzyme used as degrading agent in absence of organisms
In	Natural or synthetic medium inoculated with acclimated or unacclimated organisms; although the shake culture, BOD, and Warburg procedures fall into this category also, they have been designated individually.
Pond	Sewage oxidation pond or simulation thereof
RTF	Recycle trickling filter
RW	River water die-away
Sew	Sewage die-away
SF	Shake flask culture
Soil	Percolation through a soil medium, usually intermittently
ST	Septic tank or simulation thereof
Sunfl	Biodegradation accomplished by intact living sunflower
TF	Trickling filter
Wa	Warburg respirometer; biodegradation may be estimated from other appropriate analyses

*   Indicates field test or large scale trials in sewage treatment equipment

V. BIODEGRADATION TIME: COLUMN 4

Column 4 gives the time (d, days; h, hours) corresponding to the extent of degradation entered in column 2. In the case of continuous-flow tests this is usually the average retention time in the aeration section, while in the batch (semicontinuous) activated sludge test it is the time for one cycle.

The times cited for die-away type tests are those at which the extent of degradation levels off to a relatively constant value. If no leveling off occurs, or if the extent is low, the time and extent usually correspond to the end of the run.

No time entries are made for trickling filter and soil percolation runs, since retention times are rarely determined under those conditions.

VI. ANALYTICAL METHODS: COLUMN 5

The various analytical methods have been discussed in Chapter 3. They are abbreviated as follows in column 5:

B	Bacterial growth
BiI	Bismuth iodide
C	Organic carbon
Ch	Chelometric analysis [for chelants (!)]
COD	Chemical oxygen demand
CO_2	Carbon dioxide formation
CT	Cobalt thiocyanante
DGC	Desulfonation-gas chromatography
F	Foaming properties
GC	Gas chromatography
HgI	Mercuric iodide
IR	Infrared spectroscopy
MB	Methylene blue, methyl green, and other cationic dye methods
MP	Determination of metabolic products
NH_3	Ammonia liberation
NO_3	Nitrate production

O$_2$	Oxygen uptake
PC	Paper chromatography
PM	Phosphomolybdate
PW	Phosphotungstate
RA	Radiosulfur-tagged sulfonates or sulfates
RC	Radiocarbon tracers
RS	Radiosulfate formation from tagged sulfonates or sulfates
SMB	Sulfation-methylene blue
SO$_4$	Sulfate formation
TLC	Thin layer chromatography
UV	Ultraviolet spectroscopy
Wt	Weight of soluble organics in cell-free medium
σ	Surface tension

VII. BIODEGRADATION TABLES

A. Anionic Surfactants

A. 1. *Individual Alkylbenzenesulfonates*

TABLE 8.1

Biodegradation of ABS Individual Isomers

Substrate (see § II)	Extent (see § III)	Method (see § IV)	Time (see § V)	Analysis (see § VI)	Reference
a. *6-Carbon Alkyl Group*					
$1\phi C_6$	48;100	BOD	15;11d	O_2;MB	Hammerton, 1962
$3\phi C_6$	0	SF	7d	MB	Huddleston, 1963
C_6LAS isomers	Rate	SF	–	DGC	Huddleston, 1963[a]
b. *8-Carbon Alkyl Group*					
$1\phi C_8$	47;59	Wa	6h;8d	O_2	Ryckman, 1956, 57
	56;71	BOD	5;14d	O_2	
	97	?	?	SO_4	Ryckman, 1957
	99;71	BOD	7;14d	MB;O_2	Hammerton, 1962
	100;35	In	12d	MB;O_2	Kölbel, 1964
$2\phi C_8$	96	BAS	2d	SO_4	Ryckman, 1956, 57
	53;51	Wa	6h;8d	O_2	
	30;76	BOD	5;11d	O_2	
	100;82	In	13;30d	MB;O_2	Ruschenberg, 1963a,b
$3\phi C_8$	100	SF	7d	MB	Huddleston, 1963
$4\phi C_8$	0	In	30d	MB;O_2	Ruschenberg, 1963a,b
	50	SF	7d	MB	Huddleston, 1963[b]
C_8LAS isomers	Rate	SF	–	DGC	Huddleston, 1963[a]
$1\phi 2PrC_5$	15;10	In	30d	MB;O_2	Kölbel, 1964
c. *9-Carbon Alkyl Group*					
$3\phi C9$	0.05G	Wa	3d	O_2	Malaney, 1966
d. *10-Carbon Alkyl Group*					
$1\phi C_{10}$	94;58	RW	4;7d	MB;O_2	Hammerton, 1955, 56
	95-98	BAS	1d	SO_4	Ryckman, 1956, 57
	48	Wa	6h;8d	O_2	
	36;72	BOD	5;16d	O_2	
	98;77	BOD	4;15d	MB;O_2	Hammerton, 1962
	100;35	In	7;8d	MB;O_2	Kölbel, 1964
	66	Wa	1d	O_2	Brink, 1966
$2\phi C_{10}$	93	BAS	2d	SO_4	Ryckman, 1956, 57
	51;50	Wa	6h;8d	O_2	
	40;64	BOD	5;16d	O_2	
	100;73	In	6;30d	MB;O_2	Ruschenberg, 1963a,b
$4\phi C_{10}$	100	SF	7d	MB	Huddleston, 1963[b]
$5\phi C_{10}$	100;30	In	21;30d	MB;O_2	Ruschenberg, 1963a,b
C_{10}LAS isomers	Rate	SF	–	DGC	Huddleston, 1963[a]
	Rate	RW	–	DGC	Swisher, 1963c, 64b
$5\phi 5MeC_9$	0	SF	7d	MB	Huddleston, 1963
	5	SF	7d	MB	Allred, 1964a
$1\phi 2BuC_6$	0	In	36d	MB;O_2	Kölbel, 1964

TABLE 8.1 ABS Individual Isomers *(continued)*

Substrate (see § II)	Extent (see § III)	Method (see § IV)	Time (see § V)	Analysis (see § VI)	Reference
e. *11-Carbon Alkyl Group*					
$1\phi C_{11}$	92;50	RTF;RW	1;10d	$MB;O_2$	Burnop, 1960
	100;90	Wa	28h	MB;COD	Nelson, 1961
	40	Wa	28h	O_2	
	99;61	RW	4;20d	$MB;O_2$	Hammerton, 1962
	Rate	RW	–	DGC	Swisher, 1963b
	94	Wa	1d	O_2	Brink, 1966
C_{11}LAS isomers	Rate	RW	–	DGC	Swisher, 1963c, 64b
$1\phi8,8Me_2C_9$	100	RW	34d	MB	Swisher, 1963b
	Rate	RW	–	DGC	
	84;51	CAS	6h	RA;RS	Sweeney, 1964a
$1\phi5,7,7Me_3C_8$	56;0	CAS	6h	RA;RS	Sweeney, 1964a
	0	Wa	1d	O_2	Brink, 1966
f. *12-Carbon Alkyl Group*					
$1\phi C_{12}$	13;44	Wa;BOD	6h;5d	O_2	Bogan, 1955, Sawyer, 1956
	95	BAS	1d	SO_4	Ryckman, 1956, 57
	38;60	Wa	6h;8d	O_2	
	43;62	BOD	5;14d	O_2	
	98;70	RTF;RW	1;10d	$MB;O_2$	Burnop, 1960
	100;85	Wa	28h	MB;COD	Nelson, 1961
	45	Wa	28h	O_2	
	100;71	BOD	6;20d	$MB;O_2$	Hammerton, 1962
	100	RW	6d	MB	Swisher, 1963b
	98	In	7d	MB	Crauland, 1964
	100;42	In	7;8d	$MB;O_2$	Kölbel, 1964
	98	RW	5d	MB	Setzkorn, 1964
	100;27	CAS	6h	RA;RS	Sweeney, 1964a
	1-9M	Wa	4h	O_2	Heyman, 1967
	100;52	In	9;27d	$MB;O_2$	Kölbel, 1967
	100;76	In	7;14d	$MB;O_2$	
	100	BAS	1d	MB;UV	Swisher, 1967b
	98	In	10d	MB	Eden, 1968
$1\phi C_{12}ortho\Sigma$	100	RW	13d	MB	Swisher, 1963b
$2\phi C_{12}$	96;25	RTF;RW	1;10d	$MB;O_2$	Burnop, 1960
	100;85	Wa	28h	MB;COD	Nelson, 1961
	45	Wa	28h	O_2	
	100;66	BOD	7;20d	$MB;O_2$	Hammerton, 1962
	100;1.9G	In	6;14d	$MB;O_2$	LGC, 1962, p. 42
	100;80	In	10;21d	MB;COD	Konecky, 1963
	100;70	In	6;30d	$MB;O_2$	Ruschenberg, 1963a,b
	100	RW	6d	MB	Swisher, 1963b
	Rate	RW	–	DGC	
	100;70	In	10;21d	MB;COD	McAteer, 1964

TABLE 8.1   ABS Individual Isomers *(continued)*

Substrate (see § II)	Extent (see § III)	Method (see § IV)	Time (see § V)	Analysis (see § VI)	Reference
$2\phi C_{12}$	98	RW	5d	MB	Setzkorn, 1964
	100;50	CAS	6h	RA;RS	Sweeney, 1964a
	100	RW	5d	MB	Sweeney, 1964b
	100	BAS;RW	1;26d	MB	Cohen, 1965
	70-80	In	21d	COD	Kelly, 1965
	100;77	In	8;21d	MB;COD	Livingston, 1965
	89	Wa	30h	O_2	Brink, 1966
	80†	RW	3d	MB	Smith, 1966
	100;90	In	6;18d	MB;COD	Cordon, 1968a
$3\phi C_{12}$	99;64	BOD	6;20d	MB;O_2	Hammerton, 1962
	100;63	In	8;20d	MB;O_2	Ruschenberg, 1963b
	98	RW	5d	MB	Setzkorn, 1964
	77	Wa	1d	O_2	Brink, 1966
	100	CAS;BAS	6h;1d	MB	Swisher, 1967b
	90-94	CAS;BAS	6h;1d	UV	
	100;95	RW	7;10d	MB;UV	Swisher, 1967c
	98	In	6d	MB	Eden, 1968
	100;95	SF	3;14d	MB;UV	Swisher, 1968a
$4\phi C_{12}$	100;48	In	6;20d	MB;O_2	Ruschenberg, 1963b
	100	RW	9d	MB	Swisher, 1963b
	98	RW	7d	MB	Setzkorn, 1964
	100;48	In	6;20d	MB;O_2	Ruschenberg, 1963b
	100	RW	9d	MB	Swisher, 1963b
	98	RW	7d	MB	Setzkorn, 1964
$5\phi C_{12}$	100	SF	7d	MB	Huddleston, 1963
	100	RW	9d	MB	Swisher, 1963b
	98	RW	7d	MB	Setzkorn, 1964
	98;16	CAS	6h	RA;RS	Sweeney, 1964a
$6\phi C_{12}$	98;43	BOD	15d	MB;O_2	Hammerton, 1962
	100;42	In	21;30d	MB;O_2	Ruschenberg, 1963a,b
	100	RW	13d	MB	Swisher, 1963b
	Rate	RW	–	DGC	
	98	RW	10d	MB	Setzkorn, 1964
	90	RW	12d	MB	Sweeney, 1964b
	100	CAS;BAS	6h;1d	MB	Swisher, 1967b
	85-95	CAS;BAS	6h;1d	UV	
	100;80	RW	14;44d	MB;UV	Swisher, 1967c
	98	In	13d	MB	Eden, 1968
	100;90	SF	4;21d	MB;UV	Swisher, 1968a
C_{12}LAS isomers	Rate	SF	–	DGC	Huddleston, 1963[a]
	Rate	RW	–	DGC	Swisher, 1963b,c,64b
$1\phi 10MeC_{11}$	97	RW	6d	MB	Swisher, 1963b
	Rate	RW	–	DGC	
	Rate	RW	–	?	Sweeney, 1964a
$2\phi 2MeC_{11}$	86;29	RTF;RW	1;10d	MB;O_2	Burnop, 1960

TABLE 8.1 ABS Individual Isomers *(continued)*

Substrate (see § II)	Extent (see § III)	Method (see § IV)	Time (see § V)	Analysis (see § VI)	Reference
$2\phi2MeC_{11}$	100;85	Wa	28h	MB;COD	Nelson, 1961
	38	Wa	28h	O_2	
	100;59	BOD	6;16d	MB;O_2	Hammerton, 1962
	100;70	CAS	6h	RA;RS	Sweeney, 1964a
	83	Wa	26h	O_2	Brink, 1966
	100;30	In	16,27d	MB;O_2	Kölbel, 1967
	100;90	In	9;14d	MB;O_2	
	100;90	BAS	1d	MB;UV	Swisher, 1969
$5\phi5MeC_{11}$	50	SF	7d	MB	Huddleston, 1963[b]
$1\phi2,2Me_2C_{10}$	100;80	BAS	1d	MB;UV	Swisher, 1969
$1\phi3,3Me_2C_{10}$	100;94	BAS	1d	MB;UV	Swisher, 1969
$1\phi4,4Me_2C_{10}$	98;33	CAS	6h	RA;RS	Sweeney, 1964a
$1\phi9,9Me_2C_{10}$	95	RW	50d	MB	Swisher, 1963b
	Rate	RW	–	DGC	
	0;9	In	27d	MB;O_2	Kölbel, 1967
	8;4	In	27;28d	MB;O_2	
	0-40	BAS	1d	MB;UV	Swisher, 1969
$2\phi5,9Me_2C_{10}$	92;95	BAS;RW	1;26d	MB	Cohen, 1965
$1\phi6,8,8Me_3C_9$	30;35	Wa	28h	MB;COD	Nelson, 1961
	0	Wa	28h	O_2	
$3\phi6,8,8Me_3C_9$	66;0	RTF;RW	1;10d	MB;O_2	Burnop, 1960
	0	Wa	28h	O_2	Nelson, 1961
	4;0	BOD	20d	MB;O_2	Hammerton, 1962
$4\phi2,2,3Me_3C_9$	98	RW	13d	MB	Swisher, 1963b
	Rate	RW	–	DGC	
$1\phi5,5,7,7Me_4C_8$	85	RTF	1d	MB	Burnop, 1960
	55;65	Wa	28h	MB;COD	Nelson, 1961
	0	Wa	28h	O_2	
	93	RTF	1d	MB	Cohen, 1963
$1\phi2AmC_7$	97;52	In	36d	MB;O_2	Kölbel, 1964
$1\phi6CyC_6$	100;+	RW;BOD	15d;?	MB;O_2	Huyser, 1960
$1\phi3(3,3,5Me_3Cy)C_3$	30;10	Wa	28h	MB;COD	Nelson, 1961
	0	Wa	28h	O_2	
$1\phi2(2iPr5MeCy)C_2$	50;65	Wa	28h	MB;COD	Nelson, 1961
	0	Wa	28h	O_2	

g. *13-Carbon Alkyl Group*

Substrate	Extent	Method	Time	Analysis	Reference
$C_{13}LAS$ isomers	Rate	RW	–	DGC	Swisher, 1963c, 64b
$1\phi10MeC_{12}$	Rate	RW	–	?	Sweeney, 1964a
$2\phi2MeC_{12}$	99;21	CAS	6h	RA;RS	Sweeney, 1964a
$5\phi5MeC_{12}$	100	SF	7d	MB	Huddleston, 1963[b]
$2\phi6,10Me_2C_{11}$	94;96	BAS;RW	1;20d	MB	Cohen, 1965
$5\phi5PrC_{10}$	0	SF	7d	MB	Huddleston, 1963[b]
$2\phi2,6,8,8Me_4C_9$	6;0	BOD	30d	MB;O_2	Hammerton, 1962
	62;4	CAS	6h	RA;RS	Sweeney, 1964a

TABLE 8.1   ABS Individual Isomers *(continued)*

Substrate (see § II)	Extent (see § III)	Method (see § IV)	Time (see § V)	Analysis (see § VI)	Reference
$3\phi3,6,8,8Me_4C_9$	3;0	RW;BOD	39d;?	$MB;O_2$	Huyser, 1960
	0	In	21d	MB	LGC, 1962, p. 42
h. *14-Carbon Alkyl Group*					
$1\phi C_{14}$	12;52	Wa	1;8d	O_2	Ryckman, 1956, 57
	54-61	BOD	14d	O_2	
	99	RW	6d	MB	Swisher, 1963b
	100;40	In	10;8d	$MB;O_2$	Kölbel, 1964
	100;64	In	9;27d	$MB;O_2$	Kölbel, 1967
	100;57	In	8;10d	$MB;O_2$	
	20	In	20d	MB	Eden, 1968
$1\phi C_{14}ortho\Sigma$	90	RW	13d	MB	Swisher, 1963b
$2\phi C_{14}$	55;57	Wa	6h;8d	O_2	Ryckman, 1956, 57
	27;63	BOD	5;12d	O_2	
	100;70	In	9;30d	$MB;O_2$	Ruschenberg, 1963a,b
	80†	RW	4d	MB	Smith, 1966
$3\phi C_{14}$	40	Wa	1d	O_2	Brink, 1966
$7\phi C_{14}$	100;47	In	21;30d	$MB;O_2$	Ruschenberg, 1963a,b
	85;97	RW	20;5d	MB	Sweeney, 1964b
	50	In	20d	MB	Eden, 1968
$2\phi2MeC_{13}$	100;33	In	13;27d	$MB;O_2$	Kölbel, 1967
	100;72	In	13;16d	$MB;O_2$	
$C_{14}LAS$ isomers	Rate	RW	–	DGC	Swisher, 1963b
$1\phi11,11Me_2C_{12}$	27;17	In	27d	$MB;O_2$	Kölbel, 1967
	66;33	In	27;28d	$MB;O_2$	
$5\phi5PrC_{11}$	0	SF	7d	MB	Allred, 1964a
$1\phi2HxC_8$	0	In	36d	$MB;O_2$	Kölbel, 1964
$2\phi8CyC_8$	100;51	BOD	20d	$MB;O_2$	Hammerton, 1962
i. *15-Carbon Alkyl Group*					
$8\phi C_{15}$	95	RW	2-6d	MB	Huyser, 1960
j. *16-Carbon Alkyl Group*					
$1\phi C_{16}$	77;63	In	20d	$MB;O_2$	Kölbel, 1964
	20	In	20d	MB	Eden, 1968
$2\phi C_{16}$	100;75	In	9;30d	$MB;O_2$	Ruschenberg, 1963a,b
	80†	RW	5d	MB	Smith, 1966
$8\phi C_{16}$	0	In	30d	$MB;O_2$	Ruschenberg, 1963a,b
	0	RW	36d	MB	Sweeney, 1964b
	0	In	20d	MB	Eden, 1968
k. *18-Carbon Alkyl Group*					
$2\phi C_{18}$	80†	RW	17d	MB	Smith, 1966
l. *Unspecified Size Alkyl Group*					
1ϕ-*n*-alkane	100;+	RW;BOD	4d;?	$MB;O_2$	Huyser, 1960
1ϕ-*n*-alkane-*ortho*Σ	100;0	RW;BOD	9-11d;?	$MB;O_2$	Huyser, 1960

TABLE 8.1 ABS Individual Isomers *(continued)*

Substrate (see § II)	Extent (see § III)	Method (see § IV)	Time (see § V)	Analysis (see § VI)	Reference
2ϕ-n-alkane	100;+	RW;BOD	5-11d;?	MB;O$_2$	Huyser, 1960
2ϕ-n-alkane-$ortho\Sigma$	0	BOD	?	O$_2$	Huyser, 1960
Linear (C$_n$)$_2$Cϕ	100;0	RW;BOD	11d;?	MB;O$_2$	Huyser, 1960
Linear (C$_n$)$_3$Cϕ	30;0	RW;BOD	39d;?	MB;O$_2$	Huyser, 1960

[a]See also Allred (1964b).

[b]See also Allred (1964a).

A. 2. *Linear Alkylbenzenesulfonates*

TABLE 8.2

Biodegradation of LAS

Substrate (see § II)	Extent (see § III)	Method (see § IV)	Time (see § V)	Analysis (see § VI)	Reference
a. *Individual Chain Lengths*					
C_6LAS	20-30	SF	7d	MB	Huddleston, 1963
	82	RW	60d	MB	Swisher, 1963b
C_7LAS	90	RW	43d	MB	Swisher, 1963b
C_8LAS	70-80	SF	7d	MB	Huddleston, 1963
	100	RW	35d	MB	Swisher, 1963b
	50†	In	2.9d	MB	Tarring, 1965
	50†	?	4.5d	?	De Jong, 1967
C_9LAS	100	RW	20d	MB	Swisher, 1963b
	50†	In	3.0d	MB	Tarring, 1965
C_{10}LAS	100	SF	7d	MB	Huddleston, 1963
	100	RW	15d	MB	Swisher, 1963b
	95†	RW	4d	MB	Setzkorn, 1964
	50†	In	2.1d	MB	Tarring, 1965
	90	RW	18d	MB	Ciattoni, 1968
	50†	?	1.5d	?	De Jong, 1967
C_{11}LAS	100;60	In	5;30d	MB;O_2	Ruschenberg, 1963a,b
	96	RW	14d	MB	Swisher, 1963b
	98	SF	5d	MB	Swisher, 1966a
	2.8-3.6M	Wa	4h	O_2	Heyman, 1967
	100	RW	10d	MB	Ciattoni, 1968
	98	In	4d	MB	Halvorson, 1969b
C_{12}LAS	100	SF	7d	MB	Huddleston, 1963
	50-55	Wa	?	O_2	
	98;79	In	20d	MB;O_2	Pitter, 1963b, 64c,d
	~100	In	20d	SO_4	
	100;60	In	5;30d	MB;O_2	Ruschenberg, 1963a,b
	100	RW	6-8d	MB	Swisher, 1963b
	100	SF	3d	MB;σ	Allred, 1964b
	47	Wa	6h	O_2	
	82;58	CAS;RW	3h;5d	MB	Cordon, 1964
(ex 1-Cl-C_{12})	92-94	In	7-9d	MB	Crauland, 1964
(ex 2-Cl-C_{12})	92-94	In	7-9d	MB	Crauland, 1964
(ex α-dodecene)	94-97	In	7-9d	MB	Crauland, 1964
(ex α-dodecene)	96-97	CAS	4h	MB	Huddleston, 1964a
	98;98	BAS;SF	1;4d	MB	Orgel, 1964
	88-95	CAS	8h	MB	Pitter, 1964a
	91-96	In	20d	SO_4	Pitter, 1964c
	100	SF	4d	MB	Renn, 1964a, Orgel, 1964
	95†	RW	5d	MB	Setzkorn, 1964
	100;22	CAS	6h	RA;RS	Sweeney, 1964a
	100	RW;CAS	5d;6h	?	Sweeney, 1964b
	100	CAS	7h	MB	Swisher, 1964a

TABLE 8.2 LAS *(continued)*

Substrate (see § II)	Extent (see § III)	Method (see § IV)	Time (see § V)	Analysis (see § VI)	Reference
$C_{12}LAS$	100	RW	21d	MB;σ;F	Vath, 1964
	60	Wa	36h	O_2	
	90	In	3d	F	Bloch, 1965
(internal)	90	In	3d	F	Bloch, 1965
	92-100	SF	7d	MB	Booman, 1965
	0	CAnD	30d	MB	Maurer, 1965
	99.5	SF	8d	MB	SDA, 1965
	99.5	BAS	1d	MB	
	50†	In	2.2d	MB	Tarring, 1965
	+	In	?	B	Lambin, 1966
	99	In	2d	MB	MWB, 1966, p. 36
	80†	RW	4d	MB	Smith, 1966
(internal)	95	CAS	3h	MB	Bloch, 1967
	100	RW	8d	MB	Borstlap, 1967b
	98-99	In	7d	MB	Bunch, 1967a
	50†	?	0.7d	?	De Jong, 1967
	0-1.7M	Wa	4h	O_2	Heyman, 1967
	99;100	RTF;In	7;17d	MB	Jenkins, 1967
	78;92	CAS;BAS	6h;1d	UV	Swisher, 1967b
	100	CAS;BAS	6h;1d	MB	
	100;85	RW	7;10d	MB;UV	Swisher, 1967c
	100	RW	10d	MB	Ciattoni, 1968
	93-94	CAS	3h	MB	Janicke, 1968c
	83-96	CAS	3h	COD	
	100	BAS	1d	F;MB	SDA, 1969a,b
	100;99	RW;SF	14d	F;MB	
$C_{13}LAS$	100;60	In	5;30d	MB;O_2	Ruschenberg, 1963a,b
	94	RW	19d	MB	Swisher, 1963b
	100;38	CAS	6h	RA;RS	Sweeney, 1964a
	50†	In	2.0d	MB	Tarring, 1965
	100	RW	8d	MB	Ciattoni, 1968
	82-10	In	8-5d	MB	Nyns, 1969a
$C_{14}LAS$	50-55	Wa	?	O_2	Huddleston, 1963
	100;25	In	20;30d	MB;O_2	Ruschenberg, 1966a,b
	98	RW	21d	MB	Swisher, 1963b
	95†	RW	16d	MB	Setzkorn, 1964
	99	CAS	3h	MB	Swisher, 1964a
	50†	In	1.6d	MB	Tarring, 1965
	80†	RW	16d	MB	Smith, 1966
	99	SF	5d	MB	Swisher, 1966a
	94-98	In	7d	MB	Bunch, 1967a
	50†	?	0.7d	?	De Jong, 1967
	100	RW	10d	MB	Ciattoni, 1968
	98	In	4d	MB	Halvorson, 1969b
$C_{15}LAS$	100;20	In	30d	MB;O_2	Ruschenburg, 1963a,b

TABLE 8.2   LAS *(continued)*

Substrate (see § II)	Extent (see § III)	Method (see § IV)	Time (see § V)	Analysis (see § VI)	Reference
C_{15} LAS	93	RW	24d	MB	Swisher, 1963b
	99;10	CAS	6h	RA;RS	Sweeney, 1964a
	90;99	RW;CAS	19d;6h	?	Sweeney, 1964b
	50†	In	1.5d	MB	Tarring, 1965
	100	RW	8d	MB	Ciattoni, 1968
C_{16} LAS	100;25	In	30d	MB;O_2	Ruschenberg, 1963a,b
	100	RW	14d	MB	Swisher, 1963b
	95†	RW	16d	MB	Setzkorn, 1964
	50†	In	1.3d	MB	Tarring, 1965
	80†	RW	10d	MB	Smith, 1966
	50†	?	0.7d	?	De Jong, 1967
C_{17} LAS	50†	In	1.3d	MB	Tarring, 1965
C_{18} LAS	50-55	Wa	?	O_2	Huddleston, 1963
	100	RW	14d	MB	Swisher, 1963b
	50†	In	1.3d	MB	Tarring, 1965
	80†	RW	16d	MB	Smith, 1966
C_{21} LAS	50†	?	4.5d	?	De Jong, 1967
b. *Mixed Chain Lengths*					
Dobane JN	94	TF;CAS	–;6h	MB	Truesdale, 1959,
					Eden, 1961a
Detergent X	84	TF	–	MB	Truesdale, 1959,
					Eden, 1961a
	93;91	CAS	8;6h	MB	
ex Linear olefin	88	CAS	2-4h	MB	Bock, 1960
BI = 0.2,MG = 2.2(I)	76;38	RTF;RW	1;10d	MB;O_2	Burnop, 1960
BI = 0.3 (J)	85;30	RTF;RW	1;10d	MB;O_2	Burnop, 1960
BI = 0.8,MG = 2.1(K)	87;34	RTF;RW	1;10d	MB;O_2	Burnop, 1960
BI = 3.2 (N)	71	RTF	1d	MB	Burnop, 1960
Dobane JN	89;0.6G	BOD	15d	MB;O_2	Isaac, 1960a,b
Dobane JN(?)	83	In	6d	MB	Roberts, 1960
ex Linear olefin	97	CAS	3-4h	MB	Bock, 1961
Marlon BW2043	>80	CAS	2h	MB	Bock, 1961
Marlon BW1043	85-90	CAS	3-4h	MB	Bock, 1961
	>94	CAS	2h	MB	
Dobane JN	96-99	RW	21d	MB	Knöpp, 1961
	100	RW	21d	F	
Korenyl	95-97	RW	21d	MB	Knöpp, 1961
	100	RW	21d	F	
Dobane JN	46-69	RW	12d	MB	Gameson, 1962
	79-96	ASEff	13d	MB	
Marlon BW1043	85	*CAS	2d	MB	Huber, 1962
Marlon BW2043	69[a]	*TF	–	MB	Jendreyko, 1962
Marlon BW1043	82[b]	*TF	–	MB	Jendreyko, 1962
Dobane JN	4-7	In	2d	MB	Kimura, 1962

TABLE 8.2   LAS *(continued)*

Substrate (see § II)	Extent (see § III)	Method (see § IV)	Time (see § V)	Analysis (see § VI)	Reference
Research product	80;47	Wa	1d	MB;O$_2$	Offhaus, 1962
WAS-A	80-90	TF	–	MB	Schönborn, 1962a
	81-88	BAS	1d	MB	
WAS-K	60-84	TF	–	MB	Schönborn, 1962a
	72-90	BAS	1d	MB	
LAS	>90	BAS	6+16h	MB	Weaver, 1962
	60	ST	33d	MB	
Dobane JN(?)	92	RW	30d	MB	Weaver, 1962
DOBS JN	94	In	4d	MB	Borstlap, 1963
DOBS JN improved	99	In	4d	MB	Borstlap, 1963
BW1043	94-98	TF	–	MB	Bringmann, 1963
Korenyl (detergent A)	72-85	TF	–	MB	Husmann, 1963a, p. 34[c]
	75-82	CAS	1h	MB	Husmann, 1963a, p. 44, 46[c]
	83-91	*TF	–	MB	Husmann, 1963a, p. 60-63[c]
	73	*CAS	1¼h	MB	Husmann, 1963a, p. 64-65[c]
	69;74	In	4;9d	MB	Husmann, 1963a, p. 108[c]
LAS-C	85	Soil	–	MB	Husmann, 1963b
LAS-C;D	97;94	RW	20;35d	MB	Husmann, 1963b
10% Branched	75;93	In	7;14d	MB	Jendreyko, 1963
	47	In	22d	O$_2$	
5% Branched	85;97	In	7;14d	MB	Jendreyko, 1963
	55	In	22d	O$_2$	
0% Branched	93;100	In	7;14d	MB	Jendreyko, 1963
	70	In	22d	O$_2$	
LAS-Commercial	76;54	In	5;21d	MB;COD	Konecky, 1963
LAS-Experimental	94;69	In	13;21d	MB;COD	Konecky, 1963
Dobane JN;JNX;036	87-88	In	40d	MB	LGC, 1963, p. 48
ex Cl paraffins	90;1.7G	In	20d	MB;O$_2$	Pitter, 1963b, 64c,d
ex C$_{10-13}$ olefins	91;87	In	20d	MB;O$_2$	Pitter, 1963b, 64c,d
	~100	In	20d	MB;O$_2$	
ex C$_{11-15}$ olefins	96	In	20d	MB	Pitter, 1963b, 64c, d
C$_{11-13}$, 10% Me branch	97	In	30d	MB	Ruschenberg, 1963a
C$_{11-13}$, 5% Me branch	100	In	21d	MB	Ruschenberg, 1963a
C$_{11-13}$, 0% Me branch	100	In	9d	MB	Ruschenberg, 1963a
C$_{11-15}$ LAS	100;30	In	7;30d	MB;O$_2$	Ruschenberg, 1963a,b
C$_{11-13}$ LAS	100	In	15d	MB	Ruschenberg, 1963a,b
C$_{14-16}$ LAS	0	In	30d	MB	Ruschenberg, 1963a,b
Dobane JNX	89;58	BOD	40d	MB;O$_2$	STCSD, 1963
	91	TF;CAS	–;?	MB	
Dobane JN036	96;59	BOD	40d	MB;O$_2$	STCSD, 1963

TABLE 8.2   LAS *(continued)*

Substrate (see § II)	Extent (see § III)	Method (see § IV)	Time (see § V)	Analysis (see § VI)	Reference
Dobane JN036	93;94	TF;CAS	−;?	MB	STCSD, 1963
C_{10-13} LAS	30-40	ST	5d	RA	Straus, 1963
	90-95	ST+Soil	5d+	RA	
LAS-USA	100;22	RW;AnRW	9d	MB	Wayman, 1963a
LAS-German	92;17	RW;AnRW	9d	MB	Wayman, 1963a
ex Cl paraffin	100	SF	7d	MB	Allred, 1964a
ex Lin. α-olefins	100	SF	7d	MB	Allred, 1964a
ex.CWO	100	SF	7d	MB	Allred, 1964a
ex Olefins	97;80	RTF;RW	3;40d	MB	Berger, 1964
	90	CAS	3h	MB	
Commercial LAS	63;80	CAS;RW	3h;4d	MB	Cordon, 1964
RD036	85-93	*CAS	2d	MB	De Jong, 1964, 65
Slightly branched	70-80	*CAS	2d	MB	De Jong, 1964, 65
Ucane 12,13	100	RW	21;42d	MB;UV	Foster, 1964
Nacconol U (C_{11-14})	100	RW	14d	MB	Fuhrmann, 1964
Nacconol DN (C_{10-13})	100	RW	10d	MB	Fuhrmann, 1964
SDA	70-94	*CAS	14-24h	MB	Hanna, 1964
Nonbranched LAS	100	In	12d	MB	Hitzman, 1964
Moderately branched	87	In	13d	MB	Hitzman, 1964
Nalkylene	98;87	CAS	6;2h	MB	Huddleston, 1964a
Dobane JN	32	Wa	10d	O_2	Hunter, 1964
SDA	56;6	Alg	3d	RA;RS	Klein, 1964a[d]
	93;12	Pond	30d	RA;RS	
	9-10	ST	2d	RA	
	0-1	ST	2d	RS	
	34-44	ST	2d	MB	
	97	ST+Soil	2d+	RA	
	41-70	ST+Soil	2d+	RS	
	94-100	ST+Soil	2d+	MB	
Marlon 1033	93;55	CAS	3h	MB;UV	Krüger, 1964
LAS A,B,C	98	BAS	1d	MB	Orgel, 1964
C_{10-13} LAS	83-91	CAS	8h	MB	Pitter, 1964a
C_{10-13} LAS	0-16	BAnD	32d	MB	Pitter, 1964b,d
LAS − No. 1	67;85	RW;SF	32;4d	MB	Renn, 1964a, Orgel, 1964
	80-99	CAS	?	MB	
No. 2	97;88	RW;SF	32;4d	MB	
	50-95	CAS	?	MB	
No. 3	45;98	RW;SF	32;4d	MB	
	64-97	CAS	?	MB	
No. 4	95;96	RW;SF	32;4d	MB	
No. 5	93;95	RW;SF	32;4d	MB	
various	45-97	RW	32d	MB	
	85-96	SF	4d	MB	
	85-97	BAS	1d	MB	

TABLE 8.2   LAS *(continued)*

Substrate (see § II)	Extent (see § III)	Method (see § IV)	Time (see § V)	Analysis (see § VI)	Reference
LAS – various	50-99	CAS	5h	MB	Renn, 1964a, Orgel, 1964
LAS	98	*CAS	2d	MB	Renn, 1964a; Orgel, 1964
Ucane 12	93-97	CAS	3h	MB	Renn, 1964b
LAS	99;71	Aqu	30;184d	RA;RS	Sharman, 1964a
LAS – A;C	63;73	*CAS	?	MB	Spohn, 1964a
	78;86	*CAS+TF	?	MB	
Marlon BW2043	57-69	*TF	–	MB	Spohn, 1964b
Marlon BW1043	85-93	*TF	–	MB	Spohn, 1964b
	48-66	*CAS	1-2h	MB	
	73-85	*CAS+TF	1-2h+	MB	
Marlon BW1033	57-82	*CAS	1h	MB	Spohn, 1964b
	80-95	*CAS+TF	1h+	MB	
Korenyl	59-64	*CAS	1h	MB	Spohn, 1964b
Dobane JNX	91;93	TF;CAS	–;?	MB	STCSD, 1964
	92	In	25d	MB	
"ABS Control"	93-96	CAS	6h	RA	Sweeney, 1964a
C_{10-14} LAS	85;94	RW;CAS	9d;6h	?	Sweeney, 1964b
C_{11-15} LAS	95;98	RW;CAS	17d,6h	?	Sweeney, 1964b
C_{10-18} LAS	95;98	RW;CAS	24d;6h	?	Sweeney, 1964b
C_{11-14} LAS (high 2ϕ)	99	CAS	6h	?	Sweeney, 1964b
C_{11-14} LAS (low 2ϕ)	99	CAS	6h	?	Sweeney, 1964b
C_{10-16} LAS	100	CAS	3h	MB	Berber, 1965
SDA 3-S	95-98	SF	7d	MB	Booman, 1965
	90;15	BAS;ST	1;2d	MB	
Dobane JN	94;93	CAS;TF	6h;–	MB	Eden, 1965
	92	In	21d	MB	
Dobane JNX	91	CAS;TF	6h;–	MB	Eden, 1965
	94	In	21d	MB	
Dobane 036	94;93	CAS;TF	6h;–	MB	Eden, 1965
	95	In	21d	MB	
LAS	88	CAS	3h	MB	Fischer, 1965
ex CWO	34	*TF	–	MB	Kelly, 1965
	40-60	In	21d	COD	
SDA	85;45	TF	–	RA;RS	Klein, 1965a,b
	95-100	CAS	6h	RA	
	30-45	CAS	6h	RS	
	98	Sew	40d	RA;MB	
	96	AnSew	40d	RA;MB	
LAS	97	RW	30d	MB	Knaggs, 1965
SDA	85	*CAS	8h	MB	Knapp, 1965
SDA	75-95	*CAS	1½d	MB	Knopp, 1965
LAS	100	RW;SF	7;8d	MB	Lang, 1965
Dobane JN036	91	In	21d	MB	LGC, 1965, p. 52

TABLE 8.2   LAS *(continued)*

Substrate (see § II)	Extent (see § III)	Method (see § IV)	Time (see § V)	Analysis (see § VI)	Reference
LAS	95;54	In	6;21d	MB;COD	Livingston, 1965
Commercial LAS	0	CAnD	30d	MB	Maurer, 1965
	83	Soil	–	MB	
Dobane JN;JNX	82;88	?	2d	MB	Ōba, 1965b,e,f
Nalkylene; Ucane	91;89	?	2d	MB	Ōba, 1965b,e,f
Cracked wax	0.2-0.4B	TF	–	MB	Renn, 1965a
Ucane	0.8-1.2B	TF	–	MB	Renn, 1965a
	1.0-1.1B	CAS	?	MB	
LAS 1-1	92.1-94.8	SF	8d	MB	SDA, 1965
	95.9-98.6	BAS	1d	MB	
LAS 3-S	94.5-96.5	SF	8d	MB	SDA, 1965
	97.1-99.2	BAS	1d	MB	
LAS A	92.2-96.5	SF	8d	MB	SDA, 1965
	95.6-98.8	BAS	1d	MB	
LAS B	87.2-92.5	SF	8d	MB	SDA, 1965
	92.8-96.0	BAS	1d	MB	
LAS C	91.3-96.1	SF	8d	MB	SDA, 1965
	95.0-99.1	BAS	1d	MB	
MG 2.3-3.0	92-83	RW	35d	MB	Tarring, 1965
DMG 0.1-0.35	92-84	RW	35d	MB	Tarring, 1965
CyG 0.2-0.32	92-83	RW	35d	MB	Tarring, 1965
ex C_{12} olefins[e]	84	CAS	11h	MB	Urban, 1965
ex C_{13} olefins[e]	91	CAS	11h	MB	Urban, 1965
ex Arge olefins	99	CAS	11h	MB	Urban, 1965
Dobane JN	90	CAS	11h	MB	Urban, 1965
Experimental LAS	93;98	*TF	–	MB;IR	Urban, 1965
	100;99	*Pond	10d	MB;IR	
	36;29	*Chan	6h	MB;IR	
Dobane JNX	88	In	4-18d	MB	WPRL, 1965, p. 121
	61-66	ST	1-3d	MB	WPRL, 1965, p. 124-7
Dobane JN036	95	TF	–	MB	WPRL,1965,p.127-30
C_{5-6} LAS	11	Wa	30h	O_2	Brink, 1966
C_{7-8} LAS	0	Wa	24h	O_2	Brink, 1966
C_{11-14} LAS	103	Wa	30h	O_2	Brink, 1966
C_{11-20} LAS	98	Wa	24h	O_2	Brink, 1966
SDA	74;41	TF	–	RA;RS	Klein, 1966, McGauhey, 1966
	90-96	CAS	8h	RA	
	27-85	CAS	8h	RS	
Ucane 13	75-80	*TF	–	MB	Kumke, 1966
Dobane JNX	64-73	*TF	–	MB	Kumke, 1966
SDA 3-S	20;0	ST	3d	MB;F	Lashen, 1966
	84;86	ST+Soil	3d+	MB;F	
Dobane JNX	85-90	In	14d	MB	LGC, 1966, p. 79
Dobane JN036	90;80	In	20;60d	MB;UV	LGC, 1966, p. 80

TABLE 8.2 LAS *(continued)*

Substrate (see § II)	Extent (see § III)	Method (see § IV)	Time (see § V)	Analysis (see § VI)	Reference
SDA	100	SF	3d	MB	Long, 1966
LAS	99	?	15d	?	Manneck, 1966
Nalkylene 500	42-43	Wa	5d	MB	Marion, 1966
	55-76	Wa	5d	O_2	
Dobane JNX	90	In	9d	MB	MWB, 1966, p. 36
Dobane JNQ	95	In	8d	MB	MWB, 1966, p. 36
Dobane 055	96	In	15d	MB	MWB, 1966, p. 36
Dobane JNX	83-92	In	21d	MB	STCSD, 1966
SDA 1-1	97	SF	5d	MB	Swisher, 1966a
C_{12-14} LAS	97	SF	7d	MB	Swisher, 1966a
ex Kogasin	+	*CAS	3d	MB	Walther, 1966
Dobane JNX	88-90	In	21d	MB	WPRL, 1966a, p. 131
	91	TF	−	MB	
	93;96	CAS;BAS	?;1d	MB	
Dobane JNQ	97	In	21d	MB	WPRL, 1966a, p. 131
LAS	65	In	28d	Wt	Borstlap, 1967a
ex CWO	100	RW	20d	MB	Borstlap, 1967b
C_{10-15} LAS	90-93	In	7d	MB	Bunch, 1967a
C_{12-14} LAS	96-98	In	7d	MB	Bunch, 1967a
SDA LAS (5)	83-94	In	7d	MB	Bunch, 1967a
SDA LAS (6)	82-93	In	7d	MB	Bunch, 1967a
SDA LAS (7)	73-95	In	7d	MB	Bunch, 1967a
SDA 1-1	15-90	Alg	21d	MB	Davis, 1967
	2-100	Alg	14d	IR	
	90	Sew	18d	MB	
	100	Pond	21d	MB	
	1-18	Pond	14d	IR	
	100	BAS	21d	MB	
LAS I;III	27-70	Wa	8h	O_2	Hartmann, 1967
German LAS No. 415	95;90	CAS;BAS	3;24h	MB	Heinz, 1967
U.S. LAS No. 509	95	CAS;BAS	3;24h	MB	Heinz, 1967
German LAS	89,90	In	6d	MB	Heinz, 1967
British LAS	88	In	9d	MB	Heinz, 1967
Technical LAS	85;87	SF;In	8;5d	MB	Heinz, 1967
	94-70	In	30d	MB;O_2	
LAS	85-100	RTF	7d	MB	Jenkins, 1967
	100	In	17d	MB	
Dobane JN036	85-100	RTF	7d	MB	Jenkins, 1967
	85-96	In	17d	MB	
Dobane JNX	94	RTF	5d	MB	Jenkins, 1967
Produkt A	73-77	CAS	?	MB	Malz, 1967
	83-86	TF	−	MB	
Produkt B	64-84	CAS	?	MB	Malz, 1967
	70-73	TF	−	MB	
Produkt C	75-81	CAS	?	MB	Malz, 1967

TABLE 8.2   LAS *(continued)*

Substrate (see § II)	Extent (see § III)	Method (see § IV)	Time (see § V)	Analysis (see § VI)	Reference
LAS	20	An	3d	MB	Ōba, 1967
LAS	15	*Sewer	3h	RS	STCSD, 1967, p. 7
Marlon A350	100	BAS	3d	MB	Vaicum, 1967
Marlon A370	83-92	BAS	3d	MB	Vaicum, 1967
Dobane JN	71-83	BAS	3d	MB	Vaicum, 1967
Dobane JNQ	0	BAnD	?	?	WPRL, 1967, p. 117
	96	TF	–	MB	WPRL, 1967, p. 154
	94	*TF	–	MB	WPRL, 1967, p. 155
Dobane JNX	93	TF	–	MB	WPRL, 1967, p. 154
	91	*TF	–	MB	WPRL, 1967, p. 155
	92	In	5d	MB;UV	WPRL, 1967, p. 177
Dobane 055	96	TF	–	MB	WPRL, 1967, p. 154
Dobane JNX	48-85	In	15d	MB	Cook, 1968
	88	RW;SF	?;8d	MB	
	61;89	CAS;BAS	3h;1d	MB	
	92	RTF	?	MB	
Dobane JNQ	87-92	In	15d	MB	Cook, 1968
	93;96	RW;SF	?;8d	MB	
	66;96	CAS;BAS	3h;1d	MB	
	96	RTF	?	MB	
Dobane 055	0-66	In	15d	MB	Cook, 1968
	96;91	RW;SF	?;8d	MB	
	75;98	CAS;BAS	3h;1d	MB	
	97	RTF	?	MB	
LAS	83	In	13d	SO_4	Cordon, 1968b
Dobane JNX	92-93	In	20d	MB	Eden, 1968
	91	CAS;TF	6h;–	MB	
Dobane JN036	94-95	In	20d	MB	Eden, 1968
	94;93	CAS;TF	6h;–	MB	
Marlon BW1043;2043	81-93	*CAS	2d	MB	Huber, 1968
	33-81	*CAS	1½d	MB	
	30	*CAS	1d	MB	
Marlon BW1033	99	Soil	–	MB	Kempf, 1968
LAS	19	BOD	7d	O_2	Krone, 1968
	50;85	In	5;11d	MB	
	96	CAS	3h	MB	
	76	*TF	–	MB	
C_{10-13} LAS	83-97	In	40d	MB	Mann, 1968
	98	CAS	3h	MB	
	94	*TF	–	MB	
C_{11-15} LAS	10-88	In	40d	MB	Mann, 1968
	95-96	CAS	3h	MB	
	95	*TF	–	MB	
LAS	91-96	SF	?	MB	Ōba, 1968b
LAS	80-95	In	20d	MB	Pitter, 1968c

TABLE 8.2 LAS *(continued)*

Substrate (see § II)	Extent (see § III)	Method (see § IV)	Time (see § V)	Analysis (see § VI)	Reference
LAS	45-65	In	20d	COD	Pitter, 1968c
LAS	92;98	CAS;RW	3h;13d	MB	Rismondo, 1968
	30	ST	12h	MB	
Dobane JNX	89-91	In	21d	MB	Truesdale, 1968
	90-93	TF;RTF	−;7d	MB	
	94;97	CAS;BAS	3h;1d	MB	
Dobane JNQ	96-99	5 methods		MB	Truesdale, 1968
Dobane 055	94-98	5 methods		MB	Truesdale, 1968
$C_{10.4}$ average	86	SF	4d	MB	Gebril, 1969
$C_{12.1}$ average	88	SF	4d	MB	Gebril, 1969
$C_{14.4}$ average	89	SF	4d	MB	Gebril, 1969
C_{11-13} LAS	98	In	5d	MB	Halvorson, 1969b
$C_{11.3}$ average	98	In	4d	MB	Halvorson, 1969b
$C_{13.3}$ average	98	In	4d	MB	Halvorson, 1969b
SDA 1-1	100;97	BAS	1d	F;MB	SDA, 1969a,b
	96	BAS	17h	F	
	100;93	RW	28d	F;MB	
	91;95	SF	14d	F;MB	
	94;91	In	7d	F;MB	
LAS	90,92	SF	4d	MB	Tomiyama, 1969

[a]Range 54-85.

[b]Range 66-100.

[c]See also Jendreyko (1963).

[d]See also Klein (1965b), McGauhey (1964).

[e]Kellogg olefins.

A. 3. *TBS and Other Polypropylene Derivatives*

TABLE 8.3

Biodegradation of TBS and Related Polypropylene Products

Substrate (see § II)	Extent (see § III)	Method (see § IV)	Time (see § V)	Analysis (see § VI)	Reference
a. *Below C_{12} Alkyl Group*					
C_6TBS	2;6	BOD	5;15d	O_2	Ryckman, 1956, 57
	4	Wa	6h	O_2	
	10;0	Wa	8d	O_2;SO_4	
	5	BOD	22d	MB	
C_9TBS	2;7	BOD	5;13d	O_2	Ryckman, 1956, 57
	3	Wa	6h	O_2	
	18;0	Wa	8d	O_2;SO_4	
	5	BOD	22d	MB	
b. *Near C_{12} Alkyl Group*					
TBS	2;3	Wa;BOD	6h;5d	O_2	Bogan, 1955; Sawyer, 1956
TBS	30-50	CAS	3h	MB	Degens, 1955
TBS	52	RW	19d	MB	Hammerton, 1955
TBS	10;35	RW	20d	O_2;MB	Hammerton, 1956
TBS	80-90	*CAS	16h	RA;MB	House, 1956
	45-50	CAS	15h	RA	
	5-7	CAS	15h	RS	
	25-85	BAS	8h	RA	
Tide (50KBS:50TBS)	80;60	*TF;CAS	−;?	MB	Raybould, 1956
	25-40	*TF	−	MB	
	21	CAnD	28d	MB	
TBS	2;2	BOD	5;15d	O_2	Ryckman, 1956, 57
	2;14	Wa	6;30h	O_2	
	0	Wa	8d	SO_4	
	5	BOD	22d	MB	
TBS	60-80	RW	20d	MB	Sawyer, 1956
TBS	0.25F	Wa	6h	O_2	Barden, 1957
	0.14G	Wa	9h	O_2	
TBS	45-55	BAS;CAS	1d;6h	RA	McGauhey, 1957
	5	CAS	6h	RS	
TBS	0	BOD	10d	O_2	Isaac, 1958
TBS	0-11	CAnD	30d	MB	Johnson, 1958
TBS	19-77	CAS	4-6h	RA;MB	McGauhey, 1959a,b
	3-9	CAS	4-6h	RS	
	25	TF	−	RA;MB	
	1-3	TF	−	RS	
TBS	3-95	Wa	1d	O_2	McKinney, 1959a
TBS	60[a]	CAS	8h	MB	McKinney, 1959b
	80	CAS	5h	MB	
TBS	67	TF	−	MB	Truesdale, 1959, Eden, 1961a
	80	CAS	8h	MB	

TABLE 8.3   TBS and Related Products *(continued)*

Substrate (see § II)	Extent (see § III)	Method (see § IV)	Time (see § V)	Analysis (see § VI)	Reference
TBS	69	CAS	6h	MB	Truesdale, 1959, Eden, 1961a
TBS	22	CAS	3-4h	MB	Bock, 1960, 61
	64	CAS	?	MB	
	30-40	CAS	?	MB	
	30	*CAS	?	MB	
BI = 2.4, MG = 3.6 (P)	36;0	RTF;RW	1;10d	MB;O_2	Burnop, 1960
TBS (Q)	43;0	RTW;RW	1;10d	MB;O_2	Burnop, 1960
TBS	18	In	2d	MB	Roberts, 1960
TBS	0-35	Wa	8h	O_2;MB	Bennett, 1961
TBS	75-81	RW	21d	MB	Knöpp, 1961
	90	RW	21d	F	
Marlon TP350	37-71	TF	—	MB	Meinck, 1961
	20	BOD	5d	O_2	
	80-90	BAnD	60d	MB	
Dobane PT	16	RW	12d	MB	Gameson, 1962
	45	ASEff	18d	MB	
TBS	0-9	In	2d	MB	Kimura, 1962
TBS	20-35	Soil	—	RA	Klein, 1962, 63a
	2-30	Soil	—	RS	
	68-87	Sunfl	—	RA	
TBS	74	In	1d	MB	Offhaus, 1962
TBS	81	*CAS	2d	MB	Scherb, 1962
WAS T	45-60	TF	—	MB	Schönborn, 1962a
	54-80	BAS	1d	MB	
TBS	40-50	ST	33d	MB	Weaver, 1962
	20-30	TF	—	MB	
	40-60	BAS	6+18h	MB	
	71	RW	30d	MB	
Marlon TP350	26	Wa	1d	O_2	Winter, 1962
"ABS"	70	CAS	6h	MB	Barnhart, 1963a
"ABS"	65	BAS	1d	MB	Barnhart, 1963b
	60	CAS	?	MB	
DOBS PT	82	In	7d	MB	Borstlap, 1963
Marlon TP350	37-71	TF	—	MB	Bringmann, 1963
TBS	38	RTF	1d	MB	Cohen, 1963
TBS	55[b]	CAS	6h	MB	Eldib, 1963
	21[c]	CAS	3h	MB	
TBS	2-6	SF	3d	MB	Huddleston, 1963
	43	SF	56d	MB	
TBS	50-60	TF	—	MB	Husmann, 1963a, p. 33
	25	TF	—	MB	Husmann, 1963a, p. 35
	23	CAS	1h	MB	Husmann, 1963a, p. 42
TBS B	77	Soil	—	MB	Husmann, 1963b
	78	RW	35d	MB	

TABLE 8.3   TBS and Related Products *(continued)*

Substrate (see § II)	Extent (see § III)	Method (see § IV)	Time (see § V)	Analysis (see § VI)	Reference
TBS	3	In?	4,9d	MB	Jendreyko, 1963[d]
TBS	30-70	Soil	–	RA	Klein, 1963b, 64b
	5-38	Soil	–	RS	
	60	Sew	63d	RA	
	25-30	Sew	63d	RS	
	19	ST	20h	MB	
	2	ST	20h	RA	
SDA	45	In	?	MB	Konecky, 1963
	<8	In	?	COD	
Commercial	37	In	?	MB	Konecky, 1963
TBS	10	In	32h	Wt	Payne, 1963b
TBS	83-94	CAS	4-11h	MB	Phillips, 1963
	20-50	Wa	20h	MB	
	0.3-1.2G	Wa	10h	O_2	
TBS	0	In	4d	B	Pipes, 1963b
TBS	11;0	In	20d	$MB;SO_4$	Pitter, 1963b, 64c,d
	0.3G	In	20d	O_2	
TBS	68-98	Soil	–	MB,RA	Robeck, 1963
	29-82	Soil	–	RS	
TBS	15	CAS	3h	MB	Rohlffs, 1963
TBS	30	In	30d	MB	Ruschenberg, 1963a,b
	0	In	25d	O_2	
TBS	15	ST	5d	RA	Straus, 1963
	35	ST+Soil	5d+	RA	
TBS	80	RW	32d	MB	Swisher, 1963b
TBS	30	RW	9d	MB	Wayman, 1963a
	8	AnRW	9d	MB	
TBS	18	SF	5d	MB;σ	Allred, 1964b
	20	Wa	6h	O_2	
TBS	60	RTF	1d	MB	Berger, 1964
	27	RW	30d	MB	
TBS	20	RW	30d	σ	Cooper, 1964
TBS	0	RW	4d	MB	Cordon, 1964
	45	CAS	3h	MB	
TBS	80	In	7d	MB	Crauland, 1964
TBS	30-70	In	20d	MB	Čuta, 1964
TBS	15-40	*CAS	2d	MB	De Jong, 1964, 65
TBS	7-52	*CAS	6-18h	MB	Hanna, 1964
TBS	63	In	10d	MB	Hitzman, 1964
TBS	0;12	CAS	2;4h	MB	Huddleston, 1964a
	60-80	CAS	6h	MB	
TBS	16;2	Alg	3h	RA;RS	Klein, 1964a
	30;1	Pond	30d	RA;RS	
TBS	6-25	ST	2d	MB	Klein, 1964a[e]
	10;0	ST	2d	RA;RS	

TABLE 8.3   TBS and Related Products *(continued)*

Substrate (see § II)	Extent (see § III)	Method (see § IV)	Time (see § V)	Analysis (see § VI)	Reference
TBS	58-70	ST+Soil	2d+	MB	Klein, 1964a[e]
	55-78	ST+Soil	2d+	RA	
	13-30	ST+Soil	2d+	RS	
TBS	0-20	In	30d	MB;O_2	Kölbel, 1964
TBS	25-65	BAS	1d	MB	Orgel, 1964,
					Renn, 1964a
	17-20	SF	4d	MB	
	25	RW	32d	MB	
TBS	13-68	CAS	8h	MB	Pitter, 1964a
TBS	3-20	BAnD	25d	MB	Pitter, 1964b,d
TBS	11	In	20d	MB	Pitter, 1964c, d
TBS	0-2	?	?	SO_4	Pitter, 1964c
TBS	31-55	CAS	5h	MB	Renn, 1964a
TBS	40-50	CAS	3h	MB	Renn, 1964b
TBS	40-95	Soil	—	MB	Robeck, 1964
TBS	73	In	17d	MB	Ruschenberg, 1964
(hot sulfonation)	90	In	17d	MB	
(b.p. 280-300°C)	80	In	20d	MB	
(b.p. 285-295°C)	95	In	20d	MB	
TBS	98	Aqu	196d	RA	Sharman, 1964a
	42	Aqu	184d	RS	
TBS	90	CAS	—[f]	RA	Sharman, 1964b
TBS	27	*CAS	2d	MB	Spohn, 1964a
TBS	27-55	*TF	—	MB	Spohn, 1964b
	40-47	*CAS	1h	MB	
Alkane 56	50-75	CAS	6h	RA	Sweeney, 1964a
	8-26	CAS	6h	RS.	
TBS	55-70	RW	30d	MB	Weil, 1964
Ultrawet K	7-8	Wa	10d	O_2	Barbaro, 1965
TBS	100	RW	20d	MB;SO_4	Benarde, 1965
	70	SF	7d	MB;SO_4	
TBS	35	In	3d	F	Bloch, 1965
SDA No. 3	0-14	SF	7d	MB	Booman, 1965
	18-32	CAS	6h	MB	
TBS	4	RW	56d	MB	Cohen, 1965
	8	BAS	1d	MB	
TBS	63	RTF	2d	MB	Edeline, 1965
Dobane PT	36-42	In	21d	MB	Eden, 1965
	67	TF	—	MB	
	69	CAS	6h	MB	
TBS	35	CAS	3h	MB	Eldib, 1965
TBS	23	CAS	3h	MB	Fischer, 1965
TBS	13	*TF	—	MB	Kelly, 1965
	0-20	In	21d	COD	
TBS	35;12	TF	—	RA;RS	Klein, 1965a,b

TABLE 8.3   TBS and Related Products *(continued)*

Substrate (see § II)	Extent (see § III)	Method (see § IV)	Time (see § V)	Analysis (see § VI)	Reference
TBS	45;6	CAS	6h	RA;RS	Klein, 1965a,b
	65	Sew	40d	RA;MB	
	20	AnSew	40d	RA;MB	
TBS	30	RW	30d	MB	Knaggs, 1965
TBS	65-90	*CAS	1½d	MB	Knopp, 1965
TBS	5	In	18d	MB	LGC, 1965, p. 52
TBS	0	In	22d	MB;COD	Livingston, 1965
SDA	35	Soil	–	MB	Maurer, 1965
	0	CAnD	30d	MB	
TBS	0.3-0.6B	TF	–	MB	Renn, 1965a
TBS	9	?	2d	MB	Ōba, 1965b,f
SDA lot 3	46-70	BAS	1d	MB	SDA, 1965
	14-29	SF	8d	MB	
TBS	50	RW	30d	MB	Swenson, 1965
Alkane 56	57	CAS	11h	MB	Urban, 1965
TBS	17	*TF	–	MB	Urban, 1965
	22-35	*Pond	10d	MB	
	15	*Chan	6h	MB	
TBS	32	In	7d	MB	Brebion, 1966
TBS	44	Wa	1d	O$_2$	Brink, 1966
TBS	0	CAnD	25d	MB	Bruce, 1966
TBS	23-53	RTF	1d	MB	Cohen, 1966
TBS	19-28	TF	–	RA	Klein, 1966, McGauhey, 1966
	2-4	TF	–	RS	
TBS	0	In	?	B	Lambin, 1966
TBS	18	CAS	6h	MB	Lashen, 1966
TBS	57	?	21d	?	Manneck, 1966
Ultrawet K	46-23	Wa	5d	MB	Marion, 1966
	52-57	Wa	5d	O$_2$	
Dobane PT	44-56	In	21d	MB	WPRL, 1966a, p. 131
	74	TF	–	MB	
	75;78	CAS;BAS	?;1d	MB	
"Very hard"	0	In	21d	MB	WPRL, 1966a, p. 131
	40	BAS	1d	MB	
TBS	30	CAS	3h	MB	Bloch, 1967
TBS	24	RW	21d	MB	Borstlap, 1967b
Fisher std.	0-30	Alg	21d	MB	Davis, 1967
	75	Sew	21d	MB	
	80	BAS	21d	MB	
	85	Pond	21d	MB	
TBS	62;75	BAS;RW	1;20d	MB	Engelbrecht, 1967
TBS (I)	5-23	Wa	8h	O$_2$	Hartmann, 1967
TBS	0	In	30d	MB;O$_2$	Heinz, 1967
TBS	20-22	TF	–	MB	Malz, 1967

TABLE 8.3   TBS and Related Products *(continued)*

Substrate (see § II)	Extent (see § III)	Method (see § IV)	Time (see § V)	Analysis (see § VI)	Reference
TBS	13-24	CAS	?	MB	Malz, 1967
TBS	6	An	7d	MB	Ōba, 1967
TBS	74-77	RW	50d	MB	Snyder, 1967
	7	ASEff	10d	MB	
TBS	45	BAS	3d	MB	Vaicum, 1967
Dobane PT	71	TF	−	MB	WPRL, 1967, p. 154
TBS	0-23	In	15d	MB	Cook, 1968
Dobane PT	36-42	In	20d	MB	Eden, 1968
	64	TF	−	MB	
Dobane PT	89	RTF	?	MB	Eden, 1968, p. 123
Marlon TP	82-86	*CAS	2d	MB	Huber, 1968
	38-56	*CAS	1½d	MB	
	20	*CAS	1d	MB	
TBS	19;35	CAS	3h	MB;COD	Janicke, 1968c
Marlon TP350	98	Soil	−	MB	Kempf, 1968
TBS	12-21	SF	8d	MB	Ōba, 1968b
TBS	10	In	20d	MB	Pitter, 1968c
	10;5	In	20d	COD;O$_2$	
TBS	25;30	CAS;RW	3h;20d	MB	Rismondo, 1968
	10	ST	12h	MB	
TBS	63-74	RW	20d	MB	Snyder, 1968
Dobane PT	15-22	In	21d	MB	Truesdale, 1968
	72;88	TF;RTF	−;7d	MB	
	35	CAS	3h	MB	
	24;42	BAS	4;6h	MB	
	75	BAS	1d	MB	
TBS	0	In	5d	MB	Halvorson, 1969b
SDA #3	80;66	BAS	1d	F;MB	SDA, 1969a,b
	<60	BAS	17h	F	
	93;68	RW	28d	F;MB	
	7;19	SF	14d	F;MB	
	<60;35	In	7d	F;MB	

c. *Above C$_{12}$ Alkyl Group*

Substrate (see § II)	Extent (see § III)	Method (see § IV)	Time (see § V)	Analysis (see § VI)	Reference
C$_{13}$ TBS	(More resistant than C$_{12}$)				Huddleston, 1963
	0-7	SF	7d	MB	Allred, 1964a
(Alkane 60)	64;10	CAS	6h	RA;RS	Sweeney, 1964a
(Alkane 60)	54-77	CAS	11h	MB	Urban, 1965
C$_{15}$ TBS	0;1	BOD	5;23d	O$_2$	Ryckman, 1956, 57
	1	Wa	6h	O$_2$	
	20;0	Wa	8d	O$_2$;SO$_4$	
	5	BOD	22d	MB	
	(More resistant than C$_{12}$ TBS)				Huddleston, 1963
	0-14	SF	7d	MB	Allred, 1964a
	67	CAS	6h	RA	Sweeney, 1964a

TABLE 8.3   TBS and Related Products *(continued)*

Substrate (see § II)	Extent (see § III)	Method (see § IV)	Time (see § V)	Analysis (see § VI)	Reference
d. *Related Polypropylene Products*					
tp-C_{12} tolueneΣ	51;6	CAS	6h	RA;RS	Sweeney, 1964a
tp-$C_{12}CH_2\phi\Sigma$	82-99	RTF	1d	MB	Cohen, 1966
tp-$C_{12}\phi CH_2\Sigma$	78	RW	46d	MB	Swisher, 1960

[a]Range 15-80.

[b]Range 40-60.

[c]Range 0-30.

[d]See also Husmann (1963a, pp. 49, 60).

[e]See also McGauhey (1964) and Klein (1965b).

[f]Foam recycle.

A. 4. *Other Types of ABS*

TABLE 8.4

Biodegradation of Other Types of ABS

Substrate (see § II)	Extent (see § III)	Method (see § IV)	Time (see § V)	Analysis (see § VI)	Reference
a. *KBS (ex Kerosene)*					
Santomerse No. 1	45-65	BAS	1d	MB	Lumb, 1953
	60	TF	–	MB	
KBS	22;10	BOD;Wa	5d;6h	O_2	Bogan, 1955, Sawyer, 1956
KBS	79	RW	21d	MB	Hammerton, 1955
Santomerse No. 1	60;85	*TF;CAS	–;?	MB	Raybould, 1956
Tide (50KBS:50TBS)	80;60	*TF;CAS	–;?	MB	Raybould, 1956
	25-40	*TF	–	MB	
	21	CAnD	28d	MB	
Santomerse No. 1	15	CAnD	28d	MB	Raybould, 1956
KBS	80	RW	20d	MB	Sawyer, 1956
Santomerse No. 1	0.02G	BOD	5d	O_2	Sheets, 1956a,b
KBS	0.2-0.3F	Wa	6h	O_2	Barden, 1957
	0.4G	Wa	9h	O_2	
Dubaral	67-79;0	CAS	?	MB;SO_4	Pitter, 1961a
KBS	11-14	In	2d	MB	Kimura, 1962
KBS	36	Wa	1d	O_2	Winter, 1962
Nacconol SZA	25	In	32h	Wt	Payne, 1963a
	0.1-0.2G	Wa	2h	O_2	
Nacconol SZA	35	In	32h	Wt	Payne, 1963b
Dubaral	32;0.6G	In	20d	MB;O_2	Pitter, 1963b, 64d
	0	In	20d	SO_4	
KBS	53-64	SF	7d	MB	Allred, 1964a
KBS	30-90	In	20d	MB	Čuta, 1964
Dubaral	3	BAnD	32d	MB	Pitter, 1964b,d
KBS (30-50% linear)	48-63	In	20d	MB	Pitter, 1968c
KBS (2-5% linear)	5-23	In	20d	MB	Pitter, 1968c
b. *ex Nonlinear Olefins*					
n-Hexene dimer (Ni^a)	91	In	23d	MB	BHC, 1963
C_{6+7}Dimer (iBu_3Al^a)	90	CAS	3h	MB	Rohlffs, 1963
C_3 Dimer dimer($H_3PO_4^a$)	3-4	SF	7d	MB	Allred, 1964a
2Bu-octene ($H_2SO_4{}^b$)	53	SF	7d	MB	Allred, 1964a
2Bu-octene ($AlCl_3{}^b$)	87	SF	7d	MB	Allred, 1964a
n-C_6 Dimer (Mn^a)	96	In	23d	MB	Oldham, 1964
Cy-Cyclohexene	56;5	CAS	6h	RA;RS	Sweeney, 1964a
C_{5+6+7}CWO dimer ($LiAlH_4{}^a$)	100	RW	30d	MB	Swenson, 1965
C_{5+6+7}CWO dimer ($H_3PO_4{}^a$)	23	RW	30d	MB	Swenson, 1965
$2C_6$ Dimer ($SiAlO^a$)	90-93	RW	20d	MB	Snyder, 1967, 68
$2C_{6+8}$ Dimer ($SiAlO^a$)	39	ASEff	10d	MB	Snyder, 1967, 68
$1C_6$ Dimer ($SiAlO^a$)	91	RW	20d	MB	Snyder, 1968
$1C_{6+7}$Dimer ($SiAlO^a$)	33	ASEff	10d	MB	Snyder, 1968

TABLE 8.4   Other ABS Types *(continued)*

Substrate (see § II)	Extent (see § III)	Method (see § IV)	Time (see § V)	Analysis (see § VI)	Reference
c. *Other Characterized ABS*					
C_{11-13} high branch (L)	58	RTF	1d	MB	Burnop, 1960
Quaternary α-C^c (M)	9	RTF	1d	MB	Burnop, 1960
Low α-methyl (O)	92;0	RTF;RW	1;10d	MB;O_2	Burnop, 1960
d. *Undisclosed or Unknown Structure*					
Nacconol NR	0	BOD	?	O_2	Rudolfs, 1949
Commercial dodecyl	0	BOD	7d	O_2	Leclerc, 1952
Kreelon 8D; 8G	9;10	Wa	6h	O_2	Bogan, 1954
	21;34	BOD	5d	O_2	
Nacconol NRSE	11;18	Wa;BOD	6h;5d	O_2	Bogan, 1954
Santomerse No. 3	12;11	Wa;BOD	6h;5d	O_2	Bogan, 1954
Ultrawet DS	5;5	Wa;BOD	6h;5d	O_2	Bogan, 1954
Santomerse No. 3	0	In	10d	σ	Von Riesen, 1955
Santomerse No. 1	62	RW	32d	MB	Hammerton, 1956
Foreign ABS	64	CAS	2-4h	MB	Bock, 1960, 61
New type comml. (G)	65;33	RTF;RW	1;10d	MB;O_2	Burnop, 1960
New type comml. (H)	68	RTF	1d	MB	Burnop, 1960
Nacconol	0	BAnD	40d	MB	Manganelli, 1960
Methyl type DBS	41	Wa	1d	O_2	Winter, 1962
Detergent B; C	51;51	TF	–	MB	Husmann, 1963a, p. 36
Detergent B	43-68	CAS	1h	MB	Husmann, 1963a, p. 38
	65-75	CAS	1h	MB	Husmann, 1963a, p. 45-47
	70	In	9d	MB	Husmann, 1963a, p. 108
Detergent C	56-63	CAS	1h	MB	Husmann, 1963a, p. 38
	75-85	CAS	1h	MB	Husmann, 1963a, p. 46-47
Detergent C; D	90;80	In	9;3d	MB	Husmann, 1963a, p. 108
Detergent B; D	69;82	*TF	–	MB	Husmann, 1963a, p. 68-70
Alkylbenzene	17-30	Wa	1d	O_2	Lipman, 1963
Dodecylbenzene	0	Wa	1d	O_2	Lipman, 1963
Dodecylbenzene, comml.	0-0.07G	Wa	4h?	O_2	Sato, 1963
n-Dodecylbenzene	50-90	In	20d	MB	Čuta, 1964
ABS	0	SF	7d	MB;F	Garrison, 1964
	0;0;50	In	9d	MB;σ;F	
	0.07G	Wa	5d	O_2	
ABS-Holland	2	In	20d	MB	Pitter, 1964c
ABS-Jugoslavia	3	In	20d	MB	Pitter, 1964c
ABS-Romania	0	In	20d	MB	Pitter, 1964c
ABS 2; 3	98;98	CAS	6h	MB	Eden, 1965
	94;91	TF	–	MB	
	98;97	In	21d	MB	

TABLE 8.4   Other ABS Types *(continued)*

Substrate (see § II)	Extent (see § III)	Method (see § IV)	Time (see § V)	Analysis (see § VI)	Reference
ABS-commercial	23(0-50)	CAS	3h	MB	Ilişescu, 1966
	12-16	In	7d	MB	
ABS RD1213	33	In	21d	MB	MWB, 1966, p. 36
Dubaral (hard ABS)	71	CAS	25h	MB	Pitter, 1966b
"Difficult" ABS	33	In	21d	MB	WPRL, 1966a, p. 131
	94	TF	—	MB	
	93;97	CAS;BAS	?;1d	MB	
ABS-commercial	29-36	CAS	6h	MB	Vaicum, 1967
	23	CAS	3h	MB	
	35	BAS	3d	MB	
	30	In	?	MB	
	17;13	BOD;Wa	5d;?	MB	
ABS-synthetic	52-58	BAS	3d	MB	Vaicum, 1967
Dobane JT	60	BAS	3d	MB	Vaicum, 1967
"Difficult" ABS (RD1213)	91	TF	—	MB	WPRL, 1967, p. 154
"Difficult" ABS	6-32	In	15d	MB	Cook, 1968
	29;34	RW;SF	?;8d	MB	
	34;70	CAS;BAS	3h;1d	MB	
	83	RTF	?	MB	
ABS 1;2	98;97	In	20d	MB	Eden, 1968
	98;99	CAS	6h	MB	
	91;99	TF	—	MB	
ABS 4 ("slow to acclimatize")	28-42	In	21d	MB	Truesdale, 1968
	90	TF	—	MB	
	66-68	CAS	3h	MB	
	98-99	BAS	1d	MB	
	96-98	RTF	7d	MB	

[a]Dimerization catalyst.

[b]Alkylation catalyst

[c]Branching index 4.0

A. 5. *Other Alkyl Aromatic Sulfonates*

TABLE 8.5

Biodegradation of Other Alkyl Aromatic Sulfonates of Known Structure

Substrate (see § II)	Extent (see § III)	Method (see § IV)	Time (see § V)	Analysis (see § VI)	Reference
a. *ex Linear Di- and Polyalkylbenzenes*					
C_{11-15} toluene	90	In	20d	MB	Pitter, 1964c,d
C_{12} ethylbenzene	53	RW	21d	MB	Borstlap, 1967b
C_{12} *p*-xylene	92	RW	17-21d	MB	Borstlap, 1967b
C_{12} *m*-xylene	16	RW	21d	MB	Borstlap, 1967b
C_{12} *o*-xylene	61	RW	17-21d	MB	Borstlap, 1967b
Di-$C_6\phi$	20-35	CAS	6h	RA;RS	Sweeney, 1964a
Di-$C_6\phi$	22	RW	21d	MB	Borstlap, 1967b
p-Di-1$C_6\phi$	99;4	CAS	6h	RA;RS	Sweeney, 1964a
p-1C_6,4$C_7\phi$	35;3	CAS	6h	RA;RS	Sweeney, 1964a
Di-$C_7\phi$	60	BAS	1d	MB	Swisher, 1963a
	Rates	BAS	1d	DGC	
Di-$C_7\phi$	53	Wa	26h	O_2	Brink, 1966
b. *ex Linear Alkyl Biphenyls*					
p-1C_4-*p'*-Σ	5;0	In	30d	MB;O_2	Kölbel, 1964
p-1C_6-*p'*-Σ	12;0	In	30d	MB;O_2	Kölbel, 1964
p-1C_8-*p'*-Σ	100;30	In	25;30d	MB;O_2	Kölbel, 1964
p-1C_{10}-*p'*-Σ	100;32	In	25d	MB;O_2	Kölbel, 1964
c. *ex Alkyl Naphthalenes*					
6-1C_4-2Σ	12;0	In	30d	MB;O_2	Kölbel, 1964
6-1C_6-2Σ	100;27	In	30d	MB;O_2	Kölbel, 1964
6-1C_8-2Σ	100;34	In	15;30d	MB;O_2	Kölbel, 1964
*i*Pr (Alkanol B)	0	In	5d	B	Feisal, 1966
*i*Pr$_2$+Bu$_2$(Nekal BX)	0.01G	BOD	5d	O_2	Sheets, 1956a,b
*i*Pr$_2$+Bu$_2$(Nekal 75,78)	+	In	5d	B	Feisal, 1966
Bu$_2$	2;22	Wa	1;2d	O_2	Winter, 1962
*i*Bu (Neokal)	6	In	20d	MB	Pitter, 1964c,d
*i*Bu (Neokal)	13	BAnD	32d	MB	Pitter, 1964b,d
*i*Bu$_2$	50-90	In	20d	MB	Čuta, 1964
*i*Bu$_2$	0	Wa	8h	O_2	Hartmann, 1967
Alkyl (Naccosol A)	8;3	Wa;BOD	6h;5d	O_2	Bogan, 1954
Alkyl (Perminal BX)	F[a]	Wa	6h	O_2	Barden, 1957
d. *ex Phenylene Alkanes*					
Phenylene dodecanes[b]	99	RW	14d	MB	Swisher, 1959
e. *Linear Thiaalkylbenzenesulfonates*					
1-Thia-1C_{12}	100	SF;RW	5d	MB	Lang, 1965
	100	SF	2d	MB	Long, 1966
2-Thia-1C_{12}	100	SF;RW	9;7d	MB	Lang, 1965
	100	SF	2d	MB	Long, 1966
3-Thia-1C_{12}	100	SF;RW	15;14d	MB	Lang, 1965
	100	SF	2d	MB	Long, 1966

TABLE 8.5   Other Alkyl Aromatics *(continued)*

Substrate (see § II)	Extent (see § III)	Method (see § IV)	Time (see § V)	Analysis (see § VI)	Reference
4-Thia-1C$_{12}$	100	SF;RW	14d	MB	Lang, 1965
	100	SF	2d	MB	Long, 1966
5-Thia-1C$_{12}$	100	RW	14d	MB	Lang, 1967
6-Thia-1C$_{12}$	100	RW	14d	MB	Lang, 1967
5- to 11-Thia-1C$_{12}$	100?	SF;RW	>14d	MB	Lang, 1965
f. *ex Alkylphenols*					
Linear C$_{12}$	86	RW	21d	MB	Borstlap, 1967b
g. *ex Alkylthiophenes*					
Linear C$_{12}$	98	RW	3d	MB	Borstlap, 1967b
h. *ex Carboxyalkylbenzenes*					
Sulfophenylunde-canoic acid	44;54	BOD;Wa	19;8d	O$_2$	Ryckman, 1956
Sulfophenylunde-canoic acid	100	RW	5d	DGC	Swisher, 1964b

[a]Erratic.

[b]Mixed dialkylindane, dialkyltetralin sulfonates derived from mixed dichloro-*n*-dodecane (**6.11, 6.12**).

A. 6. *Aliphatic Hydrocarbon and Hydroxyalkane Sulfonates*

TABLE 8.6

Biodegradation of Alkane, Alkene, and Hydroxyalkane Sulfonates

Substrate (see § II)	Extent (see § III)	Method (see § IV)	Time (see § V)	Analysis (see § VI)	Reference
a. *Linear Primary Alkane Sulfonates*					
C_{10}	100;77	In	6;18d	$MB;O_2$	Kölbel, 1964
C_{12}	100;70	In	6;18d	$MB;O_2$	Kölbel, 1964
	96	In	7d	MB	Crauland, 1964
	80†	RW	3-5d	MB	Weil, 1964, 65
	95	BAS	3d	MB	Vaicum, 1967
C_{14}	100;85	In	6;18d	$MB;O_2$	Kölbel, 1964
C_{16}	100;90	In	6;18d	$MB;O_2$	Kölbel, 1964
	80†	RW	4d	MB	Weil, 1964
C_{17}	80†	RW	5d	MB	Weil, 1964
Unspecified	100;+	RW;BOD	3d;?	$MB;O_2$	Huyser, 1960
b. *Mixed Secondary Alkane Sulfonates*					
C_{13}	100;74	In	4;21d	MB;COD	McAteer, 1964
C_{15}	100;76	In	4;21d	MB;COD	McAteer, 1964
C_{16}	100;72	In	4;21d	MB;COD	McAteer, 1964
C_{17}	100;73	In	4;21d	MB;COD	McAteer, 1964
C_{19}	100;86	In	4;21d	MB;COD	McAteer, 1964
C_{14-16}	100;70	In	5;25d	$MB;O_2$	Ruschenberg, 1963a,b
C_{13-18}	99	CAS	3h	MB	Täuber, 1968
Average C_{15}	47	Wa	1d	O_2	Winter, 1962
Average C_{15} (+polysulfonates)	44	Wa	1d	O_2	Winter, 1962
ex Fischer-Tropsch	91-95	RW	8d	MB	Hammerton, 1955, 56
Avitone A	0.5F	Wa	6h	O_2	Barden, 1957
Fit[a]	55	Wa	1d	O_2	Winter, 1962
SAS	100;88	In	7;11d	MB;COD	Konecky, 1963
SAS	50-90	Wa	1d	O_2	Lipman, 1963
ex Paraffins	95	CAS	3h	MB	Berger, 1964
	91;96	RTF;RW	3;15d	MB	
ex Paraffins	95	In	7d	MB	Crauland, 1964
SAS	44	Wa	10d	O_2	Hunter, 1964
Emulgator E30	100;92	In	4;15d	$MB;O_2$	Pitter, 1964c
Mersolat H	100;91	In	5;20d	$MB;O_2$	Pitter, 1964c
SAS E411A	98;99	TF;CAS	–;?	MB	STCSD, 1964, p. 11
	99	In	2d	MB	
SAS E4215	97;99	TF;CAS	–;?	MB	STCSD, 1964, p. 11
	98	In	2d	MB	
SAS	75-85	In	21d	COD	Kelly, 1965
	70	*TF	–	MB	
SAS	96	In	7d	MB	LGC, 1965, p. 52
SAS	0	BAnD	17d	MB	Bruce, 1966
SAS	100	In	4d	MB	MWB, 1966, p. 36
Emulgator E30	+	*CAS	3d	MB	Walther, 1966

TABLE 8.6   Alkane, Alkene, Hydroxyalkane Sulfonates *(continued)*

Substrate (see § II)	Extent (see § III)	Method (see § IV)	Time (see § V)	Analysis (see § VI)	Reference
Mersolat	99	In	3d	MB	Heinz, 1967
SAS	95-100	RTF	7d	MB	Jenkins, 1967
	100	In	17d	MB	
SAS 1;2	98-99	In	20d	MB	Eden, 1968
	99;99	CAS	6h	MB	
	99;97	TF	–	MB	
SAS	45	BOD	7d	O_2	Krone, 1968
	60	Wa	3d	O_2	
	100	In	2d	MB	
	99	CAS	3h	MB	
	85	*TF	–	MB	
c. *Alkene Sulfonates*[b]					
2-Pentadecene-1-Σ	99-100	SF	8d	MB	Ōba, 1968b
C_{14} AOS	99	?	10d	?	Manneck, 1966
C_{15} AOS	Rate	SF	–	GC	Tomiyama, 1968
C_{16} AOS	95	?	7d	?	Manneck, 1966
	Rate	SF	–	GC	Tomiyama, 1968
C_{17} AOS	Rate	SF	–	GC	Tomiyama, 1968
C_{18} AOS	80†	RW	4d	MB	Weil, 1965
	98	?	7d	?	Manneck, 1966
	Rate	SF	–	GC	Tomiyama, 1968
C_{15-16} AOS	31-43	An	28d	MB	Ōba, 1967
C_{15-18} AOS	96-97	?	4d	?	Marquis, 1966
	98-100	SF	8d	MB	Ōba, 1968b
C_{15-18} AOS (C)	98	SF	8d	MB	Ōba, 1968b
(D)(15%Σ_2)	98	SF	8d	MB	Ōba, 1968b
(E)(50%Σ_2)	96	SF	8d	MB	Ōba, 1968b
C_{15-18} AOS (0%Σ_2)	99	SF	8d	MB	Tomiyama, 1968
(50%Σ_2)	95	SF	8d	MB	Tomiyama, 1968
(100%Σ_2)	95	SF	8d	MB	Tomiyama, 1968
AOS	97,99	SF	2d	MB	Tomiyama, 1969
d. *Hydroxyalkane Sulfonates*[b]					
C_{14}-3OH-1Σ	99	SF	8d	MB	Ōba, 1968b
C_{16}-1OH-2Σ	80†	RW	4d	MB	Weil, 1965
C_{16}-2OH-1Σ	80†	RW	4d	MB	Weil, 1965
C_{18}-1Σ-2Σ	80†	RW	4d	MB	Weil, 1964
$C_{15}, C_{16}, C_{17}, C_{18}$	Rates	SF	–	GC	Tomiyama, 1968

[a]Includes alkyl sulfate.

[b]Commercial-type AOS may include large amounts of hydroxyalkane sulfonates and/ or other co-products.

A. 7. *Other Sulfonated Products*

TABLE 8.7

Biodegradation of Other Sulfonated Products

Substrate (see § II)	Extent (see § III)	Method (see § IV)	Time (see § V)	Analysis (see § VI)	Reference
a. *α-Sulfo Fatty Acids and Derivatives*					
C_9 Hexyl ester	72;0	CAS;RW	3h;5d	MB	Cordon, 1964
	80†	RW	10d	MB	Weil, 1964
	83;83	In	7;19d	MB;COD	Cordon, 1968a
C_{14} Me ester	47	Wa	10d	O_2	Hunter, 1964
C_{16} Acid, salts	80†	RW	5d	MB	Weil, 1964
C_{16} Me ester	80†	RW	4d	MB	Weil, 1964, Stirton, 1965
C_{16} Et ester	80†	RW	5d	MB	Stirton, 1965
C_{16} Pr ester	80†	RW	6d	MB	Stirton, 1965
C_{16} *i*Pr ester	80†	RW	5d	MB	Weil, 1964, Stirton, 1965
C_{16} Bu ester	80†	RW	7d	MB	Weil, 1964, Stirton, 1965
C_{16} *i*Bu ester	80†	RW	7d	MB	Stirton, 1965
C_{16} *s*Bu ester	80†	RW	6d	MB	Stirton, 1965
C_{18} Acid	80†	RW	8d	MB	Weil, 1964
C_{18} TEA salt	80†	RW	4-5d	MB	Weil, 1964
C_{18} Me ester	99;91	CAS;RW	3h;4d	MB	Cordon, 1964
	80†	RW	4d	MB	Weil, 1964, Stirton, 1965
	0	CAnD	30d	MB	Maurer, 1965
	100;85	In	5;18d	MB;COD	Cordon, 1968a
C_{18} Et ester	80†	RW	5d	MB	Stirton, 1965
C_{18} Pr ester	80†	RW	5d	MB	Stirton, 1965
C_{18} *i*Pr ester	97;85	CAS;RW	3h;5d	MB	Cordon, 1964
	80†	RW	5d	MB	Weil, 1964, Stirton, 1965
	0	CAnD	30d	MB	Maurer, 1965
	100	Soil	–	MB	
	100;70	In	5;19d	MB;COD	Cordon, 1968a
C_{18} Bu ester	80†	RW	4d	MB	Stirton, 1965
C_{18} *i*Bu ester	80†	RW	4d	MB	Stirton, 1965
C_{18} *s*Bu ester	80†	RW	6d	MB	Stirton, 1965
C_{18} (2-ΣEt) ester	96;78	CAS;RW	3h;5d	MB	Cordon, 1964
	80†	RW	5d	MB	Weil, 1964
	0	CAnD	30d	MB	Maurer, 1965
	100;65	In	6d	MB;COD	Cordon, 1968a
C_{18} Ethanolamide	80†	RW	5d	MB	Weil, 1964
$9,10Cl_2C_{18}$ Acid	80†	RW	8d	MB	Weil, 1964
$9,10(OH)_2C_{18}$ Acid,					
(erythro)	80†	RW	4d	MB	Weil, 1964
(threo)	80†	RW	4d	MB	Weil, 1964
α-OctylC_{10} Acid	92;73	CAS;RW	3h;5d	MB	Cordon, 1964

TABLE 8.7 Other Sulfonated Products *(continued)*

Substrate (see § II)	Extent (see § III)	Method (see § IV)	Time (see § V)	Analysis (see § VI)	Reference
ϕC_{18} Acid	80†	RW	14d	MB	Weil, 1964
Tallow Me ester					
(Bioterg TMS)	100;55	RW;AnRW	9d	MB	Wayman, 1963a
Tallow Me ester	+	AnRW	?	MB	Knaggs, 1965
	100	BAS;RW	1;8d	MB	
b. *Fatty Acyl Isethionates*					
Oleic (Igepon AP78)	35;50	Wa;BOD	6h;5d	O_2	Bogan, 1954
Oleic (Igepon AP78)	45;67	Wa;BOD	6h;5d	O_2	Bogan, 1955, Sawyer, 1956
Igepon AP	+	In	6d	σ	Von Riesen, 1955
Oleic (Igepon A)	45;60	Wa;BOD	6h;5d	O_2	Ryckman, 1957
Oleic (Igepon A)	62	Wa	1d	O_2	Winter, 1962
Palmitic	80†	RW	1d	MB	Weil, 1964
c. *Fatty Acyl Methyl Taurides*					
Oleic (Igepon T77)	60;54	Wa;BOD	6h;5d	O_2	Bogan, 1954
Oleic (Igepon T77)	56;56	Wa;BOD	6h;5d	O_2	Bogan, 1955, Sawyer, 1956
Igepon T	0	In	10d	σ	Von Riesen, 1955
Igepon T	0.4G	BOD	5d	O_2	Sheets, 1956a
Igepon T	56;52	Wa;BOD	6h;5d	O_2	Ryckman, 1957
Oleic (Igepon T)	86	Wa	1d	O_2	Winter, 1962
Oleic (Igepon T77)	55	Wa	10d	O_2	Hunter, 1964
Oleic (Igepon T77)	37;44	Wa	10d	O_2	Barbaro, 1965
Stearic	80†	RW	3d	MB	Weil, 1964
Fatty acyl	1.3-1.5G	BOD	5d	O_2	Offhaus, 1962
d. *Sulfosuccinic Acid Diesters*					
*i*Bu	96	RW	24d	MB	Hammerton, 1956
4Me-2-pentyl	3	RW	28d	MB	Hammerton, 1956
Hx (Aerosol MA)	0	In	10d	B;σ	Von Riesen, 1955
Cyclohexyl	4	RW	28d	MB	Hammerton, 1956
n-Octyl	98-100	RW	6d	MB	Hammerton, 1956
Octyl (Aerosol OT)	0	In	10d	B;σ	Von Riesen, 1955
2Et hexyl (Manoxol OT)	91-97	RW	17d	MB	Hammerton, 1955
2Et hexyl (Manoxol OT)	98	RW	11d	MB	Hammerton, 1956
2Et hexyl (Manoxol OT)	0.3F	Wa	6h	O_2	Barden, 1957
2Et hexyl (Manoxol OT)	98-100	RW	12d	MB	Gameson, 1962
2Et hexyl (Manoxol OT)	100	RTF	12h	MB	Alexandre, 1967
Octyl	40-75	In	20d	MB	Čuta, 1964
Octyl	80†	RW	5d	MB	Weil, 1964

TABLE 8.7   Other Sulfonated Products *(continued)*

Substrate (see § II)	Extent (see § III)	Method (see § IV)	Time (see § V)	Analysis (see § VI)	Reference
3,5,5Me$_3$ hexyl	92	RW	11d	MB	Hammerton, 1956
Benzyl	100	RW	5d	MB	Hammerton, 1956
e. *Miscellaneous*					
Oleoyl-*p*-anisidine-Σ	95;45	RW	2;14d	MB;O$_2$	Hammerton, 1955, 56
C$_9$CONMeφΣ	100;30	In	14;20d	MB;O$_2$	Kölbel, 1964
C$_{11}$CONMeφΣ	100;36	In	10;20d	MB;O$_2$	Kölbel, 1964
C$_{13}$CONMeφΣ	100;42	In	8;20d	MB;O$_2$	Kölbel, 1964
C$_{15}$CONMeφΣ	100;44	In	10;20d	MB;O$_2$	Kölbel, 1964
C$_{12}$O$_2$CCΣ	80†	RW	2d	MB	Weil, 1964
C O					
C$_{9-18}$CNCCCΣ	100;83	RW	4;42d	F;COD	Sheers, 1967
	32-45	BOD	5d	O$_2$	
f. *Unspecified or Unknown Structure*					
Alkylaryl sulfonate	37-61	RW	14d	MB	Degens, 1950
	0	BOD	5d	O$_2$	
Alkylaryl alcohol sulfonate	0	In	2d	B	Van Beneden, 1952
Alkylaryl sulfonate	81-89	BAS	3h	MB	Sierp, 1954
Arctic Syntex T^a	+;30	In	10d	B;σ	Von Riesen, 1955
Cyclopon A^b	0.2G	BOD	5d	O$_2$	Sheets, 1956a
Tide	50	TF	–	MB	Barden, 1957
7-Detergent mixture	72-92	BAS	1d	MB	Mann, 1957
	53-72	TF	–	MB	
Dubosol E^c	0	In	20d	MB	Pitter, 1964c
Oleic derivative ^{d}	1.2-4.1G	Wa	8h?	O$_2$	Hartmann, 1967

^{a}"Alkyl sulfonated ethyl amide."

^{b}"Sulfonated amide of amino alcohol."

^{c}Petroleum sulfonate.

^{d}"Oleic acid hexamethylene imid sulfonate."

A. 8. *Linear Primary Alkyl Sulfates*

TABLE 8.8
Biodegradation of LPAS

Substrate (see § II)	Extent (see § III)	Method (see § IV)	Time (see § V)	Analysis (see § VI)	Reference
a. *Alkyl Group Below C_{12}*					
Octyl	96;72	RW	2;5d	$MB;O_2$	Hammerton, 1956
	+	In	?	B	Hsu, 1963
Decyl	98;66	RW	2;5d	$MB;O_2$	Hammerton, 1955, 56
	+	In	?	B	Skinner, 1959
	54	Wa	1d	O_2	Winter, 1962
	+	In	?	B	Hsu, 1963
b. *C_{12} Alkyl Group*					
Lauryl	+	In	?	B	Williams, 1949
Lauryl	+	BOD	2-7d	O_2	Leclerc, 1952
Lauryl	+	In	2d	B	Van Beneden, 1952
Lauryl	45;66	Wa;BOD	6h;5d	O_2	Bogan, 1954
n-Dodecyl	41;60	Wa;BOD	6h;5d	O_2	Bogan, 1955, Sawyer, 1956
n-Dodecyl	100	RW	1d	MB	Sawyer, 1956
Lauryl	1.4G	BOD	5d	O_2	Sheets, 1956a,b
Lauryl	100	TF	—	MB	Barden, 1957
	1.4F	Wa	6h	O_2	
	0.9G	Wa	7h	O_2	
Dodecyl	95;75	CAS	6h	RA;RS	McGauhey, 1957
Lauryl	60;57	Wa;BOD	6h;5d	O_2	Ryckman, 1957
C_{12}	+	In	?	B	Skinner, 1959
Dodecyl	74	Wa	1d	O_2	Winter, 1962
Lauryl	98	BAS	6h	MB	Barnhart, 1963b
Lauryl	100	In	1d	MB	Borstlap, 1963
Lauryl	100	CAS	3,6h	MB	Eldib, 1963
Dodecyl	+	In	?	$MB;SO_4$	Hsu, 1963
Lauryl	100	SF	1d	MB	Huddleston, 1963
Lauryl	95-100	CAS	1h	MB	Husmann, 1963a, p. 43
Lauryl	97;100	In	2;9d	MB	Jendreyko, 1963
Lauryl	100;96	In	2;3d	MB;COD	Konecky, 1963
Lauryl	0.5-0.8G	Wa	2h	O_2	Payne, 1963a
Lauryl	65	In	32h	Wt	Payne, 1963b
Dodecyl	+	In	2d	B	Pipes, 1963b
Dodecyl	100;98	In	20d	$MB;O_2$	Pitter, 1963b
Alfol 12	100;70	In	1;20d	$MB;O_2$	Ruschenberg, 1963a,b
Lauryl	0.7G	Wa	4h?	O_2	Sato, 1963
Lauryl	100	SF	1d	$MB;\sigma$	Allred, 1964b
	95	Wa	6h	O_2	
Lauryl (natural)	98;95	CAS;RTF	3h;1d	MB	Berger, 1964
	100	RW	2d	MB	
Lauryl (Ziegler)	99	RW	3d	MB	Berger, 1964
Lauryl	100	RW	5d	σ	Cooper, 1964

TABLE 8.8   Linear Primary Alkyl Sulfates *(continued)*

Substrate (see § II)	Extent (see § III)	Method (see § IV)	Time (see § V)	Analysis (see § VI)	Reference
Lauryl (pure)	100	In	10d	MB˙	Čuta, 1964
Lauryl (technical)	100	In	5-10d	MB	Čuta, 1964
Lauryl	100	In	2d	MB	Hitzman, 1964
Alfol 12	100	CAS	4h	MB	Huddleston, 1964a
Lauryl	68	Wa	10d	O_2	Hunter, 1964
Lauryl	100;98	In	1;20d	MB;O_2	Pitter, 1964c
Lauryl (Syntapon L)	100	In	10d	MB	Pitter, 1964c
Dodecyl	88-97	BAnD	32d	MB	Pitter, 1964b,d
Lauryl	98	CAS	3h	MB	Renn, 1964b
Dodecyl	80†	RW	1-2d	MB	Weil, 1964
Lauryl	100	RTF	2d	MB	Edeline, 1965
Dodecyl	100	In	1h	MB	Hsu, 1965
	100	Enz	1h	MB	
Lauryl	88	*TF	–	MB	Kelly, 1965
	85-95	In	21d	COD	
Lauryl	100	RW	3d	MB	Knaggs, 1965
Lauryl	100;96	In	3d	MB;COD	Livingston, 1965
Lauryl	97	Enz	2h	SO_4	Payne, 1965
Dodecyl	103	Wa	28h	O_2	Brink, 1966
Lauryl	+	In	?	B	Lambin, 1966
Lauryl	112-140	Wa	5d	O_2	Marion, 1966
	100	Wa	5d	MB	
Lauryl	99;100	In;BAS	20;1d	MB	WPRL, 1966a, p. 131
Lauryl	100	In	28d	Wt	Borstlap, 1967a
Lauryl	100	RW	1d	MB	Borstlap, 1967b
Dodecyl	100;105	In	5;30d	MB;O_2	Heinz, 1967
Lauryl	100	RTF;In	7;17d	MB	Jenkins, 1967
Lauryl	100	TF	–	MB	Malz, 1967
	95-100	CAS	?	MB	
Coco	100;39	An	7d	MB;COD	Ōba, 1967
	100;47	SF	15;23h	MB;COD	
Lauryl (pure)	100	CAS	3;6h	MB	Vaicum, 1967
	100	BAS	3d	MB	
	100	In	?	MB	
	100;97	BOD;Wa	5d;?	MB	
C_{12}	100	SF	8d	MB	Ōba, 1968b
n-Dodecyl	100	In	20d	MB	Pitter, 1968c
	98;80	In	20d	COD;O_2	
Lauryl	95-15	In	½-5d	MB	Nyns, 1969a
c. *Alkyl Group Above C_{12}*					
Tridecyl	100	RW	3d	MB	Berger, 1964
Tetradecyl	72	Wa	1d	O_2	Winter, 1962
Tetradecyl	+	In	?	B	Hsu, 1963
Tetradecyl (Ziegler)	95	RW	3d	MB	Berger, 1964

TABLE 8.8   Linear Primary Alkyl Sulfates *(continued)*

Substrate (see § II)	Extent (see § III)	Method (see § IV)	Time (see § V)	Analysis (see § VI)	Reference
Myristyl	98-100	In	7d	MB	Brebion, 1966
C_{15}	95	RW	3d	MB	Berger, 1964
C_{16}	68	Wa	1d	O_2	Winter, 1962
Alfol 16	100;80	In	1;25d	$MB;O_2$	Ruschenberg, 1963a,b
C_{16}	100	CAS	6h	RA	Sweeney, 1964a
C_{16}	100	RW	28d	$MB;\sigma;F$	Vath, 1964
C_{16}	103	Wa	30h	O_2	Brink, 1966
C_{16}	70	BAS	3d	MB	Vaicum, 1967
C_{18}	+	In	?	B	Skinner, 1959
C_{18}	51	Wa	1d	O_2	Winter, 1962
Stearyl	98	In	7d	MB	Crauland, 1964
Oleyl	62	Wa	1d	O_2	Winter, 1962
Oleyl	100;98	CAS;RW	3h;2d	MB	Cordon, 1964
Oleyl	100;98	In	2d	MB;COD	Cordon, 1968a
Oleyl	93	In	2d	SO_4	Cordon, 1968b
Tallow alcohol	50	In	3d	Wt	Payne, 1963b
Tallow alcohol	100	RW	11d	MB	Knaggs, 1965
Tallow alcohol (hydrogenated)	98	CAS;RW	3h;4d	MB	Cordon, 1964
	100	CAnD	30d	MB	Maurer, 1965
	100;93	In	1;19d	MB;COD	Cordon, 1968a
d. *Mixed Alkyl Groups*					
C_{12-14} (Syntapon L)	98	CAS	?	MB	Pitter, 1961a
C_{12+14} (Ziegler)	96;99	RTF;RW	1;3d	MB	Berger, 1964
C_{12-14} (fatty)	93;105	In	5;30d	$MB;O_2$	Heinz, 1967
C_{10-18}	99	In	20d	MB	Pitter, 1968c
	98;72	In	20d	$COD;O_2$	
C_{12-18} (Texapon)	60;76	Wa	1;2d	O_2	Winter, 1962
C_{12-18}	1.2G	Wa	8h	O_2	Hartmann, 1967
C_{16-18}	99	In	20d	MB	Pitter, 1968c
	95;69	In	20d	$COD;O_2$	
C_{10-20}	100	*CAS	3d	MB	Walther, 1966
Coco + tallow alc.	95	Alg	3d	RA	Klein, 1964a
Coco + tallow alc.	98;94	Pond	30d	RA;RS	Klein, 1964a, 65b
	51;61	ST	?	MB;RA	
	72;63	ST	?	MB;RA	
	92;100	ST + Soil	−	MB;RA	
	97;100	ST + Soil	−	MB;RA	
Coco + tallow alc.	99	Sew	40d	MB;RA	Klein, 1965a
	98	AnSew	40d	MB;RA	
Cetyl + oleyl	100	In	10-15d	MB	Čuta, 1964
Cetyl + oleyl (Syntapon CP)	98	In	20d	MB	Pitter, 1964c
Coco+oleyl+tallow	98	*CAS	2d	MB	De Jong, 1964, 65

TABLE 8.8   Linear Primary Alkyl Sulfates *(continued)*

Substrate (see § II)	Extent (see § III)	Method (see § IV)	Time (see § V)	Analysis (see § VI)	Reference
e. *Unspecified or Unknown Chain Length*					
Fatty alc. sulfate					
(Dreft)	0.5G	BOD	?	O_2	Rudolfs, 1949
Primary alkyl sulf.	96;+	RW;BOD	4;5d	$MB;O_2$	Degens, 1950
	96	Aqu	14d	MB	
Duponol C	45;57	Wa;BOD	6h;5d	O_2	Sawyer, 1956
Primary alkyl sulf.	+	In	?	B	Skinner, 1959
Linear primary	100;+	RW;BOD	2d;?	$MB;O_2$	Huyser, 1960
Lin. prim. un-					
saturated	+	BOD	?	O_2	Huyser, 1960
Linear primary	90†	RW	3-4d	σ	Steinle, 1964
	90†	RW	4-6d	MB	
	45-50	Wa	2d	O_2	
Fatty alc. (Zenit)	100	CAS	1d	MB	Pitter, 1966b
Alkyl sulfate	99,100	SF	1d	MB	Tomiyama, 1969

A. 9. *Other Primary Alkyl Sulfates*

TABLE 8.9

Biodegradation of Other Primary Alkyl Sulfates

Substrate (see § II)	Extent (see § III)	Method (see § IV)	Time (see § V)	Analysis (see § VI)	Reference
a. *Oxo Alcohols ex Linear Olefins*					
n-C$_{10-14}$ olefins	98	In	20d	MB	Pitter, 1963b, 64d
	85;86	In	20d	COD;O$_2$	
n-C$_{10-14}$ olefins	98;86	In	20d	MB;O$_2$	Pitter, 1964c
n-C$_{10-14}$ olefins	88	BAnD	32d	MB	Pitter, 1964b, d
n-C$_{10-14}$ olefins	98	In	20d	MB	Pitter, 1968c
	83-92	In	20d	COD	
	50-63	In	20d	O$_2$	
n-Olefins (G)	100	RW	15d	MB	Berger, 1964
n-Olefins (I)	87	RW	17d	MB	Berger, 1964
	95	CAS;RTF	3h;1d	MB	
n-Olefins[a] (H)	100	RW	11d	MB	Berger, 1964
n-Olefins[a] (J)	95	RTF	1d	MB	Berger, 1964
n-Olefins[a] (K)	86;95	CAS;RTF	3h;1d	MB	Berger, 1964
n-Olefins[a] (L)	87;95	CAS;RTF	3h;1d	MB	Berger, 1964
	91	RW	17d	MB	
n-Olefins[a] (M)	96	RTF	2d	MB	Berger, 1964
n-Olefins	90†	RW	5,9d	σ	Steinle, 1964
	90†	RW	7,10d	MB	
b. *Oxo Alcohols ex Polypropylenes, etc.*					
C$_{10}$ (ex tripropylene)	92	In	20d	MB	Pitter, 1964d
	40;47	In	20d	COD;O$_2$	
C$_{13}$ (ex tetrapropylene)	36;30	In	20d	MB;O$_2$	Pitter, 1963b, 64c,d
	55	BAnD	32d	MB	Pitter, 1964b,d
	50;25	In	20d	MB;O$_2$	Ruschenberg, 1963a
(N)	95	RTF	3d	MB	Berger, 1964
(O)	95;52	RTF;RW	3;30d	MB	Berger, 1964
	12	CAS	3h	MB	
	91	CAS	6h	RA	Sweeney, 1964a
	57-64	An	28d	MB	Ōba, 1967
	46;25	CAS	3h	MB;COD	Janicke, 1968c
	46	In	20d	MB	Pitter, 1968c
	37;26	In	20d	COD;O$_2$	Pitter, 1968c
C$_{13}$ (ex tributene)	6	An	7d	MB	Ōba, 1967
C$_{17}$ (ex tetraisobutene)	12;62	CAS;RTF	3h;3d	MB	Berger, 1964
C$_{20}$ (C$_{10}$[b] dimer)	81;94	CAS;RTF	3h;2d	MB	Berger, 1964
	55	RW	40d	MB	
c. *Branched, Cyclic, Unsaturated, or Substituted Alcohols*					
C$_8$ (2EtC$_6$)	98;69	In	20d	MB;O$_2$	Pitter, 1964c,d
C$_8$ (2,4,4Me$_3$C$_5$)	0;+	RW;BOD	18d;?	MB;O$_2$	Huyser, 1960
C$_9$ (3,5,5Me$_3$C$_6$)	4;7	RW	21d	MB;O$_2$	Hammerton, 1956
	0;+	RW;BOD	18d;?	MB;O$_2$	Huyser, 1960
C$_{9-10}$ (ex crotonaldehyde)	81;62	In	20d	MB;O$_2$	Pitter, 1964c,d

TABLE 8.9   Other Primary Alkyl Sulfates *(continued)*

Substrate (see § II)	Extent (see § III)	Method (see § IV)	Time (see § V)	Analysis (see § VI)	Reference
C_{12} (6CyC$_6$)	100;+	RW;BOD	7d;?	MB;O_2	Huyser, 1960
C_{14} (2AmC$_9$)	97	In	7d	MB	Crauland, 1964
C_{16} (Me$_2$C$_6$ dimer)	64;66	CAS;RTF	3h;3d	MB	Berger, 1964
C_{17} (8,9Cl$_2$C$_{17}$)(?)	80†	RW	3d	MB	Weil, 1964
C_{18} (9,10Cl$_2$C$_{18}$)	99;73	CAS;RW	3h;4d	MB	Cordon, 1964
	100	CAnD	30d	MB	Maurer, 1965
	100;85	In	9;16d	MB;COD	Cordon, 1968a
C_{18} (2ΣC$_{18}$)	80†	RW	4d	MB	Weil, 1964
C_{18} (3,5,5Me$_3$C$_6$ dimer)	0;0	RW;BOD	50d;?	MB;O_2	Huyser, 1960
	78	RTF	3d	MB	Berger, 1964
C_n (Linear unsaturated)	+	BOD	?	O_2	Huyser, 1960
C_n (3-*n*-alkyl-1-OH)	100;+	RW;BOD	14d;?	MB;O_2	Huyser, 1960
d. *Unspecified or Unknown Structure*					
Primary alcohol	90	BAnD	18d	MB	Bruce, 1966
Primary alcohol (odd + even)	90	BAnD	18d	MB	Bruce, 1966

[a]Includes slightly branched impurities.

[b]Oxo alcohol from tripropylene.

A. 10. *Secondary and Unknown Alkyl Sulfates*

TABLE 8.10

Biodegradation of Secondary and Unknown Alkyl Sulfates

Substrate (see § II)	Extent (see § III)	Method (see § IV)	Time (see § V)	Analysis (see § VI)	Reference
a. *Linear Secondary — Individual Isomers*					
C_x 2-sulfate	71	Wa	1d	O_2	Winter, 1962
C_{18} 9-sulfate	100	In	7d	MB	Crauland, 1964
b. *Linear Secondary — Mixed*					
C_{12}	95	BAS	3d	MB	Vaicum, 1967
C_{16}	100	RW	28d	MB;σ;F	Vath, 1964
C_{16}	100;95	In	4;5d	MB;COD	Livingston, 1965
C_{18}	100	In	7d	MB	Crauland, 1964
$C_{10.5}$ average	95	SF	2d	MB	Gebril, 1966
C_{10-13}	97;77	In	20d	MB;O_2	Pitter, 1963b, 64c,d
$C_{12.5}$ average	97	SF	2d	MB	Gebril, 1966
C_{11-15}	+	AnSew	14d	MB;F	Vath, 1964
$C_{14.5}$ average	98	SF	2d	MB	Gebril, 1966
C_{10-20}	90†	In	3d	MB	Prochazka, 1965
C_x	69	Wa	1d	O_2	Winter, 1962
C_x	90†	RW	7;8d	MB;σ	Steinle, 1964
	85-100	Wa	?	O_2	
Teepol	+	TF	—	?	Waddams, 1949
Teepol	96;+	RW;BOD	5d	MB;O_2	Degens, 1950
	96	Aqu	14d	MB	
Teepol X (C_{8-20})	94	TF	—	?	Goldthorpe, 1950
Teepol X	1.0G	BOD	5d	O_2	Goldthorpe, 1950
Teepol	99	*TF	—	MB	Hurley, 1950, 52
Teepol	100	*CAS	?	MB	Hurley, 1952
Teepol	95	CAnD	28d	MB	Hurley, 1952, p. 330
Teepol	98-100	CAS	3h	MB	Degens, 1955
Teepol	89	RW	4d	MB	Hammerton, 1955
Teepol	88	CAnD	28d	MB	Raybould, 1956
Teepol	100	TF	—	MB	Barden, 1957
	1.4F	Wa	6h	O_2	
	1.0G	Wa	7h	O_2	
Teepol	+	In	2d	F;σ;B	Skinner, 1959
Teepol	32	BAS	3d	MB	Vaicum, 1967
c. *Nonlinear Secondary*					
C_{17} (3,9Et$_2$C$_{13}$-6-OΣ)[a]	37;28	RW	47;28d	MB	Hammerton, 1955, 56
C_{24} (4[(C$_5$)$_3$C] Cy-1-OΣ)	0	In	59d	MB	Huyser, 1960
C_x (4-*t*-alkyl-Cy-1-OΣ)	10;0	RW;BOD	59d;?	MB;O_2	Huyser, 1960
d. *Unspecified or Unknown Structure*					
Alkyl sulfate	43-58	BAS	3h	MB	Sierp, 1954
Fit[b]	55	Wa	1d	O_2	Winter, 1962
C_{13} alcohol	40	In	30d	O_2	Heinz, 1967
iC_{13} alcohol	64;40	In	30d	MB;O_2	Heinz, 1967

TABLE 8.10   Secondary and Unknown Alkyl Sulfates *(continued)*

Substrate (see § II)	Extent (see § III)	Method (see § IV)	Time (see § V)	Analysis (see § VI)	Reference
Alkyl sulfate, comml.	96-100	CAS	3h	MB	Ilişescu, 1966
	95-98	In	7d	MB	
Alkyl sulfate, comml.	90;98	CAS	6;3h	MB	Vaicum, 1967
	98	BAS	3d	MB	
	94	In	?	MB	
	100;92	BOD;Wa	5d;?	MB	
Secondary alkyl sulfate	35	BAnD	?	?	WPRL, 1967, p. 117

[a]Tergitol 7.
[b]Alkyl sulfate + alkane sulfonate.

A. 11. *Alcohol Ethoxylate Sulfates*

TABLE 8.11

Biodegradation of Alcohol Ethoxylate Sulfates

Substrate (see § II)	Extent (see § III)	Method (see § IV)	Time (see § V)	Analysis (see § VI)	Reference
a. *Linear Primary Alcohols*					
$C_{12}E_3$	100	In	7d	MB	Bunch, 1967a
Lauryl E_4 (Etoxon EPA)	100;90	In	20d	MB;O_2	Pitter, 1964c
Lauryl E_5 (natural)	65-100	In	20d	MB	Čuta, 1964
(synthetic)	60-100	In	20d	MB	
$C_{12}E_8$	96;86	In	15;30d	MB;O_2	Heinz, 1967
Lauryl E_x TEA salt					
(Drene)	+	In	10d	B;σ	Von Riesen, 1955
Lauryl E_x	>96	RTF	1d	MB	Berger, 1964
Lauryl E_x	100	RW	7d	MB	Knaggs, 1965
Coco E_3	53-67	An	28d	MB	Ōba, 1967
$C_{16}E_1$	93;93	In	4;3d	MB;COD	Cordon, 1968a
$C_{16}E_3$	80†	RW	2d	MB	Weil, 1964
$C_{16}E_x$	100	RW	28d	MB;σ;F	Vath, 1964
$C_{18}E_1$	89	In	13d	SO_4	Cordon, 1968b
$C_{12+14}E_3$ (Ziegler)	100	RW	4;10d	F;MB	Myerly, 1964, Vath, 1964
	95	Wa	36h	O_2	
$C_{12+14}E_x$	97	RTF	1d	MB	Berger, 1964
$C+O+S^aE_3$(Etoxon CO)	100;84	In	10;15d	MB;O_2	Pitter, 1964c
C_xE_3	90†	RW	4;6d	σ;MB	Steinle, 1964
	90	Wa	?	O_2	
C_xE_4	90†	RW	4;7d	σ;MB	Steinle, 1964
	65	Wa	?	O_2	
b. *Oxo Primary Alcohols ex Linear Olefins*					
$C_{12-14}E_3$	100	RW	8;14d	F;MB	Myerly, 1964
$C_{12-15}E_3$ (Dobanol 25)	98	In	28d	Wt	Borstlap, 1967a
C_xE_3	90†	RW	5;8d	σ;MB	Steinle, 1964
C_xE_4	90†	RW	6d	σ;MB	Steinle, 1964
$C_xE_x^{\ b}$ (J')	98	RTF	2d	MB	Berger, 1964
$C_xE_x^{\ b}$ (L')	90	RTF	1d	MB	Berger, 1964
c. *Oxo Primary Alcohols ex Tetrapropylene*					
tp-$C_{13}E_x$ (N')	88	RTF	5d	MB	Berger, 1964
tp-$C_{13}E_x$ (O')	87	RTF	5d	MB	Berger, 1964
d. *Other Primary Alcohols*					
$2AmC_9E_2$ (9)	98	In	7d	MB	Crauland, 1964
e. *Linear Secondary Alcohols*					
$C_{16}E_x$	100	RW	28d	MB;σ;F	Vath, 1964
C_{17}-$9E_2$ (10)	100	RW	7d	MB	Crauland, 1964
$C_{18}E_2$ (11)	84	RW	7d	MB	Crauland, 1964
$C_{11-13}E_3$	100	RW	14d	MB;σ;F	Vath, 1964

TABLE 8.11 Alcohol Ethoxylate Sulfates *(continued)*

Substrate (see § II)	Extent (see § III)	Method (see § IV)	Time (see § V)	Analysis (see § VI)	Reference
$C_{11-15}E_3$	100	RW	6;11d	MB;F	Myerly, 1964
	55	Wa	3d	O_2	
$C_{11-15}E_3$	+	AnSew	14d	MB;F	Vath, 1964
$C_{11-15}E_3$	91	*CAS	2d	MB	Conway, 1965
$C_{11-15}E_3$	98	RW	14d	MB;σ	Conway, 1966
	0.7G	BOD	5d	O_2	
$C_{11-15}E_3$	96-98	In	7d	MB	Bunch, 1967a
$C_{13-15}E_5$	50	Wa	32h	O_2	Vath, 1964
C_xE_3	90†	RW	6;7d	σ;MB	Steinle, 1964
	50	Wa	?	O_2	
C_xE_5	90†	RW	6;7d	σ;MB	Steinle, 1964
	55	Wa	?	O_2	

f. *Other, Unknown or Unspecified Alcohols*

$C_{13}E_4$	25	In	30d	O_2	Heinz, 1967
$iC_{13}E_2$	53	In	30d	O_2	Heinz, 1967
$C_{12-18}E_3$	0.9-1.6G	Wa	8h	O_2	Hartmann, 1967

[a]Cetyl + oleyl + stearyl alcohols.
[b]Includes slightly branched impurities.

A. 12. *Alkylphenol Ethoxylate Sulfates*

TABLE 8.12

Biodegradation of Alkylphenol Ethoxylate Sulfates

Substrate (see § II)	Extent (see § III)	Method (see § IV)	Time (see § V)	Analysis (see § VI)	Reference
a. *Linear Primary Alkylphenols*					
$C_{12}APE_6$	45	Wa	44h	O_2	Vath, 1964
C_xAPE_4	90†	RW	4d	MB	Steinle, 1964
	35	Wa	?	O_2	
C_xAPE_6	90†	RW	10d	MB	Steinle, 1964
b. *Linear Secondary Alkylphenols*					
ortho-$2C_9APE_x$	90	In	?	?	Smithson, 1966
ortho-$5C_9APE_x$	0	In	?	?	Smithson, 1966
para-$5C_9APE_x$	0-60	In	?	?	Smithson, 1966
ortho-$2C_{10}APE_4$	94;85	CAS;In	6h;9d	MB	Smithson, 1966
$C_{10}APE_{3.5}$(nonrandom)	90	In	7d	MB	Smithson, 1966
$C_{10}APE_4$(nonrandom)	95	SF	5d	MB	Swisher, 1966a
$C_{12}APE_6$	100	RW	8d	σ;F	Vath, 1964
	65;35	RW;Wa	21;2d	MB;O_2	
$C_{11-15}APE_5$	0-50	In	7d	MB	Bunch, 1967a
C_xAPE_4	90†	RW	14d	MB	Steinle, 1964
C_xAPE_6	90†	RW	6;21d	σ;MB	Steinle, 1964
	35	Wa	?	O_2	
C_xAPE_x	68;70	CAS;In	6h;11d	MB	Smithson, 1966
c. *Linear + Branched Alkylphenols*					
C_xAPE_4 0.95MG	90†	RW	2d	σ	Steinle, 1964
C_xAPE_6 0.95MG	53	Wa	2d	O_2	Steinle, 1964
C_xAPE_4 1.17MG	37	Wa	2d	O_2	Steinle, 1964
C_xAPE_6 1.17MG	90†	RW	2d	σ	Steinle, 1964
C_xAPE_5 1.32MG	22	Wa	2d	O_2	Steinle, 1964
C_xAPE_6 1.32MG	90†;17	RW;Wa	9;2d	σ;O_2	Steinle, 1964
C_xAPE_7 1.32MG	13	Wa	2d	O_2	Steinle, 1964
C_xAPE_6 1.39MG	90†	RW	3d	σ	Steinle, 1964
C_xAPE_4 1.50MG	90†	RW	15d	σ	Steinle, 1964
C_xAPE_7 1.50MG	90†	RW	19d	σ	Steinle, 1964
C_xAPE_4 1.60MG	90†;17	RW;Wa	18;2d	σ;O_2	Steinle, 1964
C_xAPE_6 1.76MG	90†;9	RW;Wa	12;2d	σ;O_2	Steinle, 1964
C_xAPE_4 1.81MG	90†	RW	38d	σ	Steinle, 1964
C_xAPE_6 1.97MG	90†	RW	45d	σ	Steinle, 1964
d. *Branched or Unspecified Alkylphenols*					
C_9APE_x	65	RTF	3d	MB	Berger, 1964
br-C_9APE_3	20	Wa	4d	O_2	Myerly, 1964
tp-C_9APE_4	10-15	Wa	2d	O_2	Vath, 1964
br-C_9APE_4	0-24	In	7d	MB	Bunch, 1967a
br-C_9APE_x	0	In	11d	MB	Smithson, 1966
tp-$C_{12}APE_6$	4	Wa	2d	O_2	Steinle, 1964
C_xAPE_x (Triton X200)	0	In	10d	σ	Von Riesen, 1955

A. 13. *Other Sulfated Products*

TABLE 8.13

Biodegradation of Other Sulfated Products

Substrate (see § II)	Extent (see § III)	Method (see § IV)	Time (see § V)	Analysis (see § VI)	Reference
Alkyl monoglyceride (Arctic Syntex M)	+	In	6d	B;σ	Von Riesen, 1955
n-C_{14}-D-gluconamide	80†	RW	4d	MB	Weil, 1964
Coco ethanolamide (Syntopal B)	99	In	20d	MB	Pitter, 1964c
Ester (Calsolene Oil HS)	0.8F	Wa	6h	O_2	Barden, 1957
$C_{16}P_1$ (1-hexadecoxy-2-propanol)	100;75	In	7;8d	MB;COD	Cordon, 1968a
$C_{16}P_3$	90;60	In	4;8d	MB;COD	Cordon, 1968a
$C_{16}B_1$ (1-hexadecoxy-2-butanol)	93-85	In	7;11d	MB;COD	Cordon, 1968a

A. 14. *Phosphates, Phosphonates, Carboxylates, Soaps*

TABLE 8.14

Biodegradation of Phosphates, Phosphonates, Carboxylates, Soaps

Substrate (see § II)	Extent (see § III)	Method (see § IV)	Time (see § V)	Analysis (see § VI)	Reference
Primary alkyl phosphate, TEA salt	66	Wa	1d	O_2	Winter, 1962
Didecyl phosphate	100	RW	10d	σ	Cooper, 1964
α-Phosphonopalmitate *i*Pr ester	80†	RW	4d	MB	Weil, 1964
Fatty acyl sarcoside	1.3-1.5G	BOD	5d	O_2	Offhaus, 1962
Fatty acyl sarcoside	+	In	10d	B	Forsberg, 1967b
N-Coco-β-aminopropionate	95-100	CAS	3h	σ	Eldib, 1965
N-Tallow-iminodipropionate	95-100	CAS	3h	σ	Eldib, 1965
Octanoic acid	20	Wa	12h	O_2	McKinney, 1956b
Octanoate	1.2G	Wa	20h	O_2	McKinney, 1959a
Na octanoate	40-60	Wa	?	O_2	Payne, 1963a
Lauric acid	1.3G	Wa	5d	O_2	Marion, 1963b
Lauric acid	48	Wa	5d	O_2	Marion, 1966
Lauric acid	6	Wa	1d	O_2	Malaney, 1969
Na laurate	58	Wa	5d	O_2	Marion, 1966
Myristic acid	0.9G	Wa	5d	O_2	Marion, 1963b
Myristic acid	4	Wa	1d	O_2	Malaney, 1969
Palmitic acid	0.4G	Wa	5d	O_2	Marion, 1963b
Palmitic acid	3	Wa	1d	O_2	Malaney, 1969
Na palmitate	43;55	Wa	1;2d	O_2	Winter, 1962
Na palmitate	84	In	20d	O_2	Pitter, 1964c
Stearic acid	0.2G	Wa	5d	O_2	Marion, 1963b
Stearic acid	1	Wa	1d	O_2	Malaney, 1969
Na stearate	1.3G	BOD	5d	O_2	Goldthorpe, 1950
Na stearate	17;53	Wa;BOD	6h;5d	O_2	Bogan, 1955
Na stearate	0	In	?	B	Pipes, 1963b
Na oleate	1.3G	BOD	5d	O_2	Goldthorpe, 1950
Na oleate	23;64	Wa;BOD	6h;5d	O_2	Bogan, 1955
Na oleate	0.3F	Wa	6h	O_2	Barden, 1957
Na oleate	92	In	9d	COD	Konecky, 1963
Na oleate	+	In	?	B	Pipes, 1963b
Soap (Ivory Snow)	1.4G	BOD	?	O_2	Rudolfs, 1949
Soap (Lux Flakes)	1.5G	BOD	?	O_2	Rudolfs, 1949

A. 15. *MBAS in Sewage*

TABLE 8.15

Biodegradation of Sewage MBAS

Substrate (see § II)	Extent (see § III)	Method (see § IV)	Time (see § V)	Analysis (see § VI)	Reference
MBAS	67-89	*TF	–	MB	Evans, 1949
	55-71	*CAS	?	MB	
	16-58	*?	?	?	Husmann, 1956
	22-31	*CAS	?	MB	Lockett, 1956
	21;68	*CAS	5;12h	MB	Roberts, 1957
	39;56	*TF	–	MB	
	71-83	*TF	–	MB	Smith, 1957
	43	*CAS	6h	MB	Burnop, 1960
	24(0-35)	*TF	–	MB	Jendreyko, 1962[a]
	82	*CAS	2d	MB	Scherb, 1962
	72	*CAS	2d	MB	
	23	*CAS	2d	MB	
	4	*CAS	2d	MB	
	45-52	*TF	–	MB	
	26	*TF	–	MB	Jendreyko, 1963[a]
	24	*CAS	1h	MB	
	30	*CAS	6h	MB	Phillips, 1963
	5-34	*CAS	?	MB	Spohn, 1964a
	20-42	*CAS+TF	?	MB	
	1-19	*TF	–	MB	Spohn, 1964b
	5-52	*CAS	1-2h	MB	
	25-44	*CAS+TF	1-2h+	MB	
	56	*CAS	2d	MB	Conway, 1965
MBAS(LAS)	95-98	*CAS	2,3d	MB	Lashen, 1967b,c
MBAS	49	*CAS	5h	MB	Merrell, 1967
	82	*CAS+[b]	–	MB	
	60-90	*CAS	2d	MB	Huber, 1968
	22-84	*CAS	1.5d	MB	
	6	*CAS	1d	MB	
	57	*TF	–	MB	Krone, 1968
	80	Soil	–	MB	Popkin, 1968
	69	CAS	1d	MB	Köhler, 1969
	78	CAS+TF	1d+	MB	
MBAS[c], hydrolyzable	95-98	CAS	1d	MB	Köhler, 1969
	90-98	CAS+TF	1d+	MB	
MBAS[c], nonhydrolyzable	53-78	CAS	1d	MB	Köhler, 1969
	69-90	CAS+TF	1d+	MB	
MBAS[c], total	69-87	CAS	1d	MB	Köhler, 1969
	77-93	CAS+TF	1d+	MB	

[a]See also Husmann (1963a, p. 48-67).

[b]CAS + pond + soil.

[c]Spiked at times with mixture of FAS (alkyl sulfate):ABS:Mersolat D (SAS) in ratio 4:1:0.5.

B. Nonionic Surfactants

B. 1. *Linear Primary Alcohol Ethoxylates*

TABLE 8.16

Biodegradation of LPAE

Substrate (see § II)	Extent (see § III)	Method (see § IV)	Time (see § V)	Analysis (see § VI)	Reference
a. *Individual Chain Lengths*					
C_8E_4	100	RW	7d	PM	Huyser, 1960
C_8E_5	100	CAS	4h	CT	Huddleston, 1964b
	100	SF;RW	3;8d	CT	
C_8E_{14}	84	RW	34d	PM	Huyser, 1960
C_8E_{25}	35	RW	34d	PM	Huyser, 1960
$C_{10}E_7$	97	RW	6d	σ	Blankenship, 1963
	100	CAS	4h	CT	Huddleston, 1964b
	100	SF;RW	3;5d	CT	
Lauryl E_3	1.5G	Wa	5d	O_2	Garrison, 1964
	66	Wa	10d	O_2	Hunter, 1964
	75;78	Wa	10d	O_2	Barbaro, 1965
Lauryl E_4(Slovasol S)	100	CAS	10h	PW	Pitter, 1963a
	84	In	20d	O_2	Pitter, 1964c
Lauryl E_4	97	In	30d	O_2	Heinz, 1967
$C_{12}E_{4.4}$	100	CAS	4h	CT	Huddleston, 1964b
	100	SF;RW	3;4d	CT	
Lauryl E_5	25-100	In	20d	HgI	Čuta, 1964
$C_{12}E_6$	99	RW	6d	σ	Blankenship, 1963
	100	CAS	4h	CT	Huddleston, 1964b
	100	SF;RW	3;4d	CT	
	76	In	30d	O_2	Heinz, 1967
$C_{12}E_8$	96-99	RW	5-8d	σ	Blankenship, 1963
	48[a]	Wa	>3h	O_2	
	72-82	In	30d	O_2	Heinz, 1967
Lauryl E_9	56	Wa	10d	O_2	Hunter, 1964
$C_{12}E_9$	100	CAS	4h	CT	Huddleston, 1964b
	100	SF;RW	2;3d	CT	
Lauryl E_9	100;72	RW;Wa	12;15d	$\sigma;O_2$	Knaggs, 1964
	100	In	28d	Wt	Borstlap, 1967a
$C_{12}E_9$	98-100	In	7d	CT	Bunch, 1967a
$C_{12}E_{10}$	99	RW	6d	σ	Blankenship, 1963
	73	In	30d	O_2	Heinz, 1967
$C_{12}E_{16}$	62	In	30d	O_2	Heinz, 1967
$C_{12}E_{20}$	99	RW	23d	σ	Blankenship, 1963
	41	In	30d	O_2	Heinz, 1967
$C_{12}E_{30}$	95	RW	20d	σ	Blankenship, 1963
	3	In	30d	O_2	Heinz, 1967
$C_{12}E_x$ (Lissapol DS4429)	100	RTF	1d	CT;TLC	Jenkins, 1967
$C_{14}E_8$	100	RW	5d	IR	Frazee, 1964b
	95-100	In	7d	CT	Bunch, 1967a
$C_{14}E_{9.5}$	97	RW	6d	σ	Blankenship, 1963
Alfol 16 E_5	65	In	20d	O_2	Ruschenberg, 1963a
Alfol 16 E_{10}	53	In	20d	O_2	Ruschenberg, 1963a

TABLE 8.16   Linear Primary Alcohol Ethoxylates *(continued)*

Substrate (see § II)	Extent (see § III)	Method (see § IV)	Time (see § V)	Analysis (see § VI)	Reference
$C_{16}E_{10}$	100	CAS	4h	CT	Huddleston, 1964b
	100	SF;RW	2;5d	CT	
	95	RW	2d	σ;F	Weil, 1964
$C_{16}E_{10.4}$	97	RW	7d	σ	Blankenship, 1963
Alfol 16 E_{20}	33	In	20d	O_2	Ruschenberg, 1963a
$C_{16}E_{20}$	95	RW	2;>25d	σ;F	Weil, 1964
$C_{16}E_x$	97-98	RW	28d	CT;σ;F	Vath, 1964
$C_{18}E_4$	100	RW	27d	PM	Huyser, 1960
$C_{18}E_6$	100;80	RW	2;3d	F;σ	Weil, 1964
$C_{18}E_8$	100	RW	27d	PM	Huyser, 1960
Stearyl E_8	100	In	10d	TLC	Patterson, 1967
$C_{18}E_{10.5}$	100	CAS	4h	CT	Huddleston, 1964b
	100	SF;RW	2;13d	CT	
$C_{18}E_{14}$	56	RW	34d	PM	Huyser, 1960

b. *Mixed or Unknown Chain Lengths*

Substrate (see § II)	Extent (see § III)	Method (see § IV)	Time (see § V)	Analysis (see § VI)	Reference
$C_{10-12}E_6$ (Ziegler)	100	SF	7d	CT;σ;F	Huddleston, 1965b
	100	RW	2d	CT;σ;F	
	91	RW	2d	Wt	
	100	BAS	1d	CT;σ	
	95;93	BAS	1d	F;Wt	
C_{10-12} +58%E	100	CAS	4h	CT	Huddleston, 1964b
	100	SF;RW	2;5d	CT	
C_{10-12} +58%E	100	In	3d	TLC	Patterson, 1967
C_{10-12} +60%E (Alfonic 1012-6)	100	CAS;SF	4h;8d	CT	Huddleston, 1966
	100	RW	11d	CT;σ;F	
	0.7G	BOD	5d	O_2	
$C_{10-16}E_3$	90	In	20d	PW	Pitter, 1968a,b
	93;66	In	20d	COD;O_2	
$C_{10-16}E_5$	75	In	20d	PW	Pitter, 1968a,b
	84;62	In	20d	COD;O_2	
$C_{10-16}E_8$	65	In	20d	PW	Pitter, 1968a,b
	73;52	In	20d	COD;O_2	
$C_{10-16}E_{10}$	50	In	20d	PW	Pitter, 1968a,b
	64;48	In	20d	COD;O_2	
$C_{10-16}E_{14}$	48;34	In	20d	COD;O_2	Pitter, 1968a,b
$C_{10-16}E_{15}$	33	In	20d	PW	Pitter, 1968a,b
	41;32	In	20d	COD;O_2	
$C_{10-16}E_{18}$	36;23	In	20d	COD;O_2	Pitter, 1968a,b
$C_{10-16}E_{20}$	8	In	20d	PW	Pitter, 1968a,b
	27;18	In	20d	COD;O_2	
Lauryl+myristyl E_6	100	In	28d	Wt	Borstlap, 1967a
$C_{12-14}E_{7.4}$ (Ziegler)	100	RW	16d	CT;σ;F	Vath, 1964
	50	Wa	30h	O_2	

TABLE 8.16   Linear Primary Alcohol Ethoxylates *(continued)*

Substrate (see § II)	Extent (see § III)	Method (see § IV)	Time (see § V)	Analysis (see § VI)	Reference
$C_{12-14}E_9$	97;50	RW;Wa	6;3d	$CT;O_2$	Myerly, 1964
Lauryl+myristyl E_9	100	In	28d	Wt	Borstlap, 1967a
$C_{12-14}E_{10}$ (coco)	100	BAS;RW	1;14d	F;CT	SDA, 1969a,b
	100	SF	14;7d	F;CT	
	99	BAS	17h	F	
Lauryl+myristyl E_{12}	94	In	28d	Wt	Borstlap, 1967a
Lauryl+myristyl E_{15}	89	In	28d	Wt	Borstlap, 1967a
Lauryl+myristyl E_{18}	81	In	28d	Wt	Borstlap, 1967a
Lauryl+myristyl E_{30}	64	In	28d	Wt	Borstlap, 1967a
$C_{12-15}E_9$ (slight branch)	100	In	24d	TLC	Patterson, 1967
$C_{12-16}E_5+C_{12-16}E_{14}$ (synthetic)	80-90	In	20d	HgI	Čuta, 1964
$C_{12-16}E_{13}$ (synthetic)	100	In	20d	HgI	Čuta, 1964
$C_{12-18}E_8$ (Ziegler)	99-100	In	9d	$CT;\sigma;F$	Garrison, 1964
	98-100	SF	7d	$CT;\sigma;F$	
	0.9G	Wa	5d	O_2	
$C_{12-18}E_9$	0.8-1.3G	Wa	8h	O_2	Hartmann, 1967
C_{12-18} +60%E (Alfonic 1218-6)	100	CAS	4h	CT	Huddleston, 1964a
C_{12-18} +62%E	100	CAS	4h	CT	Huddleston, 1964b
	100	SF;RW	2;5d	CT	
C_{12-18} +62%E	100	In	4d	TLC	Patterson, 1967
$C_{14-16}E_8$ (Ziegler)	100	BAS	1d	SMB	Han, 1967
$C_{14-16}E_{16}$ (Ziegler)	86	BAS	1d	SMB	Han, 1967
$C_{16-18}E_5$	91;87	In	20d	PW;COD	Pitter, 1968a
$C_{16-18}E_6$	100	In	7d	TLC	Patterson, 1967
$C_{16-18}E_9$	100	In	5d	TLC	LGC, 1966, p. 142
$C_{16-18}E_9$	100	In	7d	TLC	Patterson, 1967
$C_{16-18}E_{10}$	74;70	In	20d	PW;COD	Pitter, 1968a
$C_{16-18}E_{15}$	100	In	7d	TLC	Patterson, 1967
$C_{16-18}E_{15}$	36;46	In	20d	PW;COD	Pitter, 1968a
$C_{16-18}E_{20}$	100	In	8d	TLC	LGC, 1966, p. 142
$C_{16-18}E_{20}$	100	In	9d	TLC	Patterson, 1967
$C_{16-18}E_{20}$	7;32	In	20d	PW;COD	Pitter, 1968a
$C_{16-18}E_{22}$	98	In	28d	TLC	Patterson, 1967
$C_{16-18}E_{25}$	0;13	In	20d	PW;COD	Pitter, 1968a
$C_{16-18}E_{30}$	100	In	22d	TLC	Patterson, 1967
C_{16-18} +63%E	100	CAS	4h	CT	Huddleston, 1964b
	100	SF;RW	2;5d	CT	
Tallow alcohol E_{40}	12	BAS	1d	SMB	Han, 1967
$C+O+S^b E_3$ (Slovasol CO)	88	In	20d	O_2	Pitter, 1964c
$C+O^b E_8$	60-85	In	20d	HgI	Čuta, 1964
$C+O^b E_{20}$ (Slovasol O)	0	CAS	10h	PW	Pitter, 1963a
$C+O^b E_{20}$	80	In	20d	HgI	Čuta, 1964

TABLE 8.16 Linear Primary Alcohol Ethoxylates *(continued)*

Substrate (see § II)	Extent (see § III)	Method (see § IV)	Time (see § V)	Analysis (see § VI)	Reference
$C_x E_{3-6}$	80-96	In	20d	COD	Pitter, 1968c
	61-88	In	20d	O_2	
$C_x E_8$	90†	RW	4;5d	CT;σ	Steinle, 1964
	50	Wa	2d	O_2	
$C_x E_{8-9}$	0.7-1.4G	Wa	?	O_2	Hartmann, 1967
$C_x E_9$	90†	RW	6;4d	CT;σ	Steinle, 1964
	60	Wa	2d	O_2	
$C_x E_9$	27-133	Wa	?	O_2	Hartmann, 1967
$C_x E_9$ (Empilan KM9)	98	Sew	4d	TLC	Patterson, 1966b
$C_x E_9$ (Empilan KM9)	98-99	In	21d	?	Truesdale, 1968
	98-99	BAS;RTF	1;7d	?	
$C_x E_{20}$	100	Sew	14d	TLC	Patterson, 1966b
$C_x E_{20}$ (Empilan KM20)	96-99	In;RTF	21;7d	?	Truesdale, 1968
$C_x E_{20-30}$	16-50	In	20d	COD	Pitter, 1968c
	12-27	In	20d	O_2	
$C_x E_x$	100	In	7d	?	Smithson, 1966

[a] Average asymptotic value.

[b] C+O+S = cetyl + oleyl + stearyl alcohols.

B. 2. *Other Alcohol Ethoxylates*

TABLE 8.17

Biodegradation of Other Alcohol Ethoxylates

Substrate (see § II)	Extent (see § III)	Method (see § IV)	Time (see § V)	Analysis (see § VI)	Reference
a. *Oxo Primary Alcohols ex Linear Olefins*					
$C_{11-14}E_3{}^a$	100	BAS	1d	SMB	Han, 1967
$C_{11-14}E_{13}{}^a$	92	BAS	1d	SMB	Han, 1967
$C_{12-14}E_9$	99	RW	10d	CT	Myerly, 1964
$C_{12-15}E_6{}^b$	100	In	28d	Wt	Borstlap, 1967a
$C_{12-15}E_9{}^b$	95	In	28d	Wt	Borstlap, 1967a
$C_{12-15}E_9$	99	BAS	17h	F	SDA, 1969a,b
$C_{12-15}E_{12}{}^b$	95	In	28d	Wt	Borstlap, 1967a
$C_{12-15}E_{15}{}^b$	89	In	28d	Wt	Borstlap, 1967a
$C_{12-15}E_{18}{}^b$	77	In	28d	Wt	Borstlap, 1967a
$C_{12-15}E_{30}{}^b$	68	In	28d	Wt	Borstlap, 1967a
C_xE_8	90†	RW	8;6d	CT;σ	Steinle, 1964
C_xE_9	90†	RW	6;7d	CT;σ	Steinle, 1964
b. *Oxo Primary Alcohols ex Tetrapropylene*					
$tp\text{-}C_{13}E_8$	70	In	49d	TLC	Patterson, 1967
$tp\text{-}C_{13}E_9$	11;36;0	SF	7d	CT;σ;F	Garrison, 1964
	15	In	30d	O_2	Heinz, 1967
	0-10	In	7d	CT	Bunch, 1967a
	75;63	BAS	1d	F;CT	SDA, 1969a,b
	91;100	RW	28d	F;CT	SDA, 1969a,b
	0;31	SF	14d	F;CT	SDA, 1969a,b
$C_{13}E_{14.5}$	33;86	SF;RW	4;26d	CT	Huddleston,1964b
c. *Linear Secondary Alcohol Ethoxylates*					
$C_{12}\text{-}4E_8$	99	RW	10d	σ	Blankenship, 1963
$C_{12}\text{-}6E_8$	70	RW	13d	σ	Blankenship, 1963
$s\text{-}C_{14}E_9,E_{12}$	95	CAS	3h	BiI	Wickbold, 1966
$s\text{-}C_{16}E_x$	93-98	RW	28d	CT;σ;F	Vath, 1964
$s\text{-}C_{11-13}E_6$	39;26	Wa	?	O_2;RC	Vath, 1964
$s\text{-}C_{11-13}E_8$	100	RW	17d	CT;σ;F	Vath, 1964
$s\text{-}C_{11-13}E_9$	95	BAS	1d	CT	Booman, 1965
$s\text{-}C_{11-15}E_7$	98	RW	14d	CT;σ	Conway, 1966
	65	TF	−	CT	
	0.6G	BOD	5d	O_2	
$s\text{-}C_{11-15}E_8$	92	BAS	1d	SMB	Han, 1967
$s\text{-}C_{11-15}E_9$	97;55	RW;Wa	6;3d	CT;O_2	Myerly, 1964
$s\text{-}C_{11-15}E_9$	+	AnSew	14d	CT;F	Vath, 1964
$s\text{-}C_{11-15}E_9$(Tergitol 15-S-9)	93	*CAS	2d	CT	Conway, 1965
$s\text{-}C_{11-15}E_9$(Tergitol 15-S-9)	98	RW	21d	CT;σ	Conway, 1966
	100	RW;AnRW	30d	CT;F	
	90	CAS	8h	CT;F	
	99	SF	8d	CT	
	93;94	SF	8d	σ;F	

TABLE 8.17   Other Alcohol Ethoxylates *(continued)*

Substrate (see § II)	Extent (see § III)	Method (see § IV)	Time (see § V)	Analysis (see § VI)	Reference
s-C$_{11-15}$E$_9$(Tergitol 15-S-9)	65	TF	—	CT	Conway, 1966
	0.4G	BOD	5d	O$_2$	
	+	In	?	CO$_2$	
s-C$_{11-15}$E$_9$(Tergitol 15-S-9)	92;95	CAS	3;6h	CT	Lashen, 1966
	82;81	CAS	3;6h	F	
s-C$_{11-15}$E$_9$	95	RW	21d	F	Booman, 1967
s-C$_{11-15}$E$_9$	100	In	7d	TLC	Patterson, 1967
s-C$_{11-15}$E$_9$	85-96	BAS	1d	F	SDA, 1969a,b
	93;94	·BAS	17h;1d	F;CT	
	98;100	RW	28d	F;CT	
	62;93	SF	14d	F;CT	
	85	In	7d	F	
s-C$_{11-15}$E$_{12}$	83-90	CAS	8h	CT;F	Conway, 1966
s-C$_{11-15}$E$_{13}$	70	RW	21d	F	Booman, 1967
s-C$_{11-15}$E$_{13}$	100	In	20d	TLC	Patterson, 1967
s-C$_{11-15}$E$_{15}$	+	In	?	CO$_2$	Conway, 1966
s-C$_{11-15}$E$_{15}$	57	BAS	1d	SMB	Han, 1967
s-C$_{11-15}$E$_{17}$	40	RW	21d	F	Booman, 1967
s-C$_{13-15}$E$_6$	39;26	Wa	?	O$_2$;RC	Vath, 1964
s-C$_{13-15}$E$_{9.5}$	55	Wa	30h	O$_2$	Vath, 1964
s-C$_x$E$_6$	50	Wa	?	O$_2$	Steinle, 1964
s-C$_x$E$_8$	90†	RW	6d	CT	Steinle, 1964
s-C$_x$E$_9$	35	Wa	?	O$_2$	Steinle, 1964
s-C$_x$E$_{10}$	90†	RW	6d	CT	Steinle, 1964
s-C$_x$E$_{11}$	90†	RW	6d	CT;σ	Steinle, 1964

d. *Unspecified or Unknown Structure*

Substrate (see § II)	Extent (see § III)	Method (see § IV)	Time (see § V)	Analysis (see § VI)	Reference
C$_{13}$E$_9$(Merpoxen TP90)	31;68	TF;CAS	—;3h	HgI	Schönborn, 1966, p. 117-9
C$_x$E$_7$	50-95	Alg	21d	CT	Davis, 1967
	100	Sew;BAS	18;21d	CT	
	100	Pond	14d	CT	
C$_x$E$_{7.4}$	55-100	Alg	21d	CT	Davis, 1967
	0-96	Alg	14d	IR	
	100	Sew;BAS	10d	CT	
	100	Pond	14d	CT	
	88-96	Pond	14d	IR	

[a]ex Dobanol 277.

[b]ex Dobanol 25.

B. 3. *Linear Alkylphenol Ethoxylates*

TABLE 8.18
Biodegradation of Linear Alkylphenol Ethoxylates

Substrate (see § II)	Extent (see § III)	Method (see § IV)	Time (see § V)	Analysis (see § VI)	Reference
a. *Linear Primary Alkylphenols*					
ortho-C_8APE_9	95	RW	10d	σ	Blankenship, 1963
para-C_8APE_9	98	RW	8d	σ	Blankenship, 1963
$C_{12}APE_{12}$	7	Wa	20h	O_2	Vath, 1964
C_xAPE_{10}	5	Wa	?	O_2	Steinle, 1964
C_xAPE_{12}	5	Wa	?	O_2	Steinle, 1964
b. *Linear Secondary Alkylphenols*					
s-$C_{7-9}APE_8$	95-100	SF	7d	CT	Booman, 1965
s-C_8APE_7	75	CAS	4h	CT	Huddleston, 1964b
	40;54	SF;RW	5;26d	CT	
s-C_8APE_9	71	In	28d	Wt	Borstlap, 1967a
s-C_8APE_{10}	50	RW	17d	σ	Blankenship, 1963
s-C_8APE_{15}	51	In	28d	Wt	Borstlap, 1967a
s-C_9APE_9	57;66	In	9d	CT;σ	Garrison, 1964
	75	In	9d	F	
	62;60	SF	7d	CT;σ	
	0-50	SF	7d	F	
	0.1G	Wa	5d	O_2	
s-C_9APE_9	88	CAS	4h	CT	Huddleston, 1964b
	65	SF;RW	5;26d	CT	
s-C_9APE_9	42;45	SF	7d	CT;σ	Huddleston, 1965b
	20	SF	7d	F	
	94;80	RW	20d	CT;σ	
	90;46	RW	20d	F;Wt	
	99;85	BAS	1d	CT;σ	
	75;60	BAS	1d	F;Wt	
s-C_9APE_{10}	90	RW	36d	IR	Osburn, 1966
s-C_9APE_{11}	93	RW	36d	IR	Osburn, 1966
s-C_9APE_{15} (Tergitol NP35)	26;66	TF	—	HgI[a]	Schönborn, 1966, p. 117,9
	77;75	CAS	3h	HgI[b];PC	Schönborn, 1966, p. 120
s-C_9APE_x (Igepal L0630)	88;44	CAS;SF	4h;8d	CT	Huddleston, 1966
	40	RW	11d	CT	
	35;40	RW	11d	σ;F	
	0	BOD	5d	O_2	
s-C_9APE_x	65;94	BAS	1d	F;CT	SDA, 1969a,b
	92;100	RW	28d	F;CT	
	0;68	SF	14d	F;CT	
s-$C_{9-10}APE_{10.4}$ (nonrandom)	85;99	BAS	1d	F;CT	SDA, 1969a,b
	83	BAS	17h	F	
	99;100	RW	28d	F;CT	

TABLE 8.18   Linear Alkylphenol Ethoxylates *(continued)*

Substrate (see § II)	Extent (see § III)	Method (see § IV)	Time (see § V)	Analysis (see § VI)	Reference
$s\text{-}C_{9-10}APE_{10.4}$ (nonrandom)	6;87	SF	14d	F;CT	SDA, 1969a,b
	75	In	7d	F	
ortho-2-$C_{10}APE_{9.5}$	100;95	CAS;In	6h;7d	CT	Smithson, 1966
$s\text{-}C_{10}APE_{8.5}$ (nonrandom)	91;95	CAS;In	6h;7d	CT	Smithson, 1966
$s\text{-}C_{10}APE_{10}$	100	CAS	4h	CT	Huddleston, 1964b
	92;96	SF;RW	5;26d	CT	
$s\text{-}C_{12}APE_{11}$	60	RW	36d	IR	Osburn, 1966
$s\text{-}C_{12}APE_{12}$	100	RW	28d	CT	Vath, 1964
	60;40	RW	28d	σ;F	
	15	Wa	2d	O_2	
	+	AnSew	14d	CT;F	
$s\text{-}C_{12}APE_{12}$	58;60	SF	7d	CT;σ	Huddleston, 1965b
	30	SF	7d	F	
	93;80	RW	20d	CT;σ	
	70;25	RW	20d	F;Wt	
	90;65	BAS	1d	CT;σ	
	50;5	BAS	1d	F;Wt	
$s\text{-}C_{12}APE_{17}$	100	CAS	4h	CT	Huddleston, 1964b
	90;91	SF;RW	7;26d	CT	
$s\text{-}C_{14}APE_{15}$	93	RW	26d	CT	
$s\text{-}C_xAPE_9$	60-80	In	28d	TLC	Patterson, 1968
$s\text{-}C_xAPE_{11}$	97;50	RW	28d	CT;σ	Conway, 1966
	77	CAS	8h	CT;F	
	75;48	SF	8d	CT;σ	
	0	SF	8d	F	
	0.1G	BOD	5d	O_2	
	+	In	?	CO_2	
$s\text{-}C_xAPE_{12}$	90†	RW	21;16d	CT;σ	Steinle, 1964
	15	Wa	?	O_2	
$s\text{-}C_xAPE_x$	63;60	CAS;In	6h;12d	CT	Smithson, 1966

[a]Prepurification by resin adsorption.

[b]Prepurification by butanone extraction.

B. 4. *Branched or Unspecified Alkylphenol Ethoxylates*

TABLE 8.19

Biodegradation of Branched or Unspecified Alkylphenol Ethoxylates

Substrate (see § II)	Extent (see § III)	Method (see § IV)	Time (see § V)	Analysis (see § VI)	Reference
a. *Octyl Phenols*					
t-C_8APE_5 (OPE$_5$)	7;16	Wa;BOD	6h;5d	O_2	Bogan, 1955, Sawyer, 1956
i-C_8APE_{6-7}	10,58	Wa	8h	O_2	Hartmann, 1967
t-$C_8APE_{7.5}$	100	SF	5-7d	CT	Booman, 1965
C_8APE_8 (Leuna JR51)	17	Wa	1d	O_2	Winter, 1962
t-C_8APE_9 (Triton X100)	3-4	Wa	10d	O_2	Barbaro, 1965
t-C_8APE_9 (Triton X100)	87;56	RW	35d	CT;IR	Osburn, 1966
br-C_8APE_9	0	In	7d	CT	Bunch, 1967a
br-C_8APE_9	46	In	28d	Wt	Borstlap, 1967a
t-C_8APE_{10}	97-100	SF	4-5d	CT	Booman, 1965
	92-97	CAS	6h	CT	
	50	ST	2d	CT	
i-C_8APE_{10}	8;10	SF	7d	CT;σ	Huddleston, 1965b
	25	SF	7d	F	
	95;20	RW	20d	CT;σ	
	80;53	RW	20d	F;Wt	
	97;50	BAS	1d	CT;σ	
	75;40	BAS	1d	F;Wt	
t-C_8APE_{10}	62;25	RW	28d	CT;σ	Conway, 1966
	55	CAS	8h	CT;F	
	15;11	SF	7d	CT;σ	
	0	SF	7d	F	
	0.06G	BOD	5d	O_2	
	+	In	?	CO_2	
t-C_8APE_{10} (Triton X100)	30†	RW	11d	CT	Huddleston, 1966
	0†	RW	11d	σ;F	
	10;0	SF;BOD	8;5d	CT;O_2	
t-C_8APE_{10} (Triton X100)	98;65	BAS	1d	CT;RC	Lashen, 1966
	90-100	SF	7d	CT	
	90;95	CAS	3;6h	CT;F	
	65	CAS	6h	RC	
	23-29	BOD	7d	O_2	
	58;63	ST	3d	CT;F	
	7	ST	3d	RC	
	93;84	ST+Soil	3d+	CT;F	
	46	ST+Soil	3d+	RC	
t-C_8APE_{10} (Triton X100)	90-95	RW	4-24d	F	Lashen, 1967a,b
t-C_8APE_{10} (Triton X100)	95-100	*CAS	2,3d	CT;TLC	Lashen, 1967b,c
	90-95	*CAS	2,3d	σ;F	
t-C_8APE_{10} (Triton X100)	0;95	BAS	1d	F;CT	SDA, 1969a,b
	<60	BAS	17h	F	
	80;100	RW	28d	F;CT	
	0;44	SF	14d	F;CT	

TABLE 8.19   Branched or Unspecified APEs *(continued)*

Substrate (see § II)	Extent (see § III)	Method (see § IV)	Time (see § V)	Analysis (see § VI)	Reference
t-C_8APE_{10} (Triton X100)	<60	In	7d	F	SDA, 1969a,b
C_8APE_{11} (Nonidet P80)	80	?	?	?	WPRL, 1966b
t-$C_8APE_{12.5}$	100	SF	4d	CT	Booman, 1965
br-C_8APE_{15}	49	In	28d	Wt	Borstlap, 1967a
t-C_8APE_{16}	30	RW	34d	PM	Huyser, 1960
C_8APE_x	80	TF	−	TLC	WPRL, 1965, p. 127
C_8APE_x	61	TF	−	TLC	WPRL, 1966a, p. 131
	50;47	In;BAS	21;1d	TLC	
C_8APE_x (Nonidet P80)	100	RTF;In	7d	CT	Jenkins, 1967
b. *Nonyl Phenols*					
tp-C_9APE_4	58	RW	34d	IR	Frazee, 1964b
C_9APE_4	44	BAS	1d	SMB	Han, 1967
br-C_9APE_4 (British)	50-80	In	42d	TLC	Patterson, 1968
br-C_9APE_4	53;31	In	20d	PW;COD	Pitter, 1968a
	14	In	20d	O_2	
C_9APE_6 (Merpoxen N060)	44;70	TF;CAS	−;3h	HgI[a]	Schönborn, 1966, p. 117,9
	83;90	CAS	3h	HgI[b];PC	Schönborn, 1966, p. 120
tp-C_9APE_7	24;14	CAS	3h	C;COD	Janicke, 1968a
	2	BOD	20d	O_2	
tp-C_9APE_7	28	CAS	3h	HgI	Janicke, 1968c
	30-50	CAS	3h	COD	
br-C_9APE_8	29;16	In	20d	PW;COD	Pitter, 1968a
	11;0	In	20d	O_2;UV	
br-C_9APE_9	33;10	In;SF	9;7d	CT	Garrison, 1964
	32;18	In;SF	9;7d	σ	
	0	In;SF	9;7d	F	
	0	Wa	5d	O_2	
tp-C_9APE_9	55	CAS	4h	CT	Huddleston, 1964a,b
tp-C_9APE_9	30;54	SF;RW	5;26d	CT	Huddleston, 1964b
br-C_9APE_9	10	Wa	3d	O_2	Myerly, 1964
br-C_9APE_9 (British)	50-70	In	42d	TLC	Patterson, 1968
	40-100	In	35d	TLC	
$C_9APE_{9.5}$ (Merpoxen N095)	89;85	CAS	3h	HgI;PC	Schönborn, 1966, p. 120
$C_9APE_{9.5}$	45-100	Alg	21d	CT	Davis, 1967
	0-56	Alg	14d	IR	
	100	Sew;BAS	7d	CT	
	100	Pond	21d	CT	
	28-86	Pond	14d	IR	

TABLE 8.19   Branched or Unspecified APEs *(continued)*

Substrate (see § II)	Extent (see § III)	Method (see § IV)	Time (see § V)	Analysis (see § VI)	Reference
C_9APE_{10}	100;96	BAS	3d	UV;PM	Sato, 1963
	0-0.1G	Wa	4h?	O_2	
tp-C_9APE_{10}	83;65	RW	34d	IR;UV	Frazee, 1964b
tp-C_9APE_{10}	7	Wa	1d	O_2	Vath, 1964
br-C_9APE_{10}	87;84	RW	34d	IR;UV	Osburn, 1966
	97;69	RW	34d	CT;IR	
C_9APE_{10}	29	BAS	1d	SMB	Han, 1967
br-C_9APE_{10}	11;6	In	20d	COD;O_2	Pitter, 1968a
tp-C_9APE_{11}	65	CAS	8h	CT;F	Conway, 1966
	+	In	?	CO_2	
	75;35;0	SF	7d	CT;σ;F	
br-C_9APE_{15}	0-60	In	7d	CT	Bunch, 1967a
br-C_9APE_{15}	13;6	In	20d	PW;COD	Pitter, 1968a
	3	In	20d	O_2	
br-C_9APE_{16} (British)	30-75	In	63d	TLC	Patterson, 1968
C_9APE_{20}	8	BAS	1d	SMB	Han, 1967
br-C_9APE_{20}	4;3	In	20d	PW;COD	Pitter, 1968a
	2	In	20d	O_2	
C_9APE_{30}	6;40	BAS	1d	SMB	Han, 1967
br-C_9APE_{30}	0;2	In	20d	PW;COD	Pitter, 1968a
	0	In	20d	O_2;UV	
C_9APE_x	60	CAS	6h	F	Eldib, 1963
C_9APE_x (Tergitol NPX)	75	RW;AnRW	7d	σ	Wayman, 1963a
br-C_9APE_x	80†	RW	35d	σ	Weil, 1964
C_9APE_x (Lissapol SNX)	25-50	BAnD	13d	TLC	Bruce, 1966
br-C_9APE_x (Igepal C0630)	55;30	CAS;SF	4h;8d	CT	Huddleston, 1966
C_9APE_x (1)	66;36	TF;In	–;21d	TLC	WPRL, 1966a, p. 131
(2)	72;47	TF;In	–;21d	TLC	
tp-C_9APE_x	50;95	BAS	1d	F;CT	SDA, 1969a,b
	92;98	RW	28d	F;CT	
	0;19	SF	14d	F;CT	
c. *Dodecyl Phenols*					
tp-$C_{12}APE_{10}$	20;96	BAS	1d	F;CT	SDA, 1969a,b
	87;98	RW	28d	F;CT	
	15;0	SF	14d	F;CT	
br-$C_{12}APE_{11}$	0	RW	36d	IR	Osburn, 1966
$C_{12}APE_{12}$ (Igepal CA)	0	BOD	5d	O_2	Sheets, 1956a,b
tp-$C_{12}APE_{12}$	0	Wa	2d	O_2	Steinle, 1964
d. *Di- and Polyalkyl Phenols*					
4-sBu-2-C_8APE_9	0.4G	Wa	5d	O_2	Nunn, 1967
2-sBu-4-C_8APE_9	0.4G	Wa	5d	O_2	Nunn, 1967
1,3,5-tBu$_3APE_9$	0.03G	Wa	5d	O_2	Nunn, 1967

TABLE 8.19   Branched or Unspecified APEs *(continued)*

Substrate (see § II)	Extent (see § III)	Method (see § IV)	Time (see § V)	Analysis (see § VI)	Reference
e. *Unknown or Unstated Alkyl Group*					
$br\text{-}C_xAPE_4$ (German)	50-60	In	42d	TLC	Patterson, 1968
$br\text{-}C_xAPE_9$ (Lissapol NX)	55	Sew	42d	TLC	Patterson, 1966b
$br\text{-}C_xAPE_9$ (Lissapol NX)	100;80	RTF	6;9d	CT;TLC	Jenkins, 1967
$br\text{-}C_xAPE_9$ (German)	30-50	In	42d	TLC	Patterson, 1968
C_xAPE_{10} 0.95 MG	90†	RW	6d	σ	Steinle, 1964
C_xAPE_{10} 1.17 MG	4	Wa	2d	O_2	Steinle, 1964
C_xAPE_{10} 1.50 MG	90†	RW	38d	σ	Steinle, 1964
C_xAPE_{10} 1.60 MG	3	Wa	2d	O_2	Steinle, 1964
C_xAPE_{12} 0.95 MG	3	Wa	2d	O_2	Steinle, 1964
C_xAPE_{12} 1.17 MG	90†	RW	5d	σ	Steinle, 1964
C_xAPE_{12} 1.32 MG	90†;0	RW;Wa	5;2d	$\sigma;O_2$	Steinle, 1964
C_xAPE_{12} 1.39 MG	90†	RW	5d	σ	Steinle, 1964
C_xAPE_{12} 1.76 MG	90†;3	RW;Wa	45;2d	$\sigma;O_2$	Steinle, 1964
C_xAPE_{12} 1.97 MG	90†	RW	45d	σ	Steinle, 1964
$iAPE_{4-20}$	<30;<15	In	20d	COD;O_2	Pitter, 1968c
Lissapol N	0	BOD	20d	O_2	Oldham, 1949, 58
APE_x	48;25	RW;Aqu	15;14d	σ	Degens, 1950
	0	BOD	5d	O_2	
APE_x (Lissapol N)	0	BOD	5d	O_2	Goldthorpe, 1950
APE_x	0	BOD	7d	O_2	Leclerc, 1952
APE_x (Igepal CA630)	20;6	Wa;BOD	6h;5d	O_2	Bogan, 1954, Sawyer, 1956
APE_x (Neutronyx 600)	12;5	Wa;BOD	6h;5d	O_2	Bogan, 1954, Sawyer, 1956
APE_x (Lissapol N)	40	TF	–	?	Barden, 1957
	0.4F	Wa	6h	O_2	
	0.25G	Wa	7h	O_2	
Lissapol N	<4;30	In;TF	20d;–	PM	Oldham, 1958
APE_x	0	BOD	5d	O_2	Offhaus, 1962
APE_x (Hyonic PE90)	75	RW;AnRW	7d	σ	Wayman, 1963a

[a]Prepurification by resin adsorption.

[b]Prepurification by butanone extraction.

B. 5. *Other Alkoxylate Derivatives*

TABLE 8.20

Biodegradation of Miscellaneous Alkoxylates

Substrate (see § II)	Extent (see § III)	Method (see § IV)	Time (see § V)	Analysis (see § VI)	Reference
a. *Ethoxylates of Fatty Acids and Fats*					
Lauric E_9 (Empilan AQ100)	100	BAS	6h	PC	Schönborn, 1966, p. 112
Lauric E_{10}	100	RW	2;3d	σ;F	Weil, 1964
Lauric E_{18} (Empilan AP100)	100	BAS	1½h	PC	Schönborn, 1966, p. 112
Coco E_5 (Ethofat C/15)	29;40	Wa;BOD	6h;5d	O_2	Bogan, 1954, Sawyer, 1956
Coco E_{50} (Ethofat C/60)	9;13	Wa;BOD	6h;5d	O_2	Bogan, 1954, Sawyer, 1956
Stearic $E_{2.5}$ (Cremophor AP)	+	BAS	4d	F	Schönborn, 1966, p. 112
Stearic E_6	40-100	In	20d	HgI	Čuta, 1964
Stearic E_{18} (Empilan CP100)	100	BAS	1d	PC	Schönborn, 1966, p. 112
Stearic E_{50} (Ethofat 60/60)	10;16	Wa;BOD	6h;5d	O_2	Bogan, 1955
Oleic E_6	40-90	In	20d	HgI	Čuta, 1964
Oleic E_9 (Empilan BQ100)	100	BAS	2d	PC	Schönborn, 1966, p. 112
$C_{7-15}CO_2E_3$	95;90+	In	20d	PW;COD	Pitter, 1968a
$C_{7-15}CO_2E_6$	80;90+	In	20d	PW;COD	Pitter, 1968a
$C_xCO_2E_{3-6}$	>92;80	In	20d	COD;O_2	Pitter, 1968c
$C_xCO_2E_x$ (Cirrosol FP)	0.3F	Wa	6h	O_2	Barden, 1957
Tall oil E_{12} (Merpoxen TA120)	+	BAS	6d	PC	Schönborn, 1966, p. 112
Tálový oil E_{14}	70-80	In	20d	HgI	Čuta, 1964
Spermový oil E_{20}	60-90	In	20d	HgI	Čuta, 1964
Castor oil E_{20}	100	In	10d	HgI	Čuta, 1964
b. *Ethoxylates of Fatty Amides and Ethanolamides*					
Lauric MEA E_2	100	RW	10d	σ	Knaggs, 1964
Lauric MEA E_3	54	Wa	10d	O_2	Hunter, 1964
Lauric MEA E_{10}	35	Wa	10d	O_2	Hunter, 1964
Lauric MEA E_{10}	100;76	RW;Wa	11;15d	σ;O_2	Knaggs, 1964
Coco EA E_5	100	BAS	1d	SMB	Han, 1967
Coco EA E_{14}	100	BAS	1d	SMB	Han, 1967
Coco EA E_{25}	97	BAS	1d	SMB	Han, 1967
Coco EA E_x	~100	?	?	?	WPRL, 1966b
Coco MEA E_x	100	In	3d	TLC	Patterson, 1967
Tallow EA E_8	94	BAS	1d	SMB	Han, 1967
HTA[a] E_5 (Ethomid HT/15)	20;42	Wa;BOD	6h;5d	O_2	Bogan, 1954, Sawyer, 1956
HTA[a] E_{50} (Ethomid HT/60)	0;4	Wa;BOD	6h;5d	O_2	Bogan, 1954

TABLE 8.20 Miscellaneous Alkoxylates *(continued)*

Substrate (see § II)	Extent (see § III)	Method (see § IV)	Time (see § V)	Analysis (see § VI)	Reference
$HTA^a E_{50}$ (Ethomid HT/60)	8;16	Wa;BOD	6h;5d	O_2	Bogan, 1955, Sawyer, 1956
C_{12-18} amide E_x	97-100	In;SF	9;7d	$CT;\sigma;F$	Garrison, 1964
	0.7G	Wa	5d	O_2	
Fatty EA E_x	97	TF	–	TLC	WPRL, 1965, p. 127
Tall oil amide E_9	48	Wa	10d	O_2	Hunter, 1964
Tall oil amide E_x	0.5;0.8G	Wa	1;10d	O_2	Hunter, 1964
c. *Ethoxylated Fatty Amines*					
t-Amine (Ethomeen C15)	1.0F	Wa	6h	O_2	Barden, 1957
$C_{18}NE_{20}$	25-65	In	20d	HgI	Čuta, 1964
d. *Ethoxylated Fatty Acyl Sorbitans*					
Lauroyl SoE_{20} (Tween 20)	90-95	In	20d	HgI	Čuta, 1964
Lauroyl SoE_{20} (Tween 20)	100	BAS	1d	PC	Schönborn, 1966, p. 112
Oleoyl SoE_{20} (Tween 80)	+	Wa	1d	O_2	Manganelli, 1956
Oleoyl SoE_{20} (Tween 80)	+	In	16h	O_2;B	Minami, 1958
Oleoyl SoE_{20} (Tween 80)	0.4G	BOD	5d	O_2	Sheets, 1956a,b
Stearoyl SoE_x (Tween 85)	100	BAS	6d	PC	Schönborn, 1966, p. 112
e. *Ethoxylated Sugar Derivatives*					
Tallow sucroglyceride E_x	95;65	In	7d	σ;CT	Brebion, 1964
	+	Wa;In	1;3d	O_2;B	
f. *Naphthalene Derivatives*					
Dibenzyl-β-naphthol E_{20}	30-60	In	20d	HgI	Čuta, 1964
g. *Ethylene-Propylene Oxide Derivatives*					
Pluronic F68	4;1	Wa;BOD	6h;5d	O_2	Bogan, 1954, Sawyer, 1956
EO-PO copolymer (mol. wt. 2000)	20	In	28d	TLC	Patterson, 1967
Linear alcohol $E_{6-9}P_{3-4.5}$	85	SF	7d	BiI	Weipert, 1967
h. *Poly Alkylene Oxides*					
Pluronic F68	4;1	Wa;BOD	6h;5d	O_2	Bogan, 1954, Sawyer, 1956
EO-PO copolymer, mol. wt. 2000	20	In	28d	TLC	Patterson, 1967
i. *Ethoxylate Chlorides* (e.g., $ROE_{n-1}OCH_2CH_2Cl$)					
n-$C_8E_{10}Cl$	$1.91L^b$	Wa	?	O_2	Eiseman, 1969
n-$C_8E_{14}Cl$	$1.21L^b$	Wa	?	O_2	Eiseman, 1969
Linear oxo C_8E_7Cl	$1.07L^b$	Wa	?	O_2	Eiseman, 1969
$2,3Me_2HxE_{14}Cl$	$0.09L^b$	Wa	?	O_2	Eiseman, 1969
$2EtHxE_7Cl$	$0.16L^b$	Wa	?	O_2	Eiseman, 1969

TABLE 8.20   Miscellaneous Alkoxylates *(continued)*

Substrate (see § II)	Extent (see § III)	Method (see § IV)	Time (see § V)	Analysis (see § VI)	Reference
$i\text{-}C_8E_{10}Cl$	$0.18L^b$	Wa	?	O_2	Eiseman, 1969
Ziegler $C_9E_{11}Cl$	$1.13L^b$	Wa	?	O_2	Eiseman, 1969
Ziegler $C_{10}E_{10}Cl$	$1.09L^b$	Wa	?	O_2	Eiseman, 1969
$n\text{-}C_{10}E_{14}Cl$	$1.09L^b$	Wa	?	O_2	Eiseman, 1969
Oxo $C_{10}E_{14}Cl$	$0.31L^b$	Wa	?	O_2	Eiseman, 1969
Oxo $C_{10}E_{10}P_5Cl$	$0.07L^b$	Wa	?	O_2	Eiseman, 1969
$br\text{-}C_{10}E_{10}Cl$	$0.25L^b$	Wa	?	O_2	Eiseman, 1969
$n\text{-}C_{12}E_{10}Cl$	$1.13L^b$	Wa	?	O_2	Eiseman, 1969
$n\text{-}C_{12}E_{20}Cl$	$1.18L^b$	Wa	?	O_2	Eiseman, 1969
$2,4,6Me_3C_9E_{10}Cl$	$0.03L^b$	Wa	?	O_2	Eiseman, 1969
$4Me6EtC_9E_{20}Cl$	$0.16L^b$	Wa	?	O_2	Eiseman, 1969
C_9APE_9Cl	0	Wa	?	O_2	Eiseman, 1969
j. *Unspecified Structures*					
Novidet P.80	0	In	21d	TLC	MWB, 1966, p. 36
Lissapol DS4429	100	In	7d	TLC	MWB, 1966, p. 36
k. *CTAS in Sewage*					
CTAS	60	*CAS	2d	CT	Conway, 1965

[a]HTA, hydrogenated tallow amide.

[b]L: In comparison with oxygen uptake of LAS = 1.00.

B. 6. *Sugar Derivatives*

TABLE 8.21

Biodegradation of Sugar Surfactants

Substrate (see § II)	Extent (see § III)	Method (see § IV)	Time (see § V)	Analysis (see § VI)	Reference
a. *Fatty Acyl Sucrose Derivatives*					
Laurate	63	Wa	1d	O_2	Winter, 1962
Laurate	100	In	30d	O_2	Heinz, 1967
Laurate	97;84	In	20d	$COD;O_2$	Kulovaná, 1966
Myristate	97;77	In	20d	$COD;O_2$	Kulovaná, 1966
Palmitate (Sucrodet D600)	27	BOD	14d	O_2	Isaac, 1958, 1960b
Palmitate	80†	RW	4d	σ	Weil, 1964
Palmitate	97;58	In	20d	$COD;O_2$	Kulovaná, 1966
Palmitate-stearate (SME80)	0.4G	BOD	12d	O_2	Isaac, 1960b
Stearate (Herstein)	0.8;1.1G	BOD;Sew	5;4d	O_2	Isaac, 1960a
Stearate	57	In	30d	O_2	Heinz, 1967
Stearate	93;53	In	20d	O_2	Kulovaná, 1966
Distearate	17	In	30d	O_2	Heinz, 1967
Distearate	0.5G	BOD	14d	O_2	Isaac, 1960b
Hydroxystearate	100	RW;AnRW	2d	σ	Wayman, 1963a
12-Hydroxystearate	50-86	In	30d	O_2	Heinz, 1967
Tallowate	100	RW;AnRW	2d	σ	Wayman, 1963a
Cottonseed FA (Sequol 260)	57	BOD	10d	O_2	Isaac, 1958, 60b
Sucrose ester (SE 957)	0.9G	BOD	12d	O_2	Isaac, 1960b
Hydroxy fatty acyl	100	?	?	?	Ismail, 1965
Fatty acyl	55	In	30d	O_2	Heinz, 1967
Fatty acyl	>90	In	20d	COD	Pitter, 1968c
	60-87	In	20d	O_2	
b. *Alkyl Sucrose Derivatives*					
n-C_{13} O-sucrose	100	RW	21;13d	σ;F	Swisher, 1960
tp-C_{13} O-sucrose	0;low	RW	64;69d	σ;F	Swisher, 1960
n-$C_{16-18}\pi$-sucrose[a]	100	RW	14;33d	σ;F	Swisher, 1960
tp-$C_{13}\pi$-sucrose[a]	Low	RW	69d	F	Swisher, 1960
c. *Sucroglyceride Ethoxylates*					
Tallow sucroglyceride					
E_x	95;65	In	7d	σ;CT	Brebion, 1964
	+	Wa;In	1;3d	O_2;B	

[a] π, $-OCH_2CHOHCH_2O-$.

B. 7. *Other Nonionics*

TABLE 8.22

TABLE 8.22

Biodegradation of Other Nonionics

Substrate (see § II)	Extent (see § III)	Method (see § IV)	Time (see § V)	Analysis (see § VI)	Reference
a. *Fatty Alkanolamides*					
Palm-kernel EA	0.35F	Wa	6h	O_2	Barden, 1957
Lauric MEA	86	Wa	10d	O_2	Hunter, 1964
Lauric DEA	85	Wa	10d	O_2	Hunter, 1964
Lauric DEA	100	In;SF	9;7d	CT;σ;F	Garrison, 1964
	1.7G	Wa	5d	O_2	
Lauric DEA	100;93	RW;Wa	11;15d	σ;O_2	Knaggs, 1964
Lauric DEA	100	BAS	1d	F	SDA, 1969a,b
Fatty alkanolamide (Empilan)	+	In	?	B	Skinner, 1959
Fatty "isopropylamide"	100	SF	2d	σ	Huddleston, 1965a
Fatty DEA	100	SF	2d	σ	Huddleston, 1965a
Fatty alkanolamide	0.9G	Wa	8h	O_2	Hartmann, 1967
b. *Alkanesulfonyl Polypeptides*					
Lamepon	5.0G	Wa	1d	O_2	Winter, 1962
c. *Amine Oxides*					
Lauryl dimethyl	100	SF	6d	σ	Huddleston, 1965a
Coco dimethyl	100	BAS	1d	F	SDA, 1969a,b

C. Cationic Surfactants

TABLE 8.23

Biodegradation of Cationics

Substrate (see § II)	Extent (see § III)	Method (see § IV)	Time (see § V)	Analysis (see § VI)	Reference
CTAB[a]	0	BOD	5d	O_2	Sheets, 1956a,b
CTAB[a]	0.1F	Wa	6h	O_2	Barden, 1957
CTAB[a]	100	CAS	8h	MB	Pitter, 1961b
CTAB[a]	0	BOD	5d	O_2	Winter, 1962
CPB[b]	100	TF	—	MB	Barden, 1957
	0.9-1.2F	Wa	6h	O_2	
	1.0G	Wa	6h	O_2	
CPB[b]	100	CAS	8h	MB	Pitter, 1961b
CPB[b]	64;67	Wa	10d	O_2	Barbaro, 1965
CPCl[c]	0	BOD	5d	O_2	Sheets, 1956a,b
OCE$_{20}$[d] + LEPCl[e]	60	In	10d	HgI	Čuta, 1964
DADCB[f]	0	Wa	1d	O_2	Winter, 1962
Hyamine 1622[g]	3	Wa	10d	O_2	Barbaro, 1965
t-Amine E$_x$(Ethomeen C15)	1.0F	Wa	6h	O_2	Barden, 1957
C$_{18}$ amine E$_{20}$	25-65	In	20d	HgI	Čuta, 1964
Lauryl DMAO[h]	100	SF	6d	σ	Huddleston, 1965a
Coco DMAO[h]	100	BAS	1d	F	SDA, 1969a,b
Benzalkonium chloride	0	In	?	B	Lambin, 1966
Eltren[i]	28	In	16h	TLC	Janota-Bassalik, 1969

[a]CTAB, cetyltrimethylammonium bromide.

[b]CPB, cetylpyridinium bromide.

[c]CPCl, cetylpyridinium chloride.

[d]OCE$_{20}$, oleyl-cetyl alcohol E$_{20}$.

[e]LEPCl, laurylamidoethylpyridinium chloride.

[f]DADCB, dimethylaminoacetic acid dodecylamide chlorbenzylate.

[g]Hyamine 1622, diethylbenzyl-*tert*-octylphenoxyethoxyethylammonium chloride.

[h]DMAO, dimethylamine oxide.

[i]Eltren, laurylpyridinium chloride.

D. Nonsurfactants

D. 1. *Hydrocarbons*

TABLE 8.24

Biodegradation of Hydrocarbons

Substrate (see § II)	Extent (see § III)	Method (see § IV)	Time (see § V)	Analysis (see § VI)	Reference
a. *Aliphatic*					
n-Pentane	0	Wa	7d	O_2	Marion, 1963b
	10	Wa	3d	O_2	Malaney, 1966
n-Hexane	0	Wa	7d	O_2	Marion, 1963b
	17	Wa	3d	O_2	Malaney, 1966
n-Heptane	0	Wa	7d	O_2	Marion, 1963b
	23	Wa	3d	O_2	Malaney, 1966
n-Octane	0.3G	Wa	1d	O_2	McKinney, 1959a
	0	Wa	7d	O_2	Marion, 1963b
	28	Wa	3d	O_2	Malaney, 1966
n-Nonane	0	Wa	7d	O_2	Marion, 1963b
n-Decane	0.2G	Wa	7d	O_2	Marion, 1963b
	15	Wa	3d	O_2	Malaney, 1966
n-Dodecane	4;8	Wa;BOD	6h;5d	O_2	Bogan, 1955
	0.1G	Wa	7d	O_2	Marion, 1963b
	7	Wa	3d	O_2	Malaney, 1966
	27;37	Wa	5d	O_2	Marion, 1966
n-Tetradecane	0	Wa	7d	O_2	Marion, 1963b
n-Octadecane	0	Wa	7d	O_2	Marion, 1963b
	9	Wa	3d	O_2	Malaney, 1966
b. *Alicyclic*					
Cyclohexane	+	In	3d	B	Imelik, 1948
	100	In	3d	MP	Ooyama, 1965
	0	Wa	3d	O_2	Malaney, 1966
c. *Aromatic*					
Benzene	3;2	Wa;BOD	6h;5d	O_2	Bogan, 1955
	33	Wa	12h	O_2	McKinney, 1956b
	13	Wa	8d	O_2	Malaney, 1960
	20	Wa	1d	O_2	Winter, 1962
	0-0.6G	Wa	7d	O_2	Marion, 1963a
	36	Wa	8d	O_2	Malaney, 1966
	46	Wa	5d	O_2	Marion, 1966
Toluene	47	Wa	12h	O_2	McKinney, 1956b
	0.6G	Wa	1d	O_2	McKinney, 1959a
	45	Wa	8d	O_2	Malaney, 1960
	0.1-0.9G	Wa	7d	O_2	Marion, 1963a
	48	Wa	8d	O_2	Malaney, 1966
	32	In	20d	O_2	Young, 1968
o-Xylene	0-0.2G	Wa	7d	O_2	Marion, 1963a
	40	Wa	8d	O_2	Malaney, 1966
m-Xylene	0.6G	Wa	7d	O_2	Marion, 1963a
	10	Wa	8d	O_2	Malaney, 1966

TABLE 8.24   Hydrocarbons *(continued)*

Substrate (see § II)	Extent (see § III)	Method (see § IV)	Time (see § V)	Analysis (see § VI)	Reference
p-Xylene	0-1.5G	Wa	7d	O_2	Marion, 1963a
	26	Wa	8d	O_2	Malaney, 1966
1,2,4Me$_3\phi$	0.4-2.0G	Wa	7d	O_2	Marion, 1963a
1,3,5Me$_3\phi$	0	Wa	7d	O_2	Marion, 1963a
	0	Wa	8d	O_2	Malaney, 1966
1,2,4,5Me$_4\phi$	0-0.5G	Wa	7d	O_2	Marion, 1963a
	0	Wa	8d	O_2	Malaney, 1966
Me$_5\phi$	0-0.1G	Wa	7d	O_2	Marion, 1963a
	0	Wa	8d	O_2	Malaney, 1966
Me$_6\phi$	0.1-0.4G	Wa	7d	O_2	Marion, 1963a
	0	Wa	8d	O_2	Malaney, 1966
Ethylbenzene	8;3	Wa;BOD	6h;5d	O_2	Bogan, 1955
	27	Wa	12h	O_2	McKinney, 1956b
	0	In	?	MP	Webley, 1956
	25	Wa	8d	O_2	Malaney, 1960
	45	RW	15d	CO_2	Ludzack, 1963b
	0.4-0.6G	Wa	7d	O_2	Marion, 1963a
	43	Wa	8d	O_2	Malaney, 1966
n-Propylbenzene	1;3	Wa;BOD	6h;5d	O_2	Bogan, 1955
	10	Wa	12h	O_2	McKinney, 1956b
	0	In	?	MP	Webley, 1956
	34	Wa	8d	O_2	Malaney, 1960
	0.6-1.6G	Wa	7d	O_2	Marion, 1963a
	28	Wa	8d	O_2	Malaney, 1966
i-Propylbenzene	31	Wa	8d	O_2	Malaney, 1960
	0.5-0.7G	Wa	7d	O_2	Marion, 1963a
	38	Wa	8d	O_2	Malaney, 1966
n-Butylbenzene	6;14	Wa;BOD	6h;5d	O_2	Bogan, 1965
	10	Wa	12h	O_2	McKinney, 1956b
	0	In	?	MP	Webley, 1956
	31	Wa	8d	O_2	Malaney, 1960
	14	Wa	1d	O_2	Winter, 1962
	0.3-1.9G	Wa	7d	O_2	Marion, 1963a
	48	Wa	4d	O_2	Malaney, 1966
s-Butylbenzene	1	Wa	12h	O_2	McKinney, 1956b
	31	Wa	8d	O_2	Malaney, 1960
	52	Wa	8d	O_2	Malaney, 1966
t-Butylbenzene	1;0	Wa;BOD	6h;5d	O_2	Bogan, 1955
	1	Wa	12h	O_2	McKinney, 1956b
	31	Wa	8d	O_2	Malaney, 1960
	28	RW	52d	CO_2	Ludzack, 1963b
	0-0.1G	Wa	7d	O_2	Marion, 1963a
	27	Wa	8d	O_2	Malaney, 1966
n-Amylbenzene	100	In	4d	MP	Douros, 1967

TABLE 8.24 Hydrocarbons *(continued)*

Substrate (see § II)	Extent (see § III)	Method (see § IV)	Time (see § V)	Analysis (see § VI)	Reference
s-Amylbenzene	1	Wa	12h	O_2	McKinney, 1956b
	31	Wa	8d	O_2	Malaney, 1960
	0	Wa	7d	O_2	Marion, 1963a
	56	Wa	8d	O_2	Malaney, 1966
t-Amylbenzene	1	Wa	12h	O_2	McKinney, 1956b
	22	Wa	8d	O_2	Malaney, 1960
	0-0.4G	Wa	7d	O_2	Marion, 1963a
	36	Wa	8d	O_2	Malaney, 1966
n-Nonylbenzene	100	In	6d	B;MP	Davis, 1961
n-Decylbenzene	+	In	?	MP	Webley, 1956
s-Decylbenzene	20	Wa	8d	O_2	Malaney, 1966
n-Dodecylbenzene	85	In	6d	MP	Webley, 1956
	80	In	?	B;MP	Davis, 1961
Dodecylbenzene	0.2-1.4G	Wa	7d	O_2	Marion, 1963a
Dodecylbenzene (Nalkylene 500)	6	Wa	5d	O_2	Marion, 1966
n-Octadecylbenzene	+	In	?	MP	Webley, 1956
$3\phi C_{20}$	56	In	20d	MP	Webley, 1956
Naphthalene	0	Wa	1d	O_2	Winter, 1962
1-α-Naphthyl C_{11}	17	In	13d	MP	Webley, 1956

D. 2. *Alcohols*

TABLE 8.25

Biodegradation of Alcohols

Substrate (see § II)	Extent (see § III)	Method (see § IV)	Time (see § V)	Analysis (see § VI)	Reference
a. *Aliphatic*					
Methyl	55	Wa	1d	O_2	McKinney, 1955
	0.8G	Wa	6d	O_2	Marion, 1963b
	79	In	20d	O_2	Young, 1968
Ethyl	51	Wa	1d	O_2	McKinney, 1955
	0.5G	Wa	6d	O_2	Marion, 1963b
	86	In	20d	O_2	Young, 1968
n-Propyl	55	Wa	1d	O_2	McKinney, 1955
	1.1G	RW	7d	O_2	Hammerton, 1956
	1.9G	Wa	6d	O_2	Marion, 1963b
i-Propyl	50	Wa	1d	O_2	McKinney, 1955
	0.9G	RW	8d	O_2	Hammerton, 1956
	0.3G	Wa	6d	O_2	Marion, 1963b
	70	In	20d	O_2	Young, 1968
n-Butyl	44	Wa	1d	O_2	McKinney, 1955
	1.6G	RW	4d	O_2	Hammerton, 1955
	2.0G	Wa	6d	O_2	Marion, 1963b
i-Butyl	44	Wa	1d	O_2	McKinney, 1955
	1.5G	RW	4d	O_2	Hammerton, 1955
s-Butyl	58	Wa	1d	O_2	McKinney, 1955
	1.4G	RW	5d	O_2	Hammerton, 1955
	0	Wa	6d	O_2	Marion, 1963b
t-Butyl	0.1G	RW	12d	O_2	Hammerton, 1955
	2	Wa	1d	O_2	McKinney, 1955
	0	Wa	6d	O_2	Marion, 1963b
	0	In	20d	O_2	Young, 1968
n-Amyl	1.1G	RW	5d	O_2	Hammerton, 1956
	1.4G	Wa	6d	O_2	Marion, 1963b
2Me-1-butanol	1.1G	RW	5d	O_2	Hammerton, 1956
3Me-1-butanol	1.1G	RW	5d	O_2	Hammerton, 1956
2Me-2-butanol	0.1;0.9G	RW	5;42d	O_2	Hammerton, 1956
1-Hexanol	1.2G	RW	7d	O_2	Hammerton, 1956
	+	In	1-3d	B	Payne, 1963b, Feisal, 1966
2Et-1-butanol	1.1G	RW	12d	O_2	Hammerton, 1956
2Me-2-pentanol	0.5G	RW	38d	O_2	Hammerton, 1956
1-Octanol	1.7G	RW	3d	O_2	Hammerton, 1955
	0	In	3d	B	Payne, 1963b, Feisal, 1966
2-Octanol	1.9G	RW	4d	O_2	Hammerton, 1955
2Et-1-hexanol	1.2G	RW	9d	O_2	Hammerton, 1955
1-Nonanol	1.9G	RW	4d	O_2	Hammerton, 1955
2,6Me$_2$-4-heptanol	0.2G	RW	12d	O_2	Hammerton, 1955
	0.9G	RW	24d	O_2	Hammerton, 1956
3,5,5Me$_3$-1-hexanol	0.2G	RW	10d	O_2	Hammerton, 1955

TABLE 8.25   Alcohols *(continued)*

Substrate (see § II)	Extent (see § III)	Method (see § IV)	Time (see § VI)	Analysis (see § VI)	Reference
1-Decanol	+	In	1-3d	B	Payne, 1963b, Feisal, 1966
	98	In	2d	GC	Williams, 1966
2-Undecanol	1.1G	RW	5d	O_2	Hammerton, 1956
1-Dodecanol	15;30	Wa;BOD	6h;5d	O_2	Bogan, 1955
	65	Wa	1d	O_2	Winter, 1962
	+	In	1-3d	B	Payne, 1963b, Feisal, 1966
	100	In	2d	GC	Prochazka, 1965
	78	In	2d	GC	Williams, 1966
	36	Wa	?	O_2	Prochazka, 1967
Lauryl alcohol	32	Wa	5d	O_2	Marion, 1966
1-Tetradecanol	+	In	1-3d	B	Payne, 1963b, Feisal, 1966
	70	In	2d	GC	Williams, 1966
1-Hexadecanol	80-97	In	20-48d	CO_2	Ludzack, 1957
	+	In	1-3d	B	Payne, 1963b, Feisal, 1966
	66	In	2d	GC	Williams, 1966
1-Octadecanol	+	In	1-3d	B	Payne, 1963b, Feisal, 1966
	70	In	2d	GC	Williams, 1966
b. *Aromatic*					
Benzyl	1.3G	RW	5d	O_2	Hammerton, 1956
	31;52	Wa	12h	O_2	McKinney, 1956b
	1.0G	Wa	7d	O_2	Marion, 1963a
2ϕ-Ethanol	1.3G	RW	5d	O_2	Hammerton, 1956
1ϕ-Ethanol	1.4G	RW	9d	O_2	Hammerton, 1956
3ϕ-1-Propanol	1.2G	RW	7d	O_2	Hammerton, 1956
Cinnamyl alcohol	1.1G	RW	7d	O_2	Hammerton, 1956
2ϕ-1,1Me$_2$-ethanol	0.1G	RW	28d	O_2	Hammerton, 1956

D. 3. *Glycols, Polyglycols, and Glycol Derivatives*

TABLE 8.26

Biodegradation of Glycols and Glycol Derivatives

Substrate (see § II)	Extent (see § III)	Method (see § IV)	Time (see § V)	Analysis (see § VI)	Reference
Ethylene glycol (E_1)	78	BOD	20d	O_2	Lamb, 1952
	85	BOD	10d	O_2	Mills, 1953
	67;+	Wa;In	?;7d	O_2;B	Fincher, 1962
	100	AnIn	?	B;MP	Gaston, 1963
	1.3G	Wa	7d	O_2	Marion, 1963b
	40	Wa	1d	O_2	Gerhold, 1966
	>95;70	In	20d	COD;O_2	Pitter, 1968c
	72	In	20d	O_2	Young, 1968
Diethylene glycol (E_2)	19	BOD	20d	O_2	Lamb, 1952
	0	BOD	10d	O_2	Mills, 1953
	23	BOD	15d	O_2	Mills, 1954
	68	CAnD	7d	COD	Mills, 1954
	114;+	Wa;In	?;7d	O_2;B	Fincher, 1962
	10	Wa	1d	O_2	Gerhold, 1966
	54	Wa	?	O_2	Prochazka, 1967
	95;77	In	20d	COD;O_2	Pitter, 1968a
	21	In	20d	O_2	Young, 1968
Triethylene glycol (E_3)	17	BOD	20d	O_2	Lamb, 1952
	30	BOD	10d	O_2	Mills, 1953
	60	CAnD	7d	COD	Mills, 1954
	92;+	Wa;In	?;7d	O_2;B	Fincher, 1962
	11	Wa	1d	O_2	Gerhold, 1966
	54	Wa	?	O_2	Prochazka, 1967
	95;70	In	20d	COD;O_2	Pitter, 1968a
Tetraethylene glycol (E_4)	30	BOD	10d	O_2	Mills, 1953
	99;+	Wa;In	?;7d	O_2;B	Fincher, 1962
	54	Wa	?	O_2	Prochazka, 1967
	93;65	In	20d	COD;O_2	Pitter, 1968a
PEG 200 ($E_{4.5}$)	+	In	7d	B	Fincher, 1962
PEG 300 (E_7)	+	In	7d	B	Fincher, 1962
	92;76	In	20d	COD;O_2	Pitter, 1968a,c
PEG 400 (E_9)	5	BOD	10d	O_2	Mills, 1953
	36-51	CAnD	24d	COD	Mills, 1954
	+	In	7d	B	Fincher, 1962
	68	Wa	3d	O_2	Vath, 1964
	96	Sew	14d	TLC	Patterson, 1966b
	77;100	TF;CAS	–;3h	HgI	Schönborn, 1966 pp. 117, 119
	98	In	28d	Wt	Borstlap, 1967a
	26-94	In	7d	CT	Bunch, 1967a
PEG 450 (E_{10})	100	In	13d	TLC	Patterson, 1967
PEG 600 (E_{14})	0	In	30d	B	Fincher, 1962
	0-5	In	20d	PW	Pitter, 1968a
	10;5	In	20d	COD;O_2	

TABLE 8.26   Glycols and Glycol Derivatives *(continued)*

Substrate (see § II)	Extent (see § III)	Method (see § IV)	Time (see § V)	Analysis (see § VI)	Reference
PEG 1000 (E_{23})	0	In	30d	B	Fincher, 1962
	54;79	TF;CAS	–;3h	HgI	Schönborn, 1966 pp. 117, 119
	90	In	28d	Wt	Borstlap, 1967a
	100	In	16d	TLC	Patterson, 1967
	0-5	In	20d	PW	Pitter, 1968a
	10;2	In	20d	COD;O_2	
PEG 1500 (E_{34})	0-5	In	20d	PW	Pitter, 1968a
	5;0	In	20d	COD;O_2	
PEG 1540 (E_{35})	0	In	30d	B	Fincher, 1962
PEG 1750 (E_{40})	24	BAS	1d	SMB	Han, 1967
PEG 3300 (E_{75})	30	In	28d	TLC	Patterson, 1967
PEG 3500 (E_{80})	0-5	In	20d	PW	Pitter, 1968a
	5;0	In	20d	COD;O_2	
	10	In	20d	COD	Pitter, 1968c
PEG 4000 (E_{90})	0	In	30d	B	Fincher, 1962
PEG 6000 (E_{135})	0	In	30d	B	Fincher, 1962
PEG (higher mol. wt.)	0	BOD	10d	O_2	Oberton, 1957
PEG (high mol. wt.)	20	Sew	49d	TLC	Patterson, 1967
Propylene glycol	90	In	2d	?	Ishii, 1959
	Slow	In	30d	B	Fincher, 1962
	>95;80	In	20d	COD;O_2	Pitter, 1968c
Dipropylene glycol	Slow	In	30d	B	Fincher, 1962
EO-PO copolymer (mol. wt. 2000)	20	In	28d	TLC	Patterson, 1967
2-Ethoxyethanol (EtOE_1)	52;54	Wa;BOD	6h;5d	O_2	Bogan, 1955
	96;86	In	20d	COD;O_2	Pitter, 1968c
Ethanol E_2 (EtOE_2)	40;34	Wa;BOD	6h;5d	O_2	Bogan, 1955
Phenol E_1	0	In	30d	B	Fincher, 1962
Phenol E_9	94;22	In	9d	CT;σ	Garrison, 1964
	31;9	SF	7d	CT;σ	
	0	Wa	5d	O_2	

D. 4. *Lower Sulfonates*

TABLE 8.27

Biodegradation of Lower Sulfonates

Substrate (see § II)	Extent (see § III)	Method (see § IV)	Time (see § V)	Analysis (see § VI)	Reference
Taurine ($H_2NCC\Sigma$)	50	Wa	1h	O_2	Ikeda, 1963
	100	Wa	1h	$RS;SO_4$	
	100	Wa	1h	NH_3	
Sulfoacetate	+	In	6h	RA;RS	Martelli, 1964
Me-6Σ-α-D-quinovoside	100;33	In	3h	RA;RS	Martelli, 1964
BenzeneΣ	39;75	Wa;BOD	6h;5d	O_2	Bogan, 1955
	1.1G	RW	7d	O_2	Hammerton, 1955
	0.6G	RW	7d	O_2	Hammerton, 1956
	60	BOD	6d	O_2	Ryckman, 1956
	39;50	Wa	16h;8d	O_2	
	93-100	BAS	1d	SO_4	
	0	Wa	8d	O_2	Malaney, 1960
	44	Wa	8h	O_2	Symons, 1961
	82	BOD	10d	O_2	Hammerton, 1962
	78	Wa	6h	O_2	Winter, 1962
	1.2G	In	20d	O_2	Pitter, 1963c
	67;+	Wa;In	4h;2d	O_2;B	Payne, 1963b, Feisal, 1966
	94;92	In	20d	O_2;SO_4	Pitter, 1964c
	0	Wa	3h	O_2	Tabak, 1964
	100	SF	6d	UV	Setzkorn, 1965
	100	In	16d	UV	Alexander, 1966
	0	Wa	3d	O_2	Malaney, 1966
	0	Wa	5d	O_2	Marion, 1966
	80-100	In	16d	O_2;UV	Kölbel, 1967
	98	In	20d	COD	Pitter, 1968c
p-TolueneΣ	0.9G	RW	7d	O_2	Hammerton, 1955
	0.8G	RW	7d	O_2	Hammerton, 1956
	93-100	BAS	1d	SO_4	Ryckman,1956,57
	37;59	Wa	6h;8d	O_2	
	54;65	BOD	5;10d	O_2	
	36	Wa	8h	O_2	Symons, 1961
	82	BOD	6d	O_2	Hammerton, 1962
	1.2G	In	20d	O_2	Pitter, 1963c
	91;91	In	20d	O_2;SO_4	Pitter, 1964c
	0	Wa	3h	O_2	Tabak, 1964
	100	In	3d	UV	Huddleston, 1965a
	100	SF	2d	UV	Setzkorn, 1965
	98	In	20d	COD	Pitter, 1968c
XyleneΣ	100	In	8d	UV	Huddleston, 1965a
2,3-XyleneΣ	100	SF	45d	UV	Setzkorn, 1965
2,4-XyleneΣ	100	SF	5d	UV	Setzkorn, 1965
2,5-XyleneΣ	0;100	SF;RW	45;55d	UV	Setzkorn, 1965

TABLE 8.27   Lower Sulfonates *(continued)*

Substrate (see § II)	Extent (see § III)	Method (see § IV)	Time (see § V)	Analysis (see § VI)	Reference
EthylbenzeneΣ	96-100	BAS	1d	SO$_4$	Ryckman,1956,57
	39;59	Wa	6h;8d	O$_2$	
	50;57	BOD	5;14d	O$_2$	
	100	SF	6d	UV	Setzkorn, 1965
n-PropylbenzeneΣ	97-100	BAS	1d	SO$_4$	Ryckman,1956,57
	40;41	Wa	6h;8d	O$_2$	
	2;34	BOD	5;22d	O$_2$	
i-PropylbenzeneΣ	90	BAS	2d	SO$_4$	Ryckman,1956,57
	28;40	Wa	6;16h	O$_2$	
	4;8	BOD	5;21d	O$_2$	
	0	BOD	30d	O$_2$	Hammerton, 1962
n-ButylbenzeneΣ	97	BAS	1d	SO$_4$	Ryckman,1956,57
	44;52	Wa	6h;8d	O$_2$	
	39;60	BOD	5;8d	O$_2$	
	76;97	BOD	25d	O$_2$;MB	Hammerton, 1962
s-ButylbenzeneΣ	3	BAS	2d	SO$_4$	Ryckman,1956,57
	3;10	Wa	6;16h	O$_2$	
	2;1	BOD	5;20d	O$_2$	
t-ButylbenzeneΣ	0	BAS	2d	SO$_4$	Ryckman,1956,57
	4;9	Wa	6h;8d	O$_2$	
	0;1	BOD	5;23d	O$_2$	
ButylbenzeneΣ	65	Wa	1d	O$_2$	Winter, 1962
n-AmylbenzeneΣ	100	RW	20d	MB	Swisher, 1963b
	0-1.4M	Wa	9h	O$_2$	Heyman, 1967
s-AmylbenzeneΣ	5	BAS	2d	SO$_4$	Ryckman,1956,57
	15;9	Wa	6;16h	O$_2$	
	0;1	BOD	5;20d	O$_2$	
t-AmylbenzeneΣ	0	BAS	2d	SO$_4$	Ryckman,1956,57
	4;2	Wa	6h;8d	O$_2$	
	0;1	BOD	5;22d	O$_2$	
p-ChlorobenzeneΣ	100	In	16d	UV	Alexander, 1966
p-NitrobenzeneΣ	0	In	64d	UV	Alexander, 1966
p-AnilineΣ	42	Wa	8h	O$_2$	Symons, 1961
	0	In	64d	UV	Alexander, 1966
p-PhenolΣ	45	Wa	8h	O$_2$	Symons, 1961
	100	In	32d	UV	Alexander, 1966
p-MethoxybenzeneΣ	0	In	64d	UV	Alexander, 1966
p-CarboxybenzeneΣ	0.5G	Wa	1d	O$_2$	McKinney, 1959a
	0	In	64d	UV	Alexander, 1966
	0	Wa	8h;1d	O$_2$;UV	Heyman, 1968
BenzeneΣ_2	0	In	64d	UV	Alexander, 1966
α-NaphthaleneΣ	91	In	20d	O$_2$	Pitter, 1964c
β-NaphthaleneΣ	80-100	In	16d	O$_2$;UV	Kölbel, 1967
p-BiphenylΣ	80-100	In	16d	O$_2$;UV	Kölbel, 1967

D. 5. *Miscellaneous Aliphatics*

TABLE 8.28

Biodegradation of Miscellaneous Aliphatic Compounds

Substrate (see § II)	Extent (see § III)	Method (see § IV)	Time (see § V)	Analysis (see § VI)	Reference
a. *Lower Carboxylates*					
Formic acid	66	Wa	12h	O_2	McKinney, 1956b
	87	Wa	5d	O_2	Marion, 1963b
	70	Wa	1d	O_2	Malaney, 1969
Acetic acid	50	Wa	12h	O_2	McKinney, 1956b
	43	Wa	5d	O_2	Marion, 1963b
	40	Wa	1d	O_2	Malaney 1969
Propionic acid	40	Wa	12h	O_2	McKinney, 1956b
	40	Wa	5d	O_2	Marion, 1963b
	40	Wa	1d	O_2	Malaney, 1969
Butyric acid	41	RW	5d	O_2	Hammerton, 1956
	30	Wa	12h	O_2	McKinney, 1956b
	43	Wa	5d	O_2	Marion, 1963b
	28	Wa	1d	O_2	Malaney, 1969
Trimethylacetic acid	0;51	RW	5;20d	O_2	Hammerton, 1956
b. *Ethanolamines*					
Monoethanolamine	60	BOD	20d	O_2	Lamb, 1952
	40	In	20d	O_2	Young, 1968
Diethanolamine	7	BOD	20d	O_2	Lamb, 1952
	90	BOD	10d	O_2	Oberton, 1957
Triethanolamine	6	BOD	20d	O_2	Lamb, 1952
c. *Fats and Fatty Derivatives*					
Tributyrin	0.2G	Wa	7d	O_2	Marion, 1963b
Trilaurin	0.1G	Wa	7d	O_2	Marion, 1963b
Trimyristin	0.1G	Wa	7d	O_2	Marion, 1963b
Tristearin	0	Wa	7d	O_2	Marion, 1963b
1-Monobutyrin	0.4G	Wa	7d	O_2	Marion, 1963b
1-Monostearin	0	Wa	7d	O_2	Marion, 1963b
Fatty nitrile (Arneel 18D)	0.2F	Wa	6h	O_2	Barden, 1957
Fatty amide (Armid C)	0.1F	Wa	6h	O_2	Barden, 1957
d. *Glucose*					
Glucose	21	Wa	12h	O_2	McKinney, 1956b
	77	In	22d	CO_2	Pahren, 1961
	96	In	?	COD	Konecky, 1963
	0.8G	Wa	7d	O_2	Marion, 1963b
	100	In	20d	O_2	Pitter, 1964c
	100	In	30d	O_2	Heinz, 1967
	98;80	In	20d	$COD;O_2$	Pitter, 1968c
e. *Carboxymethyl Cellulose*					
Carboxymethyl Cellulose	0.2F	Wa	6h	O_2	Barden, 1957
	+	BOD	5d	O_2	Malaney, 1957

TABLE 8.28   Miscellaneous Aliphatics *(continued)*

Substrate (see § II)	Extent (see § III)	Method (see § IV)	Time (see § V)	Analysis (see § VI)	Reference
f. *Chelants*					
Versene[a] (EDTA)	+	BOD	5d	O_2	Malaney, 1957
Versene[a]-iron	+	BOD	5d	O_2	Malaney, 1957
Nitrilotriacetate (NTA)	100	In	7d	Ch	Bunch, 1967b
	+	In	4-8d	B	Forsberg, 1967a,b
	100	CAS	3;6h	Ch	Swisher, 1967a
	85;25[b]	BAS	1d	Ch	Bouveng, 1968
	50;95	CAS	1½;6h	Ch	
	21;86	BOD	5;20d	O_2	Janicke, 1968b
	68;98	CAS	3h	Ch	
	67;96	CAS	3h	Ch	Janicke, 1968c
	70;110	CAS	3h	COD	
	86-95	BOD	20d	O_2	Pfeil, 1968
	100	BAS	12h	Ch	
	59;91	BOD	5;20d	O_2	Thompson, 1968
	90;95	In	25;20d	CO_2;NO_3	
	100	In;RW	6;13d	Ch	
	90;100	CAS;BAS	4;6h	Ch	
	90	*CAS	19h	Ch	Shumate, 1969
	60	*River	½ mile	Ch	
NTA-iron	100	CAS	6h	Ch	Swisher, 1967a

[a]Ethylenediamine tetracetic acid.

[b]At 5°C.

D. 6. *Miscellaneous Aromatics*

TABLE 8.29

Biodegradation of Miscellaneous Aromatic Compounds

Substrate (see § II)	Extent (see § III)	Method (see § IV)	Time (see § V)	Analysis (see § VI)	Reference
a. *Phenol and Derivatives*					
Phenol	32-39	Wa	12h	O_2	McKinney, 1956b
	76	Wa	1d	O_2	Winter, 1962
	18	Wa	4h	O_2	Payne, 1963b
	99	Wa	3h	AAP[a]	Tabak, 1964
	1.4G	Wa	3h	O_2	
	0.6G	Wa	8h	O_2	Malaney, 1966
	99-100	In	7d	AAP[a]	Bunch, 1967a
Phenol E_1	0	In	30d	B	Fincher, 1962
Phenol E_9	94;22	In	9d	CT;σ	Garrison, 1964
	31;9	SF	7d	CT;σ	
	0	Wa	5d	O_2	
b. *Aniline and Derivatives*					
Aniline	1.2G	RW	7d	O_2	Hammerton, 1956
	0.2-1.0G	Wa	7d	O_2	Marion, 1963a
	64	In	20d	O_2	Young, 1968
N-Methylaniline	0.1G	RW	20d	O_2	Hammerton, 1956
N-Dimethylaniline	0	RW	20d	O_2	Hammerton, 1956
Propionanilide	1.0G	RW	7d	O_2	Hammerton, 1956
p-Acetanisidide	0.9G	RW	12d	O_2	Hammerton, 1956
c. *Carboxylates*					
Benzoic acid	37	Wa	12h	O_2	McKinney, 1956b
	1.1G	Wa	7d	O_2	Marion, 1963a
	71	Wa	4h	O_2	Payne, 1963b, Feisal, 1966
	1.2G	Wa	3h	O_2	Tabak, 1964
	43	Wa	?	O_2	Prochazka, 1967
	4.6M	Wa	2-3h	O_2	Heyman, 1968
	78	In	20d	O_2	Young, 1968
Phenylacetic acid	+	In	?	B	Dagley, 1965b
	37	Wa	?	O_2	Prochazka, 1967
	4.1M	Wa	2-3h	O_2	Heyman, 1968

[a]4-Aminoantipyrine.

BIBLIOGRAPHY/AUTHOR INDEX

This index lists each reference cited in the text, (i) alphabetically by the first author's name and (ii) under each first author, chronologically according to the year of publication. Coauthors are included in the same alphabetical sequence, referenced to their publications via the first authors. In the few cases of different first authors with the same name, they are arranged together chronologically and not alphabetically by their initials.

Each publication is referenced, in italics, to the pages where it is cited in this text, so that the functions of both author index and reference index are served.

For the sake of brevity, nonstandard abbreviations have been used for several of the publications:

CA	*Chemical Abstracts*
FSA	*Fette-Seifen-Anstrichmittel*
JAOCS	*Journal of the American Oil Chemists' Society*
JISP	*Institute of Sewage Purification, Journal and Proceedings*
JWPCF	*Journal of the Water Pollution Control Federation*
Purdue Conf.	*Purdue University, Engineering Bulletin, Extension Series*, which includes the Proceedings of the annual Industrial Waste Conferences. Volume number is the number of the conference and year is the year of the conference, not of publication.
Sb. VSChT	*Sborník Vysoké školy chemicko-technologické v Praze, Technologie vody.*
Surf. Cong.	The International Congresses on Surface Activity: #1, Paris, 1954, *1er Congres Mondial de la Detergence et des Produits Tensio-Actifs*, 3 vols., Chambre Syndicale Tramagras, Paris, 1956(?). #2, London, 1957, *Proceedings of the Second International Congress of Surface Activity*, 4 vols., Butterworths, London, 1957.

#3, Cologne, 1960, *Vorträge in Originalfassung des III. Internationalen Kongresses für Grenzflächenaktive Stoffe,* 4 vols., Universitätsdruckerei, Mainz, 1961.

#4, Brussels, 1964, *Chemistry, Physics and Application of Surface Active Substances,* 3 vols., Gordon and Breach, London, 1967.

#5, Barcelona, 1968, *Chimie, Physique et Applications Pratiques des Agents de Surface,* 4 vols., Ediciones Unidas, Barcelona, 1969.

WPR. Conf. The International Conferences on Water Pollution Research:

#1, London, 1962, *Advances in Water Pollution Research,* 3 vols., Pergamon, Oxford, 1964

#2, Tokyo, 1964, *Advances in Water Pollution Research,* 3 vols., Pergamon, 1965.

#3, Munich, 1966, *Advances in Water Pollution Research,* 3 vols., WPCF, Washington, 1967.

#4, Prague, 1969.

#5, San Francisco, 1970.

AASGP (1956) (Association of American Soap and Glycerine Producers), Analytical Subcommittee. Determination of trace amounts of ABS in water. *Anal. Chem.,* **28,** 1822–1826 (1956). *63.*

AASGP (1961), Analytical Subcommittee. Determination of ABS in sewage. *JWPCF,* **33,** 85–91 (1961). *41, 63.*

Abbot (1962), D. C. Colorimetric determination of anionic surfactants in water. *Analyst,* **87,** 286–293 (1962). *52.*

Abbott (1968), B. J. and L. E. Casida, Jr. Oxidation of alkanes to internal mono-alkenes by a *Nocardia. J. Bacteriol.,* **96,** 925–930 (1968). *260, 295, 296.*

ABCM-SAC (1957), Joint Committee on Methods for the Analysis of Trade Effluents. Method for the determination of synthetic detergents. *Analyst,* **82,** 826–834 (1957). *49.*

Abeles (1961), R. H., and H. A. Lee, Jr. An intramolecular oxidation reduction requiring a Vitamin B_{12} coenzyme. *J. Biol. Chem.,* **236,** PC1 (1961). *313.*

Abeles, R. H. *See* Frey (1967).

Abou-Zeid, H. *See* Gebril (1966, 1969).

Abu-Niaaj, F. *See* Gaudy (1965).

Adamse (1968a), A. D. Formation and final composition of the bacterial flora of a dairy waste activated sludge. *Water Res.,* **2,** 665–671 (1968). *179, 180.*

Adamse (1968b), A. D. Bulking of dairy waste activated sludge. *Water Res.,* **2,** 715–722 (1968). *180.*

Afanas'ev, P. V. *See* Vas'kova (1968).

Alexander (1966), M., and B. K. Lustigman. Effect of chemical structure on microbial degradation of substituted benzenes. *J. Agr. Food Chem.*, **14**, 410–413 (1966). *406, 407.*

Alexandre (1967), D. Development of a fast test for the biodegradation of detergents. *Chim. Ind. (Paris)*, **98**, 1443–1448 (1967). *161, 365.*

Aiiso, K., See Ōba (1965f).

Allred (1964a), R. C., and R. L. Huddleston. Manufacture of microbiologically resistant alkylaryl sulfonates from sulfonate mixtures by microbiological oxidation. U.S. Pat. 3,138,543 (1964). *224, 334, 338, 339, 344, 355, 357.*

Allred (1964b), R. C., E. A. Setzkorn, and R. L. Huddleston. Detergent biodegradability as shown by various analytical techniques. *JAOCS*, **41**, 13–17 (1964). *39, 40, 42, 50, 61, 142, 213, 339, 340, 352, 367.*

Allred, R. C. See Huddleston (1963, 1964a, b, 1965b); SDA (1965); Setzkorn (1964).

Anderson (1964), D. A. Growth response of certain bacteria to ABS and other surfactants. *Purdue Conf.*, **19**, 592–601 (1964). *73, 79, 98.*

Anderson (1967), C. E. Fatty acid metabolism (oxidation and biosynethesis). In *The Encyclopedia of Biochemistry* (R. J. Williams and E. M. Lansford, Jr., eds.), Reinhold, New York, 1967, pp. 319–323. *261.*

Anderson, E. J. See Helmers (1950).

Andres, B. See Flynn (1969).

Andrews (1968), J. F. A mathematical model for the continuous culture of microorganisms using inhibitory substrates. *Biotechnol. Bioeng.*, **10**, 707–723 (1968). *183.*

Aoki (1958), K. Interactions of horse serum albumins with anionic and cationic detergents. *J. Am. Chem. Soc.*, **80**, 4904–4909 (1958). *105.*

Aoki (1959), K., and J. Hori. Interactions of egg albumin with detergents. *J. Am. Chem. Soc.*, **81**, 1885–1889 (1959). *104, 105.*

APHA (1960) (American Public Health Association, American Water Works Association and Water Pollution Control Federation). *Standard Methods for the Examination of Water and Wastewater*, 11th ed., APHA, New York, 1960. *71.*

APHA (1965), *Standard Methods for the Examination of Water and Wastewater*, 12th ed., APHA, New York, 1965. *49, 52, 54, 63, 67, 70, 132, 149.*

Archibald (1946), R. M. Determination of lipase activity. *J. Biol. Chem.*, **165**, 443–448 (1946). *326.*

Arigoni, D. See Rétey (1966).

Arkad'eva, Z. A. See Shaposhnikov (1968).

Asahara, T. See Hayano (1968).

Asaka, J.-I. See Ichikawa (1966).

ASTM (1968) (American Society for Testing and Materials). Total and organic carbon in water by combustion-infrared analysis (D2579-67T). In *1968 Book of ASTM Standards*, Part 23, ASTM, Philadelphia, 1968, pp. 836–840. *68.*

Aubert (1969), M., and J.-P. Gambarotta. Study of the effects of biodegradability of toxic chemical products on the marine biological chain. *Rev. Intern. d'Oceanographie Medicale*, **13–14**, 73–105 (1969). *75.*

Ault, W. C. See Maurer (1965).

Baars (1965), J. K. Bacterial activity in pollution abatement. *JISP*, **1965**, 36–44. *78.*

Bacon (1966), L. R. Foam generating method for evaluating biodegradability. *JAOCS*, **43**, 18–25 (1966). *44.*

Baker (1941), Z., R. W. Harrison, and B. F. Miller. Action of synthetic detergent on the metabolism of bacteria. *J. Exptl. Med.*, **73**, 249–271 (1941). *98.*

Ballinger (1962), D. G., and R. J. Lishka. Reliability and precision of BOD and COD determinations. *JWPCF*, **34**, 470–474 (1962). *68, 71.*

Banerji (1966), S. K., B. B. Ewing, R. S. Engelbrecht, and R. E. Speece. Mechanism of starch removal in activated sludge process. *Purdue Conf.*, **21**, 84–102 (1966). *11.*

Banerji, S. K. *See* Ewing (1962).

Barada, M. F. *See* Renn (1959).

Barbaro (1965), R. D., and J. V. Hunter. Effect of clay minerals on surfactant biogradability. *Purdue Conf.*, **20**, 189–196 (1965). *365, 381, 389, 398.*

Barbaro (1967), R. D., and J. V. Hunter. Surfactant adsorption on several homo-ionic forms of kaolin. *Water Res.*, **1**, 157–165 (1967). *113, 353.*

Barber, G. A. *See* Stumpf (1960).

Barden (1957), L., and P. C. G. Isaac. The effect of synthetic detergents on the biological stabilization of sewage. *Proc. Inst. Civil Engrs. (London)*, **6**, 371–405 (1957). *131, 144, 205, 330, 350, 357, 360, 362, 365, 366, 367, 373, 378, 379, 392, 393, 394, 397, 398, 408.*

Barnhart (1963a), E. L. Defining the degradability of synthetic detergents. *Wastes Eng.*, **34**, 646–648 (1963). *50, 351.*

Barnhart (1963b), E. L., and W. W. Eckenfelder, Jr. Criteria for biodegradable syndets. *Biotechnol. Bioeng.*, **5**, 247–254 (1963). *352, 367.*

Barr, E. A. *See* Stirton (1965).

Barrett, M. J. *See* Knowles (1965).

Bartels, T. J. *See* van Eyk (1968).

Barth, D. *See* Schaffer (1965).

Barth, E. F. *See* Bunch (1961).

Beaudoin, R. E. *See* Hurwitz (1960).

Beaujean, P. *See* Leclerc (1952).

Becker, H. *See* Jerchel (1954).

Beignot-Develmont, M. *See* Lambin (1966).

Bernarde (1965), M. A., B. W. Koft, R. Horvath, and L. Shaulis. Microbial degradation of the sulfonate of dodecylbenzenesulfonate. *Appl. Microbiol.*, **13**, 103–105 (1965). *229, 300, 353.*

Bendixen, T. W. *See* Popkin (1968); Robeck (1964).

Benedict, J. H. *See* Osburn (1966); SDA (1969a, b).

Bennett (1961), E. R., and D. W. Ryckman. The effect of ABS shock loadings on the activated sludge process. *Purdue Conf.*, **16**, 52–63 (1961). *351.*

Benson (1963), A. A. The plant sulfolipid. *Advan. Lipid Res.*, **1**, 387–394 (1963). *273.*

Benson, A. A. *See* Martelli (1964).

Berber (1965), J. S., R. V. Rahfuse, and H. W. Wainwright. Preparation of biodegradable synthetic detergents from low-temperature lignite tar. *Ind. Eng. Chem. Prod. Res. Develop.*, **4**, 242–247 (1965). *345.*

Berch, J. *See* Schwartz (1958).

Berg (1966), G., G. Stern, D. Berman, and N. A. Clarke. Stabilization of COD in primary wastewater effluents by inhibition of microbial growth. *JWPCF*, **38**, 1472–1483 (1966). *40.*

Berger (1964), B.   Biodegrability of anionic detergents—influence of the structure of the hydrophobic chain on degradability. *Ind. Chim.* (*Paris*), **51**, 421–431 (1964). *159, 205, 237, 239, 344, 352, 362, 367, 368, 369, 371, 372, 375, 377.*

Bergeron, B. *See* Loehr (1967).

Berman, D. *See* Berg (1966).

Bernheimer, R. *See* Kelly (1965).

Bey (1965), K.   Thin-layer chromatographic analysis in the field of surfactants. *FSA*, **67**, 217–221 (1965).   *60.*

Bhat, J. V. *See* Dias (1964, 1965).

Bhatla (1966), M. N., and A. F. Gaudy, Jr.   Studies on the causation of phasic oxygen uptake in high-energy systems. *J WPCF*, **38**, 1441–1451 (1966).   *84.*

Bhatla, M. N. *See* Gaudy (1963, 1964a, 1965).

BHC (1963) (British Hydrocarbon Chemicals, Ltd.).   Alkylbenzene detergents. Belgian Pat. 632,800 (1963); *CA*, **62**, 5455a (1955).   *357.*

Bishop (1967), D. G., L. Rutberg, and B. Samuelson.   The solubilization of the cytoplasmic membrane of *Bacillus subtilis* by SDS. *European J. Biochem.* **2**, 454–459 (1967).   *101.*

Bistline, R. G., Jr. *See* Stirton (1965).

Blakley (1967), E. R.   The metabolism of aromatic compounds with different side chains by a *Pseudomonas. Can. J. Microbiol.*, **13**, 761–769 (1967).   *267.*

Blakley, E. R. *See* Kurz (1969).

Blank, E. W. *See* Simko (1965).

Blankenship (1963), F. A., and V. M. Piccolini.   Biodegradation of nonionics. *Soap Chem. Specialties*, **39**(12), 75–78, 181 (1963).   *46, 145, 149, 244, 247, 248, 249, 315, 381, 382, 385, 387.*

Blankenship, F. A. *See* Lashen (1966).

Bloch (1965), H. S.   Alkaryl sulfonate production via *n*-olefin isomerization. U.S, Pat. 3,169,987 (1965).   *341, 353.*

Bloch (1967), H. S.   New route to linear alkylbenzenes. *Detergent Age*, **3**(8), 33–35. 105 (1967).   *341, 354.*

Bloodgood, D. E. *See* Johnson (1958); Pahren (1961).

Bloomhuff, R. N. *See* Ludzack (1959).

Bock (1960), K. J.   Biological decomposition of surface active materials. *Surf. Cong. #3*, **3**, 282–286 (1960).   *166, 342, 351, 358.*

Bock (1961), K. J.   Biological investigations on detergents. *Ber. Intern. Vortragstag PRO AQUA* (*Basel*), **1961**, 247–257.   *166, 342, 351, 358.*

Bock, K. J. *See* DAGS (1961); Jendreyko (1962).

Bogan (1954), R. H., and C. N. Sawyer.   Biochemical degradation of synethetic detergents. I. Preliminary studies. *Sewage Ind. Wastes*, **26**, 1069–1080 (1954). *205, 206, 243, 358, 360, 365, 367, 392, 393, 394.*

Bogan (1955), R. H., and C. N. Sawyer.   Biochemical degradation of synthetic detergents. II. Studies on the relation between chemical structure and biochemical oxidation. *Sewage Ind. Wastes*, **27**, 917–928 (1955).   *144, 172, 203, 205, 206, 207, 243, 269, 271, 302, 312, 335, 350, 357, 365, 367, 379, 389, 393, 394, 399, 400, 403, 405, 406.*

Bogan (1956), R. H., and C. N. Sawyer.   Biochemical degradation of synethetic detergents. III. Relationship between biological degradation and froth persistence. *Sewage Ind. Wastes*, **28**, 637–643 (1956).   *42.*

Bogan, R. H. *See* Sawyer (1956).

Böhm, W. *See* Köhler (1969).

Bolle, J. *See* Crauland (1964).

Bolton (1961), H. L., and P. J. Cooper. Analytical problems in the Luton experiment. *JISP*, **1961**, 43–47; discussion 49–56. *45, 49, 51, 63, 194*.

Bolton (1962), H. L., H. L. Webster, and J. Hilton. Collaborative work on the determination of ABS in sewage, sewage effluents, river waters and surface waters. *JISP*, **1962**, 302–308. *53*.

Boman, N. *See* Dean (1967).

Booman (1965), K. A., D. E. Daugherty, J. Dupré, and A. T. Hagler. Degradation studies on branched-chain EO surfactants. *Soap Chem. Specialties*, **41**(1), 60–63, 116, 118–119 (1965). *245, 253, 341, 345, 353, 385, 387, 389, 390*.

Booman (1967), K. A., J. Dupré, and E. S. Lashen. Biodegradable surfactants in the textile industry. *Am. Dyestuff Reptr.*, **56**(3), P82–P88 (1967). *245, 253, 386*.

Booman, K. A. *See* Lashen (1966, 1967a, b); SDA (1965, 1969a, b).

Borecký (1965), J. Identifying organic compounds. LVII. Chromatography of poly EO compounds. *Collection Czech. Chem. Commun.*, **30**, 2549–2557 (1965). *60*.

Borecký (1966), J. Use of paper chromatography in the analysis of surfactants. III Intl. Vortragstagung über grenzflächenaktive Stoffe, Berlin, March 1966; *FSA*, **68**, 572–573 (1966); *Tenside*, **3**(6), (English Supp.), 19 (1966). *60*.

Borstlap (1963), C., and P. L. Kooijman. A study of the biodegradation of anionic synthetic detergents. A new laboratory test. *JAOCS*, **40**, 78–80 (1963). *149, 200, 343, 351, 367*.

Borstlap (1964), C. Intermediate biodegradation products of anionic detergents; their toxicity and foaming properties. *Surf. Cong. #4*, **3**, 891–901 (1967). *274, 276, 277, 300, 309*.

Borstlap (1967a), C., and C. Kortland. Biodegradability of nonionic surfactants under aerobic conditions. *FSA*, **69**, 736–738 (1967). *69, 244, 245, 246, 249, 252, 312, 318, 347, 368, 375, 381, 382, 383, 385, 387, 389, 390, 404, 405*.

Borstlap (1967b), C., and C. Kortland. Effect of substituents in the aromatic nucleus on the biodegradation behavior of alkaryl sulfonates. *JAOCS*, **44**, 295–297 (1967). *223, 230, 231, 341, 347, 354, 360, 361, 368*.

Bott, R. F. *See* Merrell (1967).

Bott (1969), T. L., and T. D. Brock. Bacterial growth rates above 90°C in Yellowstone hot springs. *Science*, **164**, 1411–1412 (1969). *9*.

Bouveng (1968), H. O., G. Davisson, and E.-M. Steinberg. NTA in sewage treatment. *Vatten*, **24**, 348–359 (1968). *409*.

Boyle, O. W. *See* Pipes (1963a).

Boyle, W. *See* Crabtree (1966).

Brand (1956), B. P., and P. Johnson. Interaction of SDS with legumin. *Trans. Faraday Soc.*, **52**, 438–451 (1956). *104*.

Brebion (1964), G., R. Cabridenc, and A. Lerenard. Evaluation of the biodegradation of an ethoxylated tallow sucroglyceride. *Rev. Franc. Corps Gras*, **11**, 191–204 (1964). *325, 394, 396*.

Brebion (1966), G., R. Cabridenc, and T. Jullig. Evaluation of the biodegradability of detergents by a static method. *Tribune CEBEDEAU*, **18**(266), 13–18 (1966). *74, 78, 151, 240, 325, 354, 369*.

Brendish, K. *See* Eden (1964).

Brenner (1968), T. E.   The impact of biodegradable surfactants on water quality. *JAOCS*, **45**, 433–436 (1968).   *197.*

Brenner, T. E. *See* SDA (1965, 1969a, b).

Briggs, D. R. *See* Hill (1956).

Briggs, R. *See* Downing (1965).

Bringmann (1963), G., and W. Janicke.   Alkylaryl sufonates and biodegradation in water. *Gesundh.-Ingr.*, **84**, 330–333 (1963).   *75, 97, 100, 343, 351.*

Bringmann, G. *See* Meinck (1961).

Brink (1966), R. H., Jr., and J. A. Meyers III.   Anionic surfactant biodegradability studies by Warburg respirometry. *JAOCS*, **43**, 449–451 (1966).   *65, 142, 144, 145, 334, 335, 336, 337, 338, 346, 354, 360, 368, 369.*

Briscoe (1969), E. R. E.   Pasveer oxidation system. *Process Biochem.*, **4**(6), 32–34 (1969).   *163.*

Brock (1967), T. D.   Life at high temperatures. *Science*, **158**, 1012–1019 (1967).   *9.*

Brock, T. D. *See* Bott (1969).

Brookhart, J. D. *See* Shumate (1969).

Brooman, D. *See* Burbank (1964).

Brown (1965), T. J.   A study of protozoa in a diffused-air activated sludge plant. *JISP*, **1965**, 375–378.   *78.*

Brown (1969), M. R. W., J. H. S. Foster, and J. R. Clamp.   Composition of *Pseudomonas aeruginosa* slime. *Biochem. J.*, **112**, 521–525 (1969).   *161.*

Bruce (1966), A. M., J. D. Swanwick, and R. A. Ownsworth.   Synthetic detergents and sludge digestion: some plant observations. *JISP*, **1966**, 427–447.   *41, 100, 116, 189, 354, 362, 372, 391.*

Bruce (1969), A. M.   Percolating filters. *Process Biochem.* **4**(4), 19–23 (1969).   *156.*

Bunch (1961), R. L., E. F. Barth, and M. B. Ettinger.   Organic materials in secondary effluents. *JWPCF*, **33**, 122–126 (1961).   *4, 135.*

Bunch (1967a), R. L., and C. W. Chambers.   A biodegradability test for organic compounds. *JWPCF*, **39**, 181–187 (1967).   *40, 155, 246, 341, 347, 375, 376, 377, 381, 385, 389, 391, 404, 410.*

Bunch (1967b), R. L., and M. B. Ettinger.   Biodegradability of potential organic substitutes for phosphates. *Purdue Conf.*, **22**, 393–396 (1967).   *409.*

Bunch, R. L. *See* Murtaugh (1965); WPCF (1967).

Bungay, H. R., III. *See* Shindala (1965).

Bunker, H. J. *See* Burnop (1960).

Burbank (1964), N. C., Jr., J. T. Cookson, J. Goeppner, and D. Brooman.   Isolation and identification of anaerobic and facultative bacteria present in the digestion process. *Purdue Conf.*, **19**, 552–577 (1964).   *79.*

Burbank, N. C. Jr. *See* Cookson (1965).

Bürger (1963a), K.   Methods for the micro determination and trace detection of surfactants. III. Trace detection and determination of surface active poly EO compounds and polyethylene glycols. *Z. Anal. Chem.*, **196**, 251–259 (1963).   *57.*

Bürger (1963b), K.   Methods for the micro determination and trace detection of surfactants. IV. Thin layer chromatography method for determination of the molecular weight distribution and the degree of ethoxylation of poly EO compounds. *Z. Anal. Chem.* **196**, 259–268 (1963).   *59.*

Bürger (1964), K. Methods for the micro determination and trace detection of surfactants. V. Sedimetric determination of oxyethylates and free polyethylene glycols. *Z. Anal. Chem.*, **199**, 434–438 (1964). *57.*

Bürger (1967), K. Mechanism of aerobic biodegradation of nonionic surfactants and its analytical characterization. *Münchner Beitr. Abwasser-, Fisch.- Flussbiol.*, **9**, (2nd ed.), 56–63 (1967). *60, 323.*

Burgess (1962), S. G., and L. B. Wood. Some notes on removal and disposal of synthetic detergents in sewage effluents. *JISP*, **1962**, 158–168. *115.*

Burnette, L. W. *See* SDA (1964).

Burnop (1960), V. C. E., and H. J. Bunker. Synthetic detergents with facile biodegradation. *Centre Belge Etude Doc. Eaux, Bull. Trimestr. CEBEDEAU*, **4**(50), 262–268 (1960). *159, 335, 336, 337, 342, 351, 358, 380.*

Burns, R. O. *See* Weeks (1969).

Burris, R. H. *See* Dietrich (1967).

Burttschell (1966), R. H. Determination of EO based nonionic detergents in sewage. *JAOCS*, **43**, 366–370 (1966). *58.*

Busch (1961), A. W., and N. Myrick. Aerobic bacterial degradation of glucose. *JWPCF*, **33**, 897–905 (1961). *83, 142.*

Busch (1963), A. W. Process kinetics and design criteria for bio-oxidation of petro-chemical wastes. *J. Eng. Ind.*, **1963**, 163–172. *84.*

Busch (1966), A. W. Energy, total carbon and oxygen demand. *Water Resources Res.*, **2**, 59–69 (1966). *69.*

Busch (1968), P. I., and W. Stumm. Chemical interactions in the aggregation of bacteria. Bioflocculation in waste treatment. *Environ. Sci. Technol.*, **2**, 49–53 (1968). *162.*

Busch, A. W. *See* McLellan (1968).

Butterfield (1937), C. T., C. C. Ruchhoft, and P. D. McNamee. Studies on sewage purification. V. Biochemical oxidation by sludges developed by pure cultures of bacteria isolated from activated sludge. *Sewage Works J.* **9**, 173–196 (1937). *132, 134, 161, 171.*

Buzzell, J. C. Jr. *See* Young (1968).

Byrd, J. F. *See* WPCF (1967).

Bywaters, A. *See* Painter (1961).

Cabridenc, R. *See* Brebion (1964, 1966).

Cain (1968), R. B., and D. R. Farr. Metabolism of aryl sulfonates by microorganisms. *Biochem. J.*, **106**, 859–877 (1968). *271, 272, 292, 296.*

Calaway (1968), W. T. The metazoa of waste treatment processes—rotifers. *JWPCF*, **40**, R412–R422 (1968). *78, 162.*

Caldwell, D. W. *See* Swisher (1967a).

Campagnoli, J. M. *See* Papenmeier (1969).

Canfield, K. S. *See* Swenson (1965).

Carel, A. B. *See* Setzkorn (1963).

Carlson, S. *See* Kempf (1968).

Carr, L. P. *See* Nadeau (1964).

Carrère, C. *See* Lambin (1966).

Casida, L. E. Jr. *See* Abbott (1968).

Cassell (1966), E. A., F. T. Sulzer, and J. C. Lamb, III. Population dynamics and selection in continuous mixed cultures. *JWPCF*, **38**, 1398–1409 (1966). *178.*

Černý, J. *See* Pitter (1967b).

Chambers, C. W. *See* Bunch (1967a); Tabak (1964).

Chang, S. L. *See* Metzler (1958).

Cheeseman (1968), G. C. A preliminary study by gel filtration and ultracentrifugation of the interaction of bovine milk casein with detergents. *J. Dairy Res.*, **35**, 439–445 (1968). *104, 106.*

Chian (1968), S. K., and R. I. Mateles. Growth of mixed cultures on mixed substrates. *Appl. Microbiol.*, **16**, 1337–1342 (1968). *179.*

Chian, S. K. *See* Mateles (1969).

Chin, C. C. *See* Gorin (1967).

Chudoba, J. *See* Pitter (1966b).

Ciattoni (1968), P., and S. Scardigno. New contributions to the knowledge of the biodegradability of LAS. *Riv. Ital. Sostanze Grasse*, **45**(1), 15–26 (1968). *136, 218, 340, 341, 342.*

Cibulka (1967), J. J., and G. W. Malaney. Experimental conditions in the study of the physiological ecology of activated sludge. *Purdue Conf.*, **22**, 78–91 (1967). *125.*

Claesson, S. *See* Dean (1967).

Clamp, J. R. *See* Brown (1969).

Clarke, N. A. *See* Berg (1966); Metzler (1958).

Coackley (1969), P. Some aspects of activated sludge. *Process Biochem.*, **4** (10), 27–29, 37(1969). *162.*

Cockburn, A. *See* Curds (1968).

Cohen (1963), C. A. Alkylbenzenesulfonate detergents. U.S. Pat. 3,115,530 (1963). *337, 351.*

Cohen (1965), C. A. Perhydro bis-isoprenyl alkyl aryl sulfonates. U.S. Pat. 3,196,174 (1965). *226, 336, 337, 353.*

Cohen (1966), C. A. Alkyl aryl sulfonate detergents. U.S. Pat. 3,234,297 (1966). *354, 356.*

Cohen, J. M. *See* Robeck (1963).

Cohn (1968), M. M. The Long Island problem: detergents and cesspools. *Hydrocarbon Process.*, **47**(3), 103–108 (1968). *196.*

Coler (1969), R. A., and H. B. Gunner. Microbial populations as determinants in protozoan succession. *Water Res.*, **3**, 149–156 (1969). *78.*

Compton, J. W. *See* SDA (1965).

Comstock, R. F. *See* Gloyna (1952).

Conway (1965), R. A., C. A. Vath, and C. E. Renn. New detergent nonionics—biodegradable. *Water Works Wastes Eng.*, **2**(1), 28–31 (1965). *120, 121, 191, 376, 380, 385, 395.*

Conway (1966), R. A., and G. T. Waggy. Biodegradation testing of typical surfactants in industrial usage. *Am. Dyestuff Reptr.*, **55**(16), P607-P614 (1966). *73, 376, 385, 386, 388, 389, 391.*

Conway, R. A. *See* SDA (1965, 1969a, b).

Cook (1968), R. The bacterial degradation of synthetic anionic detergents. *Water Res.*, **2**, 849–876 (1968). *86, 126, 127, 155, 199, 348, 355, 359.*

Cookson (1965), J. T., and N. C. Burbank, Jr. Isolation and identification of anaerobic and facultative bacteria present in the digestion process. *JWPCF*, **37**, 822–841 (1965). *79.*

Cookson, J. T. *See* Burbank (1964).

Cooper, P. J. *See* Bolton (1961).

Cooper (1964), R. S., and A. D. Urfer. Sodium dialkyl phosphates: surfactant properties and use in heavy duty detergents. *JAOCS*, **41**, 337–340 (1964). *352, 367, 379.*

Cordon (1964), T. C., E. W. Maurer, J. K. Weil, and A. J. Stirton. Biodegradation of esters of α-sulfo fatty acids in activated sludge. *Develop. Ind. Microbiol.*, **6**, 3–15 (1964). *205, 340, 344, 352, 364, 369, 372.*

Cordon (1968a), T. C., E. W. Maurer, M. V. Nuñez-Ponzoa, and A. J. Stirton. Metabolism of some anionic tallow-based detergents by sewage microorganisms. *Appl. Microbiol.*, **16**, 48–52 (1968). *336, 364, 369, 372, 375, 378.*

Cordon (1968b), T. C., E. W. Maurer, O. Panasiuk, and A. J. Stirton. Analysis for sulfate ion in the biodegradation of anionic detergents. *JAOCS*, **45**, 560–562 (1968). *54, 132, 348, 369, 375.*

Cordon, T. C. *See* Maurer (1965).

Coughlin, F. J. *See* Weaver (1964).

Courtier, A. *See* Crauland (1964).

Cowgill (1968), R. W. Fluorescence and protein structure. XVI. Detergents bound to muscle proteins. *Biochim. Biophys. Acta*, **168**, 439–446 (1968). *104, 105, 106.*

Crabb (1964), N. T., and H. E. Persinger. The determination of poly EO nonionic surfactants in water at the parts per million level. *JAOCS*, **41**, 752–755 (1964). *56.*

Crabb (1968), N. T., and H. E. Persinger. A determination of the apparent molar absorption coefficients of the cobalt thiocyanate complexes of nonylphenol EO adducts. *JAOCS*, **45**, 611–615 (1968). *56.*

Crabtree (1966), K., W. Boyle, E. McCoy, and G. A. Rohlich. A mechanism for floc formation by *Zoogloea ramigera*. *JWPCF*, **38**, 1968–1980 (1966). *161, 162.*

Crauland (1964), M., A. Courtier and J. Bolle. Influence of structure on detergency and biodegradability of some alkyl sulfates, alkylbenzene sulfonates and alkane sulfonates. *Surf. Cong. #4*, **1**, 93–103 (1967). *205, 240, 335, 340, 352, 362, 369, 372, 373, 375.*

Cripps, J. M. *See* Maehler (1967).

Crisler, R. O. *See* Frazee (1964a, b).

Crosby, E. S. *See* Manganelli (1953); McGauhey (1957).

Crummett, W. B. *See* Skelly (1966).

Crutchfield, M. M. *See* Swisher (1967a).

Culbert, K. *See* Shindala (1965).

Culp, R. L. *See* Metzler (1958).

Curds (1968), C. R., A. Cockburn, and J. M. Vandyke. An experimental study of the role of the ciliated protozoa in the activated sludge process. *Water Pollution Control*, **67**, 312–329 (1968). *78.*

Čuta (1964), J., and J. Hanušová. Hygienic problems of detergents. IV. Biochemical oxidation of anionic and nonionic surfactants. *Cesk. Hyg.* **9**, 507–516 (1964). *205, 352, 357, 358, 360, 365, 368, 369, 375, 381, 383, 393, 394, 398.*

Czok (1968), R., G. Kaiser, and G. Täuber. Interaction of anionic surfactants with enzymes. *Surf. Cong. #5*, **3**, 337–346 (1969). *107.*

Dagley (1965a), S. Degradation of the benzene nucleus by bacteria. *Sci. Progr.* (*London*), **53**, 381–392 (1965). *263.*

Dagley (1965b), S., and J. M. Wood. Oxidation of phenylacetic acid by a *Pseudomonas. Biochim. Biophys. Acta*, **99**, 383–385 (1965). *267, 410.*

DAGS (1961) (Deutschen Ausschuss für Grenzflächenaktive Stoffe), Fachkommission "Abwasser." Critical observations on the utility of the Longwell-Maniece method. *Gas-Wasserfach*, **52**, 1426–1427 (1961). *53.*

Darragh, J. L. *See* House (1954).

Daugherty, D. E. *See* Booman (1965).

Davis (1961), J. B., and R. L. Raymond. Oxidation of alkyl substituted hydrocarbons by a *Nocardia* during growth on *n*-alkanes. *Appl. Microbiol.*, **9**, 383–388 (1961). *266, 401.*

Davis (1962), W. M. (Development of miniature continuous flow activated sludge unit.) Unpublished work, 1962. *169.*

Davis (1967), E. M., and E. F. Gloyna. Biodegradability of nonionic and anionic surfactants by blue-green and green algae. Report to the Center for Research in Water Resources, CRWR 20A, University of Texas, Austin, 1967. [Later versions in *JWPCF*, **41**, 1494–1504 (1969) and *JAOCS*, **46**, 604–608 (1969).] *77, 347, 354, 386, 390.*

Davisson, G. *See* Bouveng (1968).

Dawes (1962), E. A., and D. W. Ribbons. The endogenous metabolism of microorganisms. *Ann. Rev. Microbiol.*, **16**, 241–264 (1962). *141.*

Dawes (1964), E. A., and D. W. Ribbons. Some aspects of the endogenous metabolism of bacteria. *Bacteriol. Rev.*, **28**, 126–149 (1964). *81, 141.*

Dawes (1965), E. A., and D. W. Ribbons. Studies on the endogenous metabolism of *Escherichia coli*. *Biochem. J.*, **95**, 332–343 (1965). *141.*

Dawson, P. S. S. *See* Kurz (1969).

Dean (1964), A. C. R., and C. Hinshelwood. Some basic aspects of cell regulation. *Nature*, **201**, 232–239 (1964). *81, 89.*

Dean (1967), R. B., S. Claesson, N. Gellerstedt, and N. Boman. An electron microscope study of colloids in waste water. *Environ. Sci. Technol.*, **1**, 147–150 (1967). *163.*

Dean, C. L. See Shumate (1969).

De Bolt (1965), D. C. Determination of ppm of anionic surfactants by automated colorimetric analysis. *Detergent Age*, **1**(11), 18–21 (1965). *50.*

Decker (1966), R. V., and J. F. Foster. The interaction of bovine plasma albumin with detergent anions. Stoichiometry and mechanism of binding of alkylbenzenesulfonates. *Biochemistry (ACS)*, **5**, 1242–1254 (1966). *102.*

Decker (1967), R. V., and J. F. Foster. Amphoteric behavior of bovine plasma albumin and its detergent complexes. *J. Biol. Chem.*, **242**, 1526–1532 (1967). *102, 103.*

De Crombrugghe (1969), B., R. L. Perlman, H. E. Varmus, and I. Pastan. Regulation of inducible enzyme synthesis in *Escherichia coli* by cyclic adenosine 3′,5′-monophosphate. *J. Biol. Chem.*, **244**, 5828–5835 (1969). *88.*

Dee, R. J. *See* Ellerker (1968).

Degens (1950), P. N., Jr., H. van der Zee, J. D. Kommer, and A. H. Kamphuis. Synthetic detergents and sewage processing. V. Effect of synthetic detergents on certain water fauna. *JISP*, **1950**, 63–68. *12, 146, 366, 370, 373, 392.*

Degens (1953), P. N. Jr., H. C. Evans, J. D. Kommer, and P. A. Winsor. Determination of sulfate and sulfonate anion-active detergents in sewage. *J. Appl. Chem. (London)*, **3**, 54–61 (1953). *49.*

Degens (1955), P. N., Jr., H. van der Zee, and J. D. Kommer. Influence of anionic detergents on the diffused air activated sludge process. *Sewage Ind. Wastes*, **27**, 10–25 (1955). *42, 115, 158, 165, 350, 373.*

De Jong (1964), A. L. Biological degradation of detergents in a sewage treatment plant. *Air Water Pollution*, **8**, 591–608 (1964). *190, 194, 344, 352, 369.*

De Jong (1965), A. L. The biodegradation of detergents under practical conditions. *FSA*, **67**, 41–45 (1965). *190, 194, 344, 352, 369.*

De Jong (1967), A. L., and M. C. Testa. Biodegradable detergents: present position and some guidance for selection of raw materials. *Ind. Chim. Belge*, **32**, (Special No.) III, 326–330 (1967). *201, 218, 340, 341, 342.*

De Jong (1969), A. L. Determination of anionic surfactants with the Autoanalyzer. *FSA*, **71**, 567–569 (1969). *50.*

De Ley (1964a), J. *Pseudomonas* and related genera. *Ann. Rev. Microbiol.*, **18**, 17–46 (1964). *80.*

De Ley (1964b), J., and K. Kersters. Oxidation of aliphatic glycols by acetic acid bacteria. *Bacteriol. Rev.*, **28**, 164–180 (1964). *313.*

Del Valle-Rivera, L. A. *See* Symons (1962).

Denner (1969), W. H. B., A. H. Olavsen, G. M. Powell, and K. S. Dodgson. The metabolism of potassium dodecyl (^{35}S)-sulfate in the rat. *Biochem. J.*, **111**, 43–51 (1969). *307.*

Denton, R. S. *See* Painter (1968).

Desmond, C. T. *See* SDA (1964).

Dias (1964), F. F., and J. V. Bhat. Microbial ecology of activated sludge. I. Dominant bacteria. *Appl. Microbiol.*, **12**, 412–417 (1964). *78.*

Dias (1965), F. F., and J. V. Bhat. Microbial ecology of activated sludge. II. Bacteriophages, *Bdellovibrio*, coliforms, and other organisms. *Appl. Microbiol.*, **13**, 257–261 (1965). *78.*

Diddams, D. G. *See* Teletzke (1967).

Dietrich (1967), S. M. C., and R. H. Burris. Effect of exogenous substrates on the endogenous respiration of bacteria. *J. Bacteriol.*, **93**, 1467–1470 (1967). *141.*

Dodgson, K. S. *See* Denner (1969).

Dondero, N. C. *See* Prakasam (1964, 1967a, b).

Donovan, E. J. *See* McKinney (1959b).

Doudoroff, M. *See* Stanier (1966).

Douros (1967), J. D., Jr., and J. W. Frankenfeld. Fermentation process for preparing cinnamic acid and 5-phenyl valeric acid. U.S. Pat. 3,301,766 (1967). *267, 400.*

Douros (1968), J. D., Jr., and J. W. Frandenfeld. Oxidation of alkylbenzenes by a strain of *Micrococcus cerificans* growing on *n*-paraffins. *Appl. Microbiol.*, **16**, 532–533 (1968). *267.*

Dowben (1961), R. M., and W. R. Koehler. The interaction of a nonionic detergent with protein. I. Physical properties of the protein–detergent complex. *Arch. Biochem. Biophys.*, **93**, 496–500 (1961). *106.*

Dowben, R. M. *See* Koehler (1961).

Downing (1964), A. L., H. A. Painter, and G. Knowles. Nitrification in the activated sludge process. *JISP*, **1964**, 130–158. *183.*

Downing (1965), A. L., G. E. Eden, and R. Briggs. Some recently developed instruments: Can they assist sewage plant operation? *JISP*, **1965**, 74–92. *71.*

Downing (1966), A. L., and G. Knowles. Population dynamics in biological treatment plants. *WPRConf. #3*, **2**, 117–142 (1967). *177, 183.*

Downing, A. L. *See* Knowles (1965).

Drewry (1963), J. Qualitative examination of detergents by paper chromatography. *Analyst*, **88**, 225–231 (1963). *60.*

Drewry (1964), J. Examination of detergents by paper chromatography. *Analyst*, **89**, 75–76 (1964). *60.*

Drogin, R. *See* Livingston (1965).

Dronkers (1964), H., and A. P. van der Vet. Investigation into the mechanism of microbiological breakdown of dodecyl sulfate. *Surf. Cong. #4*, **3**, 841–848 (1967). *305, 306.*

Duff, R. S. *See* Webley (1956).

Dupré, J. *See* Booman (1965, 1967); Lashen (1966, 1967a).

Duthie, J. R. *See* SDA (1969a, b); Thompson (1968).

Dychdala (1968), G. R. Acid-anionic surfactant sanitizers. In *Disinfection, Sterilization and Preservation* (C. A. Lawrence and S. S. Block, eds.), Lea & Febiger, Philadelphia, 1968, pp. 253–256. *96.*

EAWAG (1968) (Eidgenössische Anstalt für Wasserversorgung, Abwasserreinigung und gewässerschutz an der ETH Zürich). Working procedure for determination of biodegradability of detergents, 5th draft, Zürich, 1968. *155.*

Eckenfelder, W. W., Jr. *See* Barnhart (1963b).

Edeline (1965), F., and G. Lambert. A rapid method for measuring biodegradability. *Tribune CEBEDEAU*, **18**, 120–126 (1965). *159, 353, 368.*

Eden (1961a), G. E., and G. A. Truesdale. Behavior of a new synthetic detergent in sewage treatment processes. *JISP*, **1961**, 30–42 (discussion 49–56). *190, 194, 342, 350, 351.*

Eden (1961b), G. E., and G. A. Truesdale. The destruction of ABS in sewage treatment processes. *Water Sewage Works*, **108**, 275–279 (1961). *194.*

Eden (1962), G. E. Preservation of samples. Discussion on Bolton (1962). *JISP*, **1962**, 306. *40.*

Eden (1964), G. E., K. Brendish, and B. R. Harvey. Measurement and significance of retention in percolating filters. *JISP*, **1964**, 513–525. *156.*

Eden (1965), G. E., and G. A. Truesdale. Synthetic detergents and water pollution. *Surface Activity and the Microbial Cell, Soc. Chem. Ind. (London), Monograph No. 19.* 1965, pp. 273–285. *157, 158, 164, 345, 353, 358.*

Eden (1967), R. E. BOD determination using a dissolved oxygen meter. *Water Pollution Control*, **66**, 537–539 (1967). *71.*

Eden (1968), G. E., G. A. Truesdale, and G. V. Stennett. Assessment of biodegradability. *Water Pollution Control*, **67**, 107–123 (1968). *154, 155, 335, 336, 338, 348, 355, 359, 363.*

Eden, G. E. *See* Downing (1965); Truesdale (1968).

Edington, M. A. *See* Jones (1960b).

Edwards (1954), G. P., and M. Ginn. Determination of synthetic detergents in sewage. *Sewage Ind. Wastes*, **26**, 945–953 (1954). *49, 53.*

Eiseman (1969), F. S., Jr., and L. M. Schenk. Low foaming biodegradable surfactant compositions. U.S. Pat. 3,426,077 (1969). *394, 395.*

Eldib (1963), I. A. Testing biodegradability of detergents. *Soap Chem. Specialties*, **39**(6), 59–63, 109–112 (1963). *351, 367, 391.*

Eldib (1965), I. A.   Biodegradability of amphoteric detergents. *Soap Chem. Specialties*, **41**(5), 77–80, 161, 163–165 (1965).   *353, 379.*

Ellerker (1968), R., H. J. Dee, F. G. I. Lax, and D. A. Sargent.   Application of chromatography in the examination of sewage, sewage sludge and sewage irrigated land. *Water Pollution Control*, **67**, 542–556 (1968).   *60.*

Elton (1949), M.   Synthetic detergents and sewage processing. I. Introduction. *JISP*, **1949**, 351–354.   *1.*

Emery, E. M. *See* Simko (1965).

Engelbrecht (1967), R. M., J. M. Schuck, and R. G. Schultz.   Polymerization process. U.S. Pat. 3,315,009 (1967).   *354.*

Engelbrecht, R. S. *See* Banerji (1966); WPCF (1967).

Epton (1948), S. R.   A new method for the rapid titrimetric analysis of sodium alkyl sulfates and related compounds. *Trans. Faraday Soc.*, **44**, 226–230 (1948).   *53.*

Essenberg, M. K. *See* Frey (1967).

Ettinger (1956), M. B.   Biochemical oxidation characteristics of steam-pollutant organics. *Ind. Eng. Chem.*, **48**, 256–259 (1956).   *7.*

Ettinger (1965), M. B.   How to plan an inconsequential research project. *J. Sanit. Eng. Div., Am. Soc. Civil. Engrs.*, **91**(SA4), 19–22 (1965).   *123.*

Ettinger, M. B.   See Bunch (1961, 1967b); Ludzack (1957, 1959, 1960b, 1961, 1963a, b, 1964); SDA (1965); WPCF (1967).

Evans (1949), H. C., and P. A. Winsor.   Processing of sewage containing wool scouring liquors, etc. *JISP*, **1949**, 365–370.   *380.*

Evans (1950), H. C.   Determination of anionic synthetic detergents in sewage. *J. Soc. Chem. Ind. (London)*, **1950** (Suppl. Issue No. 2), S76–S80.   *49.*

Evans (1963), W. C.   The microbiological degradation of aromatic compounds. *J. Gen. Microbiol.*, **32**, 177–184 (1963).   *263.*

Evans, H. C. *See* Degens (1953).

Evans, R. L. *See* Sullivan (1968).

Ewing (1962), B. B., and S. K. Banerji.   Effect of biological slime on retention of ABS on granular media. *Purdue Conf.*, **17**, 351–365 (1962).   *113, 114.*

Ewing, B. B. *See* Banerji (1966).

Eye (1966), J. D., and C. C. Ritchie.   Measuring BOD with a membrane electrode system. *JWPCF*, **38**, 1430–1440 (1966).   *71.*

Fairing (1956), J. D., and F. R. Short.   Spectrophotometric determination of ABS detergents in surface water and sewage. *Anal. Chem.*, **28**, 1827–1834 (1956).   *40, 41, 42, 49, 115.*

Fairing, J. D. *See* AASGP (1956).

Farmer, V. C. *See* Webley (1956).

Farr, D. R. *See* Cain (1968).

Feisal (1966), V. E.   Bacterial utilization of detergents and related compounds. Ph.D. Thesis, University of Georgia, Athens, 1966. (University Microfilms No. 66–13,596).   *79, 304, 360, 402, 403, 406, 410.*

Feisal, V. E. *See* Payne (1963a).

Feng (1962), T. H.   Exploration of sludge adsorption of syndets. *Water Sewage Works*, **109**, 183–185 (1962).   *44.*

Fields, R. R. *See* Foster (1964).

Fincher (1962a), E. L. Bacterial utilization of ethoxy glycols. Ph.D. Thesis, University of Georgia, Athens, 1962 (University Microfilms No. 62-5402). *311, 312, 404, 405, 410.*

Fincher (1962b), E. L., and W. J. Payne. Bacterial utilization of ether glycols. *Appl. Microbiol.*, **10**, 542–547 (1962). *311, 312, 404, 405, 410.*

Fischer (1958), W. K. Action of high concentrations of anionic surfactants on bacteria. *Arch. Mikrobiol.*, **31**, 33–49 (1958). *96, 98, 99.*

Fischer (1962), W. K. Detection methods for detergents. *Münchner Beiträge zur Abwasser-, Fischerei-, und Flussbiologie*, **9**, 168–183 (1962). *39, 41.*

Fischer (1963), W. K. The closed bottle test: a simple quantitative method for determining the biodegradability of anionic surfactants and other substances. *FSA*, **65**, 37–42 (1963). *143.*

Fischer (1965), W. K. Practical operations and equipment in the (official German) detergent test. *Vom Wasser*, **32**, 168–192 (1965). *40, 140, 167, 168, 345, 353.*

Fischer, E. *See* Knaggs (1965).

Fischer, W. K. *See* Heinz (1964, 1966, 1967, 1968); Spohn (1964b).

Flynn (1969), J. M., F. V. Padar, A. A. Guerrera, B. Andres, and W. Graner. *The Long Island Ground Water Pollution Study.* Final Report, State of New York Department of Health, April, 1969. *196.*

Follett, R. H. *See* Gaudy (1965).

Foote, J. K. *See* AASGP (1961); Sweeney (1964a).

Ford, W. *See* Fuhrmann (1964).

Forney, C., Jr. *See* Kountz (1959).

Forsberg (1967a), C., and G. Lindqvist. On biological degradation of nitrilotri-acetate (NTA). *Life Sci. (Oxford)*, **6**, 1961–1962 (1967). *409.*

Forsberg (1967b), C., and G. Lindqvist. Experimental studies on bacterial degradation of nitrilotriacetate, NTA. *Vatten*, **23**, 265–277 (1967). *379, 409.*

Foster (1954), J. F., and J. T. Yang. On the mode of interaction of surface active cations with ovalbumin and bovine plasma albumin. *J. Am. Chem. Soc.*, **76**, 1015–1019 (1954). *105.*

Foster (1962a), J. W. Bacterial oxidation of hydrocarbons. In *Oxygenases* (O. Hayaishi, ed.), Academic Press, New York, 1962, Chap. 6, pp. 241–271. *258, 263.*

Foster (1962b), J. W. Hydrocarbons as substrates for microorganisms. *Antonie van Leeuwenhoeck, J. Microbiol. Serol.* **28**, 241–274 (1962). *258.*

Foster (1964), D. J., and R. R. Fields. Synthetic detergents based on LAS. *Soap Chem. Specialties*, **40**(8), 49–52 (1964). *289, 344.*

Foster, J. F. *See* Decker (1966, 1967); Yang (1953).

Foster, J. H. S. *See* Brown (1969).

Foster, J. W. *See* Ooyama (1965).

Fowler, L. R. *See* Rammler (1964).

Francis, C. *See* SDA (1964).

Frank, J. *See* Pitter (1964b).

Franke, N. *See* SDA (1965).

Frankenfeld, J. W. *See* Douros (1967, 1968).

Franks (1955), F. Partition chromatography of synthetic detergents. *Nature*, **176**, 693–694 (1955). *60.*

Franks (1956), F.   Paper chromatography with continuous change in solvent composition. I. Separation of fatty acids. II. Separation of surface-active agents. *Analyst*, **81**, 384–393 (1956).   *60.*

Frazee (1964a), C. D., and R. O. Crisler.   Infrared determination of alkyl branching in detergent ABS. *JAOCS*, **41**, 334–335 (1964).   *63.*

Frazee (1964b), C. D., Q. W. Osburn, and R. O. Crisler.   Application of infrared spectroscopy to surfactant degradation studies. *JAOCS*, **41**, 808–812 (1964).   *56, 62, 245, 315, 321, 381, 390, 391.*

Fredricks (1966), K. M.   Experiments in bacterial adaptation. *Nature*, **212**, 539–540 (1966).   *258.*

Frey (1967), P. A., M. K. Essenberg, and R. H. Abeles.   The mechanism of hydrogen transfer in the cobamide coenzyme-dependent dioldehydrase reaction. *J. Biol. Chem.*, **242**, 5369–5377 (1967).   *313.*

Fries, B. A. *See* House (1956).

Frost, S. *See* Hughes (1968b).

Fuhrmann (1964), R., J. van Peppen, and W. Ford.   Role of external variables in degradation of straight-chain ABS. *Soap Chem. Specialties*, **40**(2), 51–53, 106 (1964).   *148, 344.*

Fuhs (1961), G. W.   Microbial degradation of hydrocarbons. *Arch. Mikrobiol.*, **39**, 374–422 (1961). (English translation available from U.S. Dept. of Commerce, Office of Technical Services, Joint Publications Research Service, Washington 25 D.C.)   *258, 263.*

Fuhs (1968), G. W.   Some factors affecting BOD as determined by manometric or manostatic devices. *Wasser- und Abwasser-Forschung*, **1968**, 161–168.   *72, 133.*

Fujiwara (1968), Y., T. Takezono, S. Kyono, S. Sakayanagi, K. Yamasato, and H. Iizuka.   Effect of alkyl chain branching on the biodegradability of alkylbenzene sulfonates. *Yukagaku*, **17**, 396–399 (1968).   *222, 223.*

Fukuzumi, K. *See* Takagi (1964).

Gaffney (1965), P. E.   Carbon dioxide effects on glucose catabolism by mixed microbial cultures. *Appl. Microbiol.*, **13**, 507–510 (1965).   *72, 86, 144.*

Gambarotta, J. -P. *See* Aubert (1969).

Gameson (1962), A. L. H., and V. H. Lewin.   The determination of anionic surface-active materials. *JISP*, **1962**, 288–301.   *39, 52, 342, 351, 365.*

Gannon (1965), J. J., J. R. Pelton, and J. Westfield.   Respirometer assembly for BOD measurement of river waters and biologically treated effluents. *Intern. J. Air Water Pollution*, **9**, 27–40 (1965).   *71.*

Gardner (1967), H. M.   A reference standard for quaternary solutions used for titrating anionic detergents. *JAOCS*, **44**, 157–158 (1967).   *52.*

Garrison (1964), L. J., and R. D. Matson.   A comparison by Warburg respirometry and die-away studies of the degradability of select nonionic surfactants. *JAOCS*, **41**, 799–804 (1964).   *252, 358, 381, 383, 385, 387, 390, 394, 397, 405, 410.*

Garrison, L. J. *See* SDA (1964, 1965, 1969a,b).

Gaston (1963), L. W., and E. R. Stadtman.   Fermentation of ethylene glycol by *Clostridium glycollicum*. *J. Bacteriol.*, **85**, 356–362 (1963).   *314, 404,*

Gaudy (1962), A. F., Jr.   Studies on induction and repression in activated sludge systems. *Appl. Microbiol.*, **10**, 264–271 (1962).   *87.*

Gaudy (1963), A. F., Jr., K. Komolrit, and M. N. Bhatla.   Sequential substrate removal in heterogeneous populations. *JWPCF*, **35**, 903–922 (1963).   *88.*

Gaudy (1964a), A. F., Jr., M. N. Bhatla, and E. T. Gaudy. Use of COD values of bacterial cells in wastewater purification. *Appl. Microbiol.*, **12**, 254–260 (1964). *74*.

Gaudy (1964b), A. F., Jr. K. Komolrit, and E. T. Gaudy. Sequential substrate removal in response to qualitative shock loading of activated sludge systems. *Appl. Microbiol.*, **12**, 280–286 (1964). *88*.

Gaudy (1965), A. F. Jr., M. N. Bhatla, R. H. Follett, and F. Abu-Niaaj. Factors affecting the existence of the plateau during the exertion of BOD. *JWPCF*, **37**, 444–459 (1965). *84*.

Gaudy (1966), A. F., Jr., and E. T. Gaudy. Microbiology of wastewater. *Ann. Rev. Microbiol.*, **20**, 319–336 (1966). *84, 88, 124, 182, 189*.

Gaudy, A. F., Jr. *See* Bhatla (1966); Komolrit (1966a, b); Krishnan (1966); Ramanathan (1969); Rao (1966); Storer (1969); Thabaraj (1969).

Gaudy, E. T. *See* Gaudy (1964a, b, 1966).

Gebhardt, H. *See* DAGS (1961).

Gebril (1966), B. A., and H. Abou-Zeid. Biodegradable alcohol sulfate surfactants from chlorinated Egyptian kerosene. *Tenside*, **3**, 150–154 (1966). *238, 239, 373*.

Gebril (1969), B. A., and H. Abou-Zeid. Biodegradable KBS surfactants from chlorinated Egyptian kerosene. *Tenside*, **6**, 194–197 (1969). *349*.

Gellerstedt, N. *See* Dean (1967).

Gellman, I. *See* Orford (1953); Rudolfs (1949).

Genetelli, E. J. *See* Hunter (1966).

Gerecht, J. F. *See* Gray (1955).

Gerhold (1966), R. M., and G. W. Malaney. Structural determinants in the oxidation of aliphatic compounds by activated sludge. *JWPCF*, **38**, 562–579 (1966). *311, 404*.

Gerhold, R. M. *See* Hanna (1964); Malaney (1969).

German Government (1962). Ordinance on the degradability of detergents in washing and cleaning agents. *Bundesgesetzblatt (Bonn) Part I*, No. **49**, 698–706 (12 Dec. 1962). *13, 131, 132, 166*.

Gibson (1968), D. T. Microbial degradation of aromatic compounds. *Science*, **161**, 1093–1097 (1968). *263*.

Gildenberg (1965), L., and J. R. Trowbridge. Gas-liquid chromatographic separation of ethylene oxide adducts of fatty alcohols via their acetate esters. *JAOCS*, **42**, 69–71 (1965). *62*.

Gilwood, M. E. *See* Hunter (1966).

Ginn, M. *See* Edwards (1954).

Gitchel, W. B. *See* Teletzke (1967).

Glassman (1948), H. N. Surface active agents and their application in bacteriology. *Bacteriol. Rev.*, **12**, 105–148 (1948). *96, 102*.

Glassman (1950), H. N. The interaction of surface active agents and proteins. *Ann. N.Y. Acad. Sci.*, **53**, 91–104 (1950). *105, 106*.

Glassman (1951), H. N., and D. M. Molnan. Precipitation and inhibition of lysozyme by surface-active agents. *Arch. Biochem. Biophys.*, **32**, 170–180 (1951). *106*.

Gloyna (1952), E. F., R. F. Comstock, and C. E. Renn. Rotary tubes as experimental trickling filters. *Sewage Ind. Wastes*, **24**, 1355–1357 (1952). *158*.

Gloyna, E. F. *See* Davis (1967).

Goeppner, J. *See* Burbank (1964).

Goldberg, M. A. *See* SDA (1964, 1965).

Goldthorpe (1949), H. H., W. H. Hillier, C. Lumb, and A. S. C. Lawrence. Problems arising from the disposal of effluents containing synthetic detergents. *Chem. Ind. (London)*, **1949**, 679–682. *1.*

Goldthorpe (1950), H. H., and J. Nixon. Further experiments with synthetic detergents at Huddersfield particularly with respect to their action on percolating beds. *J. Roy. Sanit. Inst.*, **70**(2), 116–127 (1950). *157, 373, 379, 392.*

Goodhue (1966a), C. T., and E. E. Snell. The bacterial degradation of pantothenic acid. I. Over-all nature of the reaction. *Biochemistry (ACS)*, **5**, 393–398 (1966). *298.*

Goodhue (1966b), C. T., and E. E. Snell. The bacterial degradation of pantothenic acid. III. Enzymatic formation of aldopantoic acid. *Biochemistry (AGS)*, **5**, 403–408 (1966). *298.*

Gorin (1967), G., G. Mamiya, and C. C. Chin. Urease. VIII. Its interaction with sodium dodecyl sulfate. *Experentia*, **23**, 443–445 (1967). *107.*

Gould (1962), E. An extractant for microamounts of anionic surfactant bound to large amounts of protein, with subsequent spectrophotometric determination. *Anal. Chem.*, **34**, 567–571 (1962). *42.*

Grado, C. *See* Rammler (1964).

Gram, A. *See* McKinney (1956c).

Graner, W. *See* Flynn (1969).

Gray (1928), P. H. H., and H. G. Thornton. Soil bacteria that decompose certain aromatic compounds. *Zentr. Bakteriol. Parasitenk., Abt. II*, **73**, 74–96 (1928). *132.*

Gray (1955), F. W., J. F. Gerecht, and I. J. Krems. Preparation of model long chain alkylbenzenes and study of their isomeric sulfonation products. *J. Org. Chem.* **20**, 511–524 (1955). *34, 52.*

Green, J. B. *See* Orsanco (1963).

Greenburg, A. E. *See* Maehler (1967).

Greff (1965), R. A., E. A. Setzkorn, and W. D. Leslie. A colorimetric method for the determination of parts per million of nonionic surfactant. *JAOCS*, **42**, 180–185 (1965). *56.*

Griesinger, W. K. *See* Swenson (1965).

Griffith (1957), J. C. The quantitative estimation of a nonionic detergent. *Chem. Ind. (London)*, **1957**, 1041–1042. *64.*

Griggs, S. H. *See* Schaffer (1965).

Guerrera, A. A. *See* Flynn (1969).

Gunner, H. B. *See* Coler (1969).

Hagler, A. T. *See* Booman (1965).

Hakola, H. *See* Nurmikko (1966).

Halliday, J. *See* Odgen (1961).

Halvorson (1968), H., M. Ishaque, and H. Lees. Microbiology of domestic wastes. I. Physiological activity of bacteria indigenous to lagoon operation as a function of seasonal change. *Can. J. Microbiol.*, **14**, 369–376 (1968). *95.*

Halvorson (1969a), H., M. Ishaque, and H. Lees. Microbiology of domestic wastes. II. A comparative study of the seasonal physiological activity of bacteria indigenous to a sewage lagoon. *Can. J. Microbiol.*, **15**, 563–569 (1969). *95.*

Halvorson (1969b), H., and M. Ishaque. Microbiology of domestic wastes. III. Metabolism of LAS-type detergents by bacteria from a sewage lagoon. *Can. J. Microbiol.*, **15**, 571–576 (1969). *94, 340, 341, 349, 355.*

Hammerton (1955), C.   Observations on the decay of synthetic anionic detergents in natural waters. *J. Appl. Chem. (London)*, **5**, 517–524 (1955).   *13, 143, 146, 204, 205, 206, 237, 269, 334, 350, 357, 362, 365, 366, 367, 373, 402, 406.*

Hammerton (1956), C.   Synthetic detergents and water supplies. *Proc. Soc. Water Treat. Exam.*, **5**, 145–174 (1956); reprinted in *JISP*, **1957**, 280–296.   *143, 204, 205, 206, 235, 237, 334, 350, 358, 362, 365, 366, 367, 371, 373, 402, 403, 406, 408, 410.*

Hammerton (1961), C.   Decrease in surface-active material in the River Lee. *JISP*, **1961**, 48–56.   *194.*

Hammerton (1962), C.   The biological oxidation of phenylalkyl sulfonates. Metropolitan Water Board, *40th Report*, London, 1962, pp. 36–39.   *51, 64, 226, 334, 335, 336, 337, 338, 406, 407.*

Han (1967), K. W.   Determination of biodegradability of nonionic surfactants by sulfation and methylene blue extraction. *Tenside*, **4**, 43–45 (1967).   *55, 244, 245, 248, 249, 383, 385, 386, 390, 391, 393, 405.*

Hancock, W. *See* Laws (1959).

Hanna (1964), G. P., Jr., P. J. Weaver, W. D. Sheets, and R. M. Gerhold.   A field study of LAS biodegradation. *Water Sewage Works*, **111**, 478–485, 518–524 (1964).   *190, 344, 352.*

Hanušová, J. *See* Čuta (1964).

Harada (1964a), T.   The formation of sulfatases by *Pseudomonas aeruginosa*. *Biochim. Biophys. Acta*, **81**, 193–196 (1964).   *304.*

Harada (1964b), T., and B. Spencer.   Repression and induction of arylsulfatase synthesis in *Aerobacter aerogenes*. *Biochem. J.*, **93**, 373–378 (1964).   *304.*

Harada, T. *See* Ishii (1959).

Harkness (1966), N.   Bacteria in sewage treatment processes. *JISP*, **1966**, 542–557; discussion in *Water Pollution Control*, **66**, 603–605 (1967).   *78, 79,*

Harkness, N. *See* Jenkins (1967).

Harmeson, R. H. *See* Vogel (1962).

Harrison, R. W. *See* Baker (1941).

Hartmann (1963a), L.   The removal of ABS by microorganisms. *Biotechnol. Bioeng.*, **5**, 331–345 (1963).   *41, 115, 121, 299.*

Hartmann (1963b), L.   Activated sludge floc composition. *Water Sewage Works*, **110**, 262–266 (1963).   *99, 162.*

Hartmann (1966a), L., and H. Mosebach.   Adsorption experiments with new detergents. *Tenside*, **3**, 349–354 (1966); translation in English language supplement, pp. 55–56.   *113, 116, 117.*

Hartmann (1966b), L.   Toxicity of newer surfactants to autotrophic organisms. *Gas-Wasserfach*, **107**, 251–255 (1966).   *98, 100.*

Hartmann (1967), L., P. Wilderer, and W. Staub.   Reaction-kinetic investigations into the biodegradation of modern surfactants through trickle filter organisms. *Tenside*, **4**, 138–143 (1967).   *93, 301, 317, 347, 354, 360, 366, 369, 376, 383, 384, 389, 397.*

Hartmann (1968), L., and M. E. Singrun.   Bacterial adaptation. *Water Sewage Works*, **115**, 289–294 (1968).   *91, 92.*

Harvey, B. R. *See* Eden (1964).

Hashimoto, K. *See* Sato (1963).

Hasselstrom, T. *See* Lang (1965).

Hassis, H. H. *See* Symons (1960).

Hattingh (1963), W. H. J. Activated sludge studies. I. The nitrogen and phosphorus requirements of the microorganisms. II. Influence of nutrition on the respiratory rate of the microorganisms. III. Influence of nutrition on bulking. *Water Waste Treat. J.*, **9**, 380–386, 424–426, 476–480 (1963). *137.*

Hattingh (1967), W. H. J., J. P. Kotzé, P. G. Thiel, D. F. Toerien, and M. L. Siebert. Biological changes during the adaptation of an anaerobic digester to a synthetic substrate. *Water Res.*, **1**, 255–277 (1967). *79, 189.*

Hattingh, W. H. J. *See* Toerien (1969).

Hawkes (1958), J. C., A. J. Neale, and E. R. Lynch. *t*-Butyl structures in TBS. Private communication, 1958. *228.*

Hawkes (1960), J. C., and A. J. Neale. Infrared spectra of monoalkylbenzenes. *Spectrochim. Acta*, **16**, 633–653 (1960). *228.*

Hawkes (1963), H. A. *The Ecology of Waste Water Treatment*, Macmillan, New York, 1963. *3, 78, 94, 156, 162.*

Hayano (1968), S., T. Nihongi, and T. Asahara. Thin layer chromatographic analysis of poly (oxyethylene) nonylphenol ether. *Tenside*, **5**, 80–82 (1968). *60.*

Heinerth (1966), E. The problem of determining small amounts of nonionic surfactants in water and sewage. *Tenside*, **3**, 109–114 (1966); translation in English language supplement, pp. 1–5. *55.*

Heinerth, E. *See* DAGS (1961); Reid (1967, 1968).

Heinz (1964), H. J., and W. K. Fischer. Determination of the biodegradability of detergents. Principles and discussion of old and new procedures. *FSA*, **66**, 685–691 (1964). *124.*

Heinz (1966), H. J., and W. K. Fischer. International status of methods for determining biodegradability of surfactants, and the possibilities for standardization. I. Biological and procedural principles and their application in test method development. *FSA*, **68**, 955–964 (1966). *123, 124, 229.*

Heinz (1967), H. J., and W. K. Fischer. International status of methods for determining biodegradability of surfactants, and the possibilities for standardization. II. Current test methods and suggestions for a combined international standard method. *FSA*, **69**, 188–196 (1967). *121, 124, 150, 155, 244, 347, 354, 363, 368, 369, 373, 375, 376, 381, 385, 396, 408.*

Heinz (1968), H. J., and W. K. Fischer. LAS cuts German water pollution. *Hydrocarbon Process.*, **47**(3), 96–102 (1968). *197.*

Hellwig (1964), D. H. R. Preservation of water samples. *Intern. J. Air Water Pollution*, **8**, 215–228 (1964). *40.*

Hellwig (1967), D. H. R. Preservation of wastewater samples. *Water Res.*, **1**, 79–91 (1967). *40.*

Helmers (1950), E. N., E. J. Anderson, H. D. Kilgore, Jr., L. W. Weinberger, and C. N. Sawyer. Nutritional requirements in the biological stabilization of industrial wastes. I. Experimental method. *Sewage Ind. Wastes*, **22**, 1200–1206 (1950). *172.*

Henderson, C. N. *See* Manganelli (1960).

Henderson, W. S. *See* Suffis (1965).

Hendrix, C. D. *See* SDA (1969a, b).

Henning, F. A. *See* Klein (1969).

Herbert, D. W. M. *See* Mann (1957).

Herbert, S. *See* Reynolds (1967).

Herold, B. *See* Rohlffs (1963).

Herrmann, R. *See* Ruschenberg (1964).

Hess, R. W. *See* AASGP (1956).

Hetling (1964), L. J., D. R. Washington, and S. S. Rao. Kinetics of the steady state bacterial culture. II. Variation in synthesis. *Purdue Conf.*, **19**, 687–715 (1964). *83, 84, 182.*

Hetling (1965), L. J., and D. R. Washington. Kinetics of the steady state bacterial culture. III. Growth rate. *Purdue Conf.* **20**, 254–264 (1965). *182.*

Hetling, L. J. *See* Washington (1964).

Heukelekian (1955), H., and M. C. Rand. BOD of pure organic compounds. *Sewage Ind. Wastes*, **27**, 1040–1053 (1955). *70.*

Heukelekian, H. *See* Hunter (1964, 1965); Manganelli (1960); McWhorter (1962).

Heyman (1967), J. J., and A. H. Molof. Initiation of biodegradation in surfactants. *JWPCF*, **39**, 50–62 (1967). *127, 278, 335, 340, 341, 407.*

Heyman (1968), J. J., and A. H. Molof. Biodegradation of linear alkylated sulfonates. *Environ. Sci. Technol.*, **2**, 773–778 (1968). *127, 272, 285, 286, 287, 407, 410.*

Hill (1956), R. M., and D. R. Briggs. A study of the interaction of *n*-octylbenzene-*p*-sulfonate with β-lactoglobulin. *J. Am. Chem. Soc.*, **78**, 1590–1597 (1956). *104.*

Hill, M. J. *See* Hughes (1968a).

Hillier, W. H. *See* Goldthorpe (1949).

Hilton, J. *See* Bolton (1962).

Hinshelwood, C. *See* Dean (1964).

Hitzman (1964), D. O. New accelerated test for rapid measurement of detergent biodegradability. *JAOCS*, **41**, 593–595 (1964). *151, 344, 352, 368.*

Hoadley (1965), A. W., and E. McCoy. Characterization of certain Gram negative bacteria from surface waters. *Appl. Microbiol.*, **13**, 575–578 (1965). *80.*

Hoffman, C. A. *See* Teletzke (1967).

Holley, C. W. *See* Howe (1969).

Hoover, S. R. *See* Porges (1952).

Hopkins, W. J. *See* Porges (1958).

Hori, J. *See* Aoki (1959).

Horská, M. *See* Pitter (1968d).

Horvath, R. *See* Benarde (1965).

Horwood, M. P. *See* McKinney (1952).

House (1954), R., and J. L. Darragh. Analysis of synthetic anionic detergent compositions. *Anal. Chem.*, **26**, 1492–1497 (1954). *52.*

House (1956), R., and B. A. Fries. Radioactive ABS in activated sludge sewage treatment. *Sewage Ind. Wastes*, **28**, 492–506 (1956). *40, 54, 64, 65, 170, 190, 193, 299, 350.*

House (1965a), R. (Capillary rise in dilute surfactant systems.) Personal communication, 1965. *47.*

House (1965b), R., S. H. Sharman, and D. Kyriacou. Removal from sewage of surfactants resistant to biodegradation. U.S. Pat. 3,203,893 (1965). *45, 229.*

House (1966a), R. (Note on LAS adsorption on sewage solids.) Unpublished communication to SDA Biodegradation Subcommittee, 1966. *116.*

House (1966b), R. (Sulfate formation during TBS biodegradation.) Personal communication, 1966. *299, 301.*

House, R. *See* AASGP (1956, 1961); Knight (1959); Marquis (1966); SDA (1964, 1965).

Houston, C. A. *See* SDA (1965).

Howe (1969), L. H., III, and C. W. Holley. Comparison of mercury(II) chloride and sulfuric acid as preservatives for nitrogen forms in water samples. *Environ. Sci. Technol.*, **3**, 478–481 (1969). *40.*

Howe, R. *See* Jones (1968a).

Hsie (1967), A. W., and H. V. Rickenberg. Catabolite repression in *Escherichia coli:* the role of glucose-6-phosphate. *Biochem. Biophys. Res. Commun.*, **29**, 303–310 (1967). *88.*

Hsu (1963), Y.-C. Detergent (SLS)-splitting enzyme from bacteria. *Nature*, **200**, 1091–1092 (1963). *109, 303, 367, 368.*

Hsu (1965), Y.-C. Detergent-splitting enzyme from *Pseudomonas. Nature*, **207**, 385–388 (1965). *79, 127, 303, 308, 368.*

Huang, S. K. *See* SDA (1964).

Huber (1962), L. Investigations on the decomposition of a new anionic surfactant in an oxidation channel. *Münchner Beiträge zur Abwasser-, Fisch.- Flussbiol.*, **9**, 235–265 (1962). *42, 190, 342.*

Huber (1968), L. Degradation tests with alkylbenzene sulfonates in an oxidation ditch. *Tenside*, **5**, 65–76 (1968). *42, 181, 190, 193, 348, 355, 380.*

Huddleston (1963), R. L., and R. C. Allred. Microbial oxidation of sulfonated alkylbenzenes. *Develop. Ind. Microbiol.*, **4**, 24–38 (1963). *47, 61, 65, 127, 151, 152, 206, 207, 213, 216, 217, 224, 277, 278, 279, 283, 334, 336, 337, 340, 341, 342, 351, 355, 367.*

Huddleston (1964a), R. L., and R. C. Allred. Evaluation of detergents by using activated sludge. *JAOCS*, **41**, 732–735 (1964). *50, 86, 132, 165, 340, 344, 352, 368, 383, 390.*

Huddleston (1964b), R. L., and R. C. Allred. Effect of structure on biodegradation of nonionic surfactants. *Surf. Cong. #4*, **3**, 871–882 (1967). *132, 246, 247, 250, 252, 381, 382, 383, 385, 387, 388, 390.*

Huddleston (1965a), R. L., and E. A. Setzkorn. Biodegradability of detergent hydrotropes/foam stabilizers. *Soap Chem. Specialties*, **41**, (3), 63–64, 120–121 (1965). *397, 398, 406.*

Huddleston (1965b), R. L., and R. C. Allred. Biodegradability of ethoxylated alkyl phenol surfactants. *JAOCS*, **42**, 983–986 (1965). *55, 56, 57, 250, 252, 318, 319, 323, 382, 387, 388, 389.*

Huddleston (1966), R. L. Biodegradable detergents for the textile industry. *Am. Dyestuff Reptr.*, **55**(2), P52–P54 (1966). *382, 387, 389, 391.*

Huddleston, R. L. *See* Allred (1964a, b); Setzkorn (1964, 1965).

Hughes (1968a), W. H., and M. J. Hill. Variation in distribution of inducible enzyme in pure clones of bacteria. *Nature*, **218**, 766–768 (1968). *89.*

Hughes (1968b), W., S. Frost, and V. W. Reid. Analysis of alkylbenzene sulfonates present in sewage. *Surf. Cong. #5*, **1**, 317–325 (1969). *61, 63, 217.*

Humphrey (1967), A. E. A critical review of hydrocarbon fermentations and their industrial utilization. *Biotechnol. Bioeng.*, **9**, 3–24 (1967). *258.*

Hunt, E. C. *See* Patterson (1966a).

Hunter (1964), J. V., and H. Heukelekian. Determination of biodegradability using Warburg respirometric techniques. *Purdue Conf.*, **19**, 616–627 (1964). *144, 244, 245, 344, 362, 364, 365, 368, 381, 393, 394, 397.*

Hunter (1965), J. V., and H. Heukelekian. The composition of domestic sewage fractions. *JWPCF*, **37**, 1142–1163 (1965).   *134.*

Hunter (1966), J. V., E. J. Genetelli, and M. E. Gilwood. Temperature and retention time relationships in the activated sludge process. *Purdue Conf.*, **21**, 953–963 (1966). *94.*

Hunter, J. R. *See* Postgate (1963).

Hunter, J. V. *See* Barbaro (1965, 1967).

Hurley (1950), J. The influence of synthetic detergents on sewage treatment. *JISP*, **1950**, 249–262.   *190, 373.*

Hurley (1952), J. Some experimental work on the effects of synthetic detergents on sewage treatment. *JISP*, **1952**, 306–332.   *189, 190, 373.*

Hurwitz (1960), E., R. E. Beaudoin, T. Lothian, and M. Sniegowski. Assimilation of ABS by an activated sludge treatment plant-waterway system. *JWPCF*, **32**, 1111–1116 (1960).   *5.*

Husmann (1956), W. Detergents in sewage. *Gas-Wasserfach*, **97**, 1031 (1956).   *380.*

Husmann (1962), W. Problems of biodegradation of industrial detergents and means of degradability testing. *J. Soc. Cosmetic Chemists*, **13**, 416–425 (1962).   *166, 167.*

Husmann (1963a), W., F. Malz, and H. Jendreyko. Removal of detergents from wastewaters and streams. *Forschungsber. Landes Nordrhein-Westfalen No. 1153* (1963).   *132, 133, 140, 150, 158, 165, 166, 343, 351, 356, 358, 367, 380.*

Husmann (1963b), K. Degradation of detergents by slow filtration. *Veröffentl. Inst. Siedlungswasserwirtschaft Tech. Hochschule Hannover No. 12* (1963).   *74, 343, 351.*

Husmann (1968), W. Solving the detergent problem in Germany. *Water Pollution Control*, **67**, 80–90 (1968).   *197.*

Hutchinson (1967), E., and K. Shinoda. An outline of the solvent properties of surfactant solutions. In *Solvent Properties of Surfactant Solutions* (K. Shinoda, ed.), Dekker, New York, 1967, pp. 1–26.   *19.*

Huyser (1960), H. W. Relation between the structure of detergents and their bio-degradation. *Surf. Cong. #3*, **3**, 295–301 (1960).   *58, 86, 127, 148, 205, 210, 214, 225, 233, 237, 238, 244, 247, 337, 338, 339, 362, 370, 371, 372, 373, 381, 382, 390.*

Ichikawa (1966), Y., and J.-I. Asaka. Microbial decomposition of alkylbenzene sulfonate. *Shokuhin Eiseigaku Zasshi*, **7**, 403–408 (1966).   *301.*

Iizuka, H. *See* Fujiwara (1968).

Ikeda (1963), K., H. Yamada, and S. Tanaka. Bacterial degradation of taurine. *J. Biochem. (Tokyo)*, **54**, 312–316 (1963).   *273, 406.*

Ilişescu (1966), A., and A. Mavrianopol. Relation between anionic surfactant bio-degradability and microorganism development in the activated sludge treatment. *Studii de Protecţia şi Epurarea Apelor*, **7**, 363–385 (1966).   *79, 359, 374.*

Ilişescu, A. *See* Vaicum (1967).

Imanishi (1965), A., Y. Momotani, and T. Isemura. The interaction of detergents with proteins. The effect of detergents on the conformation of *B. subtilis* α-amylase and Bence-Jones protein. *J. Biochem. (Tokyo)*, **57**, 417–429 (1965).   *104, 105, 106.*

Imelik (1948), B. Oxidation of cyclohexane by *Pseudomonas aeruginosa. Compt. Rend.*, **226**, 2082–2083 (1948).   *399.*

Impiombato, F. S. A. *See* Pitt-Rivers (1968).

Ingols, R. S. *See* WPCF (1967).

Isaac (1958), P. C. G., and D. Jenkins. Biological oxidation of sugar-based detergents. *Chem. Ind. (London)*, **1958**, 976–977.   *350, 396.*

Isaac (1960a), P. C. G., and D. Jenkins. Biological breakdown of some newer synthetic detergents. Conference on Biological Waste Treatment, Manhattan College, New York, 1960, Paper No. 5. In *Advances in Biological Waste Treatment* (W. W. Eckenfelder, Jr., and Brother Joseph McCabe, eds.), Macmillan, New York, 1963. *342, 396.*

Isaac (1960b), P. C. G., and D. Jenkins. A laboratory investigation of the breakdown of some of the newer synthetic detergents in sewage treatment. *JISP*, **1960,** 314–329. *342, 396.*

Isaac, P. C. G. *See* Barden (1957).

Isemura, T. *See* Imanishi (1963).

Ishaque, M. *See* Halvorson (1968, 1969a, b).

Ishii (1959), M., T. Harada, and Z. Nikuni. Utilization of 1,2-propylene glycol by microorganisms. *Nippon Nogei Kagaku Kaishi*, **33,** 889–893 (1959); *CA*, **57,** 2656i (1962). *405.*

Ismail (1965), R. M., and H. Simonis. Biodegradable sugar esters as detergents. *FSA*, **67,** 345–347 (1965). *396.*

Jackson, M. L. *See* KrishnaMurti (1966).

Jacoby (1964), G. A. The induction and repression of amino acid oxidation in *Pseudomonas fluorescens. Biochem. J.*, **92,** 1–8 (1964). *88.*

Jaffe (1962), H. H., and M. Orchin. *Theory and Applications of Ultraviolet Spectroscopy*, Wiley, New York, 1962, Chap. 12. *63.*

James (1965), A. M. Surface active agents in microbiology. *Surface Activity and the Living Cell*, Soc. Chem. Ind. (*London*), *Monograph No. 19*, 1965, pp. 3–23. *96.*

James, K. *See* Jenkins (1967).

Janicke (1968a), W. Indirect determination of biodegradability of nonionic surfactants by organic carbon determination. *Gas-Wasserfach*, **109,** 246–249 (1968). *66, 67, 69, 390.*

Janicke (1968b), W. Biodegradation properties of nitrilotriacetic acid. *Gas-Wasserfach*, **109,** 1181–1184 (1968). *409.*

Janicke (1968c), W. Determination of biodegradation of organic compounds through COD analysis. *Gesundh.-Ind.*, **89,** 309–314 (1968). *66, 67, 68, 341, 355, 371, 390, 409.*

Janicke (1969), W. Acclimation time in the official German test method for surfactant biodegradability. *FSA*, **71,** 843–849 (1969). *168.*

Janicke, W. *See* Bringmann (1963).

Janota-Bassalik (1969), L., C. Olezyk, and M. Kaczorowska. Degradation of some cationic detergents by *Ps. pictorum. Acta Microbiol. Polon.* **18,** 31–34 (1969). *398.*

Jasewicz, L. *See* Porges (1952, 1958).

Jendreyko (1962), H., and K. J. Bock. Pilot plant experiments on biodegradation of newer detergents on trickling filters. *Gas-Wasserfach*, **103,** 615–617 (1962). *190, 342, 380.*

Jendreyko (1963), H., and E. Ruschenberg. The behavior of biologically soft ABS in trickling filters and activated sludge plants. *Gas-Wasserfach*, **104,** 391–396 (1963). *140, 165, 190, 343, 349, 352, 367, 380.*

Jendreyko, H. *See* Husmann (1963a).

Jenkins (1967), S. H., N. Harkness, A. Lennon, and K. James. The biological oxidation of synthetic detergents in recirculating filters. *Water Res.*, **1,** 31–53 (1967). *160, 341, 347, 363, 368, 381, 390, 392.*

Jenkins, D. *See* Isaac (1958, 1960a, b); Klein (1961, 1962, 1963a).

Jenkins, G. F. *See* Lamb (1952).

Jerchel (1953), D., and H. Scheurer. Paper-electrophoretic investigation of action of cationic soaps on proteins. *Z. Naturforsch.*, **8b**, 541–547 (1953). *105.*

Jerchel (1954), D., H. Becker, and K. Schmeiser. Paper-electrophoretic investigation of the action of ¹⁴C labeled cationic soap dodecyltrimethylammonium bromide on serum albumin. *Z. Naturforsch.*, **9b**, 169–172 (1954). *105.*

Jeris, J. S. *See* McKinney (1955).

Johnson (1958), C. C., and D. E. Bloodgood. Effect of vegetable oil and ABS on anaerobic digestion of primary sludge. In *Biological Treatment of Sewage and Industrial Wastes* (Brother Joseph McCabe and W. W. Eckenfelder, Jr., eds.), Vol 2, Reinhold, New York, 1958, pp. 115–125. *189, 350.*

Johnson (1964), M. J. Utilization of hydrocarbons by microorganisms. *Chem. Ind. (London)*, **1964**, 1532–1537. *258.*

Johnson, P. *See* Brand (1956).

Jones (1968a), D. F., and R. Howe. Microbiological oxidation of long chain aliphatic compounds. I. Alkanes and alk-1-enes. II. Branched chain alkanes. III. 1-Halogenoalkanes, 1-cyanohexadecane and 1-alkoxyalkanes. IV. Alkane derivatives having polar terminal groups. V. Mechanism of hydroxylation. *J. Chem. Soc., C.*, **1968**, 2801–2833. *260, 297.*

Jones (1968b), J. G., and M. A. Edington. An ecological survey of hydrocarbon-oxidizing microorganisms. J. Gen. Microbiol., **52**, 381–390 (1968). *79.*

Jones (1968c), P. H., and D. Prasad. The effect of sterilization techniques on wastewater properties. *JWPCF*, **40**, R477–R483 (1968). *40, 134.*

Jones, K. *See* Truesdale (1959, 1962).

Jones, P. H. *See* Pipes (1963b).

Jepling, W. F. *See* Merrell (1967).

Jullig, T. *See* Brebion (1966).

Jungermann, E. *See* SDA (1965).

Kabler, P. W. *See* Tabak (1964).

Kaczorowska, M. *See* Janota-Bassalik (1969).

Kaelble (1963), E. F. Detergent component analysis. *Soap Chem. Specialties*, **39**(10), 56–59, 121, 123 (1963). *26.*

Kaelble, E. F. *See* Swischer (1961).

Kaiser, G. *See* Czok (1968).

Kallio, R. E. *See* McKenna (1964, 1965).

Kamphuis, A. H. *See* Degens (1950).

Kaplan, A. M. *See* Long (1966).

Katko, A. *See* Merrell (1967).

Kay, C. M. *See* McCubbin (1966).

Kazarovets, N. M. *See* Lipman (1963).

Kelly (1965), R. J., M. S. Konecky, J. E. Shewmaker, and R. Bernheimer. Physical and biological removals of detergent actives in a full scale sewage plant. *Surface Activity and the Living Cell, Soc. Chem. Ind. (London), Monograph No. 19*, 1965, pp. 286–304. *41, 50, 191, 336, 345, 353, 362, 368.*

Kelly, R. J. *See* Konecky (1963); Livingston (1965).

Kempf (1968), T., and S. Carlson. Chemical and microbiological decomposition processes during the passage of anionic detergents through the soil in infiltration lysimeters. *Z. Kulturtech. Flurbereinig.*, **9**, 209–226 (1968). *187, 198, 229, 300, 348, 355.*

Kersters, K. *See* de Ley (1964b).

Key (1961), A. Progress toward the solution of the synthetic detergent problem: a symposium. Introductory paper. *JISP*, **1961**, 24–26; discussion 49–56. *194.*

Khorana, H. G. *See* Moffatt (1959).

Kilgore, H. D., Jr. *See* Helmers (1950).

Kimura (1962), Y., K. Ōba, and M. Tobari. Sodium alkylbenzenesulfonate. I. Separation and properties of *o*- and *p*-sodium alkylbenzenesulfonates. *Yukagaku*, **11**, 532–538 (1962). *127, 342, 351, 357.*

Kinnard, L. M. *See* McAteer (1964).

Klein (1961), S. A., D. Jenkins, and P. H. McGauhey. Travel of synthetic detergents with percolating water. *First Annual Report*, University of California, Berkeley, 1962. *187.*

Klein (1962), S. A., D. Jenkins, and P. H. McGauhey. Travel of synthetic detergents with percolating water. *Second Annual Report*, University of California, Berkeley, 1962. *77, 187, 299, 351.*

Klein (1963a), S. A., D. Jenkins, and P. H. McGauhey. The fate of ABS in soils and plants. *JWPCF*, **35**, 636–654 (1963). *77, 113, 187, 299, 351.*

Klein (1963b), S. A., and P. H. McGauhey. The persistence of ABS in wastewaters. *Develop. Ind. Microbiol.*, **5**, 78–84 (1964). *113, 187, 299, 352.*

Klein (1964a), S. A., and P. H. McGauhey. Fate of detergents in septic tank systems and oxidation ponds. *SERL (Sanitory Engineering Research Laboratory) Report 64-1* and supplement, University of California, Berkeley, 1964. *65, 77, 187, 189, 229, 299, 344, 352, 353, 369.*

Klein (1964b), S. A., and P. H. McGauhey. Travel of synthetic detergents with percolating water. *Third Annual Report, SERL 64-2*, University of California, Berkeley, 1964. *187, 189, 299, 352.*

Klein (1965a), S. A. Travel of synthetic detergents with percolating water. *Fourth Annual Report, SERL 65-4*, University of California, Berkeley, 1965. *40, 157, 189, 229, 299, 307, 345, 353, 354, 369.*

Klein (1965b), S. A., and P. H. McGauhey. Biodegradation of biologically soft detergents by wastewater treatment processes. *JWPCF*, **37**, 857–866 (1965). *40, 157, 187, 189, 229, 299, 345, 349, 353, 354, 356, 369.*

Klein (1966), S. A., and P. H. McGauhey. Effects of LAS on the quality of wastewater effluents. *SERL Report 66-5*, University of California, Berkeley, 1966. *229, 230, 299, 346, 354.*

Klein (1969), D. A., and F. A. Henning. Role of alcoholic intermediates in formation of isomeric ketones from *n*-hexadecane by a soil *Arthrobacter*. *Appl. Microbiol.*, **17**, 676–681. (1969). *260.*

Klein, S. A. *See* McGauhey (1957, 1959a, b, 1964, 1966).

Kline, W. A. *See* AASGP (1961); Renn (1964a).

Kloubek (1969), J., and A. W. Neumann. Continuous measurement of time-dependence of surface tension of surfactant solutions. II. Aqueous solutions of sodium dodecyl sulfate. *Tenside*, **6**, 4–10 (1969). *46.*

Klug (1967), M. J., and A. J. Markovetz. Degradation of hydrocarbons by members of the genus *Candida*. II. Oxidation of *n*-alkanes and 1-alkenes by *Candida lipolytica*. *J. Bacteriol.*, **93**, 1847–1852 (1967). *260.*

Knaak (1966), J. B., S. J. Kozbelt, and L. J. Sullivan. Metabolism of 2-ethylhexyl sulfate by the rat and rabbit. *Toxicol. Appl. Pharmacol.*, **8**, 369–379 (1966). *307, 310.*

Knaggs (1964), E. A. Alkylolamides in soft detergents. *Soap Chem. Specialties,* **40**(12), 79–82, 277 (1964). *381, 393, 397.*

Knaggs (1965), E. A., J. A. Yeager, L. Varenyi, and E. Fischer. α-Sulfo fatty esters in biologically soft detergent formulations. *JAOCS,* **42**, 805–810 (1965). *345, 354, 365, 368, 369, 375.*

Knapp (1965), J. W., and J. M. Morgan, Jr. Biodegradability of detergents at Manassas Air Force Station. *Purdue Conf.,* **20**, 737–745 (1965). *40, 191, 194, 345.*

Knight (1959), J. O., and R. House. Analysis of surfactant mixtures. I. *JAOCS,* **36**, 195–200 (1959). *61.*

Knöpp (1961), H. Comparative experimental investigations on the toxicity, foaming potential and degradation of various detergents in river water. *Deutsche Gewässerkundliche Mitteilungen,* **5**, 1–8 (1961). *98, 341, 351.*

Knopp (1965), P. V., L. J. Uhren, G. A. Rohlich, and M. S. Nichols. Field study of the removal of LAS detergents by the activated sludge process. *JAOCS,* **42**, 867–873 (1965). *191, 345, 354.*

Knowles (1965), G., A. L. Downing, and M. J. Barrett. Determination of kinetic constants for nitrifying bacteria in mixed culture, with the aid of an electronic computer. *J. Gen. Microbiol.,* **38**, 263–278 (1965). *183.*

Knowles, C. M. *See* SDA (1965).

Knowles, G. *See* Downing (1964, 1966).

Knox (1961), W. E. Degradation of aromatic compounds. In *Biochemists' Handbook* (C. Long, ed.), Van Nostrand, New York, 1961, pp. 596–601. *261.*

Kobayashi, M. *See* Sato (1963).

Koehler (1961), W. R., and R. M. Dowben. The interaction of nonionic detergent with protein. II. Effects on a bovine plasma albumin-azomercurial complex. *Arch. Biochem. Biophys.,* **93**, 501–507 (1961). *106.*

Koehler, W. R. *See* Dowben (1961).

Koft, B. W. *See* Benarde (1965).

Koga, S. *See* Yano (1969).

Köhler (1969), M., W. Böhm, W. Köppe, and F. Mach. Efficiency of biological treatment systems in biodegradation of detergents. *Wasserwirtsch.-Wassertech.,* **19**, 15–19 (1969). *380.*

Kölbel (1964), H., P. Kurzendörfer, and M. Zahiruddin. Constitution and properties of surfactants. IV. Influence of structure on the aerobic biodegradation of anionic surfactants. *Tenside,* **1**, 7–18 (1964). *127, 205, 220, 232, 233, 236, 334, 335, 337, 338, 353, 360, 362, 366.*

Kölbel (1967), H., P. Kurzendörfer, and C. Werner. Constitution and properties of surfactants. V. The effect of the position of the quaternary carbon atom on the biodegradability of alkyl benzene sulfonates. *Tenside,* **4**, 33–40 (1967). *64, 127, 225, 294, 335, 337, 338, 406, 407.*

Kolbeson, R. A. *See* Moore (1956).

Kommer, J. D. *See* Degens (1950, 1953, 1955).

Komolrit (1966a), K., and A. F. Gaudy, Jr. Biochemical response of continuous-flow activated sludge processes to qualitative shock loadings. *JWPCF,* **38**, 85–101 (1966). *88.*

Komolrit (1966b), K., and A. F. Gaudy, Jr. Substrate interaction during shock loadings to biological treatment processes. *JWPCF,* **38**, 1259–1272 (1966). *88.*

Komolrit, K. *See* Gaudy (1963, 1964b).

Konecky (1963), M. S., R. J. Kelly, J. M. Symons, and P. L. McCarty. The determination of the biodegradability of detergents (Esso Research biodegradation test). Annual Meeting, Water Pollution Control Federation, Seattle, October, 1963. *150, 335, 343, 352, 362, 367, 379, 408.*

Konecky, M. S. *See* Kelly (1965); Nelson (1961).

Kooijman, P. L. *See* Borstlap (1963).

Kopp (1965), R., and J. Müller. Effects of related anionic detergents on flagellation, motility, swarming and growth of *Proteus. Appl. Microbiol.*, **13**, 950–955 (1965). *101.*

Köppe, W. *See* Köhler (1969).

Kortland, C. *See* Borstlap (1967a, b).

Kotzé, J. P. *See* Hattingh (1967).

Kountz (1959), R. R., and C. Forney, Jr. Metabolic energy balances in a total oxidation activated sludge system. *Sewage Ind. Wastes*, **31**, 819–826 (1959). *83.*

Kozbelt, S. J. *See* Knaak (1966).

Kozlova, E. I. *See* Shaposhnikov (1968).

Krems, I. J. *See* Gray (1955).

Krieg, N. R. *See* Shindala (1965).

Krieger, H. L. *See* Ludzack (1964).

KrishnaMurti (1966), G. S. R., V. V. Volk, and M. L. Jackson. Soil adsorption of linear alkylate sulfonate. *Soil Sci. Soc. Am. Proc.*, **30**, 685–688 (1966). *113.*

Krishnan (1966), P., and A. F. Gaudy, Jr. Mechanism and kinetics of substrate utilization at high biological solids concentrations. *Purdue Conf.*, **21**, 495–510 (1966). *84.*

Křížová-Chlumová, J. *See* Pitter (1964b).

Krone (1968), M., and G. Schneider. Biodegradability of secondary alkane sulfonates under laboratory and practice conditions. *FSA*, **70**, 753–757 (1968). *95, 191, 348, 363, 380.*

Krüger (1964), R. Recent studies on alkylbenzene sulfonates. *FSA*, **60**, 217–221 (1964). *276, 287, 289, 292, 344.*

Kulovaná (1966), H., and P. Pitter. The biodegradation of surfactants based on sucrose fatty acid esters. *Tenside*, **3**, 322–326 (1966). *396.*

Kumke (1966), G. W., and C. E. Renn. LAS removal across an institutional trickling filter. *JAOCS*, **43**, 92–94 (1966). *159, 191, 346.*

Kuriyama, M. *See* Nishikawa (1968).

Kurz (1969), W. G. W., P. S. S. Dawson, and E. R. Blakley. A comparative study in vivo of enzyme activities in batch, continuous and phased cultures of a pseudomonad grown on phenylacetic acid. *Can. J. Microbiol.*, **15**, 27–33 (1969). *93.*

Kurzendörfer, P. *See* Kölbel (1964, 1967).

Kyono, S. *See* Fujiwara (1968).

Kyriacou, D. *See* House (1964b); Sharman (1964b).

Labonte, R. *See* Symons (1963).

Lamb (1952), C. B., and G. F. Jenkins. BOD of synthetic organic chemicals. *Purdue Conf.*, **7**, 326–339 (1952). *70, 311, 404, 408.*

Lamb, J. C., III. *See* Cassell (1966); Lashen (1967c).

Lambert, G. *See* Edeline (1965).

Lambert, M. *See* Nyns (1969a, b).

Lambin (1966), S., C. Carrère, and M. Beignot-Devalmont. Biodegradation of synthetic detergents by wastewater microorganisms. *Ann. Pharm. Franç.*, **24**(3), 161–166 (1966). *99, 127, 341, 354, 368, 398.*

Lang (1965), D., C. Trottier, L. Long, Jr., and T. Hasselstrom. Synthesis and biodegradability of isomeric sodium *p*-(thia-*n*-dodecyl)benzene sulfonates. *JAOCS*, **42**, 1095–1097 (1965). *232, 345, 360, 361.*

Lang (1967), D., and L. Long, Jr. *p*-(Thia-*n*-dodecyl)benzene sulfonates. U.S. Pat. 3,344,173 (1967). *232, 361.*

Lashen (1966), E. S., F. A. Blankenship, K. A. Booman, and J. Dupré. Biodegradation studies on a *p*-*t*-octylphenoxypolyethoxyethanol. *JAOCS*, **43**, 371–376 (1966). *56, 65, 132, 138, 140, 189, 253, 317, 324, 354, 386, 389.*

Lashen (1967a), E. S., G. F. Trebbi, K. A. Booman, and J. Dupré. Biodegradability of nonionic detergents. *Soap Chem. Specialties.*, **43**(1), 55–58, 122–129 (1967). *253, 389.*

Lashen (1967b), E. S., and K. A. Booman. Biodegradability and treatability of alkylphenol ethoxylates—a class of nonionic surfactants. *Water Sewage Works*, **114**, (Reference No.), R155–R163 (1967). *86, 119, 191, 253, 380, 389.*

Lashen (1967c), E. S., and J. C. Lamb, III. Biodegradation of a nonionic detergent. *Water Wastes Eng.*, **4**(12), 56–59 (1967). *121, 140, 191, 253, 380, 389.*

Lashen, E. S. *See* Booman (1967).

Lau, E. P. K. *See* Sheridan (1969).

Lawrence, A. S. C. *See* Goldthorpe (1949).

Laws (1959), E. Q., and W. Hancock. Some alkyl aryl sodium sulfonates as standard substances in detergent chemistry. *Nature*, **183**, 1473–1474 (1959). *51.*

Lawson, G. R. *See* Roberts (1958).

Lax, F. G. I. *See* Ellerker (1968).

Leclerc (1952), E., and P. Beaujean. Influence of synthetic detergents on the self-purification of wastewaters. *Bull. Centre Belge Etude Doc. Eaux No. 17*, pp. 152–158 (1952). *97, 358, 367, 392.*

Lee, G. F. *See* Pfeil (1968); WPCF (1967).

Lee, H. A. Jr. *See* Abeles (1961).

Lees, H. *See* Halvorson (1968, 1969a).

Lendrum, A. C. *See* Lominski (1942).

Lennon, A. *See* Jenkins (1967).

Lerenard, A. *See* Brebion (1964).

Leslie, W. D. *See* Greff (1965).

Lewin, V. H. *See* Gameson (1962).

LGC (1962) (Laboratory of the Government Chemist). *Report of the Government Chemist 1962*, HMSO, London, 1963. *49, 335, 338.*

LGC (1963). *Report of the Government Chemist 1963*, HMSO, London, 1964. *343.*

LGC (1965). *Report of the Government Chemist 1965*, HMSO, London, 1966. *60, 345, 354, 362.*

LGC (1966). *Report of the Government Chemist 1966*, HMSO, London, 1967. *136, 316, 323, 346, 383.*

Liddicoet (1965), T. H., and L. H. Smithson. Analysis of surfactants using pyrolysis-gas chromatography. *JAOCS*, **42**, 1097–1102 (1965). *61.*

Liener, I. E. *See* Viswanatha (1955).

Lindner (1964), K. *Tenside-Testilhilfsmittel-Waschrohstoffe*, Wissenschaftliche Verlagsgesellschaft mbH, Stuttgart, 1964. *22.*

Lindqvist, G. *See* Forsberg (1967a, b).

Lipman (1963), B. L., and N. M. Kazarovets. Study of biochemical oxidation of synthetic surfactants in a Warburg apparatus. *Nauchn. Tr. Akad. Kommun. Khoz.*, **1963**(20), 3–11; *CA*, **61**, 2817g (1964). *358, 362.*

Lishka (1968), R. J., and J. H. Parker. *Water Surfactant No. 3, Study Number 32*, U.S. Public Health Service Publication No. **999-UIH-11**, Cincinnati, 1968. *53.*

Lishka, R. J. *See* Ballinger (1962).

Liu, S. K. *See* Swisher (1961).

Livingston (1965), J. R., Jr., R. Drogin, and R. J. Kelly. Detergency and biodegradation of alcohol-based secondary sulfates. *Ind. Eng. Chem., Prod. Res. Develop.*, **4**, 28–32 (1965). *336, 346, 354, 368, 373.*

Locke, J. E. *See* Mansfield (1964).

Lockett (1956), W. T. Synthetic detergents in relation to the purification of sewage. *JISP*, **1956**, 225–246. *115, 174, 380.*

Loehr (1967), R. C., and B. Bergeron. Preservation of wastewater samples prior to analysis. *Water Res.*, **1**, 577–586 (1967). *39, 40, 134.*

Loehr (1968), R. C., and J. C. Roth. Aerobic degradation of long-chain fatty acid salts. *JWPCF*, **40**, R385–R403 (1968). *255.*

Loeser (1965), G. Separation of surfactant mixtures by thin layer chromatography. *Seifen-Öle-Fette-Wachse*, **91**, 751–752 (1965). *60.*

Lominski(1942), I., and A. C. Lendrum. The effect of surfactants on *B. proteus*. *J. Pathol. Bacteriol.*, **54**, 421–433 (1942). *100.*

Long (1966), L, Jr., C. H. Trottier, M. R. Rogers, and A. M. Kaplan. Biodegradable detergents. Army Science Conference, West Point, June 1966; available as document **AD634619**, Clearinghouse for Federal Scientific and Technical Information, Washington, 14 pp. *232, 347, 360, 361.*

Long, L., Jr. *See* Lang (1965, 1967).

Longman, G. F. *See* Reid (1967, 1968).

Longwell (1955), J., and W. D. Maniece. Determination of anionic detergents in sewage, sewage effluents and river waters. *Analyst*, **80**, 167–171 (1955). *49.*

Lothian, T. *See* Hurwitz (1960).

Lowes, F. J. *See* SDA (1965).

Ludzack (1957), F. J., and M. B. Ettinger. Biological oxidation of hexadecanol under laboratory conditions. *J. Am. Water Works Assoc.*, **49**, 849–858 (1957). *73, 403.*

Ludzack (1959), F. J., R. B. Schaffer, R. N. Bloomhaft, and M. B. Ettinger. Biochemical oxidation of some commercially important cyanides. *Sewage Ind. Wastes*, **31**, 33–44 (1959). *73.*

Ludzack (1960a), F. J. Laboratory model activated sludge unit. *JWPCF*, **32**, 605–609 (1960). *166.*

Ludzack (1960b), F. J., and M. B. Ettinger. Chemical structures resistant to aerobic biochemical stabilization. *JWPCF*, **32**, 1173–1200 (1960). *70.*

Ludzack (1961), F. J., R. B. Schaffer, and M. B. Ettinger. Temperature and feed as variables in activated sludge performance. *JWPCF*, **33**, 141–156 (1961). *94, 182.*

Ludzack (1963a), F. J., and M. B. Ettinger. Estimating biodegradability and treatability of organic water pollutants. *Biotechnol. Bioeng.*, **5**, 309–330 (1963). *72, 123, 156, 163.*

Ludzack (1963b), F. J., and M. B. Ettinger. Biodegradability of organic chemicals isolated from rivers. *Purdue Conf.*, **18**, 278–282 (1963). *73, 400.*

Ludzack (1964), F. J., H. L. Krieger, and M. B. Ettinger. Interactions of waste feed, activated sludge and oxygen as traced by radioactive carbon. *JWPCF*, **36**, 782–788 (1964). *65, 73, 141, 182.*

Ludzack (1965), F. J. (Modified SDA feed for pH control.) Private communication to SDA Biodegradation Subcommittee, 1965. *132, 138.*

Lumb (1953), C. Experiments on the effects of certain synthetic detergents on biological oxidation of sewage. *JISP*, **1953**, 269–277. *157, 171, 357.*

Lumb, C. *See* Goldthorpe (1949).

Lushnikova, E. V. *See* Mosolov (1966).

Lustigman, B. K. *See* Alexander (1966).

Lynch, E. R. *See* Hawkes (1958).

Mach, F. *See* Köhler (1969).

Maehler (1967), C. A., J. M. Cripps, and A. E. Greenberg. Differentiation of LAS and ABS in water. *JWPCF*, **39**, R92–R98 (1967). *63.*

Magee (1966), P. T., and E. E. Snell. The bacterial degradation of pantothenic acid. IV. Enzymatic conversion of aldopantoate to α-ketoisovalerate. *Biochemistry (ACS)*, **5**, 409–416 (1966). *298.*

Mäkinen, K. *See* Nurmikko (1966).

Malaney (1957), G. W., and W. D. Sheets. Detergent builders and BOD. *Sewage Ind. Wastes*, **29**, 263–267 (1957). *408, 409.*

Malaney (1960), G. W. Oxidative abilities of aniline-acclimated activated sludge. *JWPCF*, **32**, 1300–1311 (1960). *399, 400, 401, 406.*

Malaney (1966), G. W., and R. E. McKinney. Oxidative abilities of benzene-acclimated activated sludge. *Water Sewage Works*, **113**, 302–309 (1966). *334, 399, 400, 401, 406, 410.*

Malaney (1969), G. W., and R. M. Gerhold. Structural determinants in the oxidation of aliphatic compounds by activated sludge. *JWPCF*, **41**, R18–R33 (1969). *379, 408.*

Malaney, G. W. *See* Cibulka (1967); Gerhold (1966); Marion (1963a, b); Sheets (1956a, b).

Malz (1967), F. The behavior of new detergents in aerobic treatment of sewage. *Münchner Beitr. Abwasser-, Fisch.- Flussbiol.*, **9**(2nd ed.), 78–90 (1967). *347, 354, 355, 368.*

Malz, F. *See* Husmann (1963a).

Mamiya, G. *See* Gorin (1967).

Manganelli (1953), R., and E. S. Crosby. Effect of detergents on sewage microorganisms. *Sewage Ind. Wastes*, **25**, 262–276 (1953). *97.*

Manganelli (1956), R. M. Effects of synthetic detergents on activated sludge. *Purdue Conf.*, **11**, 611–621 (1956). *97, 115, 394.*

Manganelli (1960), R. M., H. Heukelekian, and C. N. Henderson. Persistence and effect of an alkylaryl sulfonate in sludge digestion. *Purdue Conf.*, **15**, 199–206 (1960). *189, 358.*

Manganelli, R. *See* Rudolfs (1949).

Maniece, W. D. *See* Longwell (1955).

Mann (1957), H., and D. W. M. Herbert. Some observations of the effect of synthetic detergents on the treatment of sewage. *Water and Sanit. Engr.* **6**, 206–209 (1957). *135, 140, 157, 173, 366.*

Mann (1968), A. H.   The behavior of high molecular weight LAS in laboratory-scale biodegradation tests. *Surf. Cong. #5*, **1**, 103–114 (1969).   *136, 154, 191, 219, 348.*

Manneck (1966), H.   Preparation and use of α-olefin sulfonates. *Seifen-Öle-Fette-Wachse*, **92**, 642–644 (1966).   *347, 354, 363.*

Mansfield (1964), R. C., and J. E. Locke.   The preparation of a series of molecularly homogeneous *para-t*-octylphenoxypoly(ethoxy)ethanols. *JAOCS*, **41**, 267–272 (1964).   *60.*

Marion (1963a), C. V., and G. W. Malaney.   Ability of activated sludge microorganisms to oxidize aromatic organic compounds. *Purdue Conf.*, **18**, 297–308 (1963).   *399, 400, 401, 403, 410.*

Marion (1963b), C. V., and G. W. Malaney.   The oxidation of aliphatic compounds by *Alkaligenes faecalis*. *JWPCF*, **35**, 1269–1284 (1963).   *379, 399, 402, 404, 408.*

Marion (1966), C. V.   Oxidation of sodium lauryl sulfate, alkyl benzene sulfonate and chemically related compounds. Ph.D. Thesis, Ohio State Univ., Columbus, 1966 (University Microfilms No. 67-2488).   *116, 127, 142, 278, 303, 347, 354, 368, 379, 399, 401, 403, 406.*

Märki, *E. See* Michelsen (1961).

Markovetz, A. J. *See* Klug (1967).

Marks, A. *See* Wilmsmann (1959).

Marquis (1966), D. M., S. H. Sharman, R. House, and W. A. Sweeney.   Alpha olefin sulfonates from a commercial SO₃-air reactor. *JAOCS*, **43**, 607–614 (1966).   *234, 363.*

Martelli (1964), H. L., and A. A. Benson.   Sulfo carbohydrate metabolism. I. Bacterial production and utilization of sulfoacetate. *Biochim. Biophys. Acta*, **93**, 169–171 (1964).   *274, 406.*

Martin (1964), E. J., and D. R. Washington.   Kinetics of the steady state bacterial culture. I. Mathematical Model. *Purdue Conf.*, **19**, 724–737 (1964).   *182.*

Martin (1965), E. J., and D. R. Washington.   Kinetics of the steady state bacterial culture. IV. Transfer rates. *Purdue Conf.*, **20**, 470–500 (1965).   *182.*

Mateles (1969), R. I., and S. K. Chian.   Kinetics of substrate uptake in pure and mixed cultures. *Environ. Sci. Tehcnol.*, **3**, 569–574 (1969).   *89, 179.*

Mateles, R. I. *See* Chian (1968).

Mathews (1954), M. B.   Testicular hyaluronidase in relation to micelle formation by inactivating agents. *J. Am. Chem. Soc.*, **76**, 2948–2952 (1954).   *107.*

Matson, R. D. *See* Garrison (1964).

Maurer (1965), E. W., T. C. Cordon, J. K. Weil, M. V. Nuñez-Ponzoa, W. C. Ault, and A. J. Stirton.   The effect of tallow based detergents on anaerobic digestion. *JACOS*, **42**, 189–192 (1965).   *41, 42, 189, 205, 307, 341, 346, 354, 364, 369, 372,*

Maurer, E. W. *See* Cordon (1964, 1968a, b).

Mausner, M. *See* SDA (1965, 1969a, b).

Mavrianopol, A. *See* Ilişescu (1966).

Maxwell, P. M. *See* AASGP (1956, 1961).

Mayberry, W. R. *See* Payne (1965, 1967); Prochazka (1967); Williams (1966).

McAteer (1964), J. H., and L. M. Kinnard.   Sodium alkane sulfonates—dependence of properties on molecular weight. *Surf. Cong. #4*, **1**, 127–140 (1967).   *233, 335, 362.*

McAteer, J. H. *See* Nelson (1961).

McCarty, P. L. *See* Konecky (1963).

McCoy, E. *See* Crabtree (1966); Hoadley (1965).

McCubbin (1966), W. D., C. M. Kay, and K. Oikawa. Polarimetric studies on the interaction of muscle proteins with ionic detergents. *Biochim. Biophys. Acta*, **126**, 597–600 (1966). *104, 105.*

McDermott, G. N. *See* Schaffer (1965).

McGauhey (1957), P. H., E. S. Crosby, and S. A. Klein. The fate of alkylbenzene sulfonate in sewage treatment. *Final Report*, University of California, Berkeley, 1957. *113, 134, 165, 350, 367.*

McGauhey (1959a), P. H., and S. A. Klein. Removal of ABS by sewage treatment. *Sewage Ind. Wastes*, **31**, 877–899 (1959). *65, 113, 115, 165, 173, 177, 299, 350.*

McGauhey (1959b), P. H., S. A. Klein, and P. B. Palmer. Operating variables as they affect ABS removal by sewage treatment. *Final Report, Sanitary Engineering Research Laboratory*, University of California, Berkeley, 1959. *65, 113, 165, 173, 177, 299, 350.*

McGauhey (1964), P. H., and S. A. Klein. Travel of synthetic detergents with percolating water. *Purdue Conf.*, **19**, 1–8 (1964). *229, 299, 349, 356.*

McGauhey (1966), P. H., and S. A. Klein. Degradable pollutants—a study of the new detergents. *WPR Conf. #3*, **1**, 353–374 (1967). *229, 230, 346, 354.*

McGauhey, P. H. *See* Klein (1961, 1962, 1963a, b, 1964a, b, 1965b, 1966).

McKenna (1964), E. J., and R. E. Kallio. Hydrocarbon structure: its effect on bacterial utilization of alkanes. In *Principles and Applications of Aquatic Microbiology* (H. Heukelekian and N. C. Dondero, eds.), Wiley, New York, 1964, pp. 1–14. *267.*

McKenna (1965), E. J. and R. E. Kallio. The biology of hydrocarbons. *Ann. Rev. Microbiol.* **19**, 183–208 (1965). *258, 263, 267.*

McKenna (1966), E. J. Effect of hydrocarbon structure on mechanisms of microbial alkane metabolism. Ph.D. Thesis, University of Iowa, Iowa City, 1966. *261, 268, 296.*

McKinney (1952), R. E., and M. P. Horwood. Fundamental approach to the activated sludge process. I. Floc-producing bacteria. *Sewage Ind. Wastes*, **24**, 117–123 (1952). *125, 126, 162.*

McKinney (1953), R. E., and R. G. Wetchlein. Isolation of floc-producing bacteria from activated sludge. *Appl. Microbiol.*, **1**, 259–261 (1953). *78, 162.*

McKinney (1955), R. E., and J. S. Jeris. Metabolism of low molecular weight alcohols by activated sludge. *Sewage Ind. Wastes*, **27**, 728–735 (1955). *402.*

McKinney (1956a), R. E. Biological flocculation. In *Biological Treatment of Sewage and Industrial Wastes* (Brother Joseph McCabe and W. W. Eckenfelder, Jr., eds.), Vol. 1, Reinhold, New York, 1956, pp. 88–100. *162.*

McKinney (1956b), R. E., H. D. Tomlinson, and R. L. Wilcox. Metabolism of aromatic compounds by activated sludge. *Sewage Ind. Wastes*, **28**, 547–557 (1956). *263, 265, 379, 399, 400, 401, 403, 408, 410.*

McKinney (1956c), R. E., and A. Gram. Protozoa and activated sludge. *Sewage Ind. Wastes*, **28**, 1219–1231 (1956). *78.*

McKinney (1959a), R. E., and J. M. Symons. Bacterial degradation of ABS. I. Fundamental biochemistry. *Sewage Ind. Wastes*, **31**, 549–556 (1959). *79, 133, 223, 227, 350, 379, 399, 407.*

McKinney (1959b), R. E., and E. J. Donovan. Bacterial degradation of ABS. II. Complete mixing activated sludge. *Sewage Ind. Wastes*, **31**, 690–696 (1959). *98, 115, 133, 138, 166, 172, 350.*

McKinney (1962a), R. E. Mathematics of complete-mixing activated sludge. *J. Sanit. Eng. Div. Am. Soc. Civil Engrs.*, **88**(SA3), 87–113 (1962). *182.*

McKinney (1962b), R. E. *Microbiology for Sanitary Engineers*, McGraw-Hill, New York, 1962. *156, 162.*

McKinney (1963), R. E. Discussion of Busch (1963). *J. Eng. Ind.*, **1963**, 171–172. *83, 84.*

McKinney (1968), R. E., and W. J. O'Brien. Activated sludge—basic design concepts. *JWPCF*, **40**, 1831–1843 (1968). *3, 162.*

McKinney, R. E. *See* Malaney (1966); Mohanrao (1962); Nelson (1961); Symons (1958, 1960).

McLellan (1968), J. C., and A. W. Busch. Steady-state analysis in activated sludge process design. *Water Resources Res.*, **4**, 1307–1315 (1968). *182.*

McNamee, P. D. *See* Butterfield (1937).

McWhorter (1962), T., and H. Heukelekian. Factors influencing aerobic cell yields-and delineation of endogenous metabolism. *WPR Conf. #1*, **2**, 419–450 (1964). *82, 84.*

Meinck (1961), F., and G. Bringmann. The behavior of tetrapropylenebenzenesulfonate in sewage treatment. *Schriftenreihe Ver. Wasser-Boden-Lufthyg. No. 19*, pp. 68–81 (1961). *42, 98, 189, 351.*

Melpolder, F. W. *See* AASGP (1956, 1961); SDA (1964).

Menzel (1963), D. W., and R. F. Vaccaro. The measurement of dissolved organic and particulate carbon in sea water. *Limnol. Oceanog.*, **9**, 138–142 (1963). *68.*

Menzies, J. D. *See* WPCF (1967).

Merrell (1967), J. C., Jr., W. F. Jopling, R. F. Bott, A. Katko, and H. E. Pintler. The Santee Recreation Project, Santee, California, Final Report. Water Pollution Control Research Series Publication No. **WP-20-7**, Federal Water Pollution Control Administration, Cincinnati, 1967. *196, 380.*

Metcalf, E. S. *See* Olsen (1968).

Metzler (1968), D. F., R. L. Culp, H. A. Stoltenberg, R. L. Woodward, G. Walton, S. L. Chang, N. A. Clarke, C. M. Palmer, and F. M. Middleton. Emergency use of reclaimed water for potable supply at Chanute, Kansas. *J. Am. Water Works Assoc.*, **50**, 1021–1060 (1958). *5.*

Meusel, A. *See* Ruschenberg (1964).

Meyers, J. A. III. *See* Brink (1966).

Michael (1968), W. R. Metabolism of LAS and ABS in albino rats. *Toxicol. Appl. Pharmacol.*, **12**, 473–485 (1968). *54, 285, 301.*

Michelsen (1961), E., and E. Märki. Laboratory and field method for detection of anionic surfactants in surface-, ground- and wastewaters. *Mitt. Gebiete Lebensm. Hyg.*, **52**, 557–571 (1961). *50.*

Middleton (1954), F. M. Occurrence of excessive foam at the Wheeling West Virginia water purification plant. Environmental Health Center, Physics and Chemistry Section, Cincinnati, 1954. *4.*

Middleton, F. M. *See* AASGP (1956, 1961); Metzler (1958).

Miesch, A. T. *See* Wayman (1965).

Miholits, E. M. *See* Pipes (1963a).

Miller, B. F. *See* Baker (1941).

Mills (1953), E. J., Jr., and V. T. Stack, Jr.   Biological oxidation of synthetic organic chemicals. *Purdue Conf.*, **8**, 492–517 (1953).   *311, 404.*

Mills (1954), E. J., Jr., and V. T. Stack.   Acclimation of microorganisms for the oxidation of pure organic compounds. *Purdue Conf.*, **9**, 449–464 (1954).   *311, 404.*

Minami (1956), K.   Bactericidal action of oleic acid against *Mycobacterium avium*. *Nature*, **178**, 743 (1956).   *98.*

Minami (1958), K.   Isolation of a soil coccus capable of utilizing Tween 80 as a sole source of carbon. *Nature*, **181**, 430–431 (1958).   *79, 101, 127, 326, 394.*

Minami (1959), K., and E. Ogata.   Metabolism of Tween 80 by *Micrococcus tweenis*. *Nippon Saikingata Zasshi*, **14**, 987–992 (1959); *CA*, **55**, 22480a (1961).   *79, 326.*

Moffatt (1959), J. G., and H. G. Khorama.   The total synthesis of Coenzyme A. *J. Am. Chem. Soc.*, **81**, 1265 (1959).   *261.*

Mohanrao (1962), G. J., and R. E. McKinney.   A study of the biochemical characteristics of quaternary carbon compounds. *Intern. J. Air Water Pollution*, **6**, 153–168 (1962).   *297, 299.*

Molnar, D. M. *See* Glassman (1951).

Molof, A. H. *See* Heyman (1967, 1968).

Momotani, Y. *See* Imanishi (1965).

Montgomery (1967), H. A. C.   The determination of BOD by respirometric methods. *Water Res.*, **1**, 631–662 (1967).   *71, 72, 94, 142, 144.*

Moore (1956), W. A., and R. A. Kolbeson.   Determination of anionic detergents in surface waters and sewage with methyl green. *Anal. Chem.*, **28**, 161–164 (1956). *49.*

Morgan (1962), D. J.   The microdetermination of nonionic surfactants containing poly(ethylene oxide) linkages. *Analyst*, **87**, 223–234 (1962).   *56.*

Morgan, J. M., Jr. *See* Knapp (1965).

Mori, A. *See* Ōba (1968b); Tomiyama (1968, 1969).

Morris, J. C. *See* Weber (1962, 1963, 1964a).

Mosebach, H. *See* Hartmann (1966a).

Mosolov (1966), V. V., and E. V. Lushnikova.   Effect of surfactants on the enzymatic activity of trypsin, acetyltrypsin and succinyltrypsin. *Dokl.—Biochem. Sect.* (*English Transl.*), **167**, 92–94 (1966).   *109.*

Moss, H. V. *See* Orsanco (1963).

Müller, J. *See* Kopp (1965).

Murtaugh (1965), J. J., and R. L. Bunch.   Acidic components of sewage effluents and river water. *JWPCF*, **37**, 410–415 (1965).   *4, 134.*

MWB (1966) (Metropolitan Water Board).   Forty second report on the results of the bacteriological, chemical and biological examination of the London waters for the years 1965–1966. London, 1966.   *341, 347, 359, 362, 395.*

Myerly (1964), R. C., J. M. Rector, E. C. Steinle, C. A. Vath, and H. T. Zika.   Secondary alcohol ethoxylates as degradable detergent materials. *Soap Chem. Specialties*, **40**(5), 78–80, 82, 170–171 (1964).   *248, 375, 376, 377, 383, 385, 390.*

Myerly. R. C. *See* Steinle (1964).

Myrick, N. *See* Busch (1961).

Nadeau (1964), H. G., D. M. Oaks, Jr., W. A. Nichols, and L. P. Carr.   Separation and analysis of nonylphenoxy-EO adducts by programmed temperature gas chromatography. *Anal. Chem.*, **36**, 1914–1917 (1964).   *62.*

Nakamura (1968), H.   Genetic determination of resistance to acriflavine, phenethyl alcohol and sodium dodecyl sulfate in *Escherichia coli*. *J. Bacteriol.*, **96**, 987–996 (1968). *99*.

Neale, A. J. *See* Hawkes (1958, 1960).

Nelson (1961), J. F., R. E. McKinney, J. H. McAteer, and M. S. Konecky.   Biodegradability of ABS. *Develop. Ind. Microbiol.*, **2**, 93–101 (1961).   *44, 66, 133, 142, 143, 144, 173, 224, 226, 335, 337*.

Nichols, M. S. *See* Knopp (1965).

Nichols, W. A. *See* Nadeau (1964).

Nihongi, T. *See* Hayano (1968).

Nishikawa (1968), S., and M. Kuriyama.   Nucleic acid as a component of mucilage in activated sludge. *Water Res.*, **2**, 811–812 (1968). *161*.

Nixon, J. *See* Goldthorpe (1950).

Nord, F. F. *See* Timasheff (1951).

Nuñez-Ponzoa, M. V. *See* Cordon (1968a); Maurer (1965); Smith (1966); Stirton (1965).

Nunn (1967), L. G., Jr., and L. M. Schenk.   Biodegradable surface active agents. U.S. Pat. 3,317,612 (1967). *391*.

Nurmikko (1966), V., E. Salo, H. Hakola, K. Mäkinen, and E. E. Snell.   The bacterial degradation of pantothenic acid. II. Pantothenate hydrolase. *Biochemistry (ACS)*, **5**, 399–403 (1966). *298*.

Nyns (1969a), E. J., M. Lambert, and A. L. Wiaux.   Comparative biodegradation of detergents. I. Comparative bacterial biodegradation of lauryl sulfate and tridecylbenzene sulfonates. *FSA*, **71**, 232–236 (1969).   *128, 341, 368*.

Nyns (1969b), E. J., M. Lambert, and A. L. Wiaux.   Comparative bacterial degradation of detergents. I. Comparative bacterial biodegradation of anionic detergents and their hydrocarbon or alcohol bases. *Tenside*, **6**, 192–194 (1969).   *268, 304*.

Oaks, D. M., Jr. *See* Nadeau (1964).

Ōba (1965a), K., and S. Tomiyama.   Biochemical studies of alkyl benzene sulfonates. II. Action of ABS on microorganisms. *Yukagaku*, **14**, 364–369 (1965).   *99, 117*.

Ōba (1965b), K.   Biochemical studies of alkyl benzene sulfonates. III. Biodegradation of linear alkylbenzene sulfonates in a private cesspool. *Yukagaku*, **14**, 565–571 (1965). *346, 354*.

Ōba (1965c), K.   Biochemical studies of alkyl benzene sulfonates. IV. Methylene blue active substance mistakable for ABS in air and waters. *Yukagaku*, **14**, 625–629 (1965).   *51, 61*.

Ōba (1965d), K.   Biochemical studies of alkyl benzene sulfonates. V. Critical examination on the determination method of residual ABS in vegetables. *Nippon Nogei Kagaku Kaishi*, **39**, 508–513 (1965).   *51, 61*.

Ōba (1965e), K., and S. Tomiyama.   Biodegradation of alkylbenzene sulfonates by microorganisms in a private cesspool. *Agr. Biol. Chem. (Tokyo)*, **29**, 591–592 (1965).   *346*.

Ōba (1965f), K., and K. Aiiso.   Biological degradation of LAS by microflora in cesspool. *J. Japan Biol. Soc. Water Waste*, **2**(1), 4–6 (1965).   *80, 346, 354*.

Ōba (1967), K., Y. Yoshida, and S. Tomiyama.   Studies on biodegradation of synthetic detergents. I. Biodegradation of anionic surfactants under aerobic and anaerobic conditions. *Yukagaku*, **16**, 517–523 (1967).   *205, 237, 307, 348, 355, 363, 368, 371, 375*.

Ōba (1968a), K., Y. Yoshida, and S. Tomiyama. Studies on biodegradation of synthetic detergents. II. Determination of straight and branched chain alkylbenzene sulfonates in waters. *Yukagaku*, **17**, 455–460 (1968). *63, 197.*

Ōba (1968b), K., A. Mori, and S. Tomiyama. Biochemical studies of *n*-α-olefin sulfonates. I. Biodegradability under aerobic condition. *Yukagaku*, **17**, 517–520 (1968). *48, 233, 234, 348, 355, 363, 368.*

Ōba, K. *See* Kimura (1962); Tomiyama (1968).

Oberton (1957), A. C. E., and V. T. Stack, Jr. BOD of organic chemicals. *Sewage Ind. Wastes*, **29**, 1267–1272 (1957). *405, 408.*

O'Brien, W. J. *See* McKinney (1968).

OECD (1968) (Organisation de Cooperation et de Developement Economiques.) Provisional OECD open-flask test of biodegradability of synthetic anionic detergents. Annex to letter W/DAS/SCI/68.971, Appendix II. *150.*

Offhaus (1962), K. BOD investigations with detergents. *Münchner Beitr. Abwasser-, Fisch.-Flussbiol.*, **9**, 184–197 (1962). *115, 343, 351, 365, 379, 392.*

Ogata, E. *See* Minami (1959).

Ogden (1961), C. P., H. L. Webster, and J. Halliday. Determination of biologically soft and hard ABS in detergents and sewage. *Analyst*, **86**, 22–29 (1961). *63.*

Oikawa, K. *See* McCubbin (1966).

Olavsen, A. H. *See* Denner (1969).

Oldham (1949), L. W. Discussion of Goldthorpe (1949). *Chem. Ind. (London)*, **1949**, 680. *392.*

Oldham (1958), L. W. Investigations into the effects of a nonionic synthetic detergent on biological percolating filters. *JISP*, **1958**, 136–147. *119, 243, 246, 392.*

Oldham (1964), W. J. Biodegradable alkylbenzenes. French Pat. 1,376,000 (1964); *CA*, **62**, 7690e (1965). *357.*

Olezyk, C. *See* Janota-Bassalik (1969).

Olsen (1968), R. H., and E. S. Metcalf. Conversion of mesophilic to psychrophilic bacteria. *Science*, **162**, 1288–1289 (1968). *93.*

Ooyama (1965), J., and J. W. Foster. Bacterial oxidation of cycloparaffinic hydrocarbons. *Antonie van Leeuwenhoek, J. Microbiol. Serol.*, **31**, 45–65 (1965). *399.*

Orchin, M. *See* Jaffe (1962).

Orford (1953), H. E., M. C. Rand, and I. Gellman. A single dilution technique for BOD studies. *Sewage Ind. Wastes*, **25**, 284–289 (1953). *71.*

Orgel (1964), G., and F. A. Rupp. Evaluation of anionic surfactant biodegradability by various laboratory procedures. *Surf. Cong. #4*, **3**, 131–140 (1967). *140, 340, 344, 345, 353.*

Orgel, G. *See* Renn (1964a); SDA (1965).

O'Rourke, J. T. *See* Swisher (1964a).

Orsanco (1963) (Ohio River Valley Water Sanitation Commission) Detergent Subcommittee. Components of household synthetic detergents in water and sewage. *J. Am. Water Works Assoc.*, **55**, 369–402 (1963). *4.*

Osborn, D. W. *See* Urban (1965).

Osburn (1966), Q. W., and J. H. Benedict. Polyethoxylated alkyl phenols: relationship of structure to biodegradation mechanism. *JAOCS*, **43**, 141–146 (1966). *45, 56, 57, 62, 64, 252, 253, 319, 321, 322, 324, 387, 388, 389, 391.*

Osburn, Q. W. *See* Frazee (1964b).

Overath (1969), P., G. Pauli, and H. U. Schairer. Fatty acid degradation in *Escherichia coli*. *European J. Biochem.*, **7**, 559–574 (1969). *87, 261.*

Ownsworth, R. A. *See* Bruce (1966).

Padar, F. V. *See* Flynn (1969).

Page (1963), H. G., and C. H. Wayman. Removal of ABS and other sewage components by infiltration through soils. Eighth Annual Midwest Groundwater Conference, Hickory Corners, Dec. 1963. *Ground Water*, **3**, 10–17 (1966). *196.*

Page, H. G. *See* Wayman (1963c).

Pahren (1961), H. R., and D. E. Bloodgood. Biological oxidation of several vinyl compounds. *JWPCF*, **33**, 233–238 (1961). *73, 408.*

Painter (1961), H. A., M. Viney, and A. Bywaters. Composition of sewage and sewage effluents. *JISP*, **1961**, 302–314. *135.*

Painter (1967), H. A. (Activated Sludge bacteria.) Discussion of Harkness (1966). *Water Pollution Control*, **66**, 603–604 (1967). *91.*

Painter (1968), H. A., R. S. Denton, and C. Quarmby. Removal of sugars by activated sludge. *Water Res.*, **2**, 427–447 (1968). *83, 90, 161, 179.*

Painter, H. A. *See* Downing (1964).

Pallansch, M. J. *See* Viswanatha (1955).

Palleroni, N. J. *See* Stainer (1966).

Palmer, C. M. *See* Metzler (1958).

Palmer, P. B. *See* McGauhey (1959b).

Palumbo (1969a), S. A., and L. D. Witter. The influence of temperature on the pathways of glucose catabolism in *Pseudomonas fluorescens*. *Can. J. Microbiol.*, **15**, 995–1000 (1969). *94.*

Palumbo (1969b), S. A., and L. D. Witter. Influence of temperature on glucose utilization by *Pseudomonas fluorescens*. *Appl. Microbiol.*, **18**, 137–141 (1969). *94.*

Panasiuk, O. *See* Cordon (1968b).

Papenmeier (1969), G. J., and J. M. Campagnoli. Microcalorimetry. Thermodynamics of the reaction of an anionic detergent with a cationic detergent. *J. Am. Chem. Soc.*, **91**, 6579–6584 (1969). *33.*

Parker (1966), R. B. Continuous culture system for ecological studies of microorganisms. *Biotechnol. Bioeng.*, **8**, 473–488 (1966). *178.*

Parker, J. H. *See* Lishka (1968).

Pastan, I. *See* de Crombrugghe (1969).

Pasveer (1959), A. A contribution to the development in activated sludge treatment. *JISP*, **1959**, 436–465. *163.*

Pasveer (1960a), A. Developments in activated sludge treatment in the Netherlands. Conference on Biological Waste Treatment, Manhattan College, New York, 1960, Paper No. 34. In *Advances in Biological Waste Treatment* (W. W. Eckenfelder, Jr., and Brother Joseph McCabe, eds.), Macmillan, New York, 1963. *163.*

Pasveer (1960b), A. A simplified method for the purification of comparatively small amounts of sewage and industrial wastes. *Purdue Conf.*, **15**, 528–540 (1960). *163.*

Patterson (1966a), S. J., E. C. Hunt, and K. B. E. Tucker. Determination of commonly used nonionic detergents in sewage effluents by a TLC method. *JISP.* **1966**, 190–198. *60, 322.*

Patterson (1966b), S. J., K. B. E. Tucker, and C. C. Scott. Nonionic detergents and related substances in British waters. *WPR Conf. #3*, **2**, 103–116 (1967). *60, 384, 392, 404.*

Patterson (1967), S. J., C. C. Scott, and K. B. E. Tucker. Nonionic detergent degradation. I. Thin layer chromatography and foaming properties of alcohol polyethoxylates. *JAOCS*, **44**, 407–412 (1967). *244, 245, 246, 248, 315, 317, 382, 383, 385, 386, 393, 394, 404, 405.*

Patterson (1968), S. J., C. C. Scott, and K. B. E. Tucker. Nonionic detergent degradation, II. Thin layer chromatography and foaming properties of alkyl phenol ethoxylates. *JAOCS*, **45**, 528–532 (1968). *253, 322, 388, 390, 391, 392.*

Patterson (1970), S. J., C. C. Scott, and K. B. E. Tucker. Nonionic detergent degradation. III. Initial mechanism of the degradation. *JAOCS*, **47**, 37–41 (1970). *316, 323.*

Pauli, G. *See* Overath (1969).

Payne (1963a), W. J., and V. E. Feisal. Bacterial utilization of dodecyl sulfate and dodecylbenzene sulfonate. *Appl. Microbiol.*, **11**, 339–344 (1963). *69, 79, 97, 127, 278, 303, 357, 367, 379.*

Payne (1963b), W. J. Pure culture studies of the degradation of detergent compounds. *Biotechnol. Bioeng.*, **5**, 355–365 (1963). *69, 127, 303, 304, 312, 352, 357, 367, 369, 402, 403, 406, 410.*

Payne (1965), W. J., J. P. Williams, and W. R. Mayberry. Primary alcohol sulfatase in a *Pseudomonas* species. *Appl. Microbiol.*, **13**, 698–701 (1965). *62, 65, 303, 308, 368.*

Payne (1966), W. J., and R. L. Todd. Flavin linked dehydrogenation of ether glycols by cell-free extracts of a soil bacterium. *J. Bacteriol.*, **91**, 1533–1536 (1966). *312.*

Payne (1967), W. J., J. P. Williams, and W. R. Mayberry. Hydrolysis of secondary alcohol sulfate by a bacterial enzyme. *Nature*, **214**, 623–624 (1967). *127, 308.*

Payne, W. J. *See* Fincher (1962b); Prochazka (1965, 1967); Williams (1964, 1966).

Peck (1942), R. J. Inhibition of the proteolytic action of trypsin by soaps. *J. Am. Chem. Soc.*, **64**, 487–490 (1942). *108.*

Pensak, P. *See* SDA (1964).

Perlman, R. L. *See* de Crombrugghe (1969).

Perry, J. W. *See* Schwartz (1949, 1958).

Persinger, H. E. *See* Crabb (1964, 1968).

Pfeil (1968), B. H., and G. F. Lee. Biodegradation of nitrilotriacetic acid in aerobic systems. *Environ. Sci. Technol.* **2**, 543–546 (1968). *409.*

Phillips (1963), D. C. ABS removal by activated sludge. Ph.D. Thesis, University of Wisconsin, Madison, 1963 (University Microfilms No. 64–662). *41, 115, 301, 352, 380.*

Piccolini, V. M. *See* Blankenship (1963).

Pintler, H. E. *See* Merrell (1967).

Pipes (1963a), W. O., E. M. Miholits, and O. W. Boyle. Aerobic cell yield and theoretical oxygen demand. *Purdue Conf.*, **18**, 418–426 (1963). *74.*

Pipes (1963b), W. O., and P. H. Jones. Decomposition of organic wastes by *Sphaerotilus*. *Biotechnol. Bioeng.*, **5**, 287–307 (1963). *127, 352, 367, 379.*

Pitter (1960), P. Determination of synthetic detergents in waters. *Sb. VSChT*, **4**(1), 269–286 (1960). *52.*

Pitter (1961a), P., and J. Svitálková. Synthetic surfactants in wastewaters. II. Biodegradation of alkyl sulfate and alkylaryl sulfonates in laboratory models of activated sludge plants. *Sb. VSChT*, **5**(1), 27–43 (1961). *166, 357, 369.*

Pitter (1961b), P. Synthetic surfactants in wastewaters. III. Biodegradation of cationic agents in laboratory models of activated sludge tanks. *Sb. VSChT*, **5**(2), 25–42 (1961). *166, 398.*

Pitter (1962a), P. A new colorimetric method for determination of nonionic EO type surfactants in low concentration. *Chem. Ind. (London)*, **1962**, 1832–1833. *58.*

Pitter (1962b), P. Colorimetric determination of nonionic surfactants with hydroquinone and determination of anionic and cationic surfactants. *Sb. VSChT*, **6**(1), 547–561 (1962). *58.*

Pitter (1963a), P., and J. Trauč. Synthetic surfactants in wastewaters. IV. Biodegradation of nonionic agents in laboratory models of aeration tanks. *Sb. VSChT*, **7**(1), 201–216 (1964). *58, 133, 166, 381, 383.*

Pitter (1963b), P. Synthetic surfactants in wastewaters. V. Alkyl sulfates resistant and alkylbenzene sulfonates susceptible to biological oxidation. *Sb. VSChT*, **7**(2), 19–32 (1963). *72, 138, 205, 237, 340, 343, 352, 357, 367, 371, 373.*

Pitter (1963c), P. Biodegradability of some anionic detergents. *Chem. Prumysl*, **13**, 284–287 (1963). *406.*

Pitter (1964a), P., and J. Tutko. Synthetic surfactants in wastewaters. VI. Experiments with hard and soft alkylbenzenesulfonates in laboratory models of aeration tanks. *Sb. VSChT*, **8**(1), 167–174 (1964). *166, 340, 344, 353.*

Pitter (1964b), P., J. Frank, and J. Křížová-Chlumová. Synthetic surfactants in wastewaters. VII. The influence of surfactants on anaerobic sludge digestion. Discontinuous laboratory experiments. *Sb. VSChT*, **8**(1), 175–193 (1964). *115, 189, 307, 344, 353, 368.*

Pitter (1964c), P. Synthetic surfactants in wastewaters. VIII. Relation between molecular structure and the susceptibility of anionic surfactants to biochemical oxidation (Summary of results). *Sb. VSChT*, **8**(2), 13–39 (1964). *140, 223, 237, 309, 340, 343, 352, 353, 358, 360, 362, 366, 368, 369, 371, 373, 375, 378, 379, 381, 383, 406, 407, 408.*

Pitter (1964d), P. Alkyl sulfates resistant to biodegradation and persistence of *n*-alkylbenzene sulfonates under anaerobic conditions. *Surf. Cong. #4*, **3**, 861–869 (1967). *189, 205, 223, 237, 307, 340, 343, 344, 352, 353, 357, 360, 368, 371, 373.*

Pitter (1966a), P. An improved technique for the determination of polyoxyethylene surfactants with hydroquinone. *Chem. Ind. (London)*, **1966**, 1217–1219. *58.*

Pitter (1966b), P., and J. Chudoba. Synthetic surfactants in wastewaters. X. Treatment of laundry wastes by activated sludge process. *Sb. VSChT*, **F10**, 31–48 (1966). *359, 370.*

Pitter (1967a), P. Improved spectrometric determination of very dilute aqueous solutions of nonionic polyethoxylates with phosphotungstic acid and hydroquinone. *FSA*, **69**, 74–76 (1967). *58.*

Pitter (1967b), P., and J. Černý. Synthetic surfactants in wastewaters. XI. Determination of anionic surfactants in sludges from sewage treatment processes. *Sb. VSChT*, **F12**, 5–15 (1967). *41.*

Pitter (1968a), P. Relation between degradability and chemical structure of nonionic polyethylene oxide compounds. *Surf. Cong. #5*, **1**, 115–123 (1969). *133, 244, 245, 247, 249, 312, 316, 320, 382, 383, 390, 391, 393, 404, 405.*

Pitter (1968b), P. Biodegradability of surface-active alkyl polyethylene glycol ethers. *Collection Czech. Chem. Commun.*, **33**, 4083–4088 (1968). *244, 247, 312, 316, 382.*

Pitter (1968c), P. Synthetic surfactants in wastewaters. XII. Evaluation of surfactant biodegradability by the COD technique and ultraviolet spectra. *Sb. VSChT*, **F14**, 7–17 (1968). *222, 289, 348, 349, 355, 357, 368, 369, 371, 384, 392, 393, 396, 404, 405, 406, 408.*

Pitter (1968d), P., and M. Horská. Synthetic surfactants in wastewaters. XIII. Effect of surfactants on the dehydrogenase activity of activated sludge. *Sb. VSChT*, **F14**, 19–25 (1968). *99.*

Pitter (1968e), P. Relationship between molecular structure and the biodegradability of surfactants and their metabolism. *Sb. VSChT*, **F14**, 47–137 (1968). *v.*

Pitter, P. *See* Kulovaná (1966).

Pitt-Rivers (1968), R., and F. S. A. Impiombato. The binding of sodium dodecyl sulfate to various proteins. *Biochem. J.*, **109**, 825–830 (1968). *104.*

Polet, H. *See* Ray (1966); Reynolds (1967).

Polkowski (1959), L. B., G. A. Rohlich, and J. R. Simpson. Evaluation of frothing in sewage treatment plants. *Sewage Ind. Wastes*, **31**, 1004–1015 (1959). *44.*

Popkin (1968), R. A., and T. W. Bendixen. Improved subsurface disposal. *JWPCF*, **40**, 1499–1514 (1968). *380.*

Porges (1952), N., L. Jasewicz, and S. R. Hoover. Measurement of carbon dioxide evolution from activated sludge. *Sewage Ind. Wastes*, **24**, 1091–1097 (1952). *72.*

Porges (1958), N., A. E. Wasserman, W. J. Hopkins, and L. Jasewicz. Oxidation of radioactive glucose by aerated sludge. *Sewage Ind. Wastes*, **30**, 776–782 (1958). *82.*

Postgate (1963), J. R., and J. R. Hunter. Acceleration of bacterial death by growth substrates. *Nature*, **198**, 273 (1963). *81.*

Powell, G. M. *See* Denner (1969).

Powers, M. T. *See* Stokes (1967).

Prakasam (1964), T. B. S., and N. C. Dondero. Observations on the behavior of a microbial population adapted to a synthetic waste. *Purdue Conf.*, **19**, 835–845 (1964). *178.*

Prakasam (1967a), T. B. S., and N. C. Dondero. Aerobic heterotrophic bacterial populations in sewage and activated sludge. II. Method of characterization of activated sludge bacteria. *Appl. Microbiol.*, **15**, 1122–1127 (1967). *162.*

Prakasam (1967b), T. B. S., and N. C. Dondero. Aerobic heterotrophic bacterial populations of sewage and activated sludge. III. Adaptation in a synthetic waste. *Appl. Microbiol.*, **15**, 1128–1137 (1967). *89, 178.*

Prasad, D. *See* Jones (1968c).

Prescher, D. *See* Püschel (1968).

Prochazka (1965), G. J., and W. J. Payne. Bacterial growth as a practical indicator of extensive biodegradability of organic compounds. *Appl. Microbiol.*, **13**, 702–705 (1965). *74, 373, 403.*

Prochazka (1967), G. J., W. R. Mayberry, and W. J. Payne. Model systems for studying energy yields from synthetic organic compounds. *Develop. Ind. Microbiol.*, **8**, 167–178 (1967). *74, 305, 312, 314, 403, 404, 410.*

Püschel (1968), F., and D. Prescher. On high molecular aliphatic sulfonates. VII. Paper chromatography of sulfonates and some alkyl sulfates. *J. Chromatog.*, **32**, 337–345 (1968). *60.*

Putnam (1948), F. W. The interactions of proteins and synthetic detergents. *Advan. Protein Chem.*, **4**, 79–122 (1948). *96, 102, 103.*

Quarmby, C. *See* Painter (1968).

Rahfuse, R. V. *See* Berber (1965).

Ramanathan (1969), M., and A. F. Gaudy, Jr. Effect of high feedback on kinetic behavior of heterogeneous populations in completely mixed systems. *Biotechnol. Bioeng.*, **11**, 207–237 (1969). *182.*

Rammler (1964), D. H., C. Grado, and L. R. Fowler. Sulfur metabolism of *Aerobacter aerogenes*. I. A repressible sulfatase. *Biochemistry (ACS)*, **3**, 224–230 (1964). *304.*

Rand, M. C. *See* Heukelekian (1955); Orford (1953).

Rao (1966), B. S., and A. F. Gaudy, Jr. Effect of sludge concentration on various aspects of biological activity in activated sludge. *JWPCF*, **38**, 794–812 (1966). *179, 180.*

Rao, S. S. *See* Hetling (1964); Washington (1964).

Ray (1966), A., J. A. Reynolds, H. Polet, and J. Steinhardt. Binding of large organic anions and netural molecules by native bovine serum albumin. *Biochemistry (ACS)*, **5**, 2606–2616 (1966). *103.*

Raybould (1956), R. D., and L. H. Thompson. Some large scale investigations on the influence of alkyl benzene sulfonate detergents on sewage purification. *JISP*, **1956,** 12–35. *189, 190, 350, 357, 373.*

Raymond (1969), R. L. Biotransformations using hydrocarbons. *Process Biochem.*, **4** (9), 71–74 (1969). *265.*

Raymond, R. L. *See* Davis (1961).

Rector, J. M. *See* Myerly (1964).

Rees, H. B., Jr. *See* Williams (1949).

Reid (1967), V. W., G. F. Longman, and E. Heinerth. Determination of anionic detergents by two-phase titration. *Tenside*, **4**, 292–304 (1967). *52, 53.*

Reid (1968), V. W., G. F. Longman, and E. Heinerth. Determination of anionic detergents by two-phase titration. II. *Tenside* **5**, 90–96 (1968). *52, 53.*

Reid, V. W. *See* Hughes (1968b).

Renn (1959), C. E., and M. F. Barada. Adsorption of ABS on particulate materials in water. *Sewage Ind. Wastes*, **31**, 850–854 (1959). *113.*

Renn (1964a), C. E., W. A Kline, and G. Orgel. Destruction of linear alkylate sulfonates in biological waste treatment by field test. *JWPCF*, **36**, 864–879 (1964). *39, 50, 130, 151, 166, 191, 340, 344, 345, 353.*

Renn (1964b), C. E. Techniques for determining biodegradability of detergent bases. *Detergent Age*, **1**(2), 24–28 (1964). *345, 353, 368.*

Renn (1965a), C. E. Degradation of Ucane LAS alkylate in laboratory trickling filters. Annual Convention, Soap & Detergent Association, New York, Jan. 1965. *130, 159, 346, 354.*

Renn (1965b), C. E. Woodbridge, Virginia field tests of biodegradable LAS detergents in extended aeration activated sludge system. *Purdue Conf.*, **20**, 734–737 (1965). *191.*

Renn, C. E. *See* Conway (1965); Gloyna (1952); Kumbe (1966).

Rétey (1966), J., A. Umani-Ronchi, J. Seibl, and D. Arigoni. Mechanism of the propanediol dehydrase reaction. *Experentia*, **22**, 502–503 (1966). *313.*

Reynolds (1966), T. D., and J. T. Yang. Model of the completely-mixed activated sludge process. *Purdue Conf.*, **21**, 696–713 (1966). *182.*

Reynolds (1967), J. A., S. Herbert, H. Polet, and J. Steinhardt. The binding of divers detergent anions to bovine serum albumin. *Biochemistry (ACS)*, **6**, 937–947 (1967). *104.*

Reynolds, J. A. *See* Ray (1966).

Ribbons, D. W. *See* Dawes (1962, 1964, 1965).

Rickard (1965), M. D., and W. H. Riley. Carbon as a parameter in bacterial systems growth limitation and substrate utilization studies. *Purdue Conf.*, **20**, 98–109 (1965). *69, 74.*

Rickard, M. D. *See* Riley (1965).

Rickenberg, H. V. *See* Hsie (1967).

Riley (1965), W. H., and M. D. Rickard. The biochemical aspects of aerobic bacterial growth. *Purdue Conf.*, **20**, 235–247 (1965). *74.*

Riley, W. H. *See* Rickard (1965).

Rismondo (1968), R., and F. Zilio-Grandi. Biodegradability of anionic surfactants in an experimental cesspool model. *Riv. Ital. Sostanze Grasse*, **45**, 116–121 (1968). *349, 355.*

Ritchie, C. C. *See* Eye (1966).

Robeck (1963), G. G., J. M. Cohen, W. T. Sayers, and R. L. Woodward. Degradation of ABS and other organics in saturated soils. *JWPCF*, **35**, 1225–1236 (1963). *65, 79, 187, 188, 198, 229, 299, 352.*

Robeck (1964), G. G., T. W. Bendixen, W. A. Schwartz, and R. L. Woodward. Factors influencing the design and operation of soil systems for waste treatment. *JWPCF*, **36**, 971–983 (1964). *95, 187, 188, 353.*

Roberts (1957), F. W. The removal of anionic syndets by biological purification processes. *Water Waste Treat. J.*, **6**, 302–303 (1957). *42, 380.*

Roberts (1958), F. W., and G. R. Lawson. Some determinations of the synthetic detergent content of sewage sludge. *Water Waste Treat. J.*, **7**, 14–17 (1958). *41.*

Roberts (1960), F. W. Assessment of softness in a synthetic detergent by biological means. *Chem. Ind. (London)*, **1960**, 1282–1284. *94, 149, 342, 351.*

Roberts, H. *See* Solbe (1967).

Robertson, J. B. *See* Waymann (1963a, b, c).

Rogers, M. R. *See* Long (1966).

Rohlffs (1963), G., and B. Herold. Process for preparing biodegradable alkylaryl sulfonates. German Pat. 1,142,166 (1963). *352, 357.*

Rohlich, G. A. *See* Crabtree (1966); Knopp (1965); Polkowski (1959).

Rombach, R. *See* Thoma (1965).

Ross, J. *See* AASGP (1956).

Roth, J. C. *See* Loehr (1968).

Ruchhoft, C. C. *See* Butterfield (1937).

Rudolfs (1949), W., R. Manganelli, and I. Gellman. Effect of certain detergents on sewage treatment. *Sewage Works J.*, **21**, 605–612 (1949). *358, 370, 379.*

Rumer, R. R. Jr., *See* Weber (1965).

Rupp, F. A. *See* Orgel (1964); SDA (1965).

Ruschenberg (1963a), E. Structure elements of detergents and their influence on biodegradation. *Vom Wasser*, **30**, 232–248 (1963). *205, 206, 209, 215, 217, 222, 237, 244, 334, 335, 336, 338, 340, 341, 342, 343, 352, 362, 367, 369, 371, 381, 382.*

Ruschenberg (1963b), E. The biological behavior of surfactants. I. The relations between the constitution and the biodegradation of alkylbenzene sulfonates. *FSA*, **65**, 810–814 (1963). *205, 206, 209, 215, 217, 334, 335, 336, 338, 340, 341, 342, 343, 352, 362, 367, 369.*

Ruschenberg (1964), E., A. Meusel, and R. Herrmann. Process for preparing easily biodegradable tetrapropylene benzene sulfonates. German Pats. 1,166,185 and 1,166,186 (1964). *228, 353.*

Ruschenberg, E. *See* Jendreyko (1963).

Rutberg, L. *See* Bishop (1967).

Ryckman (1956), D. W. The significance of chemical structure in biodegradation of alkyl benzene sulfonates. Sc.D. Thesis, Massachusetts Institute of Technology, Cambridge, 1956. *54, 62, 71, 172, 275, 334, 335, 338, 350, 355, 361, 406, 407.*

Ryckman (1957), D. W., and C. N. Sawyer. Chemical structure and biological oxidizability of surfactants. *Purdue Conf.*, **12**, 270–284 (1957). *62, 71, 172, 205, 207, 236, 275, 277, 301, 302, 334, 335, 338, 350, 355, 365, 367, 406, 407.*

Ryckman, D. W. *See* Bennett (1961); Young (1968).

Sakayanagi, S. *See* Fujiwara (1968).

Sallee, E. M. *See* AASGP (1956, 1961); SDA (1964).

Salo, E. *See* Nurmikko (1966).

Salt (1968), W. G., and D. Wiseman. The uptake of cetyltrimethylammonium bromide by *Escherichia coli*. *J. Pharm. Pharmacol.*, **20** (Suppl.), 14S–17S (1968) *116.*

Samuelson, B. *See* Bishop (1967).

Sargent, D. A. *See* Ellerker (1968).

Sato (1963), M., K. Hashimoto, and M. Kobayashi. Microbial degradation of nonionic synthetic detergent, poly EO nonylphenol ethers. *Water Treat. Eng.* (*Japan*), **4**(4), 31–36 (1963). *253, 319, 358, 367, 391.*

Sauer, G. F. *See* Sheers (1967).

Sawyer (1956), C. N., R. H. Bogan, and J. R. Simpson. Biochemical behavior of synthetic detergents. *Ind. Eng. Chem.*, **48**, 236–240 (1956). *146, 203, 205, 206, 207, 227, 236, 243, 245, 301, 335, 350, 357, 365, 367, 370, 389, 392, 393, 394.*

Sawyer (1965), C. N. Milestones in the development of the activated sludge process. *JWPCF*, **37**, 151–162 (1965). *3.*

Sawyer, C. N. *See* Bogan (1954, 1955, 1956); Helmers (1950); Ryckman (1957).

Sayers, W. T. *See* Robeck (1963).

Scardigno, S. *See* Ciattoni (1968).

Schaffer (1965), R. B., C. E. van Hall, G. N. McDermott, D. Barth, V. A. Stenger, S. J. Sebesta, and S. H. Griggs. Application of a carbon analyzer in waste treatment. *JWPCF*, **37**, 1545–1566 (1965). *68.*

Schaffer, R. B. *See* Ludzack (1959, 1961).

Schairer, H. U. *See* Overath (1969).

Schenk, L. M. *See* Eiseman (1969), Nunn (1967).

Scherb (1962), K. Investigations on the decomposition of detergents in oxidation channels. *Münchner Beitr. Abwasser-, Fisch.-Flussbiol.*, **9**, 242–254 (1962). *190, 351, 380.*

Scheurer, H. *See* Jerchel (1953).

Schick (1967), M. J., ed. *Nonionic Surfactants*, Dekker, New York, 1967. *22.*

Schmeiser, K. *See* Jerchel (1954).

Schneider, G. *See* Krone (1968).

Schönborn (1962a), W. Investigations of the conditions permitting degradation of detergents in conventional sewage treatment plants. *Gas-Wasserfach*, **103**, 1133–1136 (1962). *98, 114, 115, 130, 133, 140, 158, 173, 343, 351.*

Schönborn (1962b), W. Methods for testing biodegradability of detergents. *Seifen-Öle-Fette-Wachse*, **88**, 870–875 (1962). *79, 124.*

Schönborn (1966), W. Analytical determination and biodegradation of nonionic surfactants in wastewaters. *Schriftenreiche des Deutschen Arbeitskreises Wasserforschung e.V.*, No. 12, Erich Schmidt, Berlin, 1966. *42, 55, 58, 67, 116, 119, 386, 387, 390, 393, 394, 404, 405.*

Schuck, J. M. *See* Engelbrecht (1967).

Schuett, W. R. *See* Sharman (1964b).

Schultz, R. G. *See* Engelbrecht (1967).

Schulze (1964a), K. L. A mathematical model of the activated sludge process. *Develop. Ind. Microbiol.*, **5**, 258–266 (1964). *182.*

Schulze (1964b), K. L. The activated sludge process as a continuous flow culture. I. Theory. *Water Sewage Works*, **111**, 526–538 (1964). II. Application. *Ibid.* **112**, 11–17 (1965). *182.*

Schwartz (1949), A. M., and J. W. Perry. *Surface Active Agents*, Vol. 1, Interscience, New York, 1949. *22.*

Schwartz (1958), A. M., J. W. Perry, and J. Berch. *Surface Active Agents and Detergents*, Vol. 2, Interscience, New York, 1958. *22.*

Schwartz, W. A. *See* Robeck (1964).

Schwen (1966), G. The final concentration of surfactants at the solution-air interface as a function of time. *Tenside*, **3**, 69–71 (1966). *47.*

Scott, C. C. *See* Patterson (1966b, 1967, 1968, 1970).

SDA (1964) (Soap & Detergent Association), Analytical Subcommittee. Identification and determination of trace amounts of ethoxylated alkylphenols in water and sewage. *Surf. Cong. #4*, **3**, 117–129 (1967). *64.*

SDA, (1965), Biodegradation Subcommittee. Procedure and standards for determination of biodegradability of ABS and LAS. *JAOCS*, **42**, 986–993 (1965). *13, 131, 133, 138, 151, 153, 171, 172, 341, 346, 354.*

SDA (1969a), Biodegradation Subcommittee. The status of biodegradability testing of nonionic surfactants. *JAOCS*, **46**, 432–440 (1969). *46, 57, 133, 139, 172, 205, 341, 349, 355, 383, 385, 386, 387, 388, 389, 390, 391, 397, 398.*

SDA (1969b), Biodegradation Subcommittee. The status of the biodegradability testing of nonionic surfactants. *SDA Scientific and Technical Report No.* **6**, 1969. *44, 46, 55, 57, 133, 139, 172, 205, 341, 349, 355, 383, 385, 386, 387, 388, 389, 390, 391, 397, 398.*

Sebban (1968), R. Modification of the Greff-Setzkorn method for determination of higher polyethoxylate nonionic detergents. *Chim. Anal. (Paris)*, **50**(3), 129–130 (1968). *56.*

Sebesta, S. J. *See* Schaffer (1965).

Seibl, J. *See* Rétey (1966).

Sekiguchi, H. *See* Tomiyama (1968, 1969).

Senkowski, B. Z. *See* Sheridan (1969).

Setzkorn (1963), E. A., and A. B. Carel. The analysis of alkyl aryl sulfonates by microdesulfonation and gas chromatography. *JAOCS*, **40**, 57–59 (1963). *61.*

Setzkorn (1964), E. A., R. L. Huddleston, and R. C. Allred. An evaluation of the river die-away technique for studying detergent biodegradability. *JAOCS*, **41**, 826–830 (1964). *50, 53, 148, 209, 215, 335, 336, 340, 341, 342.*

Setzkorn (1965), E. A., and R. L. Huddleston. Ultraviolet spectroscopic analysis for following the biodegradation of hydrotropes. *JAOCS*, **42**, 1081–1084 (1965). *64, 406, 407.*

Setzkorn, E. A. *See* Allred (1964b); Greff (1965); Huddleston (1965a); SDA (1964).

Shapiro, M. *See* Weeks (1969).

Shaposhnikov (1968), V. N., E. I. Kozlova, and Z. A. Arkad'eva. Microflora of petroleum-containing sewage. *Microbiology (USSR)*, **37**, 418–423 (1968). *78.*

Sharman (1964a), S. H. Extensive biodegradation of synthetic detergents. *Nature*, **201**, 704–705 (1964). *65, 229, 345, 353.*

Sharman (1964b), S. H., W. R. Schuett, D. Kyriacou, and W. A. Sweeney. Foam recycle. A method for improved removal of detergents from sewage. *Am. Chem. Soc., Div. Water Waste Chem., Preprints*, **4**(1), 175–182 (1964). *45, 229, 353.*

Sharman, S. H. *See* House (1965b); Marquis (1966).

Shaulis, L. *See* Benarde (1965).

Sheers (1967), E. H., D. C. Wehner, and G. F. Sauer. Biodegradation of a sulfonated amide surfactant. *JWPCF*, **39**, 1410–1416 (1967). *302, 366.*

Sheets (1956a), W. D., and G. W. Malaney. Synthetic detergents and the BOD test. *Sewage Ind. Wastes*, **28**, 10–17 (1956). *357, 360, 365, 366, 367, 391, 394. 398,*

Sheets (1956b), W. D., and G. W. Malaney. The COD values of syndets, surfactants and builders. *Purdue Conf.*, **11**, 185–196 (1956). *357, 360, 367, 391, 394, 398.*

Sheets, W. D. *See* Hanna (1964); Malaney (1957).

Sheridan (1969), J. C., E. P. K. Lau, and B. Z. Senkowski. Determination of nonionic surfactants by atomic absorption spectrophotometry. *Anal. Chem.*, **41**, 247–250 (1969). *58.*

Shewmaker, J. E. *See* Kelly (1965); SDA (1965, 1969a, b).

Shindala (1965), A., H. R. Bungay III, N. R. Krieg, and K. Culbert. Mixed-culture interactions. I. Commensalism of *Proteus vulgaris* with *Saccharomyces cerevisiae* in continuous culture. *J. Bacteriol.*, **89**, 693–696 (1965). *178.*

Short, E. R. *See* Fairing (1956).

Shultz, A. *See* SDA (1965).

Shumate (1969), K. S., J. E. Thompson, J. D. Brookhart, and C. L. Dean. Removal of NTA by a municipal activated sludge treatment plant—a ten-month field study. Annual Meeting, Missouri Water Pollution Control Associaion, Kansas City, Feb. 1969. *JWPCF*, in press. *409.*

Siebert (1969), M. L., and D. F. Toerien. The proteolytic bacteria present in the anaerobic digestion of raw sewage sludge. *Water Res.*, **3**, 241–250 (1969). *79.*

Sierp (1954), F., and H. Thiele. Influence of surfactants on sewage purification and on the self-purification of rivers. *Vom Wasser*, **21**, 197–246 (1954). *170, 366, 373.*

Sierra (1957), G. A simple method for the detection of lipolytic activity of microorganisms and some observations on the influence of the contact between cells and fatty substrates. *Antonie van Leeuwenhoek, J. Microbiol. Serol.*, **23**, 15–22 (1957). *326.*

Simko (1965), J. P., Jr., E. M. Emery, and E. W. Blank. Degradation of linear alkylate sulfonate in sewage. *JAOCS*, **42**, 627–629 (1965). *216.*

Simonis, H. *See* Ismail (1965).

Simpson (1964), J. R. Extended sludge aeration activated sludge systems. *JISP*, **1964**, 328–341. *83, 84, 137, 163.*

Simpson (1965), J. R. The biological oxidation and synthesis of organic matter. *JISP*, **1965**, 171–180. *84, 137.*

Simpson, J. R. *See* Polkowski (1959); Sawyer (1956).

Singrun, M. E. *See* Hartmann (1968).

Skelly (1966), N. E., and W. B. Crummett. Determination of 18 to 22 mol ethoxylates in nine mol EO adduct of *p*-nonylphenol. *J. Chromatog.*, **21**, 257–260 (1966). *60.*

Skinner (1959), F. A. Decomposition of anionic surfactants by soil bacteria. *Nature*, **183**, 548–549 (1959). *79, 127, 367, 369, 370, 373, 397.*

Smith (1957), W. S. Removal of anionic synthetic detergent by percolating filters. *JISP*, **1957**, 153–155. *380.*

Smith (1966), F. D., A. J. Stirton, and M. V. Nuñez-Ponzoa. Isomeric linear phenylalkanes and sodium alkylbenzene sulfonates. *JAOCS*, **43**, 501–504 (1966). *336, 338, 341, 342.*

Smith (1968), R. L. Water pollution control was settled in 1950. *Ind. Water Eng.*, **5**(9), 40–41 (1968). *163.*

Smith, F. D. *See* Weil (1965).

Smithson (1966), L. H. Properties of ethoxylate derivatives of nonrandom alkylphenols. *JAOCS*, **43**, 568–571 (1966). *241, 249, 250, 377, 384, 388.*

Smithson, L. H. *See* Liddicoet (1965).

Snell, E. E. *See* Goodhue (1966a, b); Magee (1966); Nurmikko (1966).

Sniegowski, M. *See* Hurwitz (1960).

Snow, C. M. *See* SDA (1965).

Snyder (1967), J. Q., R. D. Swisher, W. E. Weesner, and L. T. Wolford. Process for preparing alkaryl sulfonates by dimerizing olefins with a metal oxide promoted silica-alumina catalyst. U.S. Pat. **3,332,989** (1967). *355, 357.*

Snyder (1968), J. Q., R. D. Swisher, L. T. Wolford, and W. E. Weesner. Polymerization of olefin hydrocarbons. Canadian Pat. 786,835 (1968). *355, 357.*

Södergren (1966), A. Automatic determination of surfactants in water. *Analyst*, **91**, 113–118 (1966). *50.*

Solbe (1967), J. F. de L. G., N. V. Williams, and H. Roberts. The colonization of a percolating filter by invertebrates and their effect on settlement of humus solids. *Water Pollution Control*, **66**, 423–448 (1967). *156.*

Sorlini (1969), C., and V. Treccani. Microbial degradation of aliphatic branched compounds: isobutyric, 2,2-dimethylmalonic and 2,2-dimethylsuccinic acids. *Experentia*, **25**, 1032 (1969). *299.*

Spencer, B. *See* Harada (1964b).

Spohn (1964a), H. Biologically hard and soft detergents in sewage treatment plants. I. *Tenside*, **1**, 18–26 (1964). *190, 345, 353, 380.*

Spohn (1964b), H., and W. K. Fischer. Biologically hard and soft detergents in sewage treatment plants. II. *Tenside*, **1**, 81–88 (1964). *190, 191, 195, 345, 353, 380.*

Spohn, H. *See* DAGS (1961).

Squire (1961), G. V. V. The manufacturer's part in the Luton experiment. *JISP*, **1961**, 27–29; 49–56. *194.*

Stack, V. T., Jr. *See* Mills (1953, 1954); Oberton (1957).

Stadtman, E. R. *See* Gaston (1963).

Stander, G. J. *See* Urban (1965).

Stanier (1947), R. Y. Simultaneous adaptation: a new technique for the study of metabolic pathways. *J. Bacteriol.*, **54**, 339–348 (1947). *87.*

Stanier (1966), R. Y., N. J. Palleroni, and M. Doudoroff. The aerobic Pseudomonads: a taxonomic study. *J. Gen. Micorbiol.*, **43**, 159–271 (1966). *80.*

Starkey (1964), R. L. Microbial transformations of some organic sulfur compounds. In *Principles and Applications in Aquatic Microbiology* (H. Heukelekian and N. C. Dondero, eds.), Wiley, New York, 1964, pp. 405–429. *273.*

Staub, W. *See* Hartmann (1967).

STCSD (1960)(Standing Technical Committee on Synthetic Detergents). *Third Progress Report*, HMSO, London, 1960. *191, 194.*

STCSD (1961). *Fourth Progress Report*, HMSO, London, 1961. *129, 191, 194.*

STCSD (1962). *Fifth Progress Report*, HMSO, London, 1962. *129, 191, 194.*

STCSD (1963). *Sixth Progress Report*, HMSO, London, 1963. *343, 344.*

STCSD (1964). *Seventh Progress Report*, HMSO, London, 1964. *345, 362.*

STCSD (1966). *Supplement to the Eight Progress Report*, HMSO, London, 1966. *53, 154, 347.*

STCSD (1967). *Ninth Progress Report*, HMSO, London, 1967. *196, 348.*

Steinberg, E.-M. *See* Bouveng (1968).

Steinhardt, J. *See* Ray (1966); Reynolds (1967).

Steinle (1964), E. C., R. C. Myerly, and C. A. Vath. Surfactants containing ethylene oxide: Relationship of structure to biodegradability. *JAOCS*, **41**, 804–807 (1964). *240, 241, 242, 251, 370, 371, 373, 375, 376, 377, 384, 385, 386, 387, 388, 391, 392.*

Steinle, E. C. *See* Myerly (1964).

Stenger (1967), V. A., and C. E. van Hall. Rapid method for determination of COD. *Anal. Chem.*, **39**, 206–211 (1967). *68, 69.*

Stenger, V. A. *See* Schaffer (1965); van Hall (1964, 1967).

Stennett, G. V. *See* Eden (1968); Truesdale (1968).

Stern, G. *See* Berg (1966).

Stevenson (1954), D. G. The absorptiometric determination of a nonionic detergent. *Analyst*, **79**, 504–507 (1954). *58.*

Stirton (1965), A. J., R. G. Bistline, Jr., E. A. Barr, and M. V. Nuñez-Ponzoa. Salts of alkyl esters of α-sulfopalmitic and sulfostearic acids. *JAOCS*, **42**, 1078–1081 (1965). *364.*

Stirton, A. J. *See* Cordon (1965, 1968a, b); Maurer (1965); Smith (1966); Weil (1964, 1965).

Stokes (1967), J. L., and M. T. Powers. Glucose repression of oxidation of organic compounds by *Sphaerotilus*. *Can. J. Microbiol.*, **13**, 557–563 (1967). *88.*

Stokke (1969), O. The degradation of a branched chain fatty acid by alterations between α- and β-oxidations. *Biochim. Biophys. Acta*, **176**, 54–59 (1969). *263, 287.*

Stoltenberg, H. A. *See* Metzler (1958).

Storer (1969), F. F., and A. F. Gaudy, Jr. Computational analysis of transient response to qualitative shock loadings of heterogeneous populations in continuous culture. *Environ. Sci. Technol.*, **3**, 143–149 (1969). *87, 182.*

Straus (1963), A. E. Biodegradation of ABS in a simulated septic tank and drain field. *Science*, **142**, 244–245 (1963). *65, 189, 344, 352.*

Stumm, W. *See* Busch (1968); Weber (1962).

Stumm-Zollinger (1968), E. Substrate utilization in heterogeneous bacterial communities. *JWPCF*, **40**, R213–R229 (1968). *80, 89, 90, 126.*

Stumpf (1960), P. K., and G. A. Barber. Comparative mechanisms for fatty acid oxidation. In *Comparative Biochemistry* (M. Florkin and H. S. Mason, eds.), Vol. 1, Academic, New York, 1960, pp. 75–105. *261.*

Suess (1964), M. J. ABS adsorption on soils. *JWPCF*, **36**, 1393–1400 (1964). *113*

Suffis (1965), R., T. J. Sullivan, and W. S. Henderson. Identification of surfactants as trimethylsilyl derivatives by gas chromatography. *J. Soc. Cosmetic Chemists*, **16**, 783–794 (1965). *62.*

Sullivan (1968), W. T., and R. L. Evans. Major U.S. river reflects surfactant changes. *Environ. Sci. Technol.*, **2**, 194–200 (1968). *5.*

Sullivan (1969), W. T., and R. D. Swisher. MBAS and LAS surfactants in the Illinois River, 1968. *Environ. Sci. Technol.*, **3**, 481–483 (1969). *5, 61.*

Sullivan, L. J. *See* Knaak (1966).

Sullivan, T. J. *See* Suffis (1965).

Sulzer, F. T. *See* Cassell (1966).

Svitálková, J. *See* Pitter (1961a).

Swanwick, J. D. *See* Bruce (1966).

Sweeney (1964a), W. A., and J. K. Foote (1964a). A rapid, accurate test for surfactant aerobic biodegradability. *JWPCF*, **36**, 14–37 (1964). *40, 65, 112, 114, 115, 129, 134, 168, 182, 205, 221, 223, 224, 237, 335, 336, 337, 340, 341, 342, 345, 353, 355, 356, 357, 360, 369, 371.*

Sweeney (1964b), W. A. Aerobic biodegradability of linear alkylbenzene sulfonate. *Soap Chem. Specialties*, **30**(3), 45–47, 190 (1964). *149, 215, 336, 338, 340, 342, 345.*

Sweeney (1966), W. A. Note on straight chain ABS removal by adsorption during activated sludge treatment. *JWPCF*, **38**, 1023–1025 (1966). *116, 121.*

Sweeney, W. A. *See* Marquis (1966); Sharman (1964b).

Swenson (1965), R. L., K. S. Canfield, and W. K. Griesinger. Alkyl benzene sulfonates having high susceptibility to bacteriological degradation. U.S. Pat. 3,214,462 (1965). *354, 357.*

Swisher (1959), R. D. (Biodegradation of dialkylindane and dialkyltetralin sulfonates.) Unpublished data, 1959. *226, 360.*

Swisher (1960), R. D. (Biodegradation of *tp*-dodecylbenzyl sulfonate and alkyl sucrose surfactants.) Unpublished data, 1960. *356, 396.*

Swisher (1961), R. D., E. F. Kaelble, and S. K. Liu. Capillary gas chromatography of phenyldodecane alkylation and isomerization mixtures. *J. Org. Chem.*, **26**, 4066–4069 (1961). *29.*

Swisher (1962), R. D. (Concentration effects in river water biodegradation of ABS.) Unpublished data, 1962. *147.*

Swisher (1963a), R. D. Biodegradation rates of isomeric deheptylbenzene sulfonates. *Develop. Ind. Microbiol.*, **4**, 39–45 (1963). *61, 206, 207, 210, 211, 212, 360.*

Swisher (1963b), R. D. Biodegradation of ABS in relation to chemical structure. *JWPCF*, **35**, 877–892 (1963). *61, 148, 208, 209, 210, 213, 214, 215, 216, 220, 221, 224, 225, 335, 336, 337, 338, 340, 341, 342, 352, 407.*

Swisher (1963c), R. D. Transient intermediates in the biodegradation of LAS. *JWPCF*, **35**, 1557–1564 (1963). *61, 213, 279, 280, 281, 282, 283, 334, 335, 336, 337.*

Swisher (1963d), R. D. The chemistry of surfactant biodegradation. *JAOCS*, **40**, 648–656 (1963). *14.*

Swisher (1963e), R. D. (Degradation rates of LAS isomers in continuous activated sludge.) Unpublished data, 1963. *216.*

Swisher (1964a), R. D., J. T. O'Rourke, and H. D. Tomlinson. Fish bioassays of LAS and intermediate biodegradation products. *JAOCS*, **41**, 746–752 (1964). *50, 51, 75, 90, 169, 274, 340, 341.*

Swisher (1964b), R. D. Las: Major development in detergents. *Chem. Eng. Progr.,* **60**(12), 41–45 (1964). *61, 283, 284, 334, 335, 336, 337, 361.*

Swisher (1964c), R. D. (Adsorption of LAS by activated sludge.) Unpublished data, 1964. *185.*

Swisher (1966a), R. D. Shake culture biodegradation of LAS without inoculation. *Develop. Ind. Microbiol.,* **7**, 271–278 (1966). *152, 340, 341, 347, 377.*

Swisher (1966b), R. D. Identification and estimation of LAS in waters and effluents. *JAOCS*, **43**, 137–140 (1966). *61.*

Swisher (1966c), R. D. (Shake cultures with LAS as sole food source.) Unpublished data, 1966. *153.*

Swisher (1967a), R. D., M. M. Crutchfield, and D. W. Caldwell. Biodegradation of nitrilotriacetate (NTA) in activated sludge. *Environ. Sci. Technol.,* **1**, 820–827 (1967). *169, 174, 409.*

Swisher (1967b), R. D. Biodegradation of LAS benzene rings in activated sludge. *JAOCS*, **44**, 717–724 (1967). *64, 169, 174, 288, 289, 290, 335, 336, 341.*

Swisher (1967c), R. D. Biodegradation of surfactant benzene rings. *Ind. Chim. Belge*, **32** (Special No.) III, 718–722 (1967). *64, 289, 336, 341.*

Swisher (1968a), R. D. Biodegradation of surfactant benzene rings in shake cultures. *Develop. Ind. Microbiol.,* **9**, 270–279 (1968). *64, 153, 289, 291, 293, 336.*

Swisher (1968b), R. D. Exposure levels and oral toxicity of surfactants. *Arch. Environ. Health*, **17**, 232–246 (1968). *2.*

Swisher (1968c), R. D. (Sulfate liberation in LAS biodegradation.) Unpublished data 1968. *292.*

Swisher (1969), R. D. Benzene ring biodegradation in quaternary ABS. *Tenside*, **6**, 135–139 (1969). *64, 225 295, 337.*

Swisher, R. D. *See* SDA (1965); Snyder (1967, 1968); Sullivan (1969); WPCF (1967).

Symons (1958), J. M., and R. E. McKinney. The biochemistry of nitrogen in the synthesis of activated sludge. *Sewage Ind. Wastes*, **30**, 874–890 (1958). *83.*

Symons (1960), J. M., R. E. McKinney, and H. H. Hassis. A procedure for determination of the biological treatability of industrial wastes. *JWPCF*, **32**, 841–852 (1960). *124, 171.*

Symons (1961), J. M., and L. A. del Valle-Rivera. Metabolism of organic sulfonates by activated sludge. *Purdue Conf.*, **16**, 555–571 (1961). *269, 270, 406, 407.*

Symons (1963), J. M., and R. Labonte. A procedure for continuous nitrification corrections during Warburg respirometer studies. *Purdue Conf.*, **18**, 498–514 (1963). *142.*

Symons, J. M. *See* Konecky (1963); McKinney (1959a); Washington (1962).

Tabak (1964), H. H., C. W. Chambers, and P. W. Kabler. Microbial metabolism of aromatic compounds. I. Decomposition of phenolic compounds and aromatic hydrocarbons by phenol-adapted bacteria. *J. Bacteriol.*, **87**, 910–919 (1964). *406 410.*

Taber (1962), W. A., and B. B. Wiley. Antimicrobial activity of a monoalkylbenzene sulfonate complex. *Can. J. Microbiol.*, **8**, 621–628 (1962). *100.*

Takagi (1964), T., and K. Fukuzumi. Application of thin layer chromatography to oil chemistry. I. Chromatography of synthetic surfactants by silica plates. *Yuka-gaku*, **13**, 520–523 (1964); *CA*, **63**, 18485f (1965). *60*.

Takao, M. *See* Tomiyama (1968, 1969).

Takezono, T. *See* Fujiwara (1968).

Tamm (1962), C. Conversion of natural substances by microbial enzymes. *Angew. Chem. Intern. Ed. Engl.*, **1**, 178–195 (1962). *9*.

Tanaka, S. *See* Ikeda (1963).

Tarring (1965), R. C. The development of a biologically degradable alkylbenzene sulfonate. *Air Water Pollution*, **9**, 545–552 (1965). *218, 222, 225, 226, 340, 341, 342, 346*.

Täuber (1968), G. Paraffin sulfonate and its use as a detergent raw material. *Surf. Cong. #5*, **3**, 211–217 (1969). *362*.

Täuber, G. *See* Czok (1968).

Taylor (1965), R. C. (Quart bottle foam test.) Private communication to SDA Biodegradation Subcommittee, 1965. *44*.

Taylor, R. C. *See* SDA (1965).

Teletzke (1967), G. H., W. B. Gitchel, D. G. Diddams, and C. A. Hoffman. Components of sludge and its wet air oxidation products. *JWPCF*, **39**, 994–1005 (1967). *135*.

Teske, W. *See* DAGS (1961).

Testa (1964), M. Automatic analytical method for determining anionic detergents in polluted waters. *Föderation Europäischer Gewässerschutz, Informationsbl.*, **11**, 33–36 (1964); *CA*, **64**, 3194e (1966). *50*.

Testa, M. C. *See* de Jong (1967).

Thabaraj (1969), G. J., and A. F. Gaudy, Jr. Effect of dissolved oxygen concentration on the metabolic response of completely mixed activated sludge. *JWPCF*, **41**, R322–R335 (1969). *179, 183*.

Theron, P. F. *See* Urban (1965).

Thiel, P. G. *See* Hattingh (1967).

Thiele, H. *See* Sierp (1954).

Thijsse, G. J. E. *See* van der Linden (1965).

Thoma (1965), K., R. Rombach, and E. Ullmann. Determination of auxiliary pharmaceuticals. VI. Detection of surface active esters and ethers of polyethylene glycol by thin layer chromatography. *Arch. Pharm.*, **298**, 19–25 (1965). *60*.

Thompson (1968), J. E., and J. R. Duthie. The biodegradability of NTA. *JWPCF*, **40**, 306–319 (1968). *409*.

Thompson, J. E. *See* Shumate (1969).

Thompson, L. H. *See* Raybould (1956).

Thornton, H. G. *See* Gray (1928).

Timasheff (1951), S. N., and F. F. Nord. Investigations on proteins and polymers. V. Interaction of egg albumins with dodecylamine hydrochloride. *Arch. Biochem. Biophys.*, **31**, 309–319 (1951). *105*.

Tobari, M. *See* Kimura (1962).

Todd (1954), A. R. Water purification upset seriously by synthetic detergents. *Water Sewage Works*, **101**, 80 (1954). *4*.

Todd, R. L. *See* Payne (1966).

Toerien (1967a), D. F.   Direct isolation studies on the aerobic and facultative anaerobic bacterial flora of anaerobic digesters receiving raw sewage sludge. *Water Res.*, **1**, 55–59 (1957).   *79.*

Toerien (1967b), D. E.   Enrichment culture studies on aerobic and facultative anaerobic bacteria found in anaerobic digesters. *Water Res.*, **1**, 147–155 (1967).   *79.*

Toerien (1969), D. F., and W. H. J. Hattingh.   Anaerobic digestion. I. The microbiology of anaerobic digestion. *Water Res.*, **3**, 385–416 (1969).   *79.*

Toerien, D. F. *See* Hattingh (1967); Siebert (1969).

Tomiyama (1968), S., M. Takao, A. Mori, H. Sekiguchi, and K. Ōba.   Studies of α-olefin sulfonate. III. Biochemical behavior of α-olefin sulfonate. Annual Meeting, American Oil Chemists' Society, New York, October. 1968.   *41, 62, 116, 118, 127, 186, 234, 363.*

Tomiyama (1969), S., M. Takao, A. Mori, and H. Sekiguchi.   New household detergent based on AOS. *JAOCS*, **46**, 208–212 (1969).   *151, 205, 349, 363, 370.*

Tomiyama, S. *See* Ōba (1965a, e, 1967, 1968a, b).

Tomlinson, H. D. *See* McKinney (1956b); Swisher (1964a).

Tool (1967), H. R.   Manometric measurement of the BOD. *Water Sewage Works*, **114**, 211–218 (1967).   *72.*

Trauč, J. *See* Pitter (1963a).

Trebbi, G. F. *See* Lashen (1967a).

Treccani, V. *See* Sorlini (1969).

Trottier, C. H. *See* Lang (1965); Long (1966).

Trowbridge, J. R. *See* Gildenberg (1965).

Truesdale (1959), G. A., K. Jones, and K. G. Vankyke.   Removal of synthetic detergents in sewage treatment processes: trials of a new biologically attackable material. *Water Waste Treat. J.*, **7**, 441–444 (1959).   *157, 163, 342, 350, 351.*

Truesdale (1962), G. A., R. Wilkinson, and K. Jones.   Comparison of the behavior of various media in percolating filters. *JISP*, **1962**, 325–340.   *157.*

Truesdale (1968), G. A., G. V. Stennett, and G. E. Eden.   Assessment of biodegradability of synthetic detergents: a comparison of methods. *Surf. Cong. #5*, **1**, 91–101 (1969).   *154, 168, 174, 198, 349, 355, 359, 384.*

Truesdale, G. A. *See* Eden (1961a, b, 1965, 1968).

Tryding (1957), N., and G. Westöö.   Synthesis and metabolism of 2,2,17,17-tetramethylstearic acid. I. *Arkiv Kemi*, **11**, 291–305 (1957).   *295, 296.*

Tucker, K. B. E. *See* Patterson (1966a, b, 1967, 1968, 1970).

Tutko, J. *See* Pitter (1964a).

Uhren, L. J. *See* Knopp (1965).

Ullmann, E. *See* Thoma (1965).

Umani-Ronchi, A. *See* Rétey (1966).

Urban (1965), P. J., G. J. Stander, D. W. Osborn, P. F. Theron, and S. M. Walker.   Experiments to establish the degradability of a new biologically soft detergent. *CSIR Research Report* **231**, South African Council for Scientific and Industrial Research, Pretoria, 1965.   *40, 44, 63, 133, 138, 164, 191, 195, 346, 354, 355.*

Urfer, A. D. *See* Cooper (1964).

U.S. Public Health Service (1962).   *Drinking Water Standards, 1962, PHS Publication No. 956*, U.S. Government Printing Office, Washington, 1962, pp. 22–25.   *2.*

Vaccaro, R. F. *See* Menzel (1963).

Vaicum (1967), L., and A. Ilişescu. Biodegradability of detergents and determining methodology. *Rev. Chim.* (*Bucharest*), **18**(1), 6–12 (1967). *171, 200, 348, 355, 359, 362, 368, 369, 373, 374.*

Vaicum (1968), L. (Influence of feed composition on LAS biodegradation.) Discussion of Mann (1968). *Surf. Cong. #5*, **1**, 111–114 (1969). *137, 168.*

Vallentyne (1957), J. R. The molecular nature of organic matter in lakes and oceans, with lesser reference to sewage and terrestrial soils. *J. Fisheries Res. Board Can.*, **14**(1), 33–82 (1957). *134.*

Van Beneden (1952), G. Non-cationic surfactants: can they act as nutrient substrate for certain bacteria? *Bull. Centre Belge Etude Doc. Eauz No. 16*, pp. 138–139 (1952). *366, 367.*

Van Beneden (1964), G. Anionic surfactants: can they constitute a nutritive substrate and assure the metabolism of bacteria? *Surf. Cong. #4*, **3**, 945–955 (1967). *121.*

Van Beneden (1965), G. Study on detergents: biodegradability or destructibility? *Tribune CEBEDEAU*, **18**(254), 26–27 (1965) *121.*

Van Cauwenberghe (1969), K., M. Vandewalle, and M. Verzele. Determination of the branching degree in alkylbenzenes by pyrolysis gas chromatography. *J. Chromatographic Sci.*, **7**, 698–700 (1969). *61, 202.*

Van der Linden (1965), A. C., and G. J. E. Thijsse. The mechanisms of microbial oxidations of petroleum hydrocarbons. *Advan. Enzymol.*, **27**, 469–546 (1965). *258, 263.*

Van der Vet, A. P. *See* Dronkers (1964).

Van der Zee, H. *See* Degens (1950, 1955).

Vandewalle, M. *See* van Cauwenberghe (1969).

Vandyke, J. M. *See* Curds (1968).

Vandyke, K. G. *See* Truesdale (1959).

Van Eyk (1968), J., and T. J. Bartels. Paraffin oxidation in *Pseudomonas aeruginosa*. I. Induction of paraffin oxidation. *J. Bacteriol.*, **96**, 706–712 (1968). *93, 258.*

Van Gils (1964), H. W. Bacteriology of activated sludge. *Report No. 32*, Research Institute for Public Health Engineering (IG-TNO), The Hague, 1964. *78. 83, 162.*

Van Hall (1964), C. E., and V. A. Stenger. Use of infrared analyzer for total carbon determination. *Water Sewage Works*, **111**, 266–270 (1964). *68.*

Van Hall (1967), C. E., and V. A. Stenger. An instrumental method for rapid determination of carbonate and total carbon in solutions. *Anal. Chem.*, **39**, 503–507 (1967). *68.*

Van Hall, C. E. *See* Schaffer (1965); Stenger (1967).

Van Peppen, J. *See* Fuhrmann (1964).

Varenyi, L. *See* Knaggs (1965).

Varmus, H. E. *See* de Crombrugghe (1969).

Vas'kova (1968), L. P., and P. V. Afanas'ev. Effect of dodecyl sulfate on the structure and catalytic properties of chymotrypsinogen. *Dokl.—Biochem. Sect.* (*English Transl.*), **179**, 98–100 (1968). *110.*

Vath (1964), C. A. A sanitary engineer's approach to biodegradation of nonionics. *Soap Chem. Specialties*, **40**(2), 56–58, 182; (3) 55–58, 108 (1964). *47, 65, 74, 143, 188, 248, 252, 312, 316, 341, 369, 373, 375, 376, 377, 382, 385, 386, 387, 388, 391, 404.*

Vath, C. A. *See* Conway (1965); Myerly (1964); Steinle (1964).

Verzele, M. *See* van Cauwenberghe (1969).

Viney, M. *See* Painter (1961).

Viswanatha (1955), T., M. J. Pallansch, and I. E. Liener.   The inhibition of trypsin. II. The effect of synthetic anionic detergents. *J. Biol. Chem.*, **212**, 301–309 (1955). *108.*

Vogel (1962), O. W., and R. H. Harmeson.   ABS in the Peoria domestic water supply. *J. Am. Water Works Assoc.*, **54**, 803–810 (1962).   *5.*

Volk, V. V. *See* KrishnaMurti (1966).

Von Riesen (1955), W. L.   Studies on bacteria-surfactant relationships. I. Decomposition of anionic surfactants by bacteria. *Trans. Kansas Acad. Sci.*, **55**, 337–344 (1955).   *47, 73, 127, 358, 365, 366, 375, 377, 378.*

Voss (1963), J. G.   Effect of inorganic cations on bactericidal activity of anionic surfactants. *J. Bacteriol.*, **86**, 207–211 (1963).   *97, 118.*

Waddams (1949), A. L.   Discussion of Goldthorpe (1949). *Chem. Ind.* (*London*), **1949,** 681.   *373.*

Waddams (1950), A. L.   Synthetic detergents and sewage processing. *JISP*, **1950,** 32–52.   *97.*

Waggy, G. T. *See* Conway (1966).

Wainwright, H. W. *See* Berber (1965).

Wakil, S. J. *See* Weeks (1969).

Waldmeyer (1968), T.   Analytical records of synthetic detergent concentrations, 1956–1966. *Water Pollution Control*, **67**, 66–79 (1968).   *194, 197.*

Walker (1962), J. F.   (Preservation of sewage samples.) Discussion of Bolton (1962). *JISP*, **1962**, 306.   *40.*

Walker, S. M. *See* Urban (1965).

Walther (1966), H.-J.   Degradation of detergents in an oxidation ditch. *Wasserwirtsch.-Wassertech.*, **16**, 163–166 (1966).   *191, 347, 362, 369.*

Walton, G. *See* Metzler (1958).

Washington (1962), D. R., and J. M. Symons.   Volatile sludge accumulation in activated sludge systems. *JWPCF*, **34**, 767–790 (1962).   *83, 84, 137.*

Washington (1964), D. R., L. J. Hetling, and S. S. Rao.   Long-term adaptation of activated sludge organisms to accumulated sludge mass. *Purdue Conf.*, **19**, 655–666 (1964).   *83.*

Washington, D. R. *See* Hetling (1964, 1965); Martin (1964, 1965).

Wasserman, A. E. *See* Porges (1958).

Watts (1968), J. B.   A modified cobaltothiocyanate method for determination of low nonionic detergent concentration. *Proc. Soc. Anal. Chem.* (*Cambridge*), **5**(3), 54–55 (1968).   *56.*

Wayman (1963a), C. H., and J. B. Robertson.   Biodegradation of anionic and nonionic surfactants under aerobic and anaerobic conditions. *Biotechnol. Bioeng.*, **5**, 367–384 (1963).   *74, 94, 188, 344, 352, 365, 391, 392, 396.*

Wayman (1963b), C. H., and J. B. Robertson.   Adsorption of ABS on soil minerals. *Purdue Conf.*, **18**, 523–533 (1963).   *113.*

Wayman (1963c), C. H., J. B. Robertson, and H. G. Page.   Factors influencing the survival of *Escherichia coli* in detergent solutions. *U.S. Geol. Surv. Profess. Paper*, **475-B**, 205–208 (1963).   *99.*

Wayman (1965), C. H., and A. T. Miesch. Accuracy and precision of laboratory and field methods for the determination of detergents in water. *Water Resources Res.*, **1**, 471–476 (1965). *53.*

Wayman, C. H. *See* Page (1963).

Weatherburn (1951), A. S. A modified method for the determination of anionic surface active compounds. *JAOCS*, **28**, 233–235 (1951). *53.*

Weaver (1962), P. J. Synthetic detergents—a look into the past, the present and the future. 24th Annual Conference, Ohio Section, American Water Works Association, Cincinnati, Nov. 1962. *159, 173, 174, 189, 343, 351.*

Weaver (1964), P. J., and F. J. Coughlin. Measurement of biodegradability. *JAOCS*, **41**, 738–741 (1964). *146, 173, 174.*

Weaver, P. J. *See* AASGP (1956, 1961); Hanna (1964); Orsanco (1963); SDA (1965).

Weber (1962), W. J., Jr., J. C. Morris, and W. Stumm. Determination of ABS by ultraviolet spectrophotometry. *Anal. Chem.*, **34**, 1844–1845 (1962). *41, 63.*

Weber (1963), W. J., Jr., and J. C. Morris. Kinetics of absorption on carbon from solution. *J. Sanit. Eng. Div., Am. Soc. Civil Engrs.*, **89**(SA2), 31–59 (1963). *113.*

Weber (1964a), W. J., Jr., and J. C. Morris. Equilibriums and capacities for adsorption on carbon. *J. Sanit. Eng. Div. Am. Soc. Civil Engrs.*, **90**(SA3), 79–107 (1964). *113.*

Weber (1964b), W. J., Jr. Competitive interaction from dilute aqueous bisolute solutions. *J. Appl. Chem. (London)*, **14**, 565–572 (1964). *113.*

Weber (1965), W. J., Jr., and R. R. Rumer, Jr. Intraparticle transport of sulfonated alkylbenzenes in a porous solid: diffusion with non-linear adsorption. *Water Resources Res.*, **1**, 361–373 (1965). *113.*

Webley (1956), D. M., R. B. Duff, and V. C. Farmer. Evidence for β-oxidation in the metabolism of saturated aliphatic hydrocarbons by soil species of *Nocardia*. *Nature*, **178**, 1467–1468 (1956). *266, 400, 401.*

Webster, H. L. *See* Bolton (1962); Ogden (1961).

Weeks (1969), G., M. Shapiro, R. O. Burns, and S. J. Wakil. Control of fatty acid metabolism. I. Induction of the enzymes of fatty acid oxidation in *Escherichia coli. J. Bacteriol.*, **97**, 827–836 (1969). *261.*

Weesner, W. E. *See* Snyder (1967, 1968).

Wehner, D. C. *See* Sheers (1967).

Weil (1964), J. K., and A. J. Stirton. Biodegradation of some tallow-based surfactants in river water. *JAOCS*, **41**, 355–358 (1964). *205, 244, 353, 362, 363, 364, 365, 366, 368, 372, 375, 378, 379, 382, 391, 393, 396.*

Weil (1965), J. K., A. J. Stirton, and F. D. Smith. Sulfonation of hexadecene-1 and octadecene-1. *JAOCS*, **42**, 873–875 (1965). *234, 362, 363.*

Weil, J. K. *See* Cordon (1965); Maurer (1965).

Weinberger (1949), L. W. Nitrogen metabolism in the activated sludge process. Sc.D. Thesis, Massachusetts Institute of Technology, Cambridge, 1949. *133.*

Weinberger, L. W. *See* Helmers (1950).

Weipert (1967), E. A. Biodegradable, liquid, water-miscible alkylene oxide condensation products. U.S. Pat. 3,340,309 (1967). *394.*

Werner, C. *See* Kölbel (1967).

Westberg (1967), N. A study of the activated sludge process as a bacterial growth process. *Water Res.*, **1**, 795–804 (1967). *182.*

Westberg (1969), N. An introductory study of regulation in the activated sludge process. *Water Res.*, **3**, 613–621 (1969). *182.*

Westfield, J. *See* Gannon (1965).

Westöö, G. *See* Tryding (1957).

Wetchlein, R. G. *See* McKinney (1953).

Wiaux, A. L. *See* Nyns (1969a, b).

Wickbold (1964), R. Intermediate products in the biodegradation of a linear ABS. *Surf. Cong. #4*, **3**, 903–912 (1967). *216, 276, 287, 292.*

Wickbold (1966), R. Analysis for nonionic surfactants in water and wastewater. *Vom Wasser*, **33**, 229–241 (1966). *55, 57, 67, 316, 385.*

Wickbold, R. *See* DAGS (1961).

Wilcox, R. L. *See* McKinney (1956b).

Wilderer, P. *See* Hartmann (1967).

Wiley, B. B. *See* Taber (1962).

Wilkinson, R. *See* Truesdale (1962).

Williams (1949), O. B., and H. B. Rees, Jr. Bacterial utilization of anionic surfactants. *J. Bacteriol.*, **58**, 823–824 (1949). *79, 127, 367.*

Williams (1964), J. P., and W. J. Payne. Enzymes induced in a bacterium by growth on sodium dodecyl sulfate. *Appl Microbiol.*, **12**, 360–362 (1964). *65, 303, 305.*

Williams (1966), J. P., W. R. Mayberry, and W. J. Payne. Metabolism of linear alcohols with various chain lengths by a *Pseudomonas* species. *Appl. Microbiol.* **14**, 156–160 (1966). *304, 403.*

Williams, J. P. *See* Payne (1965, 1967).

Williams, N. V. *See* Solbe (1967).

Wills (1954), E. D. The effect of anionic detergents and some related compounds on enzymes. *Biochem. J.*, **57**, 109–120 (1954). *107.*

Wills (1955), E. D. The effect of surfactants on pancreatic lipase. *Biochem. J.* **60**, 529–534 (1955). *108.*

Wilmsmann (1959), H., and A. Marks. On the reaction of surfactants with keratin and enzymes. *FSA*, **61**, 965–973 (1959). *108.*

Wilmsmann (1963), H. Inhibition of invertase activity as a measure of the physiological mildness of anionic surfactants. *FSA*, **65**, 958–964 (1963). *108.*

Wilson, H. F. *See* WPCF (1967).

Winsor, P. A. *See* Degens (1953); Evans (1949).

Winter (1962), W. Biodegradation of detergents in sewage treatment. *Wasserwirtsch.-Wassertech.* **12**, 265–271 (1962). *205, 233, 238, 271, 352, 357, 358, 360, 362, 365, 367, 368, 369, 373, 379, 389, 396, 397, 398, 399, 400, 401, 403, 406, 407, 410.*

Wiseman, D. *See* Salt (1968).

Witter, L. D. *See* Palumbo (1969a, b).

Woelfel, W. C. *See* AASGP (1956, 1961).

Wolford, L. T. *See* Snyder (1967, 1968).

Wood, J. M. *See* Dagley (1965b).

Wood, L. S. *See* Burgess (1962).

Woodward, R. L. *See* Metzler (1958); Robeck (1963, 1964).

WPCF (1967). (Water Pollution Control Federation) Biodegradability Subcommittee. Required characteristics and measurement of biodegradability. *JWPCF*, **39**, 1232–1235 (1967). *124.*

WPRL (1962) (Water Pollution Research Laboratory). *Water Pollution Research 1962*, HMSO, London, 1963. *157, 177, 181.*

WPRL (1964). Recent developments in determination and recording of dissolved oxygen. *Notes on Water Pollution No. 26*, (1964); reprinted in *JISP*, **1965**, 183–185. *71.*

WPRL (1965). *Water Pollution Research 1965*, HMSO, London, 1966. *60, 115, 119, 136, 154, 183, 346, 390, 394.*

WPRL (1966a). *Water Pollution Research 1966*, HMSO, London, 1967. *191, 199, 200, 347, 354, 359, 368, 390, 391.*

WPRL (1966b). Synthetic detergents. II. Nonionic detergents. *Notes on Water Pollution No. 34*, (1966); reprinted in *Water Pollution Control*, **66**, 294–296 (1967). *390, 393.*

WPRL (1967). *Water Pollution Research 1967*, HMSO, London, 1968. *71, 90, 116, 156, 157, 191, 255, 348, 355, 359, 374.*

WPRL (1968). *Water Pollution Research 1968*, HMSO, London, 1969. *78, 156, 158, 162, 191, 194, 228, 253, 255.*

Wurtz-Arlet (1964), J. Disappearance of detergents in algal cultures. *Surf. Cong. #4*, **3**, 937–943 (1967). *77.*

Yamada, H. *See* Ikeda (1963).

Yamasato, K. *See* Fujiwara (1968).

Yang (1953), J. T., and J. F. Foster. Statistical or all-or-none binding of alkyl-benzene sulfonate by albumins. *J. Am. Chem. Soc.*, **75**, 5560–5567 (1953). *104.*

Yang, J. T. *See* Foster (1954).

Yang, J. T. *See* Reynolds (1966).

Yano (1969), T., and S. Koga. Dynamic behavior of the chemostat subject to substrate inhibition. *Biotechnol. Bioeng.*, **11**, 139–153 (1969). *184.*

Yeager, J. A. *See* Knaggs (1965).

Yoshida, Y. *See* Ōba (1967, 1968a).

Young (1968), R. H. F., D. W. Ryckman, and J. C. Buzzell, Jr. An improved tool for measuring biodegradability. *JWPCF*, **40**, R354-R368 (1968). *69, 71, 74, 124, 399, 402, 404, 408, 410.*

Zahiruddin, M. *See* Kölbel (1964).

Zika, H. T. *See* Myerly (1964).

Zilio-Grandi, F. *See* Rismondo (1968).

Zimmerman, M. *See* DAGS (1961).

SUBJECT INDEX

A